PRACTICAL GUIDE TO ENERGY Management for Facilities Engineers and Managers

By

Thomas E. Mull, P.E.

New York ASME Press 2001

Library of Congress Cataloging-in-Publication Data

Mull, Thomas E.
Practical guide to energy management for facilities engineers and plant managers / Thomas E. Mull.
p. cm.
ISBN 0-7918-0158-6
1. Energy conservation. 2. Plant engineering. 3. Factory management. I. Title.

TJ163 .M85 2001
658.2-dc21 00-069453

Table of Contents

Preface

The purpose of this book is to assist facilities engineers and managers in their efforts to conserve energy and reduce energy costs in commercial, institutional, and industrial facilities. Although much of the material in this book is basic and introductory in nature, the material is intended for readers who have some knowledge of, and experience with, engineering, energy consuming systems and the thermal sciences.

There are many good books and other publications on energy conservation and energy management available. However, most of these publications are primarily concerned with energy conservation for commercial and residential buildings. Very few of these publications deal with energy conservation for central plants and industrial facilities. Those publications that do discuss energy conservation at industrial and manufacturing facilities have a very narrow scope and only deal with specific processes. This book provides energy conservation guidance for the most commonly use energy consuming equipment and systems at large commercial, institutional and industrial facilities.

Chapter 1 provides an introduction to energy management. An overview of methods for conserving energy, as well as a discussion of energy codes and standards are presented. Energy use profiles for various types of facilities are discussed, including commercial buildings, institutional buildings and several large energy consuming industries.

Chapter 2 is a review of the basic scientific principles that apply to energy consuming systems and energy analyses. These basic principles include an overview of thermodynamics, heat transfer and fluid mechanics.

An important aspect of any energy analysis is an economic analysis of any proposed energy conservation improvements. Chapter 3 includes an introduction to the various methods for performing economic calculations. Financing options for energy management improvements are also given.

Chapter 4 begins the analysis of energy consuming equipment and systems with a discussion of steam and hydronic boilers. The chapter starts with a review of the basic combustion process for boilers and other combustion systems, along with factors affecting combustion efficiencies. Various types of boilers and accessories are discussed, as well as boiler performance and efficiency. Operating and maintenance procedures for improving boiler energy efficiency are discussed. Finally, energy conservation measures for new and existing boilers are given.

Chapter 5 is a discussion of steam systems. Various types of equipment used in steam systems that affect energy efficiency are discussed.

Operating and maintenance procedures for improving the energy efficiency of steam systems are presented. Energy conservation measures for new and existing steam systems are also presented.

Hydronic and pumping systems are covered in Chapter 6. Various types of pumps and pumping systems are presented. The performance of pumps and hydronic systems are included, as well as factors affecting pump energy efficiency. Operation and maintenance procedures are presented, along with energy conservation measures for new and existing pumps and hydronic systems.

Chillers and chilled water systems represent major consumers of energy at many facilities. Chapter 7 discusses the various types of chillers and their performance. Maintenance and operating procedures for improving chiller energy performance are also discussed. Energy conservation measures for new and existing chillers are included.

Many facilities use cooling towers and fluid coolers to discharge waste heat to the atmosphere. Chapter 8 discusses various types of cooling towers and fluid coolers and the factors affecting the energy performance of these devices. Operating and maintenance procedures, which can improve the energy efficiency of cooling towers and coolers, are given. Finally, energy conservation measures for cooling towers and fluid coolers are provided.

Chapter 9 includes air distribution and heating, ventilating, and air conditioning (HVAC) systems. Various types of heating, cooling and ventilation systems for buildings are discussed, including psychrometric processes, HVAC equipment and air distribution systems. Maintenance and operational procedures for improving the energy efficiency of HVAC systems are presented. Finally, energy conservation measures for new and existing HVAC systems are discussed.

Chapter 10 is a discussion of electrical and lighting systems. Basic electricity and electrical systems, as well as electric bills and billing are discussed. Various types of devices used in electrical systems, such as electric motors, conductors and transformers, are reviewed. Maintaining electrical and lighting systems for optimum energy efficiency is discussed, in addition to energy conservation improvements for new and existing lighting and electrical systems.

Compressed air systems are discussed in Chapter 11. A discussion of the basic types of air compressors are included, along with factors that affect compressor energy performance. Accessories for air compressors and compressed air systems are also discussed. Operating and maintenance procedures that improve the energy efficiency of compressed air are presented. Selection of compressors for energy efficiency is also discussed.

Chapter 12 presents a discussion of refrigeration systems. Basic refrigeration processes are analyzed, along with the performance of refrigeration systems. An overview of various refrigeration system components, condensers, evaporators and refrigeration system controls is given. Operation and maintenance procedures for improving the energy performance of refrigeration systems are discussed. Finally, energy conservation measures for new and existing refrigeration systems are discussed.

Cogeneration systems are systems that simultaneously generate electrical or mechanical power and thermal energy. Cogeneration systems are discussed in Chapter 13. Included in this chapter is an overview of the laws that regulate the operation of cogeneration systems. The various cogeneration cycles are covered, and prime movers, electrical systems and heat recovery methods are discussed. Since these systems represent a significant capital investment, conceptual design considerations for cogeneration systems are also included.

Chapter 14 presents a discussion of heat recovery and waste heat management. Many facilities discharge large amounts of waste heat to the environment. This chapter covers the basic concepts of heat recovery and waste heat management. It includes a discussion of heat recovery equipment and heat exchangers, and also discusses heat pumps, as they may be used for heat recovery applications.

Although thermal storage systems do not usually save significant amounts of energy, they can significantly reduce the energy operating costs of many facilities. Thermal energy storage systems are discussed in Chapter 15. The advantages of thermal energy systems, including applications of thermal energy storage, are also discussed. The various components found in energy storage systems are presented along with operating strategies. A quick sizing estimation method for thermal storage systems is included.

Chapter 16 is a discussion of energy management and control systems (EMCS). The basics of energy system controls is presented. There is a discussion of various control system components in addition to energy management and control system strategies for energy consuming systems and equipment.

About the Author

Thomas E. Mull has more than 25 years experience as a mechanical engineer. His extensive accomplishments cover the areas of energy management, heating/air conditioning, utilities, plumbing and fire protection. His work involves a wide range of facilities, such as medical, institutional, commercial, industrial, educational, retail and governmental. He has specialized expertise in the fields of energy management, indoor environmental control, thermodynamics, heat transfer and other thermal sciences. During his work, Mull has traveled internationally and has been involved with projects in Europe and the Pacific Rim. He is a mechanical engineer with a St. Louis, MO, based consulting engineering firm.

Mull holds Bachelors and Masters degrees in Mechanical Engineering from the University of Missouri-Rolla. He has conducted engineering research on passive solar energy and the interaction of building lighting and heating/air conditioning systems. Mull is the author of several related publications, including a book titled *HVAC Design Principles and Applications Manual*. He has also taught engineering-related courses through the Continuing Education Department at St. Louis Community College.

Mull is a Registered Mechanical Engineer in several states, including Missouri and Illinois. He is also a certified energy analyst for the State of Missouri. Mull is a member of the American Society of Mechanical Engineers and the American Society of Heating, Refrigerating and Air Conditioning Engineers. His e-mail address is mullt@asme.org.

Chapter 1

Introduction to Energy Management

The availability, cost and probable consumption of various types of energy are important considerations in the owning and operation of most types of facilities. Non-residential energy users at facilities can be grouped into two major categories: commercial and industrial/manufacturing. Most commercial use of energy is in buildings, such as office buildings, educational buildings, long-term care buildings, and hospital or medical buildings. There are five major energy intensive industries that dominate industrial energy consumption. They are, in order of consumption, chemicals and distillate products; primary metals; petroleum and coal products; stone, clay and glass products; and paper and wood products.

Frequently, energy consuming systems and equipment are selected and designed using excessively conservative methods, resulting in oversized systems. Most systems operate at maximum efficiency at, or near, their maximum load. Energy consuming systems are also frequently selected on the basis of first cost alone. From the standpoint of total costs (the costs to own and operate) of an energy consuming system, selecting equipment and systems on the basis of first cost alone is usually not the best economic choice. When considering the economic costs of energy consuming systems and equipment, it is best to look at systems on the basis of life-cycle cost. First cost, or installed cost, is only one factor in the economic equation.

1.1 ENERGY SAVINGS OPPORTUNITIES

There are three basic categories of energy saving opportunities: Operation and maintenance (O&M) improvements, energy conservation measures, and new plants and equipment. The energy savings typically expected with each type, as well as the cost to implement each type, varies.

1.1.1 Operation and Maintenance Improvements

Operation and maintenance (O&M) improvements are usually relatively low cost improvements, that are not difficult to implement. An O&M improvement is defined as any repair or modification in the maintenance and operation of energy consuming systems or equipment, and any installations that are designed to reduce energy use. Relatively speaking, O&M improvements require no significant expenditures.

O&M improvements may be of great significance in the energy consumption of a system or equipment. Depending upon how a system is operated and/or maintained, O&M improvements can save as much as 10–20%. Typical O&M improvements include:

- shutting systems off when not in use
- reducing outside air intake
- reducing heating temperatures and/or increasing cooling temperatures during unoccupied periods
- rebalancing air distribution systems
- recalibrating and adjusting controls
- resetting heating and cooling temperatures based on the actual load
- reducing illumination levels
- repairing/replacing insulation
- repairing leaks in piping systems
- cleaning lenses in lighting fixtures
- adjusting fuel/air ratios on burners
- changing filters and clean strainers
- adjusting drive on belt-driven equipment
- maintaining and/or replacing steam traps

Those list above are just a few of the more common O&M energy savings improvements; O&M energy savings improvements for specific systems and equipment are discussed in later chapters.

1.1.2 Energy Conservation Measures

Energy conservation measures (ECMs) are modifications to existing systems and equipment that have significant acquisition and installation costs. Typically, they are retrofit projects that will significantly reduce energy consumption; however, their implementation usually requires significant expenditures of capital. Since the costs are much more significant, energy consideration measures usually require some type of economic analysis to determine whether the proposed project is economically feasible.

Energy conservation measures usually have useful lifetimes of at least two years. Typical improvements include the following:

- energy management systems
- replacement of primary heating or cooling equipment
- replacement of auxiliary heating or cooling equipment
- reduction of loads for energy consuming equipment
- replacement of lighting systems with more efficient systems

The energy conservation measures listed above are just a few of the more common energy saving projects. Energy conservation measures for specific systems and equipment are discussed in later chapters.

1.1.3 New Plants and Equipment

These improvements are major capital investments for new energy consuming systems, equipment and plants. Since they are major, these types of projects usually require an economic analysis to determine whether the project is economically feasible. The economic analysis is usually part of a much larger feasibility study. These improvement projects may include:

- cogeneration systems
- new central plants
- thermal storage systems

Major energy conservation improvements are discussed in later chapters.

1.2 ENERGY CODES AND STANDARDS

There are several code and regulations at the Federal level, which govern the use and installation of energy consuming systems and equipment. In addition to Federal regulations, frequently local and state governments adopt codes for energy consuming systems. The two Federal regulations that have the most effect on energy consuming systems are the Energy Policy Act of 1992 and the Public Utilities and Regulatory Policies Act (PURPA).

In addition to codes and regulations, there are also standards for energy consuming systems and equipment. The two most common standards are published by the American Society of Heating, Refrigerating and Air Conditioning Engineers (ASHRAE): *ASHRAE Standard 90.1 Energy Efficient Design of New Buildings Except Low Rise Residential Buildings* and *ASHRAE Standard 100.3 Energy Conservation in Existing Buildings-Commercial.* While these standards are not codes and not enforceable by law, many energy codes are based on these standards.

1.2.1 Energy Policy Act of 1992

The Energy Policy Act of 1992 is a Federal law that prescribes minimum energy efficiency standards, as well as has other legislation relating to energy issues. The Energy Policy Act of 1992 primarily applies to the construction of new facilities and new energy consuming systems and equipment. The basis for the Energy Policy Act is ASHRAE Standard 90.1-1989. Some of the energy management related items covered by the Act include:

- seasonal efficiency requirements for boilers
- energy efficiency requirements for lighting

- guidelines for energy savings performance contracts
- requirements for vehicle fuel efficiency
- grants to encourage utilities to provide demand-side management, including energy conservation, energy efficiency and load management.

The Act also requires that states enact energy codes based upon ASHRAE Standard 90.1-1989.

1.2.2 Public Utilities and Regulatory Policies Act (PURPA)

The Public Utilities and Regulatory Policies Act (PURPA) was enacted in 1978. Its purpose was to encourage the development of renewable energy sources. The basis for the law was that a properly designed and operated cogeneration system would be more efficient than the alternative energy delivery system, which consisted of central plant generation and on-site, fuel-fired boilers.

PURPA removed several significant barriers that tended to limit the development of energy efficient and economic cogeneration installations. Following are some of the provisions under the law:

- Electric utilities were required to interconnect to cogenerators and small power producers (SPPs) no larger than 80 mW, and allowed them to operate in parallel with the electric utility grid.
- Electric utilities were required to purchase power from congenerators and SPPs based on the cost the utility was able to avoid by purchasing that power. The value established for power sold to a utility had to include payments for both the capacity and the energy that was provided by the purchase of cogenerated power, with a requirement that capacity and energy payments be consistent. Electric utilities were required to purchase energy from cogenerators, even if the utility did not require capacity. In this case, the cogenerator would receive a rate that was based on the fuel savings of the utility.
- Electric utilities were not required to purchase power from cogenerators if that purchase would increase the utility's costs.
- Electric utilities were required to sell supplemental, standby and maintenance power to cogenerators and SPPs at nondescrimatory rates that are based on rate-making principles and assumptions that were used to develop rates for other customers.
- PURPA extended to those cogenerators and SPPs who sold power to a utility an exemption from rate-of-return regulations by the Federal Energy Regulatory Commission (FERC) and Security and Exchange Commission (SEC) regulation as a utility.
- PURPA allowed for the voluntary wheeling of power by a utility on behalf of a cogenerator or SPP.

1.2.3 ASHRAE Standards

The American Society of Heating, Refrigerating and Air Conditioning Engineers (ASHRAE) publishes two standards that deal with energy efficiency. These standards are *ASHRAE Standard 901. Energy Efficient Design of New Buildings Except Low-Rise Residential Buildings* and *ASHRAE Standard 100.3 Energy Conservation in Existing Buildings—Commercial.* Both standards primarily deal with commercial buildings and are standards that set guidelines. They are not codes and, therefore, are not enforceable by law; however, many local and state energy codes are based on these standards.

The ASHRAE Standards prescribe energy efficiency guidelines for commercial buildings and energy consuming systems and equipment, such as:

- Building envelopes, including exterior walls, roofs, floors, windows and doors
- Heating, ventilating and air conditioning (HVAC) systems, including design conditions, operating procedures, control and efficiency standards for most heating and cooling equipment
- Lighting systems, including illumination levels, lighting power, control and efficiency standards for fixtures, lamps, ballasts, etc
- Electric power, including the design of power distribution systems and efficiency standards for equipment like transformers, motors, etc.
- Service water heating systems, including pipe insulation, operation of hot water systems, water conservation guidelines and efficiency standards for water heaters.
- Energy management and control system capabilities.

1.3 ENERGY AUDITS

Assessing the efficiency of energy use is central to an energy management program. An energy audit is the first step in assessing the energy efficiency and energy use of a facility. An energy audit denotes a study of the energy use of a facility:

1. To determine how and where energy is being used or converted from one form to another.
2. To identify opportunities to reduce energy usage.
3. To evaluate the economic and technical practicability of implementing energy conservation improvements.
4. To formulate a prioritized list and plan for implementing energy conservation improvements.

During an audit, the following information regarding energy consuming equipment and systems is either collected or generated:

- Process flow diagrams, which provide a guide to the flows of mass and energy
- Historical energy use and production data
- Batch-based variables, which may be unique to a batch process and may include batch heaters, cycle times, etc.
- Nameplate data of major energy consuming equipment. Although this information does not generally reflect the actual operating condition of the equipment, it can be quite useful in conjunction with measured quantities.
- Material properties, including thermodynamic, physical and chemical are frequently used when carrying out energy balance calculations. Properties needed for each material include specific heat, densities of fluids, thermal conductivity and corrosion characteristics.
- Equipment operating profiles.
- Fuel and electricity costs.

There are three basic types or levels of energy audits: a preliminary or walk-through audit; an energy survey and analysis, and a detailed analysis of capital-intensive modifications. Depending on the physical and energy use characteristics of a building or facility, these steps require different levels of effort.

The preliminary or walk-through audit involves developing a thorough familiarity with the energy usage of a facility, including its associated mass and energy flows. This level of audit also involves the assessment of a facility's energy costs and efficiency through the analysis of energy bills and a brief survey of the facility. This level of energy audit will identify and provide a savings and cost analysis of relatively low cost, easily identifiable energy conservation measures. It also provides a listing of potential capital improvements that merit further consideration along with an initial estimate of potential costs and savings. The information gathered during a preliminary audit can be divided into four categories:

- overall facility information
- information on energy-using systems and equipment
- types of systems and equipment at the facility
- energy billing data

An energy survey and analysis involves a more detailed survey and energy analysis. A detailed breakdown of energy use within a facility is developed. An energy analysis provides an examination of energy savings and costs of all practical energy conservation measures, as well as an economic analysis of each. The energy analysis also includes a discussion of any effect on operation and maintenance procedures. It also provides a list of potential capital-intensive improvements that require more detailed data collection and analysis, along with an initial judgment of potential costs and savings.

A detailed analysis of capital-intensive modifications focuses on potential capital-intensive projects identified in the preliminary walk-through and energy analysis audits. It requires more detailed field data gathering and engineering analysis. It provides detailed project cost and savings data with a high level of confidence sufficient for major capital investment decisions.

1.4 ENERGY USE PROFILES FOR VARIOUS FACILITIES

Non-residential buildings and facilities consume significant amounts of energy in the United States. Non-residential energy consumers can be divided into three major classifications: non-residential buildings, industrial/manufacturing facilities and transportation. Energy consumption of vehicles used in the transportation sector is beyond the scope of this book.

Space heating, ventilating, and air conditioning (HVAC) in buildings accounts for more than 20% of the energy consumed in the United States and typically accounts for up to 60% of the energy consumed in many non-residential buildings. Industrial/manufacturing activities account for approximately 36% of the total energy consumed in the US. Major energy-intensive industrial/manufacturing sectors that consume the greatest amount of energy include chemical production, primary metals, petroleum and coal, paper and wood products, and stone, clay, and glass manufacturing.

1.4.1 Non-residential Buildings

Non-residential, or commercial and institutional, buildings include office buildings, mercantile/service, warehouse/storage, educational, public assembly, lodging, health care, food service, food sales and public order/safety. Primary energy consumption intensities for each type are listed in Table 1.1.

Commercial buildings that are most likely to have large facilities, such as central plants, include office buildings, educational buildings and health care buildings.

It has been estimated that as much as 40% of the energy used to heat, ventilate, cool, and illuminate buildings and to provide service hot water could be saved by the application of energy management programs. This level of savings could be achieved without reducing building performance or occupant comfort.

Office buildings refer to buildings used for general office space, professional offices and administrative offices. These types of buildings include data processing and computer centers; financial offices and banks;

TABLE 1.1 Commercial Energy Consumption Intensities

Building Type	Consumption (10^3 Btu/SF)	Percent of Total Consumption
Mercantile and Service	155.3	19
Office	227.2	23
Warehouses and Storage	76.3	6
Educational	136.8	10
Public Assembly	169.7	6
Lodging	235.2	8
Health Care	422.6	10
Food Service	487.8	6
Food Sales	585.7	4
Public Order/Safety	142.4	2
Vacant	49.1	2
Other	281.9	3

and professional office buildings such as administration, corporate, legal, management and medical offices. Figure 1-1 shows the energy consumption by systems of a typical office building.

Educational buildings include buildings used for academic or technical classroom instruction, such as preschools, elementary schools, junior high schools, senior high schools, college or university classrooms, and vocational school buildings. Figure 1-2 shows the energy consumption by systems for a typical educational building.

Health care buildings refer to buildings used as diagnostic and treatment facilities for both inpatient and outpatient care. Inpatient facilities treat the mentally or physically ill and include buildings for overnight, or longer, stay. These include medical care hospitals, metal health facilities and rehabilitation facilities. Outpatient facilities may be medical, dental or psychiatric facilities, and may involve diagnosis and treatment services, which do not require overnight stays. These include dental clinics, medical clinics, mental health/psychiatric clinics and

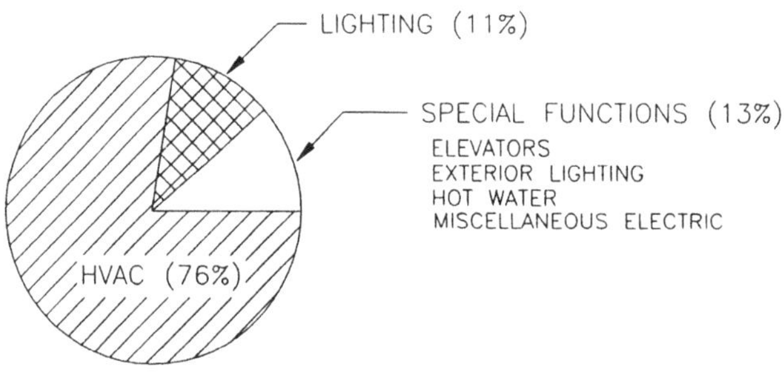

Figure 1-1. Energy consumption in office type buildings.

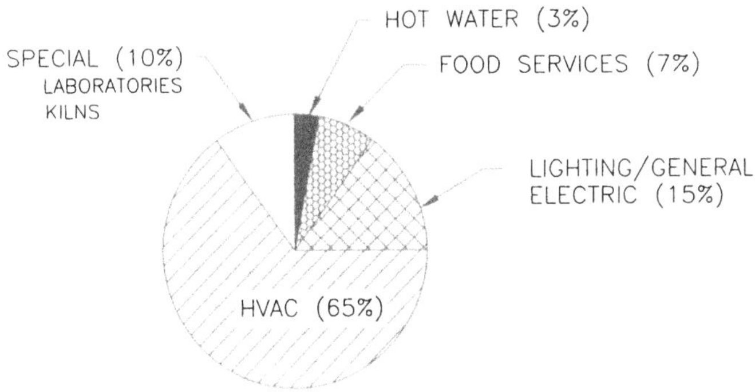

Figure 1-2. Energy consumption in educational type buildings.

veterinary facilities. Figure 1-3 shows the energy consumption by system of a typical health care building.

1.4.2 Chemical Production

Chemical and allied products include industrial organics, synthetic resins, rubber, plastics, drugs, soap and detergents, paints and other surface coating materials, inorganic industrial chemicals, agricultural chemicals and miscellaneous chemical products. The manufacture of chemicals accounts for almost 25% of the energy used by manufacturing and industrial activities in the U.S. It is the second largest energy consuming industry. About 53% of chemical industry energy use is for fuel and power and the remainder is used for feedstock. Natural gas is the dominant fuel source in the chemical industry; much of this gas is utilized as feedstock for production processes.

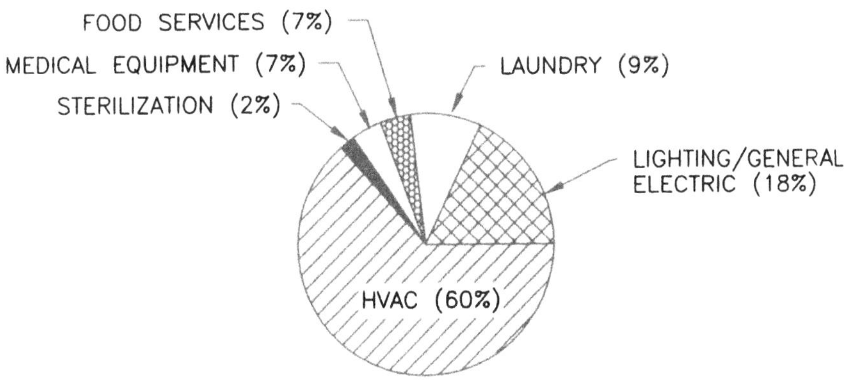

Figure 1-3. Energy consumption in health care type buildings.

The production of ethylene/propylene, ammonia, chlorine/caustic soda, ethylbenzene/sodium carbonate, cumene and phenol/acetone account for about 48% of the total energy consumed by the chemical industry. Ethylene is used as an intermediate in the production of plastics, rubber and synthetic fibers, which are, in turn, used in industrial and consumer products. It is produced by the reaction of steam and hydrocarbon feedstock, followed by thermal cracking. The resulting product mix is cryogenically cooled to -150°F and compressed to 450 – 600 psi, after which the ethylene is distilled from its feedstock and by-product materials.

Ammonia is synthesized by the reversible reaction of nitrogen and hydrogen, a reaction carried out under elevated pressures of between 80 and 1,000 atmospheres, depending on the specifics of the process, and at elevated temperatures of 750°F – 1000°F. The product mixture is then cryogenically cooled and partially distilled to remove argon, urethane and unreacted nitrogen by-products.

The production of low density polyethylene (LDPE) is one of the largest users of energy in the chemical industry. This production has a significant opportunity for energy savings with heat recovery. Almost 10% of the energy used is lost as heat of polymerization and is rejected to cooling water which is rejected to a cooling tower through an intermediate heat exchanger. The heat recovered has a temperature of about 450°F and could be used to generate low pressure steam. Heat recovery is discussed in Chapter 14.

Significant energy savings can be achieved in the production of steam in boilers and fired heaters. Computerized controls can increase the efficiency of burners. There are also a number of methods available to recover waste heat from boilers and exothermic reactors. Improvements in efficiency can be achieved by more precise monitoring of stack temperatures, fuel and air intake, as well as closed-loop process control. Significant improvements can result from using the heat from the stack gases to either preheat air or intake water with an economizer. Energy efficiency for boilers is discussed in Chapter 4.

1.4.3 Primary Metals

The primary metals industry consists of industries that manufacture aluminum, copper, iron and steel. The primary consumers of energy in the primary metals industry are steel plants, which use about 73% of the energy utilized by this sector. Aluminum, copper and iron foundries account for an additional 19% of the energy consumed.

In 1994, the U.S. steel industry consumed a gross total of 2.3 quads of energy, including the generation of electricity and transmission losses. Nearly half of the industry's energy is derived from coal, most of which is used to produce coke for blast furnaces. Electricity and natural gas are other major sources of energy.

A steel plant takes in iron ore and scrap steel, raw materials such as coal and limestone; fuels and electricity, and produces finished steel products. A steel plant may also produce intermediate steel products, such as ingots, billets and blooms. The major energy consuming processes in steelmaking are coking, coal, agglomerating iron ore; ironmaking; steelmaking; soaking ingots; reheating blooms, billets and slabs; and heat-treating/forging. Over 80% of the energy consumed is in these operations.

Energy is consumed in steel production during the processes of six major stages: preparation of raw materials, ironmaking, steelmaking, primary finishing, secondary finishing and heat-treating. The raw materials that must be prepared in the production of steel include iron ore, coal and limestone. These materials are combined in a blast furnace, where the iron ore is smelted to metallic ore. During the materials preparation process, there are two processes that are particularly energy-intensive: the benefication of iron ore and coking. Benefication of iron ore is a process in which iron ore chunks are first crushed and ground, then refined. They are then agglomerated (heated to form a mass) and formed into marble-size pellets. The agglomeration processes are particularly energy-intensive. Coke is made from coal to produce 90% pure carbon, which is what remains when coal is heated at 1650° – 2000°F for 12 – 18 hours to boil off its volatile components.

In the ironmaking process, iron ore must first be transformed to metallic iron by the reduction of the iron oxides in a blast furnace. To make iron, iron bearing materials (iron ore, sinter, pellets, mill scale, slag iron or steel scrap, etc.), fuel (coke) and flux (limestone and/or dolomite) are charged into the top of the blast furnace. Heated air (blast) is blown in at the bottom. The hot air blast burns the coke to heat, reduce and melt the charge as it descends into the furnace.

Steelmaking is a refinement process whereby undesirable amounts of other chemical elements (such as carbon) are reduced and removed from the pig iron and other elements (fluxes and alloy materials) are added to produce the desired steel properties. Steel is made in three types of furnaces: the basic oxygen furnace, the electric arc furnace and the open hearth furnace.

Primary finishing includes the operations of casting—pouring liquid steel into its first solid form (raw steel) and then converting the raw steel into semi-finished shapes, such as slabs, blooms and billets. Secondary finishing includes the operations of reheating the slabs, blooms and billets produced in primary finishing, and then transforming them into final products through hot and cold rolling. Reheat furnaces are used to heat steel shapes to temperatures of 2300° – 2400°F prior to rolling operations.

The final step in the finishing operations is heat-treating. Cold-working steel results in a highly stressed product with low ductility. The purpose of heat-treating is to relieve stresses, obtain full recrystallization to a more uniform grain structure and improve ductility.

All the stages in the production of steel use energy to alter the chemical makeup of the metal or to work the metal into useful forms and shapes. Under the most ideal circumstances, the energy required to produce solid iron from iron oxide cannot be less than 7 million Btu per ton. When process losses are included, the average energy required to produce solid iron is between 9 and 14 million Btu per ton.

There are significant opportunities for energy savings in steelmaking due to the high temperatures involved in the process, and also because by-product gases can be burned in a furnace. The major opportunities for energy efficiency improvement is in the area of process change, in particular, the replacement of ingot casting by continuous casting as well as the substitution of the electric arc furnace production (using scrap metal feedstocks) for the blast furnace/ basic oxygen furnace combination (using iron ore feedstocks). With continuous casting, there is significantly more energy saved than with batch operations.

An electric arc furnace (EAF) saves energy by allowing the substitution of scrap metal for iron ore. Continuous casting is more energy efficient for two reasons. First, the use of continuous casting eliminates the need for ingot stripping, reheating and primary rolling. Second, the yield of slabs and billets from continuous casting is much greater than from ingot casting, since less metal must be returned to the steelmaking process in the form of waste and unfilled ingot molds. The use of continuous casting results in a savings of about 2 million Btu per ton in reheating and about 2.5 million Btu per ton in the general plant.

1.4.4 Petroleum Refining Industry

The petroleum refining industry is defined as the group of establishments engaged in refining petroleum, producing paving materials and manufacturing lubricating oils. Petroleum refining involves both chemical and mechanical energy changes. Although some conversion processes are exothermic, most are endothermic and require large amounts of heat. Such processes include distillation, cracking, hydrotreating, reforming and alkylation. A large amount of mechanical energy is also needed for pumping, transportation, compression of gases, etc.

The petroleum refining industry is the single largest energy consuming industry. In 1994, it consumed a total of 5.9 quads. About 50% of the energy consumed by petroleum refining was used to manufacture feedstock, such as asphalt, lubricants and waxes. The rest, about 3 quads, was used for heat and power. More than half of the energy used in petroleum refining is by-product refinery waste gas and petroleum coke.

The principal energy consuming operations in refining are crude oil distillation, desulfurization processes, catalytic cracking, naphtha reforming, alkylation, aromatic extraction and coking. They consume about 80% of the energy required in the refining industry. Natural gas, refinery produced gas, petroleum coke and fuel oil are the primary sources of energy.

The principal processes in refining include atmospheric distillation, vacuum distillation, fluid catalytic cracking, catalytic reforming, coking, alkylation, hydrocracking, hydrotreating and residual desulfurizing. In the atmospheric distillation process, incoming crude oil is first treated to remove organic salts and then dehydrated. Under a slight pressure, it is then heated to a boil in a column where the various components of the crude oil are separated according to their boiling temperature.

During the vacuum distillation process, some crude oil components are too heat sensitive or have boiling points that are too high to be distilled at atmospheric pressure. In such cases, the so-called topped crude (material from the bottom of the atmospheric column) are distilled in a column operating under a vacuum. This process lowers the boiling point of the material and thereby allows distillation of heavier fractions without excessive thermal decomposition.

In the fluid catalytic cracking process, crude petroleum, whose lighter fractions were removed by atmospheric or vacuum distillation, is entrained in a hot moving catalyst and chemically converted to lighter materials. The catalyst is then separated and regenerated, while the reaction products are fractionated into their various compounds by distillation.

Catalytic reforming is a catalytic process that takes low octane materials and increases the octane to approximately 100. Although several chemical reactions take place, the predominant one is the removal of hydrogen from naphthenes to aromatics (benzene ring compounds). This significantly increases the octane number.

Petroleum coke is used as a residual fuel, although it may have a high sulfur content. Typically, the residual bottoms from the crude unit are blended with lighter oils. Since these residuals do contain lighter fractions of naphthas and gas oils, they may be recovered if the residual oil is coked at high temperatures.

In the alkylation process, isobutane (a low molecular weight gas) is chemically added to the carbon-to-carbon double bonds that occur in certain hydrocarbons. The resulting product, which contains many isobutyl side groups, has a much higher octane number.

Hydrocracking is a catalytic, high-pressure process that converts a wide range of hydrocarbons to a lighter, cleaner, more valuable product. The process catalytically adds hydrogen under very high pressure. The process increases the ratio of hydrogen to carbon in the feed and produces low boiling point material.

Hydrotreating uses the catalytic addition of hydrogen to remove sulfur compounds from naphthas and distillates (light and heavy gas and oils). Removal of the sulfur protects the catalyst in subsequent processes (such as catalytic reforming).

Residuum desulfurization is a process that is applied to the material remaining after atmospheric and vacuum column distillation. These processes operate at pressures and temperatures between low-severity hydrotreating and the much more severe hydrocracking. The desulfurization residual can be used as a blending component of low-sulfur fuel oils or as feedstock for a coke producing unit.

In typical refining processes, feed streams are heated to cause a physical separation or to provide energy for a heat absorbing reaction, such as catalytic reforming. Although heat exchange is used to preheat feed streams, additional heat is usually required. Petroleum refining processes use energy in the form of fuel, steam or electrical energy for the following functions:

- To heat crude units and other process feed streams.
- To make steam for mechanical drive turbines to power major compressors and some large pumps for process heating, steam-stripping and steam jet vacuum ejectors.
- To heat reboilers (steam-fired or fuel-fired).
- To power most pumps and fans.

Energy losses in refining operations are the result of the following:

- Heat rejected by air- and water-cooled heat exchangers used to cool recycled and product streams, and which accounts for approximately 50% of refinery heat losses.
- Unrecovered heat in flue gases from furnaces and steam boilers; this is approximately 25% of refinery energy losses.
- Convection and radiation losses from hot equipment and piping.
- Steam system losses.

Additional heat recovery between process units is probably the largest potential opportunity for energy savings. Additional heat can be exchanged between process streams instead of dumping waste heat to air or cooling water streams. However, recovering additional heat requires relatively large capital investments in heat exchange equipment. Heat recovery systems are discussed in Chapter 14.

Process heaters and steam boilers offer many opportunities for energy savings, including the following:

- Reducing excess air and improving combustion with stack gas analyzers and better combustion controls.
- Reducing stack gas temperature with preheaters to heat combustion air.
- Installing convection sections at heater outlets to heat incoming feed or to generate steam.

Boilers and combustion systems are discussed in Chapter 4.

In most refineries, steam is generated and distributed at pressures around 600 psig, as well as 150 and 50 psig. The steam is used for heating and for mechanical drives (usually steam turbines) at many locations throughout a refinery. Ideally, the steam is generated, distributed and used at the same pressure levels and is reduced only while doing useful work. At the lowest pressure, all steam is used for heating so that no steam is wasted by being vented to atmosphere. Ideally, all of the steam condensate is collected and returned to the boiler for reuse. Typically, steam systems are not operated in an ideal manner, and significant opportunities for energy savings are usually available. Steam systems are discussed in Chapter 5.

Most refineries use a large number of pumps for pumping various liquids. Usually, the pump drivers (electric motors and mechanical drive steam turbines) operate at a constant speed. The output from the pumps is typically achieved with throttling control valves. As a result, the pump wastes energy by discharging against a partially closed control valve, while the unnecessarily large motor driver operates well below its rated output, which also wastes energy. The proper selection and application of pumps and pump impellers can save significant amounts of pumping energy. Pumping systems are discussed in Chapter 6.

1.4.5 The Paper and Wood Products Industry

The paper and wood products industry is the 11th largest manufacturing industry in the U.S. industrial sector; however, it is the third largest energy consumer. In 1994, this industry consumed about 3.3 quads of energy. Of the total used by the industry, the lumber and wood products segment utilized about 520 trillion Btu, and the pulp and paper segment used about 2.8 quads.

Almost 85% of the energy consumed by the paper and wood products industry is in the production of pulp and paper. On average, about 31 million Btu are presently used to produce one ton of paper. There are three energy-intensive steps, which account for 83% of the energy consumed by this segment. Pulping accounts for about 26% of the usage, chemical recovery requires about 19% and papermaking consumes about 38%.

Paper is made by separating the cellulosic fibers in wood and then removing the lignin that binds the fibers (pulping). The cellulose fibers are usually further conditioned by bleaching and refining before being interlaced in sheets. Finally, water is removed from the sheets by mechanical pressing and the application of heat, leaving the final product—paper.

Pulping is a major energy consuming process in papermaking. Commercial pulping operations are of three major types, depending upon the input (kind of trees) and the desired output (products). The principal pulping methods are mechanical, full chemical and semi-chemical.

In mechanical pulping, wood is reduced to fibrous states by mechanical means such as grinding. In a stone ground pulping operation, logs are first ground to a pulp by large, revolving grindstones, while water is sprayed against the stone to control the temperature and to carry away the resulting pulp. A variation of pure mechanical pulping is thermomechanical pulping (TMP), which uses pressurized disk refiners, in addition to heat and chemicals, to produce pulp.

Full chemical pulping employs chemicals to separate cellulose fibers from other wood components. Wood chips are cooked with chemicals in an aqueous solution, usually at temperatures around 350°F and elevated pressures, to dissolve lignin and other compounds while leaving the cellulose intact and in a fibrous form.

Semi-chemical pulping, which is relatively new, is typically used on wood from deciduous trees. It involves softening the wood with mild chemical action and then mechanically grinding it into a pulp.

Wood pulp must be bleached before it can be used to make white paper. Almost all bleaching is carried out with chlorine or chlorine compounds, leaving an effluent containing high levels of chemicals that must be biologically degraded in a wastewater treatment plant.

Refining is a stock preparation process that occurs after bleaching but before actual papermaking. In this process, the proper mixture of pulp types is blended. Recycled waste paper is also added to the paper stream for repulping. During the refining process, the unmodified cellulose fibers (from the pulping process) are separated, crushed, frayed, fibrillated and cut. Water is added and absorbed by the fibers, causing them to become more flexible and pliable. In the refining process, the major energy consumers are electric motors.

The final stage of the process is papermaking, during which 95% of the water must be removed for the cellulose mixture. This process consumes up to 40% of the energy in the total papermaking process and is the single most energy-intensive process. Paper sheets are made by depositing a cellulose mixture, with a consistency of less than 1% cellulose solids suspended in water, on a continuously moving screen removing the water. Water may be removed by one of three methods: the Foundrinier process, the cylinder machine or a twin wire former.

The types of purchased energy used by paper mills are natural gas, coal and residual oil. These fuels are employed to generate steam or electricity, most of which is used to produce paper. Most processes in papermaking occur at relatively low temperatures; exceptions are the lime kiln and the black liquor recovery furnace where temperatures may reach 1500°F.

Major opportunities for energy savings occur in the operations of lime kilns and steam boilers. The kiln is a rotary furnace used for drying and calcining the lime mud. Both natural gas and fuel oil are used to fire lime kilns. Significant energy savings can be achieved in lime kilns by controlling excess air in the combustion process. Energy savings for combustion processes are discussed in Chapter 4.

Significant energy saving opportunities also are available in the operation of steam boilers. Flue gas heat recovery, boiler combustion controls and preheating of boiler feedwater offer opportunities for energy savings. Energy savings opportunities for boilers are discussed in Chapter 4.

1.4.6 Stone, Clay and Glass Manufacturing

In the stone, clay and glass industry, the production of hydraulic cement is the greatest consumer of energy. Hydraulic cement is a powder formed when limestone, sand, clay and other materials are heated in a kiln. The kiln produces an amalgam of these materials, which is called clinker. The clinker is ground to form powdered cement.

The primary energy consumer in the production of cement is the kiln in which calcining of raw materials takes place. The calcining process that produces the clinker involves an endothermic chemical reaction between the raw materials at high temperatures.

A kiln is a long cylindrical furnace that rotates slowly and is mounted at a slight incline to the horizontal. The rotational speed of a kiln is about one revolution per minute. The raw materials gradually move down the kiln due to the inclination.

The kiln burner is located at the discharge end of the kiln. As the raw materials move down the kiln, they become progressively hotter as they approach the burner. At lower temperatures, water vapor and CO_2 are driven off as the material is heated. In the higher temperature areas of the kiln, temperatures approach 2700°F and a chemical reaction between the lime, silica, aluminum and iron takes place. The resulting clinker exits the kiln as lumps and pellets.

In the wet cement making process, the rotary kiln consumes over 85% of the entire process. One way to reduce energy consumption is to convert from a wet process to a dry process. The dry process is approximately 15% more efficient than the wet process.

Another opportunity for energy savings is in the kiln combustion process. Combustion controls for the furnace and preheating combustion air with recovered waste heat can produce significant energy savings. Combustion processes are discussed in Chapter 4.

Suspension preheaters may be used to provide more heat transfer to the incoming raw feed materials. Typical suspension preheaters heat the feed to about 1400°F. As a result, chemical reactions also occur and about 30 – 40% of the calcination occurs before the material goes into the rotary kiln. Typical preheaters have up to four stages of heat. The addition of a suspension preheater converts the process from a wet process to a dry process.

Chapter 2

Basic Scientific Principles

Analyses of energy consuming equipment and systems involves the application of basic scientific principles such as physics, chemistry, thermodynamics, heat transfer and fluid mechanics. The reader, and anyone conducting an energy audit, should have a background in the thermal-fluid sciences. The material presented in this chapter is an overview of the thermal-fluid sciences. The application of the principles are presented in later chapters that deal with actual energy consuming systems and equipment.

2.1 THE CONCEPT OF HEAT AND WORK

Energy can be defined as the capacity to produce an effect or the capacity for doing work. It can also be defined as the potential for providing useful work or heat. Energy can neither be created nor destroyed; it can only change from one form to another. The lowest form of energy is heat.

There are two general classes of energy: energy in transition and stored (or internal) energy. Stored energy is energy that resides in the matter itself, either in a state of aggregation or motion. An example of stored energy is a coiled spring with potential energy stored in its coils. Examples of energy in transition are work and heat.

There are essentially six basic forms of energy: they are heat (thermal energy) work, kinetic energy, potential energy, chemical energy and electro-magnetic energy.

2.1.1 Heat

Heat is a form of energy that, when added to a substance, causes the substance to rise in temperature, fuse, evaporate, or expand. Heat itself can not be seen or observed. It is only possible to observe the effects of heat. The primary effect of adding heat to a substance is an increase in the vibration of the molecules of the body. Heat always flows from the high state (high temperature) to a lower state (lower temperature). In other words, heat always flows from a hotter area to a colder one. The cold does not flow to the warmer area rather, heat flows to the colder area. An analogy would be a bucketful of water with a leak. The water flows out of the bucket, the emptiness does not flow in.

The standard unit of measurement for heat is the British Thermal Unit, or Btu. A Btu is the amount of heat required to raise the temperature of one pound of water from 59.5 °F to 60.5 °F. Another commonly used measurement of heat in S.I. units is the kilowatt-hour, which is equal to 3412 Btu.

Heat occurs in two forms: sensible heat and latent heat. Sensible heat is heat that changes the temperature of a substance. An example of sensible heat is a substance, such as water, air or a solid, increasing or decreasing in temperature. Latent heat is heat that produces a change in state or phase change of a material. Examples of latent heating of a substance are evaporation, condensation, melting and boiling. In each case, heat is added to or removed from the substance, the temperature of the material remains constant as the phase change occurs.

2.1.2 Work

Work is another important concept in energy analyses. Work is a transient form of mechanical energy. Work is done on an object whenever a force is exerted on the object and it is displaced or accelerated in the direction of the force. The commonly used units for work are foot-pounds per minute. Power is the time rate at which work is done and is measured in horsepower. Examples of work commonly found in energy consuming systems include fans circulating air, pumps circulating liquids and compressors compressing gases.

$$\text{Work} = \text{Force} \times \text{Distance} \qquad \text{Eq. (2.1)}$$

The force acting may be constant or variable, The work required to lift a 200 pound weight a distance of 10 feet would be equal to 2,000 ft.-lbs. (neglecting any friction).

Heat may be transformed into work. Such an example of this would be an internal combustion engine where fuel is burned and converted to work at a drive shaft or flywheel. Work may also be transformed into heat either directly, as in the case of a heat pump, or indirectly through friction. In all cases, some energy will be lost due to inefficiency of the process.

Power is the rate at which work is done and is usually measured in horsepower (HP).

Examples of work found in energy using systems include fans circulating air, pumps circulating fluids and compressors compressing gasses.

2.1.3 Kinetic Energy

The energy a quantity of matter has, or posses, by virtue of its motion is called kinetic energy. Every body in motion that has mass posses kinetic energy. Whenever a body is in motion, it is able to exert a force and do work in coming to rest.

$$\text{Kinetic Energy} = \frac{mV^2}{2g} \qquad \text{Eq. (2.2)}$$

where m = the mass of the object
V = the velocity of the object
g = the acceleration due to gravity (32.2 ft./sec.^{2})

2.1.4 Potential Energy

The energy that matter has by virtue of its position or configuration is called potential energy. A body that has been moved or distorted to a new position is able to do work because of its position or configuration.

When a body undergoes a displacement (change in elevation or configuration), a force does work on the body. An example of vertical displacement would be the lifting of a weight. An example of a change in configuration would be the compression of a spring. In each case, a force does work on a body, resulting in stored potential energy.

2.1.5 Chemical Energy

Chemical energy is stored energy by virtue of the chemical composition of the material. An example of chemical energy is gasoline, which, when burned, releases thermal energy. Another example of chemical energy is atomic energy, which exists due to the atomic makeup of certain materials.

2.1.6 Electro-magnetic Energy

Electro-magnetic energy is potential energy resulting from magnetism. Electrical energy can be generated by rotating coils in a magnetic field.

2.2 THERMODYNAMICS

An important principle in energy analysis is the First Law of Thermodynamics and the concept of a control volume. A control volume can be any part of the system in which we want to examine the flow of energy into and out of it. An example might be a room, which is part of the system as a whole, that we wish to know the difference between the heat flow into the room and the heat flow out. The room is the control volume and the walls, ceiling and floor are boundaries of the control volume. We may wish to examine the heat flowing out through the walls in order to determine the amount of heat that must be added to maintain the room temperature. Another control volume may be the air handling system itself. Heat is added to the control volume by a heating device, such as a coil. Work or mechanical energy is added by a fan and energy flows out of the control volume through air grills and registers. Another control volume that we may wish to consider is the heating coil. If the coil is a steam heating coil, heat flows into the control volume in the form of steam and out as heat is transferred to the air stream. Control volumes are used frequently when studying energy using systems. A control volume is used whenever it is necessary to isolate a particular component in order to analyze the flow of energy into and out of the control volume or component.

The science of thermodynamics is based upon two laws: The first law deals with the quantities involved in energy transformations, and the second law with the direction in which transformations take place. The reader should consult Sontag and Van Wylen (1971) for a more detailed discussion of thermodynamics.

2.2.1 The First Law of Thermodynamics

Basically, the First Law of Thermodynamics (the conservation of energy law) states that the energy flowing into a control volume is equal to the energy flowing out of a control volume plus the energy stored within the control volume. A simple mathematical expression for the First Law of Thermodynamics is:

$$\text{Energy In} = \text{Energy Out} + \text{Energy Stored} \qquad \text{Eq. (2.3)}$$

This is shown schematically in Figure 2-1, where the dashed box indicates the boundaries of the control volume.

Energy can cross the system boundaries in any form, i.e., heat, work, chemical or potential energy.

Using the example of a room as a control volume, let's look at a typical application. Suppose a room had an outside wall exposed to a cold outside temperature and it also had a radiator to supply heat. Energy in the form of heat would flow out of the room, or control volume, and heat energy would flow into the room from the radiator. If the heat flowing out of the room through the wall (heat loss) were equal to the heat flowing into the room (heat gain), no energy would be stored and the room temperature would remain constant. If, however, the heat supplied by the radiator were

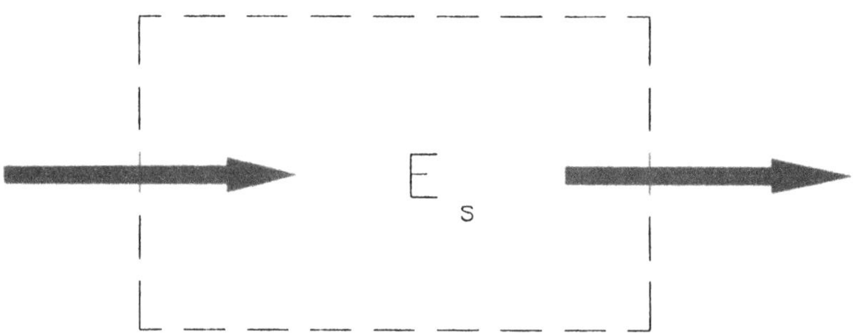

Figure 2-1. Control volume for First Law of Thermodynamics.

greater than the heat loss, energy would be stored in the room in the form of heat. The result of this addition of stored heat would result in an increase in room air temperature. Conversely, if the heat loss were greater than the heat gain from the radiator, a net heat loss would occur, resulting in negative heat storage and a resultant decrease in the room air temperature.

The First Law of Thermodynamics and the concept of the control volume are applied over and over again in energy analyses. A control volume can be as large as an entire building, it can be an entire system, such as an air conditioning system, or it can be as small as an individual component in the system. In each case, the First Law of Thermodynamics is used to perform an energy balance on the control volume in question.

As an example, consider handling unit with a hot water heating coil and a fan, as shown in Figure 2-2.

Heat energy enters the controlled volume as heat in the form of heated water to the heating coil. Mechanical energy (work) enters the controlled volume via a shaft driving the fan. The fan converts the mechanical energy into potential energy (pressure). Heat and potential energy leave the controlled volume as heated pressurized air. Heat energy may also be lost through the casing of the air handling unit if it is not well insulated. Some of the mechanical energy will be converted to heat energy due to friction losses in the fan drives. Energy in the form of heat will also be stored within the air handling unit as the components inside the unit heat up. All of the energy flowing into the unit is either stored or flows out of the unit, even though it may change forms.

2.2.2 Forms of Heat

Heat may be stored in substances in two forms: sensible heat and latent heat. Sensible heat is most recognizable type of heat. A change in the

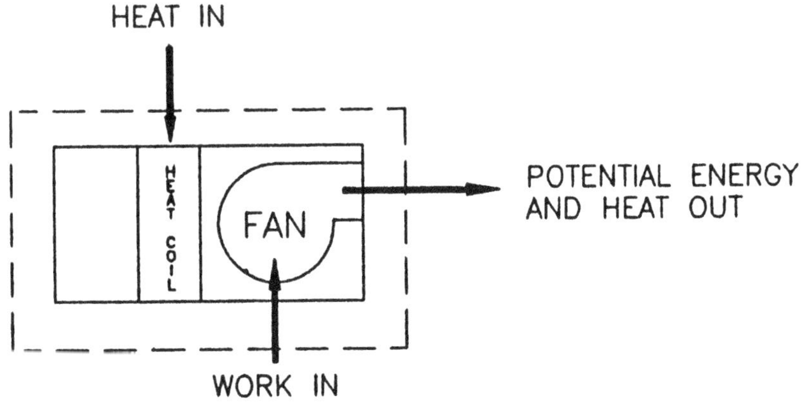

Figure 2-2. Energy balance for an air handling unit.

sensible heat content of a substance causes a change in its temperature. As an example, heat applied to water will cause an increase in the temperature of water. The change in temperature caused by sensible heat is measured by an ordinary thermometer. When measuring the temperature change of air due to a change in sensible heat, the temperatures measured are called "dry bulb" temperatures.

Latent heat is heat associated with a phase change of a substance. A phase of a substance may be defined as a quantity of the substance that is homogeneous throughout. As for example, water may be in the liquid phase, the vapor phase (steam) or the solid phase (ice). If water were in the vapor phase, it would be completely steam, with no liquid or solid water present. A change of phase can be from a liquid to a vapor or gas (evaporation), from a vapor or gas to a liquid (condensation), from a liquid to a solid (solidification) or from a solid to a liquid (melting). A substance may also change directly from a solid to a gas or vapor (sublimation). In each case, heat is either added to or removed from the substance as it changes from one phase to another. Heat that is added to or removed from a substance during a change of phase is called latent heat. During the change of phase, the temperature of the substance remains constant as heat is added or removed until all of the substance has completely changed phase. As an example, when water is boiling, the temperature of the water will remain constant at 212 °F as heat is added until all of the water has boiled away and has become steam.

2.2.3 Enthalpy

Enthalpy is a thermal property of a substance and consists of the internal energy of the substance plus energy in the form of work due to flow into or out of a control volume. The energy due to flow is the product of the pressure times the volume. Enthalpy is a measure of the total energy content of the substance measured from an arbitrary datum level. Internal energy, which is the major component of enthalpy, is energy associated with motions and configurations of a substance's interior particles, primarily due to temperature as the result of the addition of heat.

Enthalpy is used frequently in HVAC calculations, although it is actually the total heat content of a substance. The flow work component in HVAC applications is usually ignored since it is relatively small. HVAC calculations are primarily concerned with the total heat of air, including any suspended water vapor in the form of latent heat. When the term "enthalpy of the air" is used, it is really referring to the total heat content of the air/vapor mixture.

2.2.4 Specific Heat

Specific heat, or heat capacity as it is sometimes called, is the quantity of heat required to change one unit mass of a substance one degree. Specific heat may be expresses mathematically as

$$q = mc(T_2 - T_1) \qquad \text{Eq. (2.4)}$$

where q = quantity of heat added or removed from the substance, in Btu
m = mass of the substance, in pounds
c = specific heat of the substances, in Btu/lb °F
$T_2 - T_1$ = temperature change, in °F

Since water is a common substance, it has arbitrarily been chosen as the standard measure of specific heat. In the English units system, water has a specific heat of 1.0. The heat required to raise one pound mass of water from 59.5 °F to 60.5 °F is One British Thermal Unit (Btu). Therefore, the specific heat of water is 1.0 Btu/lb - °F.

As a comparison, the specific heat of air is 0.24 Btu/lb-mass °F. It can be observed that it takes more than four times as much heat to raise one pound mass of water 1°F as it does to raise one pound of air 1°F.

In general, the specific heat of a substance remains constant. Specific heat does vary slightly with temperature, but for most calculations it may be assumed a constant.

Example 2.1. Dry air has a specific heat of 0.24 Btu/lb.-°F and pure water has a specific heat of 1.0 Btu/lb.-°F. How much heat is required to raise one pound of each substance, 1°F?

Using Equation (2.4),

$$\text{For the air}: q = (1 \text{ lb})(0.24 \text{ Btu/lb} - {}^\circ\text{F})(1^\circ\text{F}) = 0.24 \text{ Btu}$$

$$\text{For the water}: q = (1)(1)(1) = 1.0 \text{ Btu}$$

2.2.5 The Second Law of Thermodynamics

There is a universal law of nature that differences of energy levels will always tend to disappear. All energy levels tend to level out. Energy will always flow from a region with a higher energy state to a region with a lower energy state. Heat will always flow from an area of higher temperature to an area of lower temperature. Heat cannot naturally flow in the other direction. Potential energy, such as pressure, will flow naturally from an area of high pressure to an area of low pressure.

The Second Law of Thermodynamics is also known as the Law of Degradation of Energy. The basic underlying principle of the second law is that heat cannot be extracted from a body at a low temperature and delivered to another at a higher temperature unless work is done to accomplish this result. If energy were to flow from a region of lower energy to a higher energy region, the First Law of Thermodynamics would still have been obeyed; however, such a flow does not occur in nature and would be a violation of the second law. The second law is the basic principle that makes mechanical refrigeration necessary.

A reversible process is one that can be made to traverse completely, in the reverse order, the same steps or states it passed through during the direct process and, in the end, the condition of the working substance is restored to its initial state. In order for a process to be reversible, the following conditions must be completely and precisely met.

1. The process may proceed in either direction.
2. The process may be reversed by infinitely small changes in the values of the conditions controlling it.
3. The working substance is always in a state of equilibrium throughout the process.
4. No energy transformations occur as the result of friction, i.e., there are no losses due to friction.
5. All energy transformations that occur as the process proceeds in one direction must equal, in both form and amount, those which take place in the reverse order.
6. The working substance must be returned to the initial condition by retracing every step of the original process in reverse order.

It is obvious that it is impossible to build a process or machine that meets the requirements for a reversible process. It is not possible to construct a completely reversible process nor is it possible to construct a process that is 100% efficient. There will always be energy losses, such as friction losses and heat losses, in every process.

In the HVAC system design, it is the second law that makes it necessary to have refrigeration systems to cool buildings. For example, consider a space that is to be maintained at 75°F with an exterior wall to an outside temperature of 95°F. The natural flow of heat is from the region of higher energy to the region of lower energy, i.e., from outside to the inside through the exterior wall. In order to maintain the space at 75°F, it is necessary to transfer heat from the cooler inside to the warmer outside. To accomplish this process, it is necessary to use a refrigeration process (an air conditioner) to transfer the heat. The refrigeration process cannot transfer the heat from the cooler inside to the warmer outside without the addition of mechanical energy in the form of work. A compressor is used to add mechanical energy to the refrigeration system. A simple refrigeration system schematic is shown in Figure 2-3.

During the refrigeration process, some of the energy is lost due to friction in the compressor and the refrigerant flow in the piping, etc. Energy may also be lost due to uncontrolled heat transfer from the system. These losses make the refrigeration process irreversible. It is, therefore,

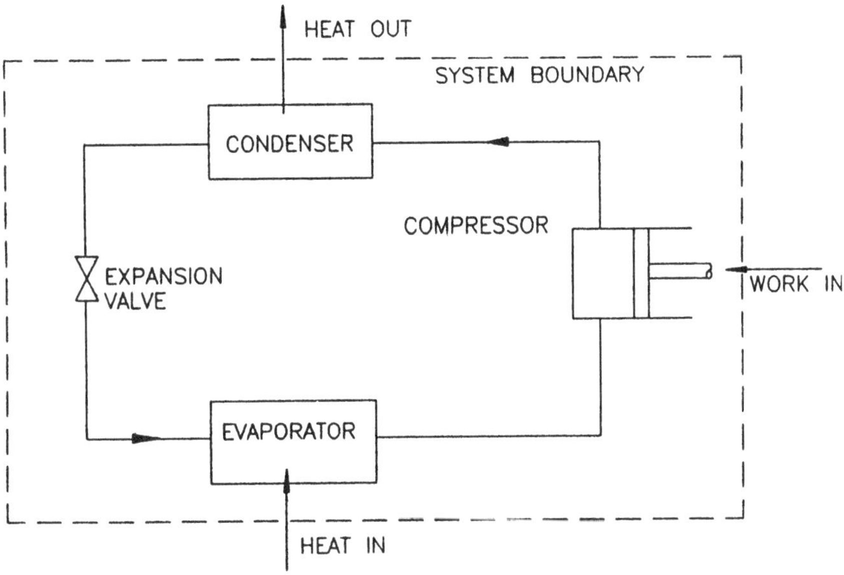

Figure 2-3. Refrigeration system schematic.

necessary to add more energy to the system than would ideally be necessary due to the inherent inefficiencies of the system.

2.2.6 Psychrometrics

Psychrometrics can be defined as the science involving the thermodynamic properties of moist air and the effect of atmospheric moisture on materials and human comfort. All HVAC heating, air-cooling, humidification and dehumidification processes involve the application of psychrometrics. It is especially important in air cooling and dehumidification processes. Psychrometric processes for HVAC systems are discussed in more detail in Chapter 9.

2.3 HEAT TRANSFER

Heat transfer involves the transfer or flow of energy in the form of heat. Whenever a temperature difference (or gradient) exists within a substance, from one side of a substance to another, or whenever a temperature difference exists between two bodies, heat will transfer from the high temperature region to the low temperature region. Heat transfer is energy transfer in the form of heat, which takes place between two bodies, or through a single body, solely due to a temperature difference.

The study of heat transfer attempts to predict the rate at which heat is transferred through a substance or from one body to another. Unlike thermodynamics, heat transfer does not always involve the transfer or flow of mass in order for heat to flow from the high temperature region to the low temperature region. There are three methods by which heat is transferred through a substance or between bodies: They are conduction, connection and radiation.

2.3.1 Conduction Heat Transfer

Conduction heat transfer occurs through, and by means of, intervening matter. It does not involve any obvious motion of the matter. It is a mechanism of internal energy exchange from one part of a substance to another part by the exchange of energy between molecules of the substance by direct communication. The molecules physically touch one another and transfer heat as they touch.

Consider a long metal rod with one end placed in a flame and the other out of the flame. In time, the end of the rod that is out of the flame will get hotter even though it is not in direct contact with the flame. Heat reaches the cooler end by conduction through the material of the rod. Microscopically, the molecules at the hot end of the rod increase in energy and increase the energy of their vibrations as the temperature increases. The heated molecules then interact with their more slowly moving neighbors farther from the flame. As they interact, they share some of their energy with these neighbors, who, in turn, pass energy along to those still farther from the flame. In this manner, energy associated with thermal motion is

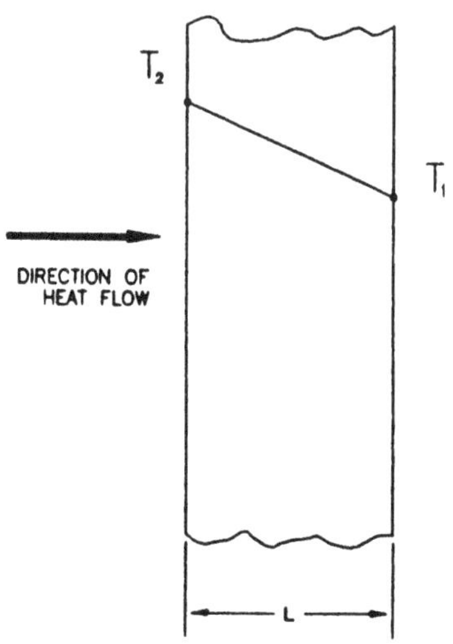

Figure 2-4. Heat conduction through a wall.

passed from molecule to molecule, while each individual molecule remains in its original position.

As another example, suppose a thick homogenous concrete wall had a hot plate applied to one side and a cold plate applied to the other side, as shown in Figure 2-4. Assume also that the hot and cold plates had been applied for a long period of time. Heat would flow through the concrete wall from the hot plate to the cold plate at a constant rate. The rate of heat flow would be a function of the wall material in this case concrete and the thickness of the material.

The concrete offers some resistance to the heat as it flows. Some materials transmit or conduct heat more readily than others. As an example, glass is frequently used for cookware since it conducts heat rather quickly and easily. Metals are also excellent conductors of heat. But some other materials are poor conductors of heat. Wood and polystyrene (styrofoam) are poor conductors of heat and, therefore, are frequently used as insulators. Thermal conductance is a measure of the ease with which heat travels through material of a specified thickness. Knowing the thermal conductance, or just conductance, of a given material, the rate at which heat is transferred can be determined. Equation (2.5) may be used to predict the rate at which heat is conducted through a given material of a given thickness.

$$q = CA(T_2 - T_1) \qquad \text{Eq. (2.5)}$$

where q = rate of heat transfer, in Btu/hr.
C = thermal conductance of material of specified thickness, in Btu/hr.-ft^2- °F
A = surface area of the material, in square feet
$T_2 - T_1$ = difference in temperature, in °F.

Equation (2.5) is used to predict the steady state conduction of heat. In other words, it is assumed that the temperature difference $T_2 - T_1$ has existed for a long period of time and the rate of heat transfer is a constant (not changing with time).

Example 2.2. Assume that 100 square feet of 17/2 inch thick gypsum board (C = 2.2 Btu/hr.-ft.^{2}- °F) has a temperature of 70°F on one side and 100°F on the other side. At what rate is heat conducted through the gypsum board?

By using Equation (2.5):

$$q = (2.2 \ \text{BTU/hr.} - \text{ft.}^2 - °\text{F})(100 \ \text{ft.}^2)(70 - 10)°\text{F} = 13,230 \ \text{Btu/hr.}$$

Thermal conductance is a measure of the ease with which a material of a given thickness transmits heat. Conversely, the reciprocal of the conductance gives the resistance to heat flow (R).

$$R = \frac{1}{C} \qquad \text{Eq. (2.6)}$$

The higher the R value, the greater the resistance to the flow of heat.

Thermal conductance is provided for a material of a given thickness. Frequently, though, tables give the conductance per unit thickness, which is known as thermal conductivity of the material.

$$C = \frac{k}{x} \qquad \text{Eq. (2.7)}$$

where k = thermal conductivity, in Btu/hr - ft.^{2} - °F - in or Btu/hr-ft^2-°F-ft.
x = material thickness in inches or feet

Note: Care must be taken to ensure that all units are consistent.

Example 2.3. What is the thermal conductance of a 1/2-inch gypsum board that has a thermal conductivity of k = 0.0926 Btu/hr. - ft.^{2} - °F - in.

$$x = (1/2 \text{ inch})(1 \text{ ft.}/12 \text{ in}) = 0.0417 \text{ ft.}$$

from Equation (2.7)

$$C = \frac{k}{x} = (0.0926)(1/0.0417) = 2.22 \text{ Btu/hr.} - \text{ft.}^2 - {}^\circ\text{F}$$

For a given material of a given thickness, Equations (2.5) and (2.7) may be combined to yield Equation (2.8). Equation (2.8) is used frequently for calculating the rate of conduction heat transfer through building exterior surfaces, such as walls and roofs.

$$q = \frac{kA(T_2 - T_1)}{L} \qquad \text{Eq. (2.8)}$$

where q = rate of conduction heat transfer, in Btu/hr.
k = the thermal conductivity of the material, in Btu/hr.-ft.-°F
A = the area of heat transfer, in ft.^{2}
L = the thickness of the material, in feet
T_2 = the high temperature side of the material, in °F
T_1 = the low temperature side of the material, in °F

The thermal conductivity of most common materials may be obtained from ASHRAE (1997) and Holman (1972) or any heat transfer text book.

Example 2.4. A 75-square-foot concrete wall that is 8 inches thick, has a temperature of 95°F on one side and a temperature of 25°F on the other side. The thermal conductivity of concrete is 0.17 Btu/hr.-ft.-°F. Assuming the temperatures have been maintained for a long period of time, what is the rate of heat transferred through the wall?

Using Equation (2.5),

$$q = (0.17\ \text{Btu/hr.} - \text{ft.} - {}^\circ\text{F})(75\ \text{ft.}^2)(95 - 25\ {}^\circ\text{F})(12/8) = 1,339\ \text{Btu/hr.}$$

Note: All units must be consistent, i.e., all feet or all inches.

Example 2.5. Assume the wall in Example 2.4 had 2 inches of polystyrene insulation with a thermal conductivity of 0.18 Btu/hr.-in.-°F applied to one side. What is the rate of heat transfer through the composite wall?

For the polystyrene,

$$k = (0.18\ \text{Btu/hr.} - \text{in.} - {}^\circ\text{F})(12\ \text{in./ft.}) = 2.16\ \text{Btu/hr.} - \text{ft.} - {}^\circ\text{F.}$$

Modifying Equation (2.8),

$$q = \frac{75(95 - 25)}{(1/0.17)(8/12) + (1/2.16)(2/12)} = 1,313\ \text{Btu/hr.}$$

In the above examples, it was assumed that the conditions (the temperature difference) had existed for a long period of time. This assumption implies that the heat flow had reached a steady state condition. In other words, the temperatures were not varying and there was no thermal storage within the wall itself. The reader may want to refer to ASHRAE (1997) and Holman (1972) for more information about non-steady heat flow and thermal storage.

2.3.2 Convection Heat Transfer

Convection is a heat transfer mechanism that is caused by the motion of a fluid (liquid or gas), within the fluid. Heat is transferred as it flows by the mixing of the molecules within the fluid. An example of convection heat transfer would be cold air flowing over a warm surface. As the air flows, the air close to the surface is heated as it comes in contact with the surface. As the air flows, it also mixes, and the air heated at the surface mixes with the free stream air.

Convection heat transfer, or just convection, can be due to either forced flow or natural flow. In the above example, the flow of the air over the surface could be forced flow from a fan.

An example of natural convection would be a cold fluid next to a warm vertical surface. As the fluid next to the surface is heated, the natural buoyancy forces due to the change in density of the fluid, would cause the warmed fluid to rise. This would cause a natural flow of fluid past the surface. Conversely, if the fluid were cooled by the surface, the fluid next to the surface would drop.

Whenever a fluid moves past a solid surface, it may be observed that the fluid velocity varies from zero immediately adjacent to the surface, to a "free stream" velocity some distance away. The flow of the fluid can be forced flow from a fan or pump, or it can be a natural flow. Natural convection occurs due to changes in density of the fluid at different temperatures.

The resistance to heat transfer at the solid-fluid boundary is a function of the convection, or film, coefficient, h. The units for the convection are Btu per hour-square foot of surface area per °F temperature difference. The equation for convection heat transfer is known as Newton's Law of Cooling and may be expressed as

$$q = hA(T_s - T_f) \qquad \text{Eq. (2.9)}$$

where q = rate of heat transfer, in Btu/hr.
h = convection heat transfer coefficient, in Btu/hr. - ft.^{2} - °F
A = surface area, in ft.^{2}
T_s = surface temperature, in °F
T_f = free stream fluid temperature, in °F

The convection coefficient is a function of the fluid velocity (if any), the difference in temperature between the fluid and the surface, the orientation of the surface (vertical, horizontal, etc.), the roughness of the surface and the viscosity of the fluid. Fluid viscosity is discussed later in this chapter. Convection coefficients are frequently determined experimentally, although they may also be calculated mathematically from empirical formulas.

Example 2.6. Air at 50°F if flowing over a flat surface that is maintained at 130°F. The surface is 8 feet by 10 feet. The convection coefficient has been determined to be 5 Btu/hr.-ft.^{2}-°F. What is the rate of heat transfer between the surface and the air?

Using Equation (2.6),

$$\begin{aligned} q &= (8 \text{ ft.} \times 10 \text{ ft.})\,(5 \text{ Btu/hr.} - \text{ft.}^2 - °\text{F})\,(130 - 50\ °\text{F}) \\ &= 32{,}000 \text{ Btu/hr.} \end{aligned}$$

Values for the convection heat transfer coefficient, **h**, are determined experimentally. In a manner similar to the resistance for thermal conductance, the resistance for the convection coefficient is given as

$$R = \frac{1}{h} \qquad \text{Eq. (2.10)}$$

2.3.3 Radiation Heat Transfer

Radiation heat transfer is heat that is transferred via electromagnetic waves. In order for radiation heat transfer, or just radiation, to occur between two bodies, it is not necessary that the two bodies be in direct physical contact with each other. Also, it is not necessary that there be a medium through which heat is transferred as radiation can pass through a complete vacuum. An example of radiation is the heating of an object by another object, such as the sun heating a surface on earth.

All objects radiate energy, regardless of their temperature. The greater the temperature of the object, the greater the energy radiated. All objects also receivc radiant energy from all other objects. The radiant exchange goes on continuously. When an object reaches a constant temperature, it does not stop radiating energy, but it is receiving energy at the same rate that it is giving off energy.

The rate at which a body emits radiant energy increases very rapidly as the temperature increases. The amount of radiant energy emitted is proportional to the fourth power of the body's absolute temperature. Examples of radiant energy include visible light, infrared radiation and ultra-violet radiation. The type of radiation is a function of the frequency and wavelength of the electro-magnetic radiation, which are functions of the absolute temperature of the radiating object. As the temperature increases, the wavelength of the radiation shortens. At 800°C, a body would emit enough visible radiation to be self-luminous and appear to be "red hot." At 3000°C, the temperature would be hot enough for the object to be "white hot" and most of the radiant energy would be in the visible light spectrum. The thermal radiation spectrum is in the range of 0.1 to 100 μm, and the visible is from 0.39 to 0.78 μm. (1 μm = 10^{-6} meters).

The ability of a surface to emit and absorb radiant energy is referred to as the emittance or emissivity of a surface. Emittance is a dimensionless quantity that varies from 0 to 1.0, with 1.0 being a perfect emitter. A body that is a good emitter of energy is also a good receiver of radiant energy. An object with a black rough surface is an excellent emitter (as well as an excellent absorber) of radiant energy. A highly polished surface is a poor emitter and absorber of radiant energy since it has a low emittance.

The radiant energy exchange between two objects is also strongly dependent on the geometry of the two objects. In order for objects to exchange radiant energy, they must "see each other." If objects are shielded by another object, they cannot exchange radiant energy unless the shield is transparent to the radiation. If the objects are partially shielded, the radiant energy will be reduced proportionally. This is accounted for by a shape factor (configuration factor) between the objects. A shape factor of 0 means that the objects cannot see each other at all. A shape factor of 1.0 indicated that the objects only "see each other" and nothing else (all radiant energy leaving one object falls on the other and nothing else). The distance between the objects also affects the shape factor.

All of the radiation falling on a surface will be either absorbed by the surface, reflected by the surface or transmitted through the surface. Figure 2-5 shows the distribution of radiant energy falling on a surface. The relative amounts of absorbed, reflected or transmitted energy are a function of the surface absorptance, reflectance and transmittance.

For any given surface, the relative amounts are also a function of the angle of incidence of the radiation. The more directly the radiation falls on a surface, the greater its intensity and its absorptance.

The amount of radiant exchange between two objects may be predicted by Equation (2.11), which is known as the Stefan-Boltzman Law.

$$q = \sigma A F_e F_s (T_1^4 - T_2^4) \qquad \text{Eq. (2.11)}$$

where
q = the rate of heat transfer, in Btu/hr.
σ = a constant, 0.1714×10^{-8}, Btu/hr.- ft.^{2} -°R^4
F_e = absorption factor
F_s = shape or configuration factor
A = surface area
$T_1 T_2$ = the absolute temperature of the bodies, in °R

Note: The relative temperature of 0°F is equal to the absolute temperature of 460°R.

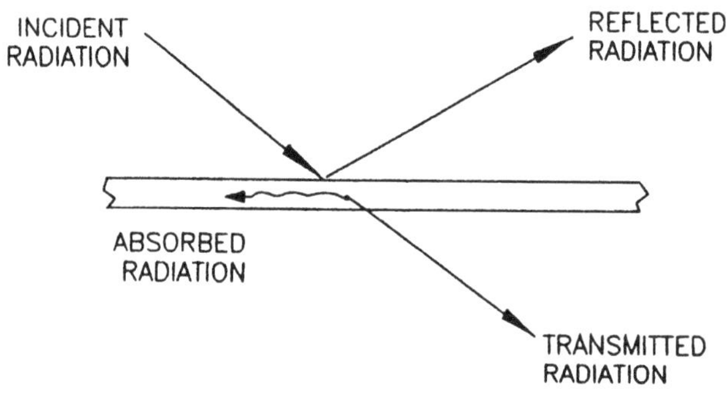

Figure 2-5. Distribution of radiant energy on a surface.

Example 2.7. Two flat black plates that are infinitely large in each direction are separated by a space that is occupied by a vacuum. One plate is maintained at 0°F and the other at 150°F. What is the radiant heat transfer per square foot between the plates?

Since the plates are infinitely long in each direction, they can only see each other and therefore their shape factor is equal to 1.0. Since they are flat black, their emissivity of each can be assumed to be 1.0.

To convert the temperatures of the two plates to absolute temperatures, their temperatures must be converted from degrees Fahrenheit to degrees Rankine. To make the conversion, it is necessary to add 460°F to their relative temperatures. The temperature of the cold plate would be 460°R and the hotter plate would be 610°R.

Using Equation (2.11)

$$\frac{q}{A} = (0.1714\text{x}10^{-8}\ \text{Btu/hr.} - \text{ft.}^2 - {}^{\circ}\text{R}^4)(1)(1)\left[(610^{\circ}\text{R})^4 - (460^{\circ}\text{R})^4\right]$$
$$= 160.6\ \text{Btu/hr.} - \text{ft.}^2$$

2.3.4 Transient Heat Transfer

Previously, all of the heat transfer examples were steady state, one directional. In other words, the rate of heat transfer was assumed to be constant over a long period of time and assumed the heat to flow in only one direction. In steady state heat transfer, heat entering an object, or control volume, is equal to the heat leaving, and this has been the case for a long period of time. No heat is stored. Frequently, in actual heat transfer applications, this is not the case.

Consider a situation where a concrete wall is shaded from the sun's rays by some external object. Assume that the object is suddenly moved and the sun's rays shine on the outside surface of the wall. The exterior surface immediately begins to increase in temperature, but the interior surface does not. There will be a considerable time lag between the time that the sun's rays initially strike the exterior surface and when the inside surface begins to increase in temperature. Thus, the rate of heat transfer through the wall is not constant, but varies with time.

The time lag between the application of heat to one surface and heat's appearance on the other side is due to thermal or heat storage of the material. The difference between the instantaneous heat gain on a building and the actual cooling load the air conditioning system must absorb is illustrated graphically in Figure 2-6.

Note in Figure 2-6 that the instantaneous heat gain peaks sooner and is greater than the actual cooling load. Only where the instantaneous heat and the cooling load curves cross is the heat gain equal to the cooling load. Prior to the time when the curves cross, the heat gain is greater than the cooling load and heat is stored by the building mass. After the curves cross,

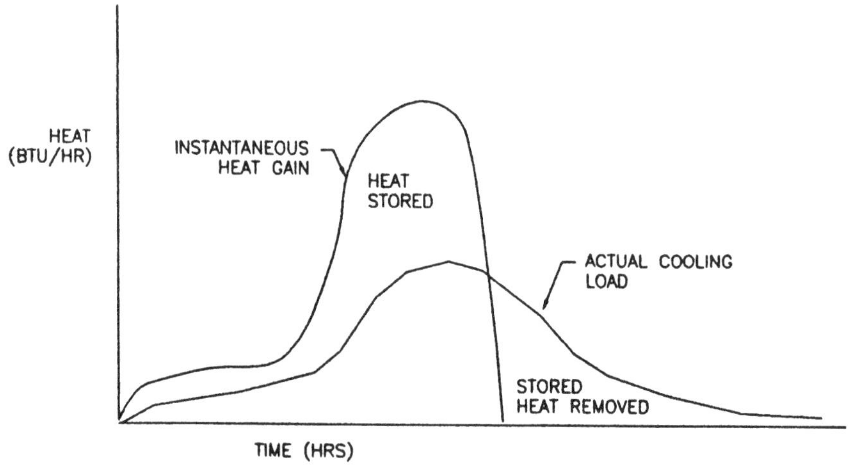

Figure 2-6. The effect of thermal storage on cooling loads.

the building is no longer storing heat. The building mass is rejecting heat to the cooling system.

The thermal mass of a building directly affects the time lag between the peak heat gain and peak cooling load, and inversely affects the intensity of the peak cooling load. The greater the thermal mass of the building, the greater the time lag and the lower the peak intensity of the cooling load. This is illustrated graphically in Figure 2-7.

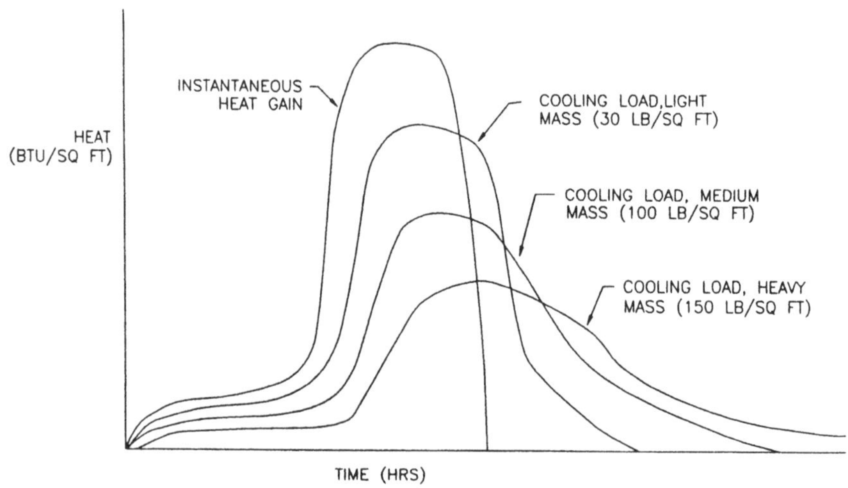

Figure 2-7. The effect of building mass on cooling loads.

In some cases, the time lag between the heat gain and the cooling load is so great, the peak cooling load for the air conditioning system occurs at night when the building is empty.

2.3.5 Log Mean Temperature Difference

Heat exchange devices (heat exchangers) are used frequently in energy consuming systems. Examples of heat exchange devices include heating and cooling coils, furnace heat exchangers, which exchange heat between flue gases and the supply air, steam to heating water heat exchangers, etc. Whenever heat is transferred between two fluid streams that do not intermix, a heat exchanger is used.

The amount of heat that is transferred between two fluids passing through a heat exchanger is a direct function of the temperature difference between the two fluid streams. The log mean temperature difference is the effective temperature difference between the two fluid streams. For an example, consider a hot water heating coil in an air handling unit. Assume the air enters the coil at 40°F and leaves at 100°F. Assumc also that the heating water enters the coil at 200°F and leaves at 170°F. What is the effective temperature difference between the air and the water in the coil? It is necessary to determine the log mean temperature difference in order to answer this question and similar questions involving heat exchange devices.

The log mean temperature difference (LMTD) is the temperature difference at one end of the heat exchanger less the temperature difference at the other end of the heat exchanger, divided by the natural logarithm of the ratio of these two temperature differences; it is expressed mathematically by Equation (2.12).

$$LMTD = \frac{(T_{h2} - T_{c2}) - (T_{h1} - T_{c1})}{ln\ [(T_{h2} - T_{c2})/(T_{h1} - T_{c1})]} \qquad \text{Eq. (2.12)}$$

where T_{h1} = entering temperature of the hotter fluid
T_{h2} = leaving temperature of the hotter fluid
T_{c1} = entering temperature of the cooler fluid
T_{c2} = leaving temperature of the cooler fluid

A flow schematic of a heat exchanger is shown in Figure 2-8.

Equation (2.12) is based on three important assumptions:

1. the specific heat of the two fluids remains constant
2. the convection heat transfer coefficients of the two fluids are constant.
3. the flow rates of the two fluid streams remain constant.

From Figure 2-7, it may be observed that the schematic diagram shows a counterflow heat exchanger. That is, the flow of the two fluid streams inside the heat exchanger run opposite, or counter, to each other. In this

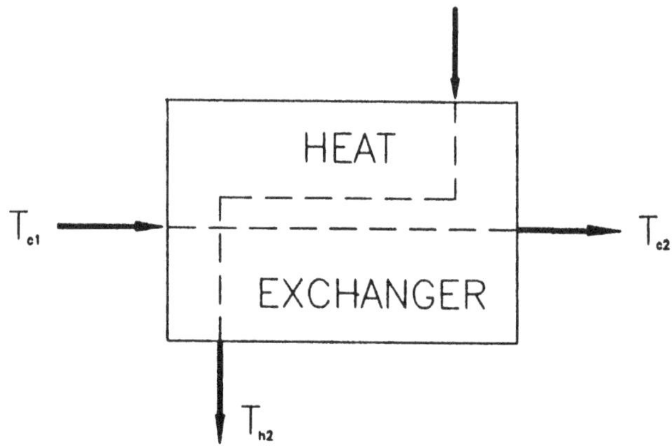

Figure 2-8. Heat exchanger flow schematic.

case, the entering cold fluid (which is the coldest) meets the leaving hot fluid (which is also the coolest). Had the hot fluid entered at the top left of the heat exchanger schematic and exited at the bottom right, the heat exchanger would have been a parallel flow heat exchanger since the flow streams inside it would be parallel.

Example 2.8. Air enters a heating coil at 40°F and is heated to 100°F. The coil has heating water entering at 200°F and leaving at 170°F. Assuming constant flow and fluid properties for both streams, what is the effective temperature difference between the air and water?

Using Equation (2.12)

$$LMTD = \frac{(170-100)-(200-40)}{ln\left[\frac{(170-100)}{(200-40)}\right]} = \frac{-90}{ln\ (0.4375)} = 108.9°F$$

2.4 CONCEPTS OF FLUID MECHANICS

A fluid may be defined as a substance that deforms continuously when subjected to a shear stress, no matter how small that stress may be. A fluid is a substance that is capable of flowing, having particles that easily move and change their relative position without separation of the mass, and that yields easily to pressure. A fluid may be a liquid, such as water, or it may be a gas, such as air.

Gases are very readily compressible, whereas liquids are only slightly compressible. For most energy analysis applications, the pressures are low enough that compressibility of air may be ignored. In other words, the density of air may be considered to be a constant without any significant error in the calculations.

Fluid statics deals with fluids at rest while fluid dynamics deals with fluids in motion. Fluid statics involves the forces exerted by a body of fluid at rest, such as fluid pressure, on a submerged object. Fluid dynamics deals with the motion of a fluid and the forces that result from the motion of that fluid.

2.4.1 Basis Concepts in Fluid Mechanics

2.4.1.1 Pressure

Fluid pressure is a concept that most people are familiar with. Any fluid at rest exerts a perpendicular, or normal, force on its boundary and any submerged objects, which is known as pressure; this type of pressure is called static pressure. The static pressure of a fluid is a result of the weight of the fluid itself. The static pressure in a fluid is directly proportional to its depth. The deeper the location within the fluid, the greater the static pressure. As an example, the pressure exerted by the atmosphere is greater at sea level than it is on high mountains. The forces exerted by static pressure are equal in all directions.

When performing calculations or making measurements involving static pressure in a fluid, the fluid pressure is usually stated with respect to a reference pressure. Normally, the reference pressure is atmospheric pressure and the resulting measured pressure is excess pressure above atmospheric pressure, or gage pressure. Pressure measured relative to a perfect vacuum (absolute zero pressure) is called absolute pressure. The relationship between pressure measurements is shown in Figure 2-9.

The relationship between gage pressure and absolute pressure is given by Equation (2.13).

$$P_{abs} = P_{gage} + P_{atm} \qquad \text{Eq. (2.13)}$$

where P_{abs} = absolute pressure
P_{gage} = gage pressure
P_{atm} = atmospheric pressure

Vacuum is the depression of pressure below atmospheric pressure. Standard atmospheric pressure is assumed to be 14.7 psia (29.92 inches mercury). Measurements of gage pressure are normally in pounds per square inch (psi) or feet of water column. Low pressures, less than 1 psi, are usually expressed in inches of mercury or inches of water gage (wg).

A fluid in motion also exerts an additional pressure due to its momentum. This pressure is exerted in the direction of flow. The pressure due to fluid motion is known as velocity pressure. For any given fluid, the magnitude of the velocity pressure is a direct function of the velocity of the fluid. The greater the velocity of the fluid, the greater the velocity pressure.

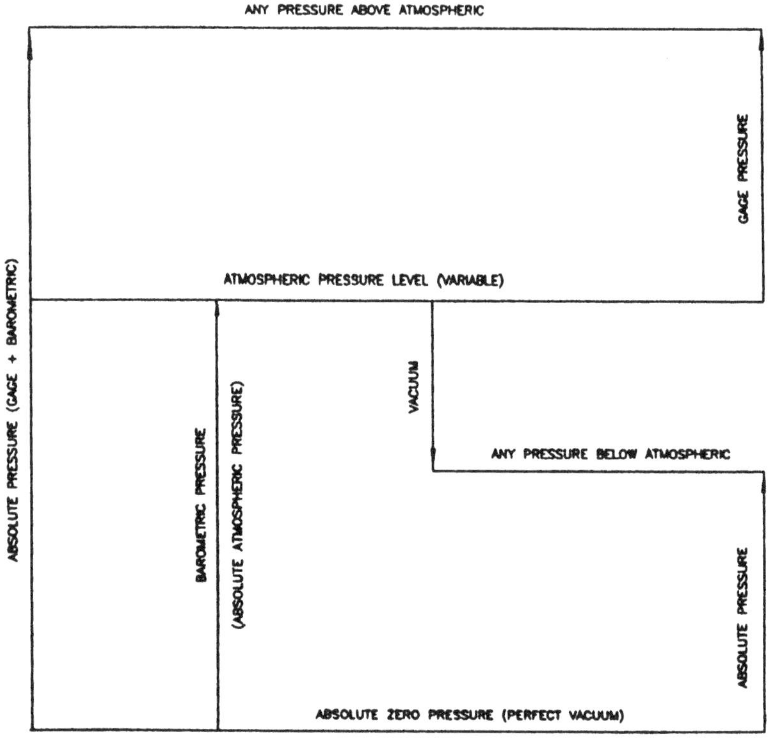

Figure 2-9. Relationship between pressure measurements.

2.4.1.2 Density

The density of a fluid is its mass per unit volume. A homogenous material has the same density throughout. The density of a liquid is given by Equation (2.14).

$$\sigma = \frac{m}{V} \qquad \text{Eq. (2.14)}$$

where m = mass, in pounds
V = volume, in ft^3.

For liquids, which are incompressible for the most part, the density may be assumed as a constant. The density of most liquids does, however, vary to some extent with temperature. Unless liquid temperatures are quite high, the density of a liquid may be assumed a constant without any significant error.

The density of gases and vapors cannot be assumed to be constant. The density of any given gas varies with the absolute pressure and absolute temperature. The density of a gas may be calculated using Equation (2.15).

$$\sigma = \frac{P}{RT} \qquad \text{Eq. (2.15)}$$

where P = the absolute pressure, in psia.
R = a gas constant for the particular gas, in ft.-lbf./lbm.-°R
T = the absolute temperature, in °R.

Equation (2.15) is applicable only for ideal or perfect gases, such as air.

2.4.2 Viscosity

Viscosity is a measure of the ease with which a fluid flows. It is that property of a fluid, which causes resistance to relative motion within the fluid and is observed only when it is in motion. Viscosity enables a fluid to develop and maintain an amount of shearing stress dependent upon the velocity of the flow, and then to offer continued resistance to flow. Energy is lost because of the friction in the flowing fluid due to the viscosity.

Viscosity is caused by the molecular structure of the fluid. In the liquid state, the molecules are packed closely together and the viscosity is a result of the cohesiveness of the molecules. As the temperature of a liquid increases, the cohesion decreases and the viscosity decreases. For a gas, the molecules are spaced much farther apart and the viscosity is due to the activity of the molecules, which strike against one another. As the temperature increases, the activity of the molecules increase, the collisions increase and the viscosity of the gas increases.

The absolute viscosity of a fluid is a measure of its resistance to internal deformation of shear. Kinematic viscosity is the ratio of the absolute viscosity to the mass density.

The effect of pressure on the viscosity of liquids and gases is so small that it may be ignored without causing any significant error. With an increase in temperature, the viscosity of most liquids decreases, while the viscosity of most gases increases.

When performing calculations involving fluid flow, it is important to determine the character of the flow. Under some conditions, the fluid will appear to flow in layers in a smooth and regular manner. This type of flow is referred to as laminar flow. In laminar flow, the layers of fluid slide over one another. If the flow is not regular and smooth, it is referred to as turbulent flow. In turbulent flow there is flow in random directions within the general flow stream.

Energy is lost due to friction in a flowing fluid as the result of fluid viscosity. The magnitude of the energy loss for any given fluid is directly dependent on whether the flow is laminar or turbulent.

2.4.3 The Continuity Equation

According to the continuity equation, which is known as the conservation of mass for fluid flow, no fluid is created or destroyed as fluid flows in a closed conduit. In more basic terms, it states that the fluid flow out of a system is equal to the fluid flow into the system, plus any storage within the system. For incompressible fluids, the mass flow past one point in a closed conduit must be equal to the flow past another point in the conduit if there is no storage within. If the cross-sectional area of the conduit changes, the fluid velocity must change accordingly. The continuity equation is important in the study of the flow of fluids in closed conduits such as pipes and ducts. Equation (2.16) is a mathematical expression of the continuity equation.

$$Q = A_1 V_1 = A_2 V_2 = \text{constant} \qquad \text{Eq. (2.16)}$$

where Q = the rate of fluid flow, in cfm, gpm, ft.^{3}/sec., etc.
A = the cross-sectional area of the conduit, in ft.^{2}
V = the velocity of the fluid, in ft./min. or ft./sec.

Equation (2.16) assumes steady state flow, that is, a constant flow rate and a non-compressible fluid.

Example 2-9. Figure 2.10 shows a pipe that increases from an inside diameter of 8 inches at point 1 to an inside diameter of 12 inches at point 2. Water at 60°F is flowing through the pipe. If the water velocity is 5.72 ft./sec. at point 1, how much water is flowing through the pipe and what is the velocity at point 2?

$$A_1 = (1/4)(\pi)(8^2)(1/144) = 0.35 \text{ ft.}^2$$

$$A_2 = (1/4)(\pi)(12^2)(1/144) = 0.79 \text{ ft.}^2$$

Using Equation (2.16), the flow rate is

$$Q = A_1 V_1 = (0.35 \text{ ft.}^2)(5.72 \text{ ft./sec}) = 2.0 \text{ ft.}^3\text{/sec.}$$

Again, using Equation (2.16), the velocity at point 2 is

$$V_2 = Q/A_2 = \frac{(2.0 \text{ ft.}^3\text{/sec})}{(0.79 \text{ ft.}^2)} = 2.53 \text{ ft./sec.}$$

Equation (2.16) is applicable only to non-compressible fluids and technically it is not applicable to calculations involving a compressible fluid, such as air. However, for most calculations, the pressures and velocities are so low that air may be assumed to be incompressible, and Equation (2.16) may be used without any significant error. For applica-

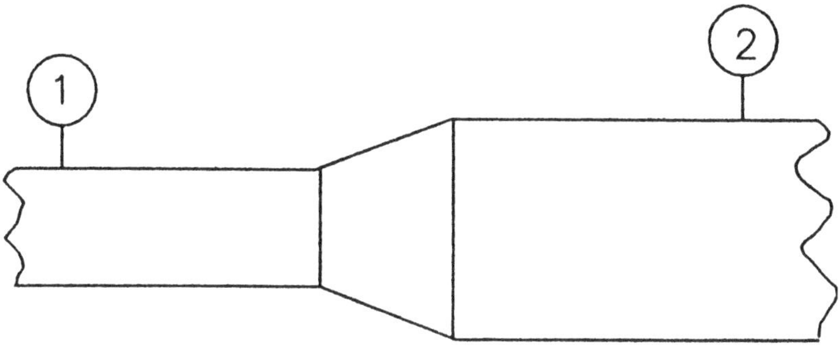

Figure 2-10. Pipe for example 2.9.

tions that do involve higher pressures and velocities, Equation (2.17) may be used for the continuity equation. It is applicable to any fluid flow at velocities less than sonic velocity.

$$Q = \sigma_1 A_1 V_1 = \sigma_2 A_2 V_2 = \text{constant} \qquad \text{Eq. (2.17)}$$

where σ = the fluid density, in lbm./ft.^{3}

2.5.3 Bernoulli's Equation

Fluid flow also follows the law of conservation of energy, i.e., energy within a system is neither created nor destroyed. It can, however, be transformed from one form to anther. Bernoulli's Equation is a means of expressing and applying the law of conservation of energy to the flow of fluids in a closed system such as a conduit. It relates the pressure, flow velocity and height for flow of a fluid.

In order for fluid to flow through a conduit, energy must be applied to the fluid in the form of work (pressure). Bernoulli's Equation states mathematically that as an incompressible fluid flows, the total head (pressure) remains constant. In other words, the energy (work) remains constant as the fluid flows.

There are three forms of energy that should be considered in any fluid flow analysis. They are potential energy, kinetic energy, and flow energy or work. They may be expressed as follows:

1. Potential Energy. Energy due to the elevation of the fluid relative to some reference elevation.

$$\text{P.E.} = wZ \qquad \text{Eq. (2.18)}$$

where w = the weight of the fluid, in pounds
Z = the elevation of the fluid above a reference elevation, in feet.

2. Kinetic Energy. Energy due to the velocity and momentum of the fluid

$$\text{K.E.} = 1/2(wV^2/g) \qquad \text{Eq. (2.19)}$$

where V = the velocity of the fluid, in ft./sec.
g = the acceleration due to gravity (32.2 ft./sec.^{2})

3. Flow Energy or Work. Also called pressure energy or flow work, it represents the amount of work necessary to move the fluid against a pressure.

$$\text{F.E.} = wP/\gamma \qquad \text{Eq. (2.20)}$$

where P = the pressure, in lbs./ft.^{2}
γ = the specific weight of the fluid, in lbs./ft.^{3}

The total amount of energy of these three forms of energy possessed by the fluid as it flows is expressed in Equation (2.21).

$$E = \text{F.E.} + \text{P.E.} + \text{K.E.} = \text{w}P/\gamma + \text{w}Z + \text{w}V^2/2g \qquad \text{Eq. (2.21)}$$

Since energy is conserved within the system, the energy between two points in the system must remain constant. Assuming the mass flow remains constant, Equation (2.21) may be expressed as

$$P_1/\gamma + Z_1 + V_1^2/2g = P_2/\gamma + Z_2 + V_2^2/2g \qquad \text{Eq. (2.22)}$$

Equation (2.22) is referred to as Bernoulli's Equation.

In the analysis of energy consuming systems, the potential energy due to elevation changes can usually be neglected since elevation changes are relatively small. In the analysis of most energy consuming systems, potential energy and kinetic energy cannot be ignored. Since potential energy can usually be neglected in the design of air systems, Equation (2.22) may be rewritten as Equation (2.23).

$$(1/\gamma)(P_2 - P_1) + (1/2g)(V_2^2 - V_1^2) = 0 \qquad \text{Eq. (2.23)}$$

Equation (2.23) states mathematically that the only transformation of energy in an air system is due to changes in pressure. The changes in pressure in air distribution are due to friction losses and changes in velocity. Energy may be lost from a system since the friction converts pressure energy to heat.

2.4.4 Fluid Flow Friction Losses

In "Section 2.3.2: Convection Heat Transfer", it was noted that whenever a fluid flows in a closed conduit or along a solid surface, the fluid right next

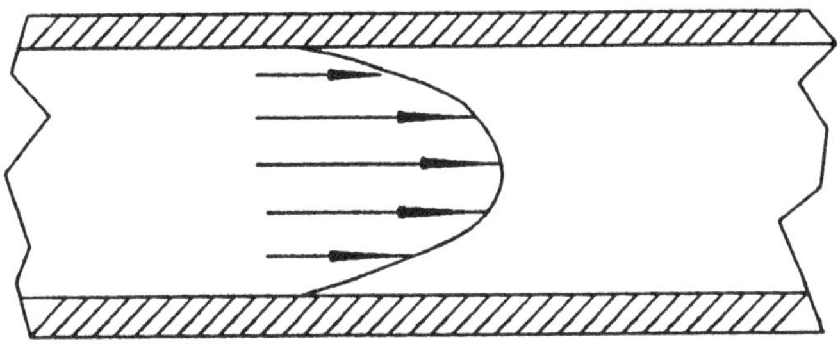

Figure 2-11. Fluid velocity profile.

to the surface is actually not moving and has zero velocity. This is due to friction between the flowing fluid molecules and the solid surface. Figure 2.11 shows a typical profile for a fluid in a conduit such as a duct.

In Figure 2-11, the arrows represent relative velocities of the fluid, with the longer arrows being higher velocities and the shorter arrows indicating lower velocities. It may be observed from Figure 2.11 that, while the fluid near the walls of the conduit has a velocity of zero, the velocity near the center of the conduit is moving at a much greater velocity. The velocity near the center is known as the free stream velocity.

The velocity gradient exists between the fluid at the walls and the center due to friction between the individual molecules of the fluid. Friction loss also results from the friction between the fluid and the walls of the conduit. The energy loss due to friction may be expressed as

$$h_L = C_L V^2/2g \qquad \text{Eq. (2.24)}$$

where h_L = the energy loss due to friction
C_L = a loss coefficient or friction factor
V = the velocity of the fluid
g = the acceleration due to gravity

The loss coefficient or friction factor is a function of the fluid velocity, fluid viscosity and the roughness of the conduit walls.

Chapter 3
Economics for Energy Management

Almost all energy conservation improvements involve some costs and investment of capital funds. The aim of the improvements is to reduce operating costs or production costs by reducing energy consumption and energy costs. Typically, the implementation of energy conservation improvements involve an initial outlay of capital: The investment must be made early on, while the benefits of energy cost savings do not occur until some time in the future. In order to aid in the decision to implement energy conservation improvements, an economic analysis, or evaluation, must be made for each proposed improvement. The economic evaluation provides a method of comparing present costs with future monetary returns.

3.1 BASIC ECONOMIC INVESTMENT ANALYSIS

3.1.1 Benefits and Cost

In general, the motivation for facility owners and managers to invest in and implement energy conservation projects is that they expect the resulting benefits to exceed the investment costs. Typical potential benefits for implementing energy conservation projects include:

- energy cost savings
- reduced size, hence lower capital cost of energy consuming equipment
- reduced maintenance costs for energy consuming equipment
- reduced costs for production labor
- improved product quality
- revenue from sales of recovered heat or energy
- reduced environmental pollution

The major cost associated with the implementation of energy conservation improvements is the initial investment, although there may be other costs, which are discussed later in this chapter. The initial investment includes not only the price and installation costs of equipment and materials, but also the ancillary costs of the finished project. These ancillary costs may include the transportation, installation, licensing fees, permits and, in some cases, the working capital required to finance the project. Other costs that may be associated with an energy conservation project include engineering planning and design costs, space costs, costs of production downtime during installation, system commissioning and "debugging," taxes and changes in hazard insurance costs.

After an energy conservation project is implemented, it is hoped that the benefits listed above will result in a reasonable return on investment. When an investment is made in an energy conservation project, returns will be realized over the lifetime of the project. The objective is to gain a net return of cash from the project and use the money to amortize the investment. Net returns are discrete or continuous. Returns are typically credited at discrete times (monthly, yearly, etc.) rather than on a continuous basis.

3.1.2 Time Value of Money and Equivalence

The implementation of energy conservation projects frequently result in financial consequences, which occur over a substantial period of time. When money consequences occur over a short period of time, the various sums of money are simply added up to obtain a net result. However, it is not appropriate to treat money the same way when the time span is greater.

The use of money is a valuable asset—so valuable that people are willing to pay to have money available for their use. Money can be rented in roughly the same way that a building is rented; however, with money, the charge for its use is called interest instead of rent. If interest is charged only on the amount of the principal, the interest is called simple interest. Money is rarely rented, or loaned, with simple interest. Charging interest on unpaid interest is called compound interest. Money is almost always loaned with compound interest.

Equivalence is a concept used in economics that implies that different repayment plans are equivalent if they give the borrower the same amount of money at the beginning of a loan. It is an essential factor in engineering economic analysis. When we are indifferent whether we have a sum of money in the future, or series of future sums of money, we say the present sum of money is *equivalent* to the future sum or series of future sums. Equivalence is very dependent upon the interest rate and time period in an economic analysis. Because of the time value money, one cannot simply add up the sums of money at different points in time directly. This means that a comparison between alternatives cannot be made in actual dollars at different points in time, but must be made in some equivalent comparable sums of money. It is necessary that we compute a value that represents the entire period of the loan. Typically, two payment streams are converted to "today's dollars." By computing equivalent sums at the same point in time ("now"), we will have values that may be validly compared. However, the technique of equivalence is not limited to a present comparison. Instead, it is possible to compare the equivalent sum of a cash flow at any point in time. It is possible to compare alternatives in equivalent (year 10) dollars rather than present (year 0) dollars.

3.1.3 Compound Interest

The interest rate is the ratio of the interest payable at the end of a time period to the money or principal owed at the beginning of the period. The interest period is generally a one year time period, although it may be payable at other times.

As indicated above, interest may be calculated in one of two ways: as simple or as compound interest. Simple interest is calculated over the life of the loan and is based only upon the principal of the loan. Compound interest involves the calculation of payable interest based upon the amount owed at the beginning of the appropriate time period for repayment. Compound interest charges interest on the unpaid balance, which includes the unpaid principal plus unpaid interest. It is common practice to use a yearly repayment system and to apply compound interest.

For a loan with compound interest, a single payment at the end of the loan, Equation (3.1), may be used to calculate the future payment.

$$F = P(1 + i)^n \qquad \text{Eq. (3.1)}$$

where i = the interest rate for the period
n = the number of interest periods
P = a present sum of money
F = a future sum of money

A functional notation that may be used to indicate Equation (3.1) is

$$F = P(F/P,\ i\%,\ n)$$

Equation (3.1) may be rearranged to calculate a present sum of money, P, given a future sum for a given compound interest rate, which results in Equation (3.2).

$$P = F(1 + i)^{-n} \qquad \text{Eq. (3.2)}$$

Equation (3.2) is referred to as the *single payment present worth* formula and it may be functionally abbreviated as

$$P = F(P/F,\ i\%,\ n)$$

Example 3.1 If $500 were deposited in a bank savings account, how much would be in the account three years hence if the bank paid 4% interest compounded annually?

The present sum of money is $500. The interest rate is 4% and in three years there are three interest periods. The future sum of money is to be calculated.

The functional notation is (*F/P*, 4%, 3). Using Equation (3.1):

$$F = P(1 + i)^n = 500(1 + 0.04)^3 = \$562.50$$

Example 3.2 If you wanted to have \$800 in a savings account at the end of four years, and 5% interest was paid annually, how much should you put into the savings account now?

$$F = \$800 \quad i = 0.05 \quad n = 4 \quad P = \text{unknown}$$

The functional notation would be (*P/F*, 5%, 4). Using Equation (3.2):

$$P = F(1 + i)^{-n} = 800(1 + 0.05)^{-4} = \$658.16$$

In the above examples, the compound interest was compounded annually, or once a year. Frequently, the interest is compounded for intervals less than one year, in which case n would not equal the number of years, but the number of interest periods.

Example 3.3 Suppose the bank changed its policy in Example 3.1 to "4% interest, compounded quarterly." For this situation a \$500 deposit now would result in how much money in the account at the end of three years? There are two elements involved with the 4% interest compounded quarterly:

---

1. 4% interest: Unless otherwise described, it is customary to assume the state interest is for a one year period. If the stated interest is for other than one year, then it must be indicated.
2. Compounded quarterly: This indicates there are four interest periods per year; an interest period is three months long.

The 4% interest is an annual rate; if it were anything different, it must be stated. Since there are four interest periods per year, the interest rate per period is 1% and for the total three year duration, there are 12 interest periods.

$$P = \$500 \quad i = 0.01 \quad n = 4 \times 3 = 12 \quad F = \text{unknown}$$

$$F = P(1 + i)^n = \$500(1 + 0.01)^{12} = \$563.50$$

Equations (3.1) and (3.2) may be used for compound interest, which is compounded for a discrete number of compounding periods. In many applications, however, the interest is continuously compounded and the number of interest periods approaches infinity. For continuous compounding, the single payment compound amount formula is given by

Equation (3.3) and the single payment present worth formula is given by Equation (3.4).

$$F = Pe^{rn} \qquad \text{Eq. (3.3)}$$

$$P = Fe^{-rn} \qquad \text{Eq. (3.4)}$$

where e = the exponential function = 2.71828
r = the nominal interest rate

Example 3.4 If $2,000 were deposited in a bank that pays 4% interest, compounded continuously, how much would be in the account at the end of two years?

Using Equation (3.3) where

r = nominal interest rate = 0.04
n = the number of years = 2

$$F = 2{,}000\ e^{(0.04 \times 2)} = \$2{,}166.00$$

Example 3.5 A bank offers to sell savings certificates that will pay the purchaser $5,000 at the end of 10 years, but will pay nothing to the purchaser in the meantime. If interest is computed at 6%, compounded continuously, at what price is the bank selling the certificates?

Using Equation (3.4) where

$$F = \$5{,}000 \qquad r = 0.06 \qquad n = 10 \text{ years}$$

$$P = 5{,}000\ e^{-(0.06 \times 10)} = \$2{,}744$$

Therefore, the bank is selling the $5,000 certificates for $2,744.

3.1.4 Economic Life and Depreciation

All equipment and systems have a finite economic life. With proper repair and maintenance, almost anything can be kept operating indefinitely. While it may be possible to keep a given piece of equipment or system going indefinitely, the cost may be excessive. Rather than basing decisions on what the remaining life of equipment could be, it is more important to consider its economic life. The economic life, which is used in the analysis of investment projects, is usually the best estimate of time that equipment can be economically used. The estimated service life of various pieces of equipment is discussed later in this chapter.

In addition to projected physical lifetimes for equipment, there are other limits on economic life. These limits have been taken into account by the Internal Revenue Service (IRS) in stipulating the economic life for certain investments. The economic life may also be mutually agreed upon by contract, in a lease or franchise, etc.

Depreciation is an accounting technique that assigns a portion of the total cost for tangible fixed assets as if their cost were a prepaid expense spread over their lives. The word depreciation is defined as a decrease in value. Depreciation may be defined in two ways: decrease in market value, or decrease in value to the owner. The accounting profession defines depreciation in a third way: The systematic allocation of the cost of an asset (less its salvage value) over its useful life. Therefore, there are three distinct definitions of depreciation:

- decline in market value of an asset
- decline in value of an asset to its owner
- systematic allocation of the cost of an asset (less its salvage value) over its useful life

Depreciable fixed assets consist of buildings, equipment, machinery and other tangible items used in the production of goods and services. There are several common accounting methods used to depreciate an asset, which include straight line depreciation, declining balance depreciation and sum-of-the-years depreciation. In all depreciation methods, the book value of the asset is the cost of the asset minus the cumulative depreciation. Therefore,

$$\text{Book value} = \text{cost} - \text{depreciation charges made}$$

Straight line depreciation is the simplest depreciation method. In this method, a constant depreciation charge is made for each year. The total amount to be depreciated is divided by the useful life in years to obtain the annual depreciation charge. The annual depreciation charge may be calculated using Equation (3.5).

$$\text{Annual straight line depreciation charge} = 1/N(P - F) \qquad \text{Eq. (3.5)}$$

where N = useful life in years
P = cost of asset
F = end of useful life salvage value

Example 3.6 If an asset has an initial cost of \$900, has a useful life of 5 years and an end of useful life salvage value of \$70, what is the annual straight line depreciation charge and what is the book value at the beginning and end of each year for the life of the asset?

Using Equation (3.5),

$$\text{Annual depreciation charge} = \frac{1}{5}(900 - 70) = \$166$$

Year	Book value before depreciation charge	Depreciation for year	Book value after depreciation charge
1	cost = $900	$166	$734
2	734	166	568
3	568	166	402
4	402	166	236
5	236	166	70 = salvage value
	total depreciation =	$830	

Declining balance depreciation is one of several accelerated write-off methods allowed by the IRS for assets with a life of three or more years. Under this method, a constant depreciation rate is applied to the book value of the asset. The depreciation rate allowed by the IRS depends on the type of asset and is stated in terms of the alternate straight line depreciation, which is $1/N$. For most equipment used for energy conservation projects, double the straight line rate is used, or $2/N$. Therefore

$$\text{Double declining balance depreciation in any year} = 2/N(\text{book value}) \qquad \text{Eq. (3.6)}$$

Example 3.7 Use the double declining balance depreciation method to calculate the depreciation for each year of the asset described in Example 3.6.

Cost of asset, P = $900
Useful life, N = 5 yrs.
End of useful life salvage value, F = $70
Using Equation (3.6)

$$\text{DDB depreciation} = \frac{2}{N}(P - \text{depreciation charge to date})$$

$$\text{1st year} = \frac{2}{5}(900 - 0) = \$360$$

$$\text{2nd year} = \frac{2}{5}(900 - 360) = 216$$

$$\text{3rd year} = \frac{2}{5}(900 - 576) = 130$$

$$4\text{th year} = \frac{2}{5}(900 - 706) = 78$$

$$5\text{th year} = \frac{2}{5}(900 - 784) = 46$$

$$\text{Total Depreciation} = \overline{\$830}$$

Another accelerated write–off method is the sum-of-the-years depreciation method. This method results in larger than straight line depreciation charges during the early years in the life of an asset, and necessarily smaller charges as the asset nears the end of its estimated useful life. Each year the depreciation charge is computed as the remaining useful life at the beginning of the year divided by the sum of the years digits for the remaining useful life; this ratio is multiplied by the total amount to be depreciated (P–F). The sum-of-the-years digits depreciation charge may be calculated using Equation (3.7).

$$\text{Sum–of–the–years digits depreciation} = L_R/L_S\ (P - F) \qquad \text{Eq. (3.7)}$$

where L_R = remaining useful life at beginning of year
L_S = sum-of-the-digits for total useful life

The sum-of-the-years digits may be calculated using Equation (3.8).

$$\text{Sum–of–the–years digits (sum)} = \frac{N}{2}(N + 1) \qquad \text{Eq. (3.8)}$$

Example 3.8 Use the sum-of-the-years digits (SYOD) depreciation method to calculate the depreciation for each year of the asset described in Example 3.6.

Cost of asset, P = 900
Useful life, N = 5 yrs.
End of useful life salvage value, F = 70

Using Equation (3.8) to calculate the sum-of-the-years digits,

$$\text{Sum} = \frac{5}{2}(5 + 1) = 15$$

Using Equation (3.7) to calculate the depreciation

$$1\text{st year SYOD depreciation} = \frac{5}{15}(900 - 70) = 277$$

$$2\text{nd year} = \frac{4}{15}(900 - 70) = 221$$

$$3\text{rd year} = \frac{3}{15}(900 - 70) = 166$$

$$4\text{th year} = \frac{2}{15}(900 - 70) = 111$$

$$5\text{th year} = \frac{1}{15}(900 - 70) = 55$$

$$\text{Total Depreciation} = \overline{\$830}$$

3.2 OWNING AND OPERATING COSTS

3.2.1 Owning Costs

Most energy conservation projects require an investment of capital funds. Depending upon the size of the proposed project, the initial investment could range from a few hundred dollars to millions of dollars. The initial investment includes not only the price of the equipment, material and installation, but also other ancillary costs. Other costs may include engineering studies, design costs, permits, taxes, insurance and rent. In addition to the costs listed above, the owning cost of a proposed project includes the opportunity cost or the cost of capital for the project.

3.2.2 Operating Costs

Operating costs include costs associated with the installation of a system or equipment. These costs may include fuel costs, electricity costs, as well as maintenance and utilities, such as water and sewer costs.

The total cost of electrical energy typically includes two types of charges for most commercial and industrial users: energy consumption charges and demand charges. The energy charge is basically the cost of the electricity generated and includes, in part, the capital costs of the generation equipment and fuel costs. Most utility rates have step rate schedules for consumption, and the last unit of energy consumed may be substantially different from that of the first.

Most electric utilities have added what is referred to as a fuel adjustment to the energy charge. Fuel adjustment charges result from variations in fuel prices used to generate electricity. This adjustment may not be reflected in the rate schedule. Typically, this is a direct charge passed on to the customer that reflects the escalation of fuel costs included in the initial setting of the energy charge.

Electrical demand charges are based on the customers peak kilowatt demand. The demand charge reflects the rate at which electrical energy is used. Typically, the demand represents the peak rate of energy use averaged over a specific time period, usually 15, 30 or 60 minutes. Additional information regarding electrical energy costs and electric billing can be found in Chapter 10.

Natural gas rates are usually a combination of two main components: (1) utility rate for gas consumption and (2) purchased gas adjustment (PGA) charges. Gas rate schedules are similar to electricity rate schedules in that they usually have step rate schedules for consumption. Although gas prices may include a PGA charge, they do not include demand charges. Gas is metered by volume; however, it is often sold by energy content (therm). Large facilities that have the ability to temporarily burn other fuels, such as light fuel oil, can qualify for special interruptible gas rates. During periods of peak gas usage, interruptible customers may be curtailed by the gas utility and have to temporarily switch to the alternate fuel. The utility rate and PGA are usually considerably cheaper for interruptible customers than for firm rate (noninterruptible) customers. Customers who use large amounts of natural gas may also purchase gas supplies on the open market. Natural gas purchased on the open market is called transportation gas. The customer contracts with a gas producer or broker at the source. Transportation fees are then paid to a pipeline company carrying the gas to the customer's local gas utility.

Other fossil fuels include coal, fuel oil, propane and diesel oil. Since these fuels are sold in bulk, they must be transported to the customers site and stored. Transportation costs are usually included in the fuel price. The cost for storage of these fuels must be included in the cost of these fuels.

3.3 SIMPLE ECONOMIC EVALUATION METHODS

These methods are the simplest and easiest methods used to evaluate alternative projects. These methods are called partial methods because they do not fully assess the economic desirability of alternatives. Despite their shortcomings, the partial techniques of analysis can serve a useful purpose. They can provide a first level measure of profitability, which is quick, simple and inexpensive to calculate. Simple economic methods of evaluation include the payback method and return on investment method.

3.3.1 Payback Method

The payback (payout) method determines the number of years required for the invested capital to be offset (paid back) by the resulting benefits. It is the number of years for the recovery of the invested capital, or to break even, and is defined as

$$\text{Payback} = \frac{\text{initial investment}}{\text{net annual benefit}} \qquad \text{Eq. (3.9)}$$

The advantage of the payback is its simplicity. However, it does not always measure profitability because the cash flows after the payback period are ignored. There are four important points to be understood about payback calculations.

- This method is an approximate rather than an exact economic calculation.
- All costs and all income or savings of the investment prior to payback are included without considering differences in their timing.
- All economic consequences beyond the payback period are completely ignored.
- Being an approximate calculation, payback period may or may not select the correct alternative.

Example 3.9 A recuperator for a boiler costs $10,000 to purchase and install. On the average, it costs $300 per year to maintain and it is expected that it will save $1,400 per year in gas costs by preheating the combustion air. What is the payback period for the recuperator?

Using Equation (3.9).

$$\text{Payback} = \frac{\$10{,}000}{\$1{,}400 - 300} = 9.1 \ \text{years}$$

3.3.2 Return on Investment

The return on investment (ROI) calculates the net average annual benefits as a percentage of the original invested capital. Equation (3.10) may be used to calculate the return on investment.

$$\text{ROI} = \frac{\text{Average Annual Net Benefit}}{\text{Original Capital Investment}} \times 100 \qquad \text{Eq. (3.10)}$$

Example 3.10 A waste heat economizer is installed on a boiler and the installed cost is $15,000. The economizer has an expected life of 10 years with no salvage value, will have an annual maintenance costs of $200 and is expected to save $5,000 in annual fuel cost. Assuming straight line depreciation, what is the annual return on investment?

Using Equation (3.5)

$$\text{Annual depreciation} = \frac{1}{10}(15{,}000 - 0) = \$1{,}500 \text{ per year}$$

Using Equation (3.10)

$$\text{ROI} = \frac{\$\,5{,}000 - (1{,}500 - 200)}{\$15{,}000} \times 100 = 22\%$$

3.4 COMPREHENSIVE ECONOMIC EVALUATION METHODS

The comprehensive economic evaluations use the basic principles of present value analysis to account for the time value of money. Since investments in energy conservation projects will generally require a number of expenditures of money spread over a period of time and will result in cost savings (or revenue receipts) also spread over time, dollars will be spent, saved or received at different times and, therefore, will not be of equal value.

Since costs, savings and receipts of money occur at different times and are, therefore, not of equal value, comprehensive evaluation methods using present worth analysis provide more accurate evaluations. The total present value (present worth) for any analysis is determined by summing the individual present worth of all individual items under consideration, including both future single payment items and series of equal future payments.

3.4.1 Single Payment Present Worth Analysis

The most commonly used economic analysis method is present worth. This technique compares the equivalent cash needed on hand to own and operate a system over an entire selected time period. It produces a single investment value in dollars at the beginning of a project. Equation (3.2) may be used to calculate the present worth for a future sum of money. Equation (3.11), which is called the series present worth formula, may be used to calculate the present worth of a uniform series of payments.

$$P = A\left[\frac{(1+i)^n - 1}{i(1+i)^n}\right] \qquad (3.11)$$

where A = an end of period cash receipt or disbursement in a uniform series for n periods

The functional notation for the series present worth is

$$P = A(P/A,\ i\%\ n)$$

Example 3.11 Two energy conservation projects are being considered. Project A has an initial cost of \$4,000 and the estimated annual energy cost savings is \$110. Project B has an initial cost of \$3,000 and the estimated annual energy cost savings is \$100. If the useful life of each project is 20 years and the interest rate is 8%, which project has the lower present worth or present cost?

For Project A, the initial cost = \$4,000.

Using Equation (3.11), the present worth of the energy cost savings is

$$P = \left[\frac{(1+i)^{20}-1}{0.08(1+.08)^{20}}\right] = \$100(9.818) = \$1{,}079.98$$

The present worth of Project A is

$$PW_A = \$4{,}000.00 - 1{,}079.98 = \$2{,}920.02$$

For Project B, the initial cost = \$3,000

The present worth of the energy cost savings is

$$P = \$100(9.818) = \$981.80$$

The present worth of Project B is

$$PW_B = \$3,000.00 - 981.80 = \$2,018.20$$

Project B has the lowest present worth or present cost.

3.4.2 Internal Rate of Return

The internal rate of return (IRR) method calculates a return on investment over a specified period of time. The IRR is a single figure of merit that is easily understood. It provides a basic comparison of return on investments. This method assumes that savings are reinvested at the same rate of return.

The internal rate of return analysis calculates the rate of return an investment is expected to yield. It expresses each investment alternative in terms of a rate of return (a compound interest rate). The criterion for selection among alternatives is to choose the investment with the highest rate of return. The annual savings and costs are not discounted, and a cash flow is established for each year of the analysis period. The cash flow is then discounted until the discount rate that yields a net present value of zero is found. The rate of return for competing projects may then be directly compared.

Equation (3.11) may be used to calculate the rate of return for a proposed project. However, the net present value and end of period cash receipts or disbursements are known, and an interest rate is calculated. The rate of return (interest rate) is usually calculated by a trial-and-error process, whereby the net cash flow is computed for various discount rates until the value is reduced to zero.

Example 3.12 A heat exchanger costs $10,000 to install, will last 10 years and result in a fuel savings of $3,000 per year. What is the internal rate of return for the heat exchanger?

Using Equation (3.11), an interest rate is assumed and the net present value is calculated. The process is repeated until the net present value is zero.

Assuming an interest rate of 25%

$$NPV = \$3{,}000[(1 + 0.25)^{10} - 1/.25(1 + 0.25)^{10}] - 10{,}000$$
$$= 10{,}713 - 10{,}000 = -\$713$$

Assuming an interest rate of 30%

$$NPV = \$3{,}000[(1 + 0.30)^{10} - 1/.30(1 + 0.30)^{10}] - 10{,}000$$
$$= 9276 - 10{,}000 = -\$724$$

For i = 25%, the net present value is positive and for i = 30%, the net present value is negative. Thus, for some discount rate between 25% and 30%, the present value benefits are equal to the present value costs and the net present value is zero. Guess an interest rate of 27% and calculate the net present value.

$$NPV = \$3{,}000[(1 + 0.27)^{10} - 1/.27(1 + 0.27)^{10}] - 10{,}000$$
$$= 10{,}093 - 10{,}000 \approx 0$$

Therefore, if the heat exchanger were installed, the internal rate of return would be about 27%.

3.4.3 Uniform Annualized Cost

This method may be used to analyze a project in terms of a uniform (e.g., average annual) cost over an analysis period. This method compares the cost of both investment and periodic costs on a periodic basis, usually an annual basis. The basic procedure for determining uniform annualized costs is to first determine the present worth of all costs and then apply the capital recovery factor to determine equal payments over the analysis period. Equation (3.12) may be used to calculate a uniform annualized cost for a proposed project.

$$A = P[\, i(1+i)^n/(1+i)^n - 1] \qquad \text{Eq. (3.12)}$$

The functional notation for a uniform annualized cost is

$$A = P(A/P,\ i\%,\ n)$$

Example 3.13 A piece of equipment has an installed cost of $1,000, will last 10 years and the interest rate is 7%. What is the uniform annualized cost for the equipment?

Using Equation (3.12)

$$A = \$1{,}000[0.07(1+.07)^{10}/(1+.07)^{10} - 1] = 1{,}000(0.1424)$$
$$= \$142.40 \text{ per year}$$

3.4.4 Benefit/Cost Ratio

The benefit/cost ratio method expresses benefits as a fraction of the costs, where benefits and costs have been discounted to either a present value or an annual cost equivalent. The ratio of the equivalent annual benefit to the equivalent annual cost must be greater than 1.0 for the proposed project to be economically advisable. Equation (3.11) may be used to determine the present worth of the annual benefits.

Example 3.14 A firm is trying to decide which of two energy saving devices to install. Both devices have a useful life of five years and no salvage value. Device A will have an installed cost of $1,000, which will result in an annual savings of $300. Device B has an installed cost of $1,300, which will result in an annual savings of $400. The interest rate is 7%. Using the benefit/cost ratio method, determine which device should be installed.

For device A

$$PW\ \text{cost}_A = \$1{,}000$$

Using Equation (3.11) to determine the present worth of the benefit

$$\text{PW benefit}_A = \$300\left[(1+0.07)^5 - 1/.07(1+0.07)^5\right] = 300(4.10) = \$1{,}230$$

$$B/C_A = PW\ \text{benefit}/PW\ \text{cost} = \$1{,}230/1{,}000 = 1.23$$

For device B

$$PW\ \text{cost}_B = \$1{,}300$$

The present worth of the benefit is

$$PW\ \text{benefit}_B = \$400(4.10) = \$1{,}640$$

$$B/C_B = \$1{,}640/\$1{,}300 = 1.26$$

Device B has the higher benefit/cost ratio and, therefore, is the better economic choice.

3.5 SPECIAL FACTORS

3.5.1 Taxes

Energy conservation projects usually result in benefits in the form of energy cost savings. The after-tax value of these benefits will be lessened by the loss of any related income tax deductions, such as for operating costs. Taxes can reverse the relative profitability of projects. When determining the present worth of annual energy cost savings, the net savings (gross savings less any increase in taxes) should be used.

Depreciation of capital improvements for energy savings projects must also be considered. System equipment costs may result in tax deductions in terms of depreciation allowances and/or investment tax credits. Depreciation was discussed earlier in this chapter. The choice of a depreciation method that yields higher depreciation allocations in the early years will increase the present value of depreciation deductions, thereby decreasing the present value of income tax payments and increasing the profitability of an investment to the firm.

3.5.2 Inflation

Inflation can have a significant impact on economic analyses of energy conservation improvements. Inflation is the gradual increase in the cost of goods and services over time. Inflation makes future dollars less valuable than present dollars, and can result in increased energy costs, as well as increased equipment and labor costs.

The prices of different items change at different rates due to the complexities of a free competitive economy. However, an approach often followed in investment analysis is to assume that all costs and revenues inflate at the same general rate, and that they, therefore, remain constant in relative terms. With this assumption, renewal costs and other future expenses and benefits are evaluated at present prices. Assuming constant future prices can generally simplify an analysis and, in many applications, result in reasonably accurate results.

Generally speaking, inflation benefits long-term borrowers of money since they will repay their debt in dollars with reduced purchasing power. Inflation is clearly unfavorable for long-term lenders of money.

3.5.3 Uncertainty

Uncertainty can be broadly defined as disparity between the predicted and actual occurrence. It encompasses two specific concepts: "risk" and "uncertainty." Risk involves an event whose probability of occurrence can be predicted. Uncertainty involves an event whose predicted chance of occurrence cannot be predicted.

Among the key factors that affect the outcome of economic evaluations of energy conservation projects are the following:

- The cost estimates, purchasing, installing, operating and maintaining energy conservation improvements.
- The additional cost imposed by the project, such as downtime and production loss
- The future rate of real price inflation in energy costs.
- The amount of actual energy saved.
- The economic lives of system components, the length of intended use of the system and the salvage values at termination.
- The interest rate used to convert future costs and benefits to a common time.

There are three commonly used general approaches to dealing with uncertainty in these types of situations: (1) probability analyses, (2) sensitivity analysis, and (3) break-even analysis. Probability analysis is generally used for situations in which the probability of an expected occurrence can be estimated, i.e., for evaluating risk. By multiplying the probability that an event will occur by the resulting dollar value if it does occur, it is possible to express costs and benefits as "expected values" rather than simply as "point estimates."

Sensitivity analysis can be used to assess the consequences of assuming alternative values for the significant variables in the analysis. By determining the effect on the outcome of potential variation in a factor, the analyst identifies the degree of importance of that estimate or assumption and can then seek more information about it if desired.

Break-even analysis is another method for dealing with uncertainty. It focuses on a single key variable, which is regarded as uncertain, and calculates the minimum (or maximum) value of the variable that is required to achieve a given outcome.

3.6 FINANCING OPTIONS FOR ENERGY MANAGEMENT IMPROVEMENTS

There are several options available for financing energy conservation improvements and projects. These include internal capital budget financing, performance contracts, leverage purchase and leasing. The method of financing a project will directly affect its financial evaluation due to the time value of money. It will be necessary to perform a cash flow analysis if a financing method that involves any cash flow (costs or benefits) other than a single lump sum is chosen.

3.6.1 Internal Capital Budget Financing

The most common form of energy conservation project financing uses the corporate capital budget. The facility owner or manager makes a decision to finance the improvements internally, taking advantage of all the savings and assuming all the risk. All funds necessary to finance the project come from within the company. The project is considered to be equivalent to other non-energy-related projects, vying for a share of the annual budget. If the financial ratios for the energy conservation project are acceptable, if it fits into the corporate business plan and if the project is judged to be technically sound, it is allocated funds for implementation by the corporate management.

Since the financing for this method is a single lump sum, funds allocated from the capital budget are considered, for analysis purposes, to have not time costs. However, the financial return from the project must be more advantageous than simply investing in a less risky investment.

Corporate project financing provides a relatively low risk and the best cash flow of any of the financing methods. Since no money is borrowed and there is no obligation to make payments, there is little financial risk and there is no increase in corporate indebtedness. Generally speaking, an energy conservation project is its own source of funds—repaying capital costs from energy cost savings.

3.6.2 Leveraged Purchase

A leveraged purchase uses borrowed money to pay for energy conservation projects instead of money from the owners capital budget. The amount borrowed may be for all, or part, of the funds needed for the project, as required. The borrowing is arranged through a financial institution, such as a bank, and becomes part of the corporate indebtedness.

A leveraged purchase is not normally used to finance energy conservation projects since it has several disadvantages. The risk is at a maximum because the corporate financial credit is used for the loan. Any failure on the part of the energy improvement to repay its cost must be made up from corporate resources. In addition, when the loan is repaid to the lender, only the interest portion of the payment is tax deductible. A leveraged purchase should be considered only when no other form of energy project financing is available, and the energy cost savings is extremely attractive.

6.3.3 Performance Contracting

In performance contracting, an energy service company (ESCO) performs the energy audit, purchases and installs the energy conservation project. The ESCO bears most of the risk for the performance of energy conservation improvements. There are two common project financing methods used in performance contracting: shared savings and guaranteed savings contracts.

When projects are financed with a shared savings contract, typically independent contractors install the energy conservation projects at their cost. The owner of the facility then pays a prearranged fee to the contractor, which is usually related to the energy cost savings achieved by the project. The fee is frequently a portion, or share, of the savings.

The facility owner has no cash investment, usually no maintenance costs and receives a positive cash flow immediately from the project. The ESCO also receives a positive cash flow from the project and also receives the tax advantages that result from the cost of the equipment and installation. In exchange, the facility owner gives up part of the energy cost savings, and does not have control of the energy saving equipment and systems.

The ESCO typically conducts an energy audit of the owner's facility without cost to the owner. It then prepares a proposal for the owner, which recommends energy cost reduction measures and estimates of the energy cost savings. One, or more, energy projects are selected for shared savings financing. The basis for the savings is explained and the method of calculating the sharing explained. Guarantees are given for maintenance, replacement and the amount of savings. Upon signing the contract, the ESCO installs the energy project and maintains it for the life of the contract. At the end of the contract period, the owner has the option of renewal, but at the market value, or removal of the equipment. Normal terms for a shared savings plan involve a 50-50 split between the owner and the ESCO over a contract period of five to seven years.

Guaranteed savings contracts are similar to shared savings contracts, except that the cost of the energy savings improvements are specified up-front. The facility owner can pay for the improvements at the beginning of the contract, finance it over a period of years or have the ESCO finance it with repayment out of the savings. The ESCO then guarantees a minimum amount of energy cost savings. The energy savings are guaranteed, usually on an annual basis. If savings greater than the guaranteed amount are achieved, some contracts require that a bonus be paid to the ESCO. If less than the guaranteed amount are realized, the owner is reimbursed for the difference by the ESCO.

3.6.4 Leasing

An alternative to purchasing energy conservation projects is leasing, which is a form of rental with the owner of the energy project as the lessor (usually a financial institution) and the user of the project being the lessee.

Under a leasing arrangement for energy improvements, the lessee pays the lessor a fixed amount each month or year for the use of the project. This amount can be tailored to the energy cost savings, which results in a positive cash flow for the lessee. The lease is for a fixed length of time. At the end of the lease, the lessor can extend the lease, or sell the project to the lessee at fair market value.

Leasing has the advantage that the lessee makes no initial payment on the project. The corporate balance sheet shows no loan or capital expense, and the corporate borrowing ability is not directly affected by the lease. Lease payments are made before taxes as an operating expense, and are deductible from before-tax income, providing certain requirements are met. At the end of the lease period, the lessor can remove the energy project or replace it with a more efficient design. Alternatively, the lessor can sell the energy improvement to the lessee at fair market value, or renegotiate the lease at a lower rate for an extended period.

3.6.5 Comparison of Financing Methods

A general comparison with the advantages and disadvantages of the various financing methods is tabulated below.

- Internal Capital Budget Financing
  - Best net cash flow
  - Poor net present value
  - Good internal project control
  - No indebtedness
- Leveraged Purchase
  - Worst net cash flow
  - Worst net present value
  - Good internal project control
  - Worst indebtedness
- Performance Contracting
  - Good net income
  - Good net present value
  - No internal project control
  - No indebtedness

- Leasing
  - Good net income
  - Best net present value
  - Good internal project control
  - Lease payments, no indebtedness

Chapter 4

Combustion Systems and Boilers

It has been estimated that over 50% of the energy consumed by industry is used to generate steam or hot water in boilers and other combustion processes, such as ovens, kilns and furnaces. Combustion is a complex process whereby chemically bound energy is released in an exothermic reaction of carbon and hydrogen to produce heat, carbon dioxide (CO_2) and water vapor (H_2O). Since combustion systems consume such a significant percentage of energy consumed by industry, improvements in the performance of fuel-burning systems can have a significant impact on facility operating costs.

The most common type of combustion system is a boiler. A boiler is a pressure vessel designed to transfer heat from the combustion process to a fluid. The fluid produced by boilers is usually steam or hot water. Since boilers are the most common combustion systems, the majority of the discussion in this chapter centers around boilers, although most of the information contained in this chapter is applicable to other combustion systems.

This chapter deals primarily with energy management for the heat producing combustion systems. "Chapter 5: Steam Systems" and "Chapter 6: Hydronic and Pumping Systems" deal with energy management for the heated fluid distribution systems.

4.1 GENERAL COMBUSTION SYSTEM OPERATION AND THEORY

The actual process of burning fuel, or combustion, is the heart of boilers, furnaces, ovens, kilns, etc. In order to analyze these systems from an energy management standpoint, it is first necessary to have an understanding of the combustion process itself. This section introduces the basic principles of fuels and the combustion process.

4.1.1 Definitions and Terms Used for Combustion Systems and Boilers

There are a number of terms that are unique to boilers and combustion systems. See Glossary for definition of terms that are commonly used when discussing boilers and combustion systems.

4.1.2 An Overview of the Combustion Process

Since combustion is a integral part of the operation of a combustion system, it is important to have a basic understanding of the combustion process. Combustion of flue gases represent the primary source of inefficiency and energy losses for a boiler or process heater. This section provides an overview of the basic principals involved in the combustion process. The following concepts are important when analyzing combustion.

Conservation of matter: All matter that enters a combustion system must also leave the system or be stored inside. Matter is never destroyed. Therefore, the total mass of the fuel plus the mass of the air, the reactants (usually expressed in pounds), is equal to the total mass of the flue gas, the products, exiting through the stack. In addition, individual atomic species also remain constant, i.e., every atom of carbon entering the combustion system also leaves as an atom of carbon. Every atom of oxygen that enters also leaves as an atom of oxygen, etc.

First Law of Thermodynamics: This is also known as the Law of Conservation of Energy. Essentially, it states that during a thermodynamic process, such as combustion, energy is neither created nor destroyed, although it may be stored within the process and it may also change from one form to another. A simplified version of the First Law is stated in equation form below:

$$\text{Energy In} = \text{Energy Out} + \text{Energy Stored} \qquad \text{Eq. (4.1)}$$

In the combustion process, the two main forms of energy, considered are chemical energy and thermal energy (heat).

Ideal (Perfect) Gas Law: Although some fuels used in combustion processes begin as solids or liquids (such as coal or fuel oil), the combustion process actually occurs with all constituents in the gaseous phase. All solids and liquid fuels must be vaporized before combustion can take place. Once the constituents are in the gaseous phase, they tend to behave like an ideal gas and follow the ideal gas law, which is stated mathematically in Equation (4.2).

$$PV = mRT \qquad \text{Eq. (4.2)}$$

where P = the absolute pressure
V = the volume
m = the mass
R = a constant
T = the absolute temperature

Dalton's Law of Partial Pressures states that the total pressure of an ideal gas is equal to the sum of the partial pressures of the constituent gases. Equation (4.3) is a mathematical statement of Dalton's Law.

$$P = P_a + P_b + P_c + \ldots\ldots\ldots\ldots \qquad \text{Eq. (4.3)}$$

Amagat's Law of Partial Volumes is similar to Dalton's Law. It states that the total volume of a mixture of an ideal gas is equal to the sum of the partial volumes of the constituent gases if each gas existed alone at the total pressure and temperature of the mixture. Amagat's Law is expressed mathematically in Equation (4.4).

$$V = V_a + V_b + V_c + \ldots\ldots \qquad \text{Eq. (4.4)}$$

Avagadro's Law states that equal volumes of different gases at the same pressure and temperature contain the same number of molecules.

An important concept in chemistry, which is also an important concept in the combustion process, is the mole. The mass of a substance in pounds equal to its molecular weight is called a pound-mole of the substance. A pound-mole is usually just called a mole. As an example, carbon (C) has a molecular weight of 12. Therefore, a mole of carbon has a weight of 12 pounds. For gases, the volume occupied by a mole is referred to as the *molal volume* and it has a constant volume of 394 ft^3 for an ideal gas at 80°F and 14.7 psia. The concept of a mole and molar volume is used frequently in combustion calculations and analyses.

In order for combustion to take place, three elements must be present: fuel, oxygen and heat. The oxygen and fuel must be available in gaseous form, be in the proper quantities and be adequately mixed together. The fuel must be in a gaseous form, or it must be vaporized to adequately mix with the oxygen. The mixture of fuel and oxygen must be heated to a temperature high enough for combustion to occur. In most combustion processes, oxygen is supplied in the form of air, which is a mixture of a number of gases, including oxygen, nitrogen and water vapor.

Most fuels used for combustion processes are hydrocarbon fuels. Almost all fuels consist of carbon, hydrogen and small amounts of sulfur. For practical purposes, the oxidation of only these elements (hydrogen, carbon and sulfur) is involved in the combustion process. The three primary chemical reactions involved in the combustion process are:

$$C + O_2 \rightarrow CO_2 + \text{Heat}$$

$$H + O_2 \rightarrow H_2O + \text{Heat}$$

$$S + O_2 \rightarrow SO_2 + \text{Heat}$$

Table 4.1 lists the common chemical reactions for hydrocarbon fuels in the combustion process.

TABLE 4.1 Common Chemical Reactions of Combustion

Combustible	Reaction	Moles	Pounds	Heat of combustion (high) (Btu/lb$_m$ of fuel)
Carbon (to CO)	$2C + O_2 = 2CO$	2 + 1 = 2	24 lb$_m$ + 32 lb$_m$ = 56 lb$_m$	4,000
Carbon (to CO_2)	$C + O_2 = 2CO_2$	1 - 1 = 1	12 + 32 = 44	14,100
Carbon monoxide	$2CO + O_2 = 2CO_2$	2 + 1 = 2	56 + 32 = 88	4,345
Hydrogen	$2H_2 + O_2 = 2H_2O$	2 + 1 = 2	4 + 32 = 36	61,100
Sulfur (to SO_2)	$S + O_2 = SO_2$	1 + 1 = 1	32 + 32 = 64	3,980
Methane	$CH_4 + 2O_2 = CO_2 + 2H_2O$	1 + 2 = 1 + 2	16 + 64 = 80	23,875
Acetylene	$2C_2H_2 + 5O_2 = 4CO_2 + 2H_2O$	2 + 5 = 4 + 2	52 + 160 = 212	21,500
Ethylene	$C_2H_4 + 3O_2 = 2CO_2 + 2H_2O$	1 + 3 = 2 + 2	28 + 96 = 124	21,635
Ethane	$2C_2H_6 + 7O_2 = 4CO_2 + 6H_2O$	2 + 7 = 4 + 6	60 + 224 = 284	22,325
Hydrogen sulfide	$2H_2S + 3O_2 = 2SO_2 + 2H_2O$	2 + 3 = 2 + 2	68 + 96 = 164	7,100

The heat of combustion listed in Table 4.1 is the heat released during the combustion process. The values listed for the heat of combustion assume that all of the fuel is oxidized and complete combustion of the fuel has occurred.

When carbon is burned, or oxidized, the oxygen combines with the carbon to form carbon dioxide (CO_2). If sufficient oxygen is supplied for all of the carbon to oxidize to CO_2, the process will release approximately 14,540 Btu/lb of carbon. When hydrogen is burned in the presence of oxygen and then cooled to its initial temperature, the result is water (H_2O) and approximately 61,000 Btu/lb of hydrogen is released. Many fuels also contain sulfur as a free element. Sulfur will burn to produce sulfur dioxide (SO_2) and will release about 4,050 Btu/lb of sulfur.

The above processes assume ideal, or perfect, combustion, which is also called stoichiometric combustion. In stoichiometric combustion of a hydrocarbon fuel, the fuel is reacted with the exact amount of oxygen required to oxidize all carbon, hydrogen and sulfur in the fuel to CO_2, H_2O and SO_2. The combustion process reactants contain no incomplete oxidized fuel constituents and no unreacted oxygen.

Oxygen for most combustion processes is typically supplied by ambient air drawn into the combustion process. For stoichiometric combustion, the amount of air supplied is called theoretical air, which is the minimum amount of air that supplies sufficient oxygen for complete combustion of all hydrogen, carbon and any other combustible reactants. When complete combustion is achieved with theoretical air, the products of combustion contain no oxygen. The oxygen in the air combines with the combustible reactants to form CO_2, H_2O and SO_2.

Although the air used in combustion contains the oxygen necessary for the combustion process, oxygen is not the primary constituent of air. Air is a mixture of many components, primarily nitrogen. Typically, air contains about 78% nitrogen, by volume, and about 21% oxygen, with about 1% argon and traces of other gases. The ratio by weight is about 76.8% nitrogen and 23.2% oxygen. During the combustion process, air also combines with nitrogen to form nitrogen oxides or NO_x. Nitrogen oxides are important in the combustion process from an air pollution standpoint. Atmospheric air also contains small amounts of water in the form of vapor. The average amount of water vapor in air at 0°F is about 0.08% by weight and about 1.56% at 70°F.

Example 4.1. Natural gas is burned in a combustion process. The natural gas has the following volumetric analysis at 60°F.

CO_2	0.40%
CH_4	92.10%
C_2H_6	4.10%
N_2	3.40%

What is the weight of the oxygen and what is the weight of the air required for stoichiometric combustion?

The composition of the fuel is given on a volumetric basis, which is typically the case. Therefore, it is necessary to convert from a volumetric analysis to a mass or weight analysis. This is done by multiplying the percent volume for each gas by its densi ty at 60°F. The densities are obtained from Table 4.1 and the conversions are tabulated below.

Constituent Product	Volume (%)	Density (lb/ft^3)	Component Weight (lb/ft^3)
CO_2	0.40	0.1161	0.0004644
CH_4	92.10	0.0423	0.0389583
C_2H_6	4.10	0.0792	0.0032472
N_2	3.40	0.0739	0.0025026
	100.00		0.0451725

Therefore, the percentage of each constituent product on a weight basis is as follows:

Constituent Product		Weight (%)
CO_2	$\frac{0.0004644}{0.0451725}$ =	1.03
CH_4	$\frac{0.0389583}{0.0451725}$ =	86.25
C_2H_6	$\frac{0.0032472}{0.0451725}$ =	7.18
N_2	$\frac{0.0025026}{0.0451725}$ =	5.54
		100.00%

Since both N_2 and CO_2 are essentially inert, they do not react and, therefore, they pass through the combustion process unchanged. Referring to Table 4.1, the molecular weights of the remaining constituents are obtained and listed below.

$O_2 = 32$	$CO_2 = 44$
$C = 12$	$CH_4 = 16$
$H = 2$	$C_2H_6 = 30$
$H_2O = 18$	

Writing a combustion equation for the CH_4, using the unknown coefficients A and B to mass balance the oxygen, the following equation is obtained:

$$CH_4 + AO_2 \rightarrow CO_2 + BH_2O$$

Performing a mass balance on the hydrogen, the equation is solved for $B = 2$. Performing another mass balance on the oxygen, the equation is solved for $A = 2$. The completed combustion equation for CH_4 is

$$CH_4 + 2O_2 \rightarrow CO_2 + 2H_2O$$

Performing a mass balance on all the reactants and products, using the molecular weight of each, the total molecular weight for burning the CH_4 is

$$16 + 2(32) = 44 + 2(18) = 64$$

Writing a combustion equation for the C_2H_6, using the unknown coefficients D and E to mass balance the hydrogen and oxygen, the following equation is obtained:

$$C_2H_6 + DO_2 \rightarrow 2CO_2 + EH_2O$$

Doing a mass balance on the H and O_2, $E = 3$ and $D = \left(\frac{7}{2}\right)$. The completed combustion equation is

$$C_2H_6 + \tfrac{7}{2}O_2 \rightarrow 2CO_2 + 3H_2O$$

Performing a mass balance on the reactants and products of combustion results in

$$30 + \tfrac{7}{2}(32) = 2(44) + 3(18) = 112$$

To determine the mass of the oxygen required for the two reactants, CH_4 and C_2H_6, the percent by weight of each is multiplied by the ratio of the total weight to the weight of the combustible reactants.

$$\begin{array}{ll} CH_4\text{:}\ \ 0.8625\left(\frac{64}{16}\right) = & 3.4500 \\ CH_4\text{:}\ \ 0.0718\left(\frac{112}{30}\right) = & \underline{0.2681} \\ \text{External } O_2 \text{ required per lb of fuel} = & 3.7181 \end{array}$$

Therefore, for each pound of fuel, 3.7181 pounds of oxygen is required for stoichiometric combustion. Since O_2 for the combustion process is obtained from air and air is primarily a mixture of nitrogen and oxygen, it is necessary to use the weight ratio of nitrogen to oxygen to determine the weight of the nitrogen in the combustion air.

$$\text{Weight of } N_2 = (3.7181 \text{ lb } O_2)\left(\frac{0.768}{0.232}\right) = 12.3082 \text{ lbs of } N_2$$

The weight of the combustion air required per pound of fuel is

$$\text{Weight of Air} = 3.7181 + 12.3082 = 16.0263 \text{ lb of air per lb of fuel}$$

In the previous example, the combustion equations were developed and mass balances were performed for all the constituents in the reactants and products of combustion. This process will work well for all fuels used in the combustion process. However, there are fuels that are commonly used in combustion systems. The procedure outlined in Example 4.1 can be a tedious and laborious process. Since most fuels used in combustion systems are frequently used repeatedly, tables have been developed for common fuels to simplify and streamline combustion analysis process.

TABLE 4.2 Data for Performing Simplified Calcualtions for Commonly used Fuels

No.	Substance	Formula	Molecular Weight	lb per cu ft	cu ft per lb	Sp Gr Air = 1.0000	Heat of Combustion: Btu per cu ft Gross (High)	Btu per cu ft Net (Low)	Btu per lb Gross (High)	Btu per lb Net (Low)	For 100% Total Air Moles per mole of Combustible or cu ft per cu ft of Combustible: Required for Combustion O_2	N_2	Air	Flue Products CO_2	H_2O	N_2	For 100% Total Air lb per lb of Combustible: Required for Combustion O_2	N_2	Air	Flue Products CO_2	H_2O	N_2
1	Carbon	C	12.01	–	–	–	–	–	14.093	14.09	31.0	3.76	4.76	1.0	–	3.76	2.66	8.86	11.53	3.66	–	8.86
2	Hydrogen	H_2	2.016	0.0053	187.723	0.0696	325	275	61.095	51.623	0.5	1.88	2.38	–	1.0	1.88	7.94	26.41	34.34	–	8.94	26.41
3	Oxygen	O_2	32.00	0.0846	11.819	1.1053	–	–	–	–	–	–	–	–	–	–	–	–	–	–	–	–
4	Nitrogen (atm)	N_2	28.01	0.0744	13.443	0.9718	–	–	–	–	–	–	–	–	–	–	–	–	–	–	–	–
5	Carbon monoxide	CO	28.01	0.0740	13.506	0.9672	321	321	4.347	4.347	0.5	1.88	2.38	1.0	–	1.88	0.57	1.90	2.47	1.57	–	1.90
6	Carbon dioxide	CO_2	44.01	0.1170	8.548	1.5282	–	–	–	–	–	–	–	–	–	–	–	–	–	–	–	–
Paraffin series																						
7	Methane	CH_4	16.04	0.0425	23.552	0.5543	1,012	911	23.875	21.495	2.0	7.53	9.53	1.0	2.0	7.53	3.99	13.28	17.27	2.74	2.25	13.28
8	Ethane	C_2H_6	30.07	0.0803	12.455	1.0488	1,773	1,622	22.323	20.418	3.5	13.18	16.68	2.0	3.0	13.18	3.73	12.39	16.12	2.93	1.80	12.39
9	Propane	C_3H_8	44.09	0.1196	8.365	1.5617	2,524	2,322	21.669	19.937	5.0	18.82	23.82	3.0	4.0	18.82	3.63	12.07	15.70	2.99	1.63	12.07
10	*n*-Butane	C_4H_{10}	58.12	0.1582	6.321	2.0665	3,271	3,018	21.321	19.678	6.5	24.47	30.97	4.0	5.0	24.47	3.58	11.91	15.49	3.03	1.55	11.91
11	Isobutane	C_4H_{10}	58.12	0.1582	6.321	2.0665	3,261	3,009	21.271	19.628	6.5	24.47	30.97	4.0	5.0	24.47	3.58	11.91	15.49	3.03	1.55	11.91
12	*n*-Pentane	C_5H_{12}	72.15	0.1904	5.252	2.4872	4,020	3,717	21.095	19.507	8.0	30.11	38.11	5.0	6.0	30.11	3.55	11.81	15.35	3.05	1.50	11.81
13	Isopentane	C_5H_{12}	72.15	0.1904	5.252	2.4872	4,011	3,708	21.047	19.459	8.0	30.11	38.11	5.0	6.0	30.11	3.55	11.81	15.35	3.05	1.50	11.81
14	Neopentane	C_5H_{12}	72.15	0.1904	5.252	2.4872	3,994	3,692	20.978	19.390	8.0	30.11	38.11	5.0	6.0	30.11	3.55	11.81	15.35	3.05	1.50	11.81
15	*n*-Hexane	C_6H_{14}	86.17	0.2274	4.398	2.9704	4,768	4,415	20.966	19.415	9.5	35.76	45.26	6.0	7.0	35.76	3.53	11.74	15.27	3.06	1.46	11.74

Olefin series																					
16 Ethylene	C_2H_4	28.05	0.0742	13.475	0.9740	1,604	1,503	21.636	20.275	3.0	11.29	14.29	2.0	2.0	11.29	3.42	11.39	14.81	3.14	1.29	11.39
17 Propylene	C_3H_6	42.08	0.1110	9.007	1.4504	2,340	2,188	21.048	19.687	4.5	16.94	21.44	3.0	3.0	16.94	3.42	11.39	14.81	3.14	1.29	11.39
18 *n*-Butene	C_4H_8	56.10	0.1480	6.756	1.9336	3,084	2,885	20.854	19.493	6.0	22.59	28.59	4.0	4.0	22.59	3.42	11.39	14.81	3.14	1.29	11.39
19 Isobutene	C_4H_8	56.10	0.1480	6.756	1.9336	3,069	2,868	20.737	19.376	6.0	22.59	28.59	4.0	4.0	22.59	3.42	11.39	14.81	3.14	1.29	11.39
20 *n*-Pentene	C_5H_{10}	70.13	0.1852	5.400	2.4190	3,837	3,585	20.720	19.359	7.5	28.23	35.73	5.0	5.0	28.23	3.42	11.39	14.81	3.14	1.29	11.39
Aromatic series																					
21 Benzene	C_6H_6	78.11	0.2060	4.852	2.6920	3,752	3,601	18.184	17.451	7.5	28.23	35.73	6.0	3.0	28.23	3.07	10.22	13.30	3.38	0.69	10.22
22 Toluene	C_7H_8	92.13	0.2431	4.113	3.1760	4,486	4,285	18.501	17.672	9.0	33.88	42.88	7.0	4.0	33.88	3.13	10.40	13.53	3.35	0.78	10.40
23 Xylene	C_8H_{10}	106.16	0.2803	3.567	3.6618	5,230	4,980	18.650	17.760	10.5	39.52	50.02	8.0	5.0	39.52	3.17	10.53	13.70	3.32	0.85	10.53
Miscellaneous gases																					
24 Acetylene	C_2H_2	26.04	0.0697	14.344	0.9107	1,477	1,426	21.502	20.769	2.5	9.41	11.91	2.0	1.0	9.41	3.07	10.22	13.30	3.38	0.69	10.22
25 Naphthalene	$C_{10}H_8$	128.16	0.3384	2.955	4.4208	5,884	5,654	17.303	16.708	12.0	45.17	57.17	10.0	4.0	45.17	3.00	9.97	12.96	3.43	0.56	9.97
26 Methyl alcohol	CH_3OH	32.04	0.0846	11.820	1.1052	868	767	10.258	9.066	1.5	5.65	7.15	1.0	2.0	5.65	1.50	4.98	6.48	1.37	1.13	4.98
27 Ethyl alcohol	C_2H_5OH0	46.07	0.1216	8.221	1.5890	1,600	1,449	13.161	11.917	3.0	11.29	14.29	2.0	3.0	11.29	2.08	6.93	9.02	1.92	1.17	6.93
28 Ammonia	NH_3	17.03	0.0456	21.914	0.5961	441	364	9.667	7.985	0.75	2.82	3.57	–	1.5	3.32	1.41	4.69	6.10	–	1.59	5.51
													SO_2						SO_2		
29 Sulfur*	S	32.06	–	–	–	–	–	3.980	3.980	1.0	3.76	4.76	1.0	–	3.76	1.00	3.29	4.29	2.00	–	3.29
30 Hydrogen sulfide	H_2S	34.08	0.0911	10.979	1.8095	646	395	7.097	6.537	1.5	5.65	7.15	1.0	1.0	3.65	1.41	4.69	6.10	1.88	0.53	4.69
31 Sulfur dioxide	SO_2	64.06	0.1733	5.770	2.2640	–	–	–	–	–	–	–	–	–	–	–	–	–	–	–	–
32 Water vapor	H_2O	18.02	0.0476	21.017	0.6215	–	–	–	–	–	–	–	–	–	–	–	–	–	–	–	–
33 Air	–	–	0.0766	13.063	1.0000	–	–	–	–	–	–	–	–	–	–	–	–	–	--	–	–

*Carbon and sulfur are considered as gases for molal calculations only.

Table 4.2 provides data for performing simplified calculations for commonly used fuels.

Example 4.2. Use Table 4.2 to determine the stoichiometric combustion air for the fuels listed in Example 4.1.

In Example 4.1, the percentage by weight of the combustible constituents was determined as follows:

CH_4	86.25%
C_2H_6	7.18%

Referring to Table 4.2, the air required for stoichiometric combustion can be determined as follows:

$$\begin{aligned} \text{Weight of Air} &= 0.8626\left(\frac{17.27 \text{ lb air}}{\text{lb fuel}}\right) + 0.0718\left(\frac{16.12 \text{ lb air}}{\text{lb fuel}}\right) \\ &= 16.053 \text{ lb air per lb fuel} \end{aligned}$$

The slight difference between the results of the two examples is due to rounding off errors. This simplified method should be more than adequate for combustion ca'.:ulations.

The amount ot air required for complete combustion in the previous examples was for theoretical air, which is the amount of air exactly sufficient for complete burning of all combustibles in the fuel. As the term theoretical air implies, this is the amount of air that will product perfect, or stoichiometric, combustion. In actual combustion processes, this situation is not possible. As the result of incomplete mixing, as well as other physical limitations, more air than theoretical air must be supplied to ensure complete combustion. The amount of air required above theoretical air is called excess air. Excess air is required due to incomplete mixing and also for safety reasons. The possibility of unburned fuel in the flue stack could present a hazardous condition. As a result, the amount of excess air is frequently more than actually required for complete combustion of all the fuel. When complete combustion is achieved, the products of combustion will contain no pure oxygen and no unreacted (unburned) fuel.

The amount of air supplied to a combustion process is expressed in terms of percent theoretical air or it may be expressed in terms of excess air. The excess air is the amount over and above the theoretical air. Therefore, 150% theoretical air is equivalent to 50% excess air.

As will be discussed later, the amount of excess air supplied to a combustion process can have a significant effect on the efficiency of a combustion process. Determining the actual amount of combustion air flow by measuring the air flow as it enters the combustion process is difficult and not very accurate. It is usually easier to measure excess air by analyzing the products of combustion in the flue gas. Equation (4.5) may be used to determine the amount of excess air if the products of combustion in the flue gas are known.

$$\text{Excess air, } (\%) = \frac{100[O_2 - (CO/2)]}{0.264N_2 - [O_2 - (CO/2)]} \qquad \text{Eq.(4.5)}$$

where O_2, CO and N_2 are percentages by volume from the flue gas analysis. Methods for performing flue gas analyses will be discussed later in this chapter.

4.1.3 Fuels for Boilers and Combustion Systems

Fuels used for combustion in boilers and other combustion systems are referred to as hydrocarbon fuels. These are fuels with combustible chemicals primarily consisting of hydrogen and carbon. Sulfur is also a combustible and is present in many hydrocarbon fuels. In addition to combustible elements, fuels may also contain noncombustible elements such as nitrogen. The physical and chemical properties of fuels have a significant effect on the efficiency of combustion.

Fuels may be classified as either gaseous, liquid or solid (coal). Regardless of the fuel type, the fuel must be in a gaseous or vapor form and mixed with oxygen in sufficient quantities in order for combustion to occur. Liquid fuel burners must atomize or vaporize the fuel into small droplets or a vapor and must be adequately mixed with the combustion air. Solid fuel combustion equipment must: (1) heat the fuel to vaporize sufficient volatiles and to initiate and sustain combustion; (2) provide residence time to complete combustion; and (3) provide space for ash containment.

The heating value of a fuel is the quantity of heat generated or released during the combustion process and is a constant value. The heating value is also referred to as the heat of combustion or the caloric value of the fuel. Table 4.2 lists two heating values for each type of fuel. The higher heating value, gross heating value or total heating value, includes the latent heat of vaporization and is determined when water vapor in the fuel combustion products is condensed. The lower heating value or net heating value is the heat that results when the latent heat of vaporization is not included. Heating values are usually expressed in Btu/ft^3 for gaseous fuels, Btu/gal for liquid fuels and Btu/lb for solid fuels. Heating values are always given in relation to a certain reference temperature and pressure, usually 60, 68 or 77°F and 14.7 psia.

Gaseous fuels are commonly burned in boilers and combustion systems. Smaller packaged boilers frequently are gas fired. Gaseous hydrocarbon fuels are a mixture of many different hydrocarbons. The two primary sources of gaseous fuels are natural gas wells and certain chemical manufacturing processes. The major constituent of natural gas is methane, which distinguishes it from manufactured gas. Other gaseous fuels include butane and propane. Liquid propane (LP) must be vaporized prior to burning. Gaseous fuels require little preparation. They mix readily with combustion air and burn with a low luminous flame. The most common gaseous fuel is natural gas. The composition of natural gas depends upon the geographical location of its source. Heating values for natural gas vary from 900 to 1,200 Btu/ft^3, usually 1,000 to 1,050 Btu/ft^3 at sea level. Liquid petroleum gases consist primarily of propane and butane. Although these fuels are stored as a liquid under pressure, they become gaseous at usual atmospheric pressures and temperatures. Commercial propane has a heating value of about 21,560 Btu/lb or about 2,500 Btu/ft^3 of gas. Commercial butane has a heating value of about 21,180 Btu/lb or about 3,200 Btu/ft^3.

Liquid hydrocarbon fuels consist of various grades of fuel oil, which are broadly classified as distillate fuel oils (lighter oils) or residual fuel oils (heavier oils). Grades No. 1 and 2 are distillate fuel oils; Grades 4 (light), 5 (heavy) and 6 are residual fuel oils. Characteristics that determine fuel oil grade classification and suitability for given applications include: (1) viscosity; (2) flash point; (3) pour point; (4) water and sediment content; (5) carbon content; (6) ash; (7) distillation qualities or distribution temperature ranges; (8) specific gravity; (9) sulfur content; (10) heating value; and (11) carbon-hydrogen content. Viscosity is a measure of the oil's resistance to flow. Flash point is the lowest temperature to which an oil must be heated for its vapors to ignite in a flame. Specific gravity is the ratio of fuel oil to the density of water at a specific temperature. The density of a fuel is a good indicator of its grade and heating value. Figure 4-1 shows the relationship between the higher heating value and density of a fuel.

Viscosity of a fuel is a very important property, since it affects both the ability to pump the oil to the burner, as well as the atomizing characteristics. Fuel oil will not burn as a liquid. It must be converted to a vapor and then mixed with air. Burner nozzles are designed for oil of a specific viscosity range and variations from those viscosities will result in poor atomization. The viscosity of fuel oils is dependent upon the grade of the fuel oil and its temperature. Typical viscosities of fuel oils at various temperatures are shown in Figure 4-2.

In order for some heavier fuel oils to be pumped to the burner and atomized, it is necessary to preheat the oil. The oil temperature must be carefully controlled. If the preheat is too low, the viscosity of the oil will be too high, resulting in improper atomization and droplet sizes that are too large. If the preheat is too high, the oil viscosity is lower than recommended and poor, fluctuating atomization occurs.

Solid fuels include coal, coke, wood, and waste products of industrial and agricultural operations. Only coal will be discussed here since this is the most widely used solid fuel. Coal has a complex composition of

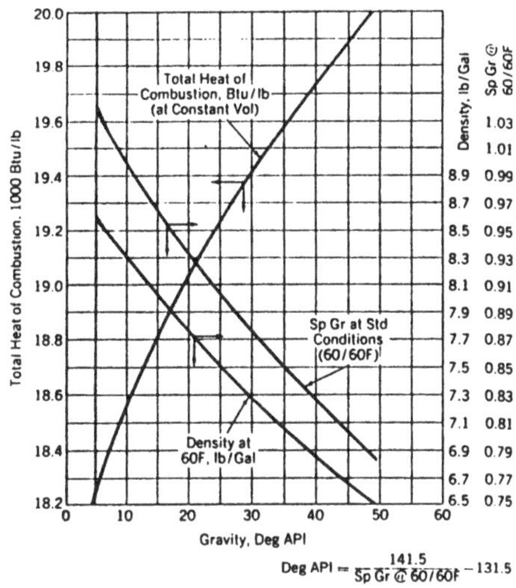

Figure 4-1. Heating value and density of fuel oils.

hydrocarbons that makes it difficult to classify. Chemically, coal consists of carbon, hydrogen, oxygen, nitrogen, sulfur, mineral residue and ash.

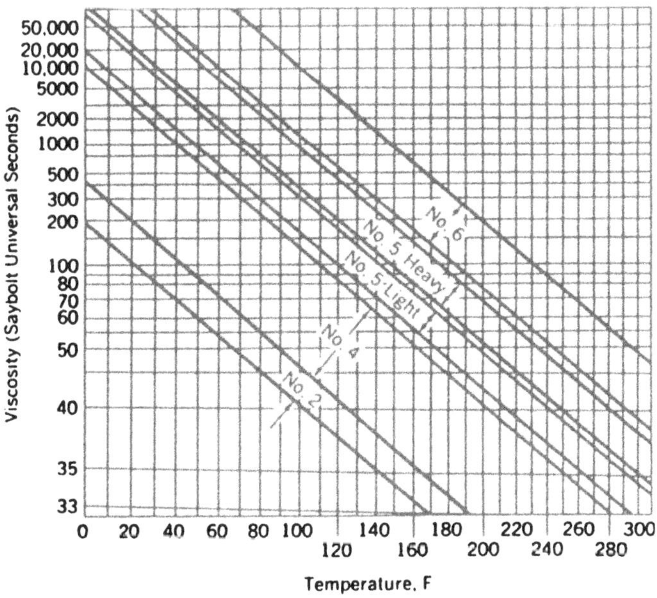

Figure 4-2. Approximate viscosity of fuel oil at various temperatures.

Common classifications of coal include anthracite, semianthracite, bituminous, semibituminous, subbituminous and lignite. Anthracite coal is a clean, dense hard coal that creates little dust and is hard to ignite but burns freely once started. Semianthracite has a higher volatile content than anthracite, is not as hard and ignites more easily. Anthracite coals tend to be high in fixed carbon and low in volatile matter. Bituminous coals have a higher proportion on volatile matter and a proportionally lower fixed carbon content. The lower grade coals, such as subbituminous, have lower heating values. Lignite is woody in structure, very high in moisture content and has a low heating value. High grade coals have a heating value ranging from approximately 14,000 to 15,000 Btu/lb. Lower grade coals have a lower carbon content and have heating values ranging from about 7,000 to 14,000 Btu/lb. Table 4.3 classifies coals by rank (type).

In order for combustion of coal to occur, it must also be mixed with combustion air to provide the oxygen required for combustion. To facilitate the mixing of coal and air, coal may be pulverized and burned in air suspension (fluidized bed), or it may be burned in chunks on a traveling grate. Burners that consume less than 200,000 lb per hour are typically stoker (grate) fired. Larger units utilize suspension firing of pulverized coal.

4.1.4 Factors Affecting Combustion Efficiency

Since the combustion process is the heart of boilers, furnaces, ovens, kilns, etc., improving the efficiency of the combustion process can have a significant effect on the overall efficiency of the combustion equipment listed above. The combustion analyses and predicted heat release discussed above assumed 100% combustion efficiency. The combustion process, however, is not 100% efficient. Basically, the efficiency of a combustion process is the ratio of the useful heat to the heat input or the heating value of the fuel. Equation (4.6) defines the efficiency of a combustion process.

$$\eta_T = \frac{q_g - q_{ft}}{q_g} \times 100\% \qquad \text{Eq. (4.6)}$$

where η_T = thermal efficiency, %
q_g = gross heating value of the fuel per unit volume
q_{ft} = heat losses per unit volume of fuel gas

Earlier, the combustion process was discussed and the heat released for various fuels was also listed in Tables 4.1, 4.2 and 4.3. The tabulated heating values are gross without any losses. To determine the net useful heat from a combustion process, it is necessary to deduct the heat losses

TABLE 4.3 Classification of Coals by Rank

Class	Group	Fixed Carbon Limits, % (Dry, Mineral-Matter-Free Basis)		Volatile Matter Limits, % (Dry, Mineral-Matter-Free Basis)		Calorific Value Limits, Btu/lb (Moist,[b] Mineral-Matter-Free Basis)		Agglomerating Character
		Equal or Greater Than	Less Than	Greater Than	Equal or Less Than	Equal or Greater Than	Less Than	
I. Anthracitic	1. Meta-anthracite	98	—	—	2	—	—	Nonagglomerating
	2. Anthracite	92	98	2	8	—	—	
	3. Semianthracite[c]	86	92	8	14	—	—	
II. Bituminous	1. Low volatile bituminous coal	78	86	14	22	—	—	Commonly agglomerating[e]
	2. Medium volatile bituminous coal	69	78	22	31	—	—	
	3. High volatile A bituminous coal	—	69	31	—	14,000[d]	—	
	4. High volatile B bituminous coal	—	—	—	—	13,000[d]	14,000	
	5. High volatile C bituminous coal	—	—	—	—	11,500	13,000	
						10,500[e]	11,500	Agglomerating
III. Subbituminous	1. Subbituminous A coal	—	—	—	—	10,500	11,500	Nonagglomerating
	2. Subbituminous B coal	—	—	—	—	9,500	10,500	
	3. Subbituminous C coal	—	—	—	—	8,300	9,500	
IV. Lignitic	1. Lignite A	—	—	—	—	6,300	8,300	
	2. Lignite B	—	—	—	—	—	6,300	

[a] This classification does not include a few coals, principally non-banded varieties, which have unusual physical and chemical properties and which come within the limits of fixed carbon or calorific value of the high-volatile bituminous and subbituminous ranks. All of these coals either contain less thah 48% dry, mineral-matter-free fixed carbon or have more than 15,500 moist, mineral-matter-free British thermal units per pound.

[b] Moist refers to coal containing its natural inherent moisture but not including visible water on the surface of the coal.

[c] If agglomerating, classify in low-volatile group of the bituminous class.

[d] Coals having 69% or more fixed carbon on the dry, mineral-matter-free basis shall be classified according to fixed carbon, regardless of calorific value.

[e] It is recognized that there may be nonagglomerating varieties in these groups of the bituminous class, and there are notable exeptions in high volatile C bituminous group.

from the gross heating value to get a net useful heat value. The heat losses in the combustion process is the total of a number of heat losses. The types of heat that can be lost during the combustion process include:

- loss of sensible heat in the dry flue gases
- heat loss in water vapor in the products of combustion formed by the combustion of hydrogen
- heat loss in water vapor in the combustion air
- heat loss due to incomplete combustion
- heat loss from unburned carbon in the ash or residue

Probably the largest source of heat loss in a combustion process is the loss of sensible heat in the dry flue gases. Most combustion systems, such as boilers, have very large flue gas losses. The flue gas losses are primarily a function of the amount of excess combustion air supplied to the burner and the temperature of the exiting flue gases.

Depending upon the fuel burned, the leaving flue gases can have temperatures in the range of 400 to 600°F. High flue gas temperatures are an indication that heat is not being transferred as useful heat to the process fluid, such as steam, hot water, air, etc. Sensible heat losses can be reduced by removing as much heat as possible from the flue gases, lowering the exit temperature. There are, however, limitations on the extent to which flue gases may be cooled, as will be discussed later.

The amount of excess air also has a significant effect on combustion efficiency. As the air is drawn or blown into the combustion process, it must be heated to the combustion temperature as part of the process. Large amounts of excess air require significant amounts of heat to raise the air temperature. The impact of excess air is discussed later in this chapter.

4.2 BOILER TYPES AND CHARACTERISTICS

Although there are many types of combustion systems in use, boilers represent the largest number of hydrocarbon fuel fired combustion systems in use. Boilers have been in use since early in the eighteenth century; they are actually pressure vessels designed to transfer heat (produced by combustion) to a fluid. In most boilers, the fluid is usually water in liquid or steam form. It has been estimated that over 50% of the energy consumed by industry is used to generate steam in boilers. Since boilers are in such widespread use, they will be discussed in some detail here. Most, if not all, of the energy conservation opportunities available to boilers are also applicable to other types of hydrocarbon combustion systems.

4.2.1 Boiler Classifications

Boilers may be grouped into classes based upon working pressure and temperature, amount of fuel burned and material of construction. They may also be classified by shape, size, usage (such as heating or process), steam, or hot water or by design.

When boilers are classified by working pressure and temperature, there are two classifications: low pressure and medium/high pressure. Low pressure boilers are constructed for maximum working pressures of 15 psig steam and up to 160 psig hot water. Low pressure hot water boilers are limited to a 250°F operating temperature. Medium/high pressure boilers are designed to operate at pressures greater than 15 psig or above 160 psig hot water or above 250°F.

Stream boilers are available in standard sizes up to and above 100,000 lb steam per hour (60,000 to over 100,000 Btu/hr.). Water boilers are available in standard sizes from 50,000 to over 100,000 Btu/hr.

Boilers may also be classified as field-erected or packaged. A packaged boiler is shipped from the manufacturer as a complete unit, ready for installation and use. Packaged boilers typically include the burner, combustion space or firebox, convection section, combustion air fan, instrumentation and controls—all in one complete package. A field-erected boiler, as the name implies, must have all its components installed at the facility where it is to be installed. Package boilers are available in capacities up to 800 boiler horsepower. Field-erected boilers can have much greater capacities.

4.2.2 Boiler Designs

There are three basic boiler designs: firetube (Scotch Marine), firebox and watertube. These boiler designs are available as either packaged or field erected boilers.

Firetube boilers are probably the most widely used. In a firetube boiler, the flame is produced in a furnace and the combustion gases flow through tubes to the vent stack while the water to be heated surrounds the tubes. The furnace, the tubes water and steam are contained in a vessel. In the firetube design, the tubes direct the hot combustion gases from the furnace through the boiler to the flue stack. The tubes are submerged in the boiler water and heat is transferred from the hot flue gases to the water through the wall of the tube. A typical packaged firetube boiler is shown in Figure 4-3.

Firetube boilers are usually classified by the number of passes in them. The passes refers to the number of times the flue gases flow through the water in the boiler before exiting the boiler. Designs are available with 2, 3 or 4 gas passes. A cutaway of a package 4 pass firetube boiler is shown in Figure 4-4.

Firetube boilers are also available in dryback or wetback designs. Dryback boilers have a refractory lined chamber outside of the firebox that

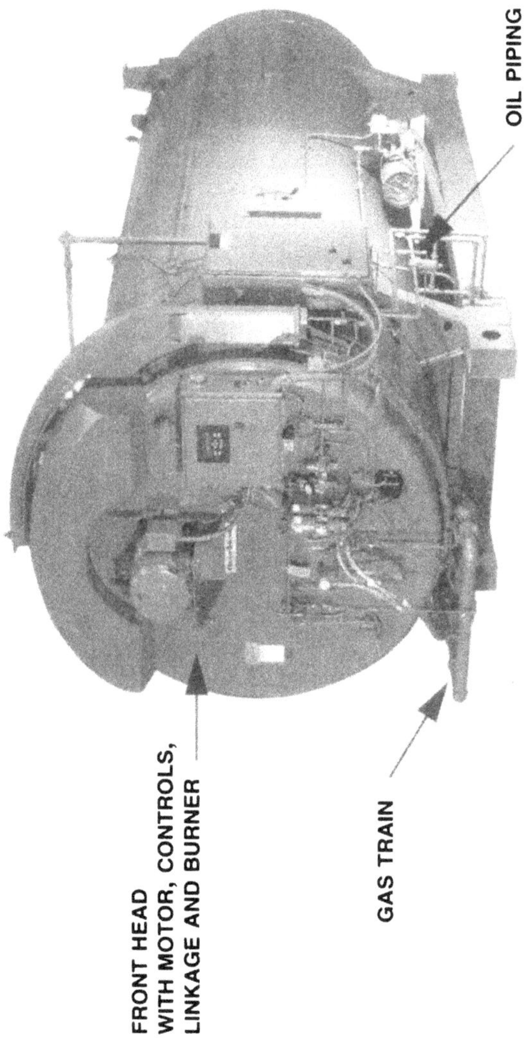

Figure 4-3. Typical packaged firetube boiler (*Courtesy of Cleaver Brooks*).

Figure 4-4. Cutaway of a four pass firetube boiler (*Courtesy of Cleaver Brooks*).

directs the combustion gases from the furnace to the tube banks. The wetback boiler design has a water-cooled turn-around chamber, which directs the flue gases from the furnace to the tube banks. A wetback firetube boiler is shown in Figure 4-5.

Firetube boilers contain large amounts of water that allow them to respond to load changes with little variation in steam pressure or hot water temperature. Steam pressures in firetube boilers is generally limited to around 350 psig.

Firebox boilers are similar in design to firetube boilers. In firetube boilers, the combustion chamber is cylindrical in shape; firebox boilers have a combustion chamber that is not cylindrical in shape. Firebox boilers are typically limited to low pressure steam or hot water applications.

Boilers may also have a watertube design. In watertube boilers, the water to be heated flows through tubes while the combustion gases flow around the outside of the tubes. Typically, a watertube boiler design consists of an upper drum (steam drum) and a lower drum (mud drum) with interconnecting water tubes. The boiler water flows by convection from the lower drum to the upper drum. The drums and water tubes of a typical water boiler are shown in Figure 4-6.

Watertube boilers typically have fast steaming capabilities. Steam or hot water can be generated very rapidly due to the relatively low volume of water. Industrial watertube boilers are capable of generating steam pressures in excess of 350 psig. They generate either saturated or superheated steam. Steam can be generated at operating conditions at flow rates as high as 134,000 lb/hr.

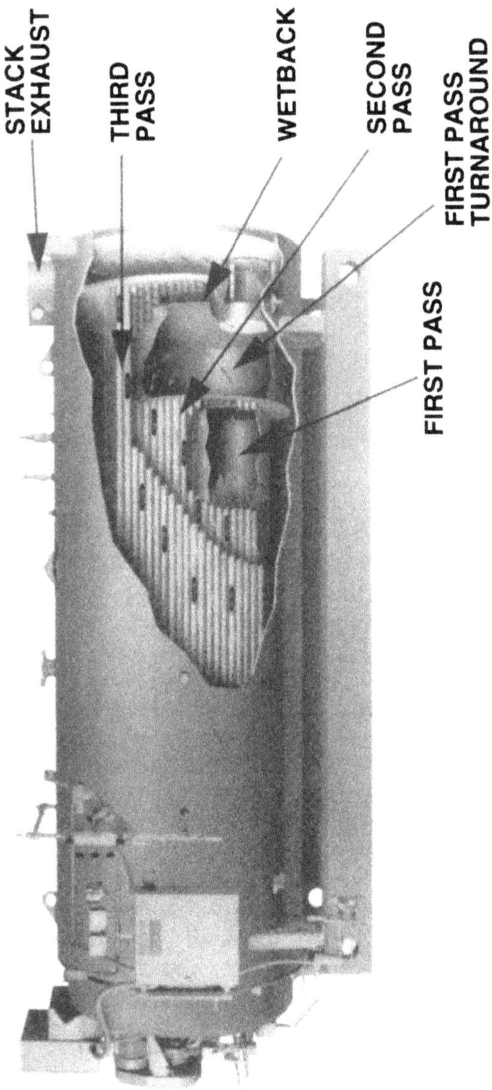

Figure 4-5. Wetback firetube boiler (*Courtesy of Cleaver Brooks*).

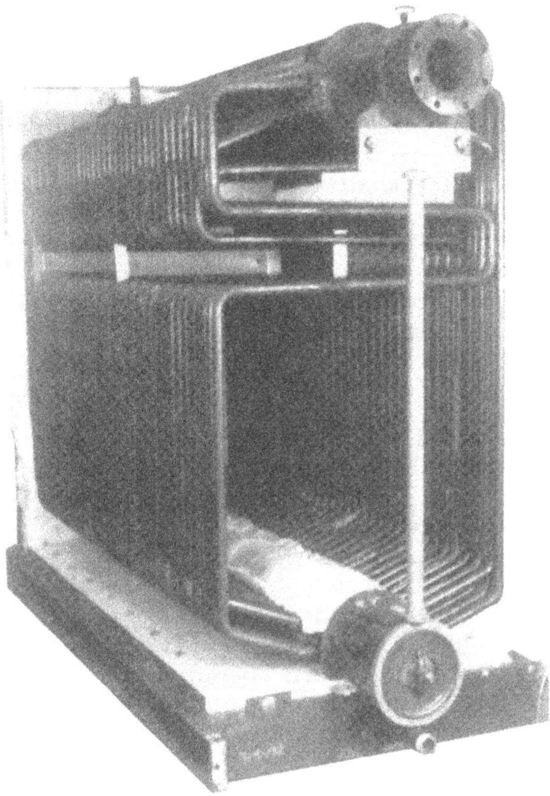

Figure 4-6. Tubes and drum of a watertube boiler (*Courtesy of Cleaver Brooks*).

Earlier, the need for oxygen in the combustion process was discussed. Oxygen for the combustion process is normally supplied by bringing air into the combustion chamber. The term draft is commonly used to designate the air static pressure in a furnace. Depending upon the design of the furnace, air may either be drawn into the combustion chamber or forced into it. Burners may be one of two basic types: atmospheric burners and power burners. As the name implies, atmospheric burners operate with the combustion chamber at, or near, atmospheric pressure. Combustion air is naturally drawn in as a result of a low pressure created by the combustion process. Power burners use a fan to supply combustion air. Power burners can be natural draft or forced draft. In natural draft burners, a chimney must draw the products of combustion through the boiler or furnace. The burner fan supplies only enough power to move the air through the burner. In forced draft burners, the fan not only moves through the burner, but also forces it through the boiler. Combustion occurs under pressure in controlled air flow. The power burner usually has superior combustion control.

4.3 BOILER CONTROLS

Boiler controls and instrumentation have two main function: They operate the boiler within safe limits and operate the boiler at the desired output or capacity. Safety controls and interlocks are intended to operate the boiler within safe limits and operating controls operate the boiler at desired capacities.

Safety controls and interlocks are primarily provided to protect against explosions and other hazards, which may result in unsafe conditions and/or cause damage to the boiler or other equipment. These controls shut the fuel off or prevent firing if an unsafe condition, such as the following, exist:

- Flame safeguard: The purpose of the flame safeguard controls is to monitor the burner start-up and control the firing rate during normal operation. The flame safeguard also controls the shutdown sequence.
- Gas pilot flame adjustment: This is an adjustment to set the size of the pilot flame so it is large enough to ignite the main flame and also be detected by the flame sensor.
- High limit control: A pressure (or temperature) limiting control that is set above the normal operating limit control and, in the event that the pressure (or temperature) in the boiler rises above the normal operating limit, the control shuts down the boiler operation.
- Low fire hold control: This control limits the burner to the low fire position until the pressure (or temperature) in the boiler is high enough that the boiler has sufficient time to warm up.
- Low water cutoff: A control that is set to de-energize and shut down the burner upon detecting a low water condition in the boiler.
- Auxiliary low water cutoff: An additional supplementary control, which is wired in series with the primary low water cutoff.
- Low gas pressure switch: It closes the main gas valve whenever the gas pressure drops below a preset minimum.
- High gas pressure switch: It closes the main gas valve whenever the gas pressure rises above a preset maximum.
- Combustion air proving switch: A pressure actuated switch, which shuts down the boiler if the is inadequate air pressure from a forced draft combustion air fan.
- Low oil temperature switch: When heavier grade fuel oils are burned, the oil must be heated in order for it to properly atomize. The low oil temperature control will shut the burner down if the oil temperature is below that required for proper atomization.
- High oil temperature switch: If the fuel oil is heated to a higher temperature than acceptable for proper atomization, the high oil temperature control will shut the boiler down.

- Low oil pressure switch: In the event that the oil pressure drops below a level required for proper burner operation, the control will shut the boiler down.

Operating controls initiate the normal starting, stopping and operation of the burner to maintain set operating conditions in the boiler and to control its output capacity. Steam boilers are controlled by boiler mounted pressure actuated controls, which vary the input of fuel to the burner in response to the steam pressure. Common control sequences include on-off, high-low-off and modulating control. Modulating control varies the fuel input in an infinite number of settings from 100% down to a selected minimum point.

Hot water boilers are controlled by temperature-actuated controls that are usually mounted on the boiler. Control sequences for hot water boilers are the same as for steam boilers, i.e., on-off, high-low-off and modulating.

The control system must also provide combustion control, which is accomplished by providing the correct amounts of combustion air and fuel depending upon the boiler control sequence. These controls are also referred to as oxygen trim controls. An optimum range of excess air exists for each combustion of fuel, firing mode and furnace condition. Too little air flow results in unburned fuel in the flue gas—combustible emissions that are hazardous, polluting and can contain considerable amounts of unused energy. Too little excess air results in higher NO_X emission levels and increased energy losses.

Conventional trim control systems are typically a mechanical jackshaft system. A typical jackshaft burner control system is shown in Figure 4-7.

There are two basic methods by which trim controls provide combustion control for burners, series control and parallel control. Figure 4-8 shows block diagrams of the basic combustion control methods. Parallel positioning control is probably the most common method for boilers under 50,000 Btu/hr. This method of control uses a signal from a steam pressure sensor (or water temperature sensor), which simultaneously goes to the fuel control devices and air flow regulating devices. Changes in steam pressure (or water temperature) result in a change in both fuel flow and combustion air flow. Parallel positioning control systems respond quickly to load changes, have simplicity of operation and are very reliable. The primary disadvantage of the parallel positioning system is that the control of the individual air and fuel regulators is basically without feedback. Therefore, there is no assurance that the fuel/air ratio is within the desired range. The jackshaft burner control system shown in Figure 4-7 is an example of a parallel positioning system.

In series positioning control, either the fuel flow or air flow is monitored and the other is adjusted accordingly. The control signal from the steam pressure sensor (or hot water sensor) is not simultaneously sent to the fuel

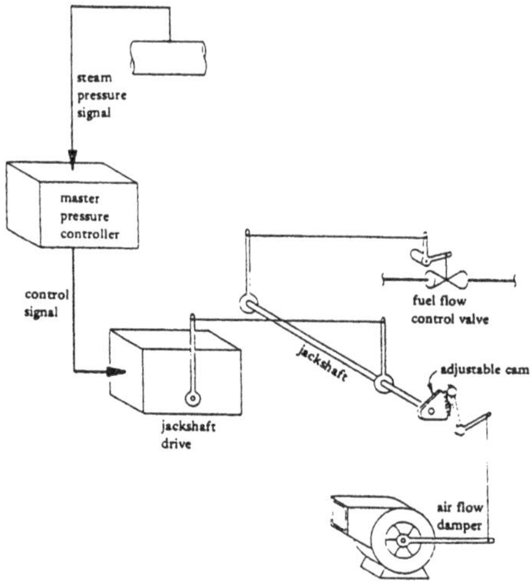

Figure 4-7. Jackshaft burner control system.

and air regulating devices. Rather, the signal is sent to only one of the regulating devices. The displacement of the first device is then measured and used to control the position of the second device. A series positioning control system is shown in Figure 4-9.

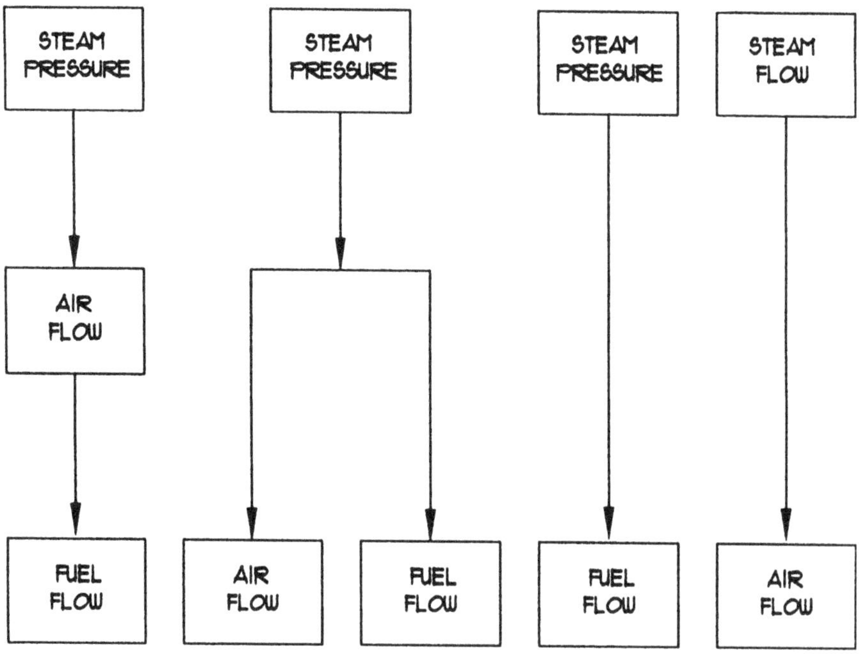

Figure 4-8. Basic combustion control methods.

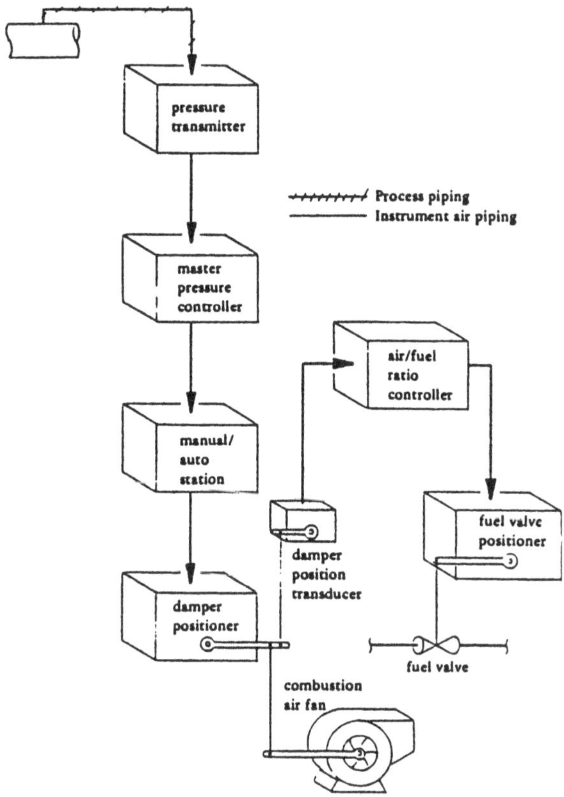

Figure 4-9. Series positioning control system (air leading fuel).

The series positioning method is also without feedback (open loop) and, therefore, there is no assurance that the air/fuel ratio is within the desired range. The series method is slower to react to rapid load changes than the parallel.

Combustion controls may also be classified as positioning or metering controls. Positioning controls respond to system load changes by moving to a preset position. Metering systems respond to actual measurements of the fuel and/or air flow rates. Metering control systems overcome the major shortcomings of series and parallel control systems. With metering control systems, the control of the fuel and air regulators is a closed loop. The steam pressure (hot water temperature) is measured and feedback is provided to a master controller, which adjusts the fuel and air flows. The master controller then adjusts the respective fuel and air flow control devices. Metering control systems are typically used on large boilers due to the added complexity and cost of these systems. A metering control system is shown in Figure 4-10.

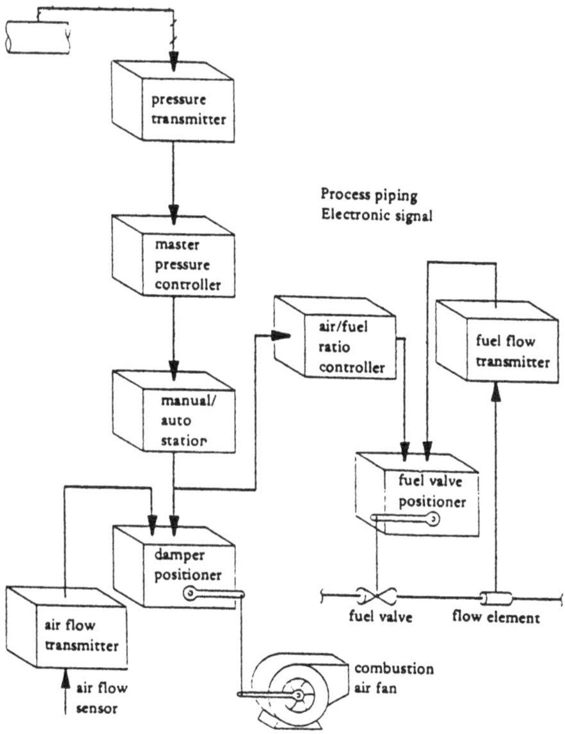

Figure 4-10. Metering control system.

Other common controls found on boilers include feedwater control and blowdown control. Feedwater control is necessary to maintain the proper water level in a boiler despite variations in load and despite the fact that it may take some time for steam condensate to return to the boiler. The simplest boiler feed control system uses only the boiler water level as the input to control the feed system. This arrangement can cause make-up water to be unnecessarily brought into the system during periods of high loads. This may result in too much water in the system during light load periods. Too much water may result in excess water being discharged from the system. Controllability of feedwater systems can be improved by adding steam flow sensors to monitor the steam flow rate and control the boiler make-up water flow rate.

Boiler blowdown is a method for maintaining water quality in the boiler by periodically removing, or blowing down, contaminated water from the boiler. As water in the boiler is heated and evaporated to produce steam, the concentration of impurities in the water, such as dissolved minerals or organic compounds, increases. Boiler blowdown releases hot water containing impurities from the boiler. Blowdown separators are sometimes used to separate flash steam in from the blowdown water and to cool the water before it is discharged to a sewer system. Automatic blowdown controllers typically use electromagnetic cells to measure the concentrations of impurities by measuring the electrical conductivity and pH levels. The electrical signal corresponding to the impurity level is compared to a setpoint of the controller and when the concentration of impurities exceeds the setpoint, blowdown is initiated.

4.4 BOILER ACCESSORIES

In addition to the boiler itself, along with its controls, there are a number of accessories required for proper boiler operation. Accessories that may be typically used with boilers include deareators, feedwater systems, make-up water conditioning equipment and blowdown systems.

4.4.1 Deaerators

Water, which is fed into a boiler (feedwater) to generate steam, contains two dissolved gases that are harmful to boilers and the steam distribution system. These gases are oxygen (O_2) and carbon dioxide (CO_2). Raw, untreated make-up water can contain up to 10 cc of O_2 per liter. If the dissolved gases are not removed before entering the boiler, they will be liberated when the water is heated. A deareator is a mechanical device that uses heat to remove dissolved gases from feedwater before it enters the boiler. A deaerator is shown in Figure 4-11.

A deaerator consists of a pressurized ASME tank, transfer pumps and controls. Steam condensate returning from the steam distribution system, as well as any required make-up water, is fed into the pressurized tank. As the condensate and/or water enters the tank, it passes through a spray manifold causing it to fall into the tank as a mist. Steam is injected into the bottom of the tank, which heats the water to its saturation temperature. The combination of the spraying and heating drives the dissolved gases out of the water. The gases are released through a vent in the top of the tank. Controls on the deaerator include a float switch and a low water cutoff. After the water is deaerated, it flows by gravity to transfer pumps where it is usually transferred to a boiler feed unit.

4.4.2 Feedwater Systems

In any steam distribution system, there is a time lag between when the steam leaves the boiler and it returns in the form of condensate. The greatest time lag occurs during a cold start-up of the system. The total volume of steam must come from the boiler during this period; this causes

Figure 4-11. *Deaerator.*

the water level in the boiler to drop. Unless condensate is available to raise the level of water in the boiler, raw make-up water must be drawn into the system by the boiler. This can result in too much water in the system, which is discharged later. In order to compensate for normal operations with fluctuating loads, boiler feed systems are used to handle the fluctuations and maintain proper water level in the boiler. A typical feedwater system is shown in Figure 4-12.

A feedwater system consists of a collection tank and transfer pumps. The collection tank is sized to store sufficient volumes of water to keep an adequate level of water in the boiler without bringing additional water into the system. The tank also serves to store large volumes of returning condensate, reducing the need to discharge excess water from the system.

The water flows by gravity from the tank to the boiler feed pumps, which transfer the water to the boiler when the water level drops. The boiler feed pumps are controlled by a level switch on the boiler.

Some boiler feed units also preheat the water before it enters the boiler, particularly on systems that require large amounts of make-up water. The water is usually preheated with steam from the boiler. Preheating the make-up water reduces the thermal shock to the boiler.

Figure 4-12. Feedwater system.

4.4.3 Make-up Water Conditioning Equipment

As discussed earlier, make-up water contains significant amounts of impurities. In addition to dissolved gases, such as oxygen and carbon dioxide, make-up water usually contains large amounts of suspended solids and chemicals. Unless theses solids are removed, significant corrosion and scale buildup will occur in the boiler and steam distribution system. Corrosion can lead to premature failure of steam system components. A buildup of scale will significantly reduce the heat transfer within the boiler, which will significantly reduce efficiency. To reduce the problems associated with untreated raw make-up water, the water is usually treated by water conditioning equipment before entering the system. A typical water conditioning system is shown in Figure 4-13.

A water conditioning system typically includes some sort of filtration and a water softener. Filters are available which will remove free chlorine, some dissolved organic materials and sediment. They can also remove suspended solids, colloidal matter, sand and iron. Water softeners are used to remove calcium salts and magnesium salts, as well as reduce the hardness of water.

The most common softening method is a sodium zeolite process. Although softening of the water reduces many harmful chemicals from make-up water, chemicals must still be added to a boiler and steam system to control corrosion and scale.

4.4.4 Expansion Tanks

Expansion tanks are required on boilers that produce hot water for process or heating purposes. An expansion tank is a pressurized ASME tank. The

Figure 4-13. Water conditioning system.

primary function of an expansion tank is to provide a means for the water in a closed system to expand as it is heated, without significantly changing its pressure. Essentially, water is an incompressible fluid, although it does expand as it is heated. Unless a means were provided for expansion as the water is heated, the system would rupture.

Expansion tanks also maintain a minimum pressure, or charge, on the system. Water will boil at 212°F and a pressure of 14.7 psia. If the water is heated to, or above, 212°F, the system must be pressurized to keep the water from boiling. The higher the temperature, the greater the required pressure.

Finally, expansion tanks may also be used to remove air from the hot water system. Expansion tanks are frequently equipped with air vents, which allows trapped air to escape from the system.

4.4.5 Blowdown Systems

The need for boiler blowdown was briefly discussed earlier in this chapter. As water is heated, the suspended solids in the water become less soluble, tend to crystallize and concentrate on boiler heat transfer surfaces as scale. Scale has a low heat transfer coefficient and, therefore, becomes an insulating barrier on the heat transfer surface. To reduce the concentration of impurities in a boiler and reduce scale buildup, small amounts of water containing the impurities are drained (blown down) from the boiler. There are two primary methods by which boilers are blown down: manual bottom blowdown and surface blowdown.

Bottom blowdown is typically a manual method. As the name implies, it removes water from the bottom of the boiler. Bottom blowdown also removes sludge and total dissolved solids, which collect in the "mud drum" of the boiler. Since bottom blowdown is a manual method, it is done on an intermittent basis.

Surface blowdown is a constant, or continuous, removal of impurities from the top surface of the water in the boiler. It is a continuous, low flow skimming process that skims sediment, oil and other impurities floating on the surface. Surface blowdown is regulated by a flow control valve, which is adjusted according to the total dissolved solids level in the boiler.

4.5 BOILER EFFICIENCY

Before implementing energy efficiency improvements to increase boiler efficiency and performance, it is first necessary to understand what boiler efficiency is and what affects the performance and efficiency of boilers. For any given boiler, its efficiency basically is the difference between the energy input and the energy output divided by the energy input. The efficiency of a boiler may be defined in three ways: combustion efficiency, overall efficiency and seasonal efficiency.

Combustion efficiency was discussed earlier in this chapter. It is a measure of the effectiveness of the burner only and is related to the burner's ability to completely burn the fuel. Combustion efficiency is the energy input minus the flue gas losses divided by the energy input. The boiler itself has little bearing on combustion efficiency. For non-condensing boilers, combustion efficiencies typically range from 75 to 86%. Approximate efficiencies for non-condensing boilers can be determined under operating conditions by measuring the operating flue gas temperature and percentage of O_2 and CO_2 in the flue gas.

Overall efficiency is the gross energy output divided by the energy input. Gross energy output is measured in the steam or hot water leaving the boiler. It is also called the "fuel-to-steam" efficiency. The overall efficiency depends on the construction of the boiler and also on individual installation characteristics. Overall efficiency is a heat balance that considers stack and temperature losses, excess air levels, and radiation and convection losses for the boiler. Losses may also include off-cycle energy losses for applications where the boiler cycles on and off. Overall efficiency is lower than combustion efficiency by the percentage of heat loss for the above losses. It can be approximated by determining the combustion efficiency and subtracting the percent heat losses for radiation losses, flue gas losses and off-cycle losses.

Seasonal efficiency is the actual operating efficiencies that a boiler will achieve during an entire season of operation. Since many boilers operate at part load at various times, there can be a significant difference between overall efficiency and seasonal efficiency for the same boiler. For boilers with variable loads, seasonal efficiency accounts for their part load efficiency, as well as heat losses when the boiler is off.

In addition to the above methods for calculating boiler efficiency, there are several ways of defining boiler efficiency, including:

- As found efficiency, which is the efficiency of installed boilers in the field in the present state of repair or maintenance. This is used as a baseline for any modifications that would improve efficiency.
- Tuned-up efficiency, which is the efficiency after operating adjustments (excess air adjustment) and minor repairs have been made to the boiler.
- Maximum attainable efficiency is the efficiency that results from adding currently available efficiency improvements without regard to cost.
- Maximum economically achievable efficiency, which is similar to the maximum attainable efficiency except that it includes only improvements that are justifiable from an economic standpoint.

Table 4.4 lists the average "as-found" operating efficiencies of various types of boilers for comparison purposes.

4.5.1 Factors Affecting Boiler Performance and Efficiency

For any given boiler, there are a number of factors that can have an impact on its efficiency. For the most part, these factors have to do with the combustion process. Factors affecting boiler efficiency include the following.

Waste heat losses in the flue gases include the dry (sensible) flue gas loss and the moisture (latent) loss. Dry loss is equal to the heat required to raise the fuel gas from room temperature to the exiting flue gas temperature measured just after the last boiler heat transfer surface. Heat is also lost in the form of moisture or latent heat. Latent heat is heat that would be available if all the water vapor in the flue gas were condensed into liquid water. Flue gas temperatures reflect the energy that did not transfer from the combustion products to the steam or hot water in the boiler. The lower the leaving flue gas temperature, the more effective the heat exchange and the higher the operating efficiency of the boiler. Minimum flue gas temperatures are limited by the possibility of condensation in the flue gases, which would result in sulfuric acid and corrosion.

Incomplete combustion results in combustible material, such as carbon, hydrogen and carbon monoxide, in the flue gases. This combustible material is unburned fuel, which has not been oxidized and, therefore, has released no heat. Excess air is typically supplied to the burner to reduce the possibility of incomplete combustion.

Radiation heat losses from a boiler include heat losses through the exterior surfaces of the boiler. Heat is lost in the form of radiant heat exchange with the boiler room and convection heat transfer to the ambient air in contact with the boiler surfaces. Radiation heat losses represent a

TABLE 4.4 Average As-Found Operating Efficiency of Boilers

	Rated Capacity Range (MBtu/hr)							
	10–16		16–100		100–250		250–500	
Category/Fuel	**Measured**	**Calculated**	**Measured**	**Calculated**	**Measured**	**Calculated**	**Measured**	**Calculated**
Watertube								
Gas	(78.0)	79.9	79.5	79.9	81.2	80.9	(82.9)	81.2
Oil	(81.5)	83.7	82.8	83.7	(83.4)	84.6	(82.7)	85.3
Coal -Stoker	*	81.0	76.6	81.2	82.2	81.8	*	82.5
Pulverized	*	83.2	*	83.3	(86.6)	86.1	(85.3)	86.3
Firetube								
Gas	(81.0)	79.9	79.5	79.9	NA		NA	
Oil	(86.3)	83.7	(85.8)	83.7	NA		NA	

relatively small percentage of efficiency losses; however, radiation heat losses are essentially constant throughout the firing range of the boiler. Therefore, they are a more significant loss at low fire than at high fire. Boilers are insulated to reduce radiation heat losses. The boiler design can have an effect on radiation heat losses. Radiation and convection losses are also affected by ambient air temperatures and the velocity of air flowing across a boiler. Lower ambient temperatures and high air velocities increase radiation losses.

The boiler firing rate also affects the efficiency of a boiler. As the load on the boiler decreases, the flue gas temperature also drops. As indicated above, low flue temperatures result in increased efficiency. However, since radiation losses are constant for all boiler loads, the improvement in efficiency is somewhat offset by the constant radiation losses.

The amount of excess air supplied for combustion has a significant effect on boiler efficiency for any given boiler. Since ambient air is brought into the burner at room temperature (unless preheated), it must be heated to the maximum combustion temperature by the combustion process. This usually requires significant amounts of heat from the combustion process. Combustion air quantities greater than actually required for complete combustion requires significant amounts of heat, which directly affects boiler efficiency. The type of fuel burned has an effect on boiler efficiency. In the case of gaseous fuels, the greater the hydrogen content, the more water vapor will be formed in the combustion process. The water vapor absorbs heat as it changes phase in the combustion process. The ratio of hydrogen to carbon in the fuel also affects boiler efficiency. Carbon tends to burn at a higher temperature and releases less heat per cubic foot than hydrogen as it is burned. When evaluating boiler efficiency, knowing the actual fuel specification is important.

Ambient air temperatures in the boiler room can have an effect on boiler efficiency. Most boiler efficiency calculations are based on an ambient air temperature of 80°F. As indicated above, higher ambient air temperatures reduces the amount of heat required to raise the temperature of combustion air and also reduces radiation losses for the boiler.

4.6 MAINTAINING AND OPERATING BOILERS FOR OPTIMUM ENERGY EFFICIENCY

For boilers and combustion systems to operate at peak efficiency and minimum energy consumption, it is necessary that they receive regular maintenance and service. Significant energy savings can be achieved by routinely making adjustments to the combustion controls. Proper routine maintenance should be the first step in any energy management program for boilers and combustion systems. Since the combustion process is the heart of boilers and combustion systems, most energy conservation maintenance is directed at the burner and controls. The following are operations and maintenance procedures, which will result in optimum energy efficiency for most boilers and combustion systems.

Maintain Accurate Boiler Logs

It is important to monitor the performance of a boiler on a regular basis. It is also important to record any modifications, adjustments or routine maintenance performed on a boiler. The objective of a boiler log is to monitor and document any deviations from the desired performance as a function of time, as well as a history of boiler maintenance. The usual practice is to check and record boiler performance on an hourly basis. Efficiency related performance items that should be checked and recorded include:

- General data to establish output
  - steam pressure or water supply temperature
  - steam or water flow rate
  - superheated steam temperature (if applicable)
  - feedwater temperature
- Firing system data
  - fuel type (in multifuel boilers)
  - fuel oil flow rate
  - oil or gas supply pressure
  - pressure at burners
  - fuel temperature
  - burner damper settings
  - windbox to furnace pressure differential
- Air flow indication
  - air temperature at burner inlet
  - stack gas O_2
  - fan damper position
- Flue gas temperatures
  - boiler outlet gas
  - economizer or air heater outlet
  - air temperature entering preheater
- Unburned combustible indication
  - CO measurement
  - stack appearance
  - flame appearance
- Air and flue gas pressures
  - forced draft fan discharge
  - boiler outlet
  - economizer differential
  - air heater and gas side differential
- Unusual conditions
  - steam or hot water leaks
  - abnormal vibrations

- equipment malfunction
- excessive make-up water
- Blowdown operation
- Sootblower operation

Flue Gas Analysis

Combustion control is the foundation of boiler energy management since it is a major factor in determining boiler efficiency. Flue gas loss is the largest single energy loss in a combustion process. The density of air varies with its temperature. A proper air/fuel ratio in the summer will result in too much air in winter. Conversely, a proper air/fuel ratio in the winter will result in incomplete combustion in the summer. The best method of determining the combustion efficiency of a boiler is to measure and analyze the products of combustion in the flue gases.

Flue gas temperatures are also important indicators of boiler efficiency. Optimum boiler efficiency occurs when flue gas temperatures are at a minimum (without condensation in the flue gas). Higher than normal flue gas temperatures are typically an indication of fouled heat transfer surfaces or high excess air, which reduces combustion chamber heat transfer. Lower than normal flue gas temperatures may be an indication of incomplete combustion, possible ruptured boiler tubes or cold air infiltrating the combustion chamber.

It is important that the combustion efficiency of a boiler be periodically checked with a flue gas analyzer of some sort. Combustion analyses should be performed at least twice a year as a minimum; monthly analyses are better.

Portable flue gas analyzers may be used to measure the combustion products in the flue gas. They must be able to measure oxygen, carbon monoxide and carbon dioxide. Excess O_2 and CO_2 in flue gases are the key constituents to be measured in a flue gas analysis. Oxygen is the most frequently measured flue gas component. It is an excellent indicator of excess air level (and, therefore, combustion efficiency). Figure 4-14 graphically shows relationship between the percent excess air and percent O_2 and CO_2 for various fuels.

There are four methods presently used to measure O_2, CO and CO_2 in flue gases: Orsat test, paramagnetic O_2 sensors, wet electrochemical cells and zirconium oxide cells. Orsat testing is the only portable method for testing boilers without permanent flue gas sampling. In the Orsat test, a sample of flue gas, which has been conditioned (cleaned, dried and cooled) is passed through a series of pipets, each of which contains a separate chemical reagent. Each reagent absorbs a different chemical in the flue gas, usually O_2, CO and CO_2. After the gas has been passed through each reagent, the volume of the gas is accurately measured. The reduction in volume indicates the amount of gas that was absorbed from the original sample.

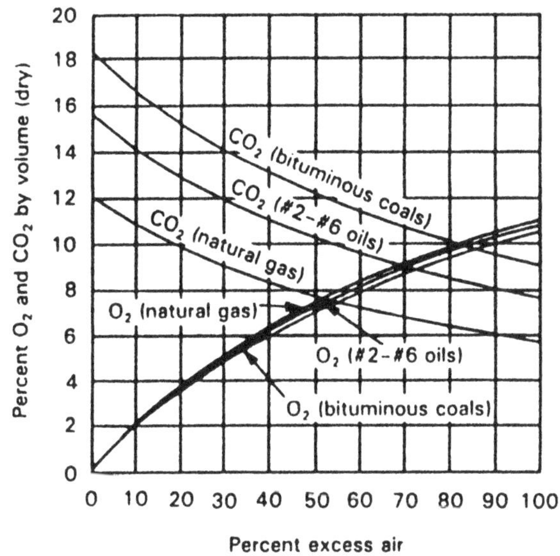

Figure 4-14. Oxygen and carbon dioxide versus excess air.

The initial step in improving the combustion efficiency of a boiler is to determine the "as is" or present condition and performance of the combustion system. The procedure for performing flue gas tests is as follows:

1. Operate the boiler on low fire, allow time for the system to stabilize (about 5 minutes), then measure and record the O_2, CO, CO_2 and flue gas temperatures.
2. Increase the load on the boiler to about 40% and allow the boiler to stabilize (about 5 minutes), then measure and record O_2, CO, CO_2 and flue gas temperatures.
3. Repeat step 2 with the boiler loaded at 60%, 80% and 100%.
4. Repeat steps 2 and 3 in reverse order (100%, 80%, 60%, 40% and low fire).
5. Compare the measurements from the two sets. If the difference between the two sets of data is significant, the boiler controls should be repaired to remove sloppy conditions, or play, in the controls.
6. Record all pertinent data for future comparisons.

There are four basic combustion zone conditions that can result in inefficient combustion:

- Insufficient combustion air applied to the flame to permit complete combustion.
- Non-uniform delivery of fuel or combustion air in the combustion zone.

- Insufficient temperature in the combustion zone to permit proper burning of the fuel.
- Insufficient flame turbulence or inadequate mixing of fuel and air during vaporization and combustion.

The disadvantages of the Orsat test include the length of time involved in the analysis, the fact that the accuracy is dependent upon the purity of the reagents and the skill of the operator. Orsat analyses should be performed by experienced boiler service personnel. Also, there is no way to provide an automatic signal for a recorder or control system.

The potential savings of about 2–20% can be achieved depending upon the condition of the boiler in the "as is" condition. All potential energy conservation improvements should be based on the boiler performance in the tuned-up condition.

Reduce Excess Air

Earlier in this chapter, the need for the proper quantity of excess air was discussed. Too little excess air will result in incomplete combustion. Too much excess air will result in inefficiency. The air brought in for the combustion process must be heated by the process to the combustion temperature. Too much air will result in more heat being used to heat the air and less heat being transferred to the steam or hot water. Too much excess air will also result in reduced heat transfer through the tubes to the steam or hot water. Maximum combustion efficiency occurs when the correct amount of excess air is supplied, such that both the unburned fuel loss and flue gas heat loss are minimized, as shown in Figure 4-15.

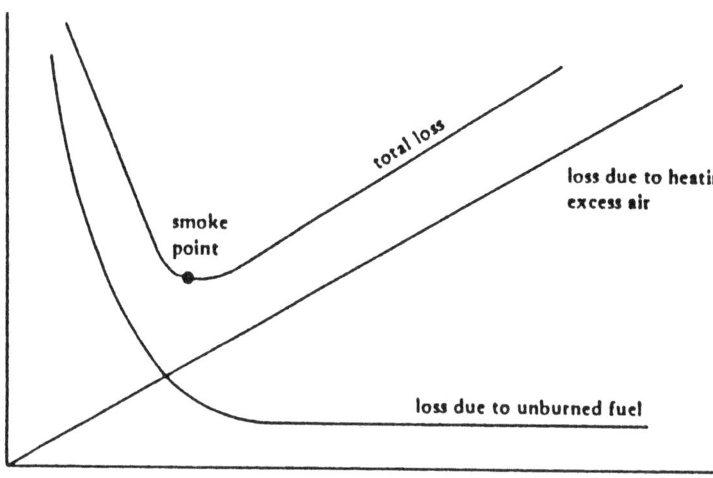

Figure 4-15. Boiler flue gas losses.

TABLE 4.5 Optimum Excess Air for Various Fuels

Fuel Type	Firing Method	Optimum Excess Air (%)	Equivalent O_2 (by volume)
Natural gas	-	5–10	1–2
Propane	-	5–10	1–2
Coke oven gas	-	5–10	1–2
No. 2 fuel oil	steam-atomized	10–15	2–3
No. 6 fuel oil	steam-atomized	10–15	2–3
Coal	pulverized	15–20	3–3.5
Coal	stoker	20–30	3.5–5

In addition to checking for "sloppy" combustion controls, the flue gas analysis (test) should also be used to determine the excess air levels for a boiler. When flue gas measurements are made, a low CO_2 or high O_2 reading indicates the presence of too much excess air in the combustion process. Reducing excess air to the proper levels can result in significant improvements in combustion efficiency.

The main causes of too much excess air are air leaks, improper trim control and faulty burner operation. Low O_2 in the combustion chamber and high O_2 in the flue gas indicates leaks in the furnace casing. High O_2 in the combustion chamber and flue gas indicates improper trim control. CO_2 readings would be opposite the O_2 readings in the above scenarios. Measuring O_2 is preferable to measuring CO_2.

Maximum combustion efficiency is achieved when the correct amount of excess air is supplied such that the sum of both unburned fuel loss and flue gas heat loss is minimized, as shown in Figure 4-15. The common practice in the industry is to adjust the combustion air flow rate to provide 50–100% air above that required for complete combustion. This significantly reduces boiler efficiency. Most boilers can be adjusted to provide 10–20% air above that required for complete combustion without any compensation for changing environmental conditions. Table 4.5 lists the optimum excess air for various fuels.

Example 4.3. A gas (methane) fired boiler has a gas input of 16.6×10^6 Btu/hr. A flue gas test was performed on the boiler and the test indicated that the flue gas had 7.8% oxygen and was at a temperature of 600°F. Combustion air is supplied at 70°F. The boiler is operated for 8,760 hours per year.

How much excess air is being used in the combustion process? If the controls on the burner were adjusted so there was 2% oxygen in the flue gases, what would the annual reduction in gas consumption be? Assume natural gas has a heating value of 1,000 Btu/ ft^3 and that the combustion air is at 70°F and 14.7 psia and has a specific heat of 0.24 Btu/lb-°F.

From Table 4.2, methane gas (CH_4) has a density of 0.0425 lb/ft^3. The weight of flow of the natural gas in the combustion process is

$$W_g = (16.6 \times 10^6 \text{ Btu/hr})(1 \text{ ft}^3/1{,}000 \text{ Btu})(0.0424 \text{ lb/ft}^3)$$
$$= 705.5 \text{ lb/hr}$$

From Table 4.2, the weight of air required for complete combustion of methane gas (100% excess air) is 17.27 lb-air/lb-gas. The weight of the air flow at 100% excess air is

$$W_{a-100\%} = (705.5 \text{ lb/hr})(17.27 \text{ lb - air/lb - gas})$$
$$= 12{,}183.99 \text{ lb - air/hr}$$

The O_2 measured in the flue gas was 7.8%. Figure 4-12 shows that the amount of excess air is 150%. The weight of the present air flow (150% excess air) is:

$$W_{a-150\%} = (12{,}183.99 \text{ lb - air/hr})(150\%/100)$$
$$= 18{,}275.99 \text{ lb - air/hr}$$

The burner trim controls are adjusted and the O_2 in the flue gas is measured at 2%. From Figure 4-14, the amount of excess air at 2% O_2 for natural gas is 110%. The weight of the air flow after the adjustment to the trim controls is

$$W_{a-110\%} = (12{,}183.99)(110\%/100) = 13{,}402.39 \text{ lb - air/hr}$$

The reduction in the air flow rate between 150 % and 110% is

$$w = 18{,}275.99 - 13{,}402.39 = 4{,}873.6 \text{ lb - air/hr}$$

The air has a specific heat of 0.24 Btu/lb-°F and is heated from 70°F to 600°F in the combustion process. The heat savings (reduction in heat required to heat the air) for reducing the excess air from 150% to 110% is

$$q = (4{,}873.6 \text{ lb - air/hr})(0.24 \text{ Btu/lb} - {}^\circ\text{F air})(600 - 70)^\circ\text{F}$$
$$= 619{,}922 \text{ Btu/hr}$$

Which is about a 4% reduction in the required heat input. The annual hours of operation are 8,760; therefore, the annual heat savings are

$$q/\text{yr} = (619{,}922 \text{ Btu/hr})(8{,}760 \text{ hr/yr}) = 5{,}430 \times 10^6 \text{ Btu/yr}$$

A therm of gas is equal to 1x10^5 Btu. The annual gas savings is

$$\text{Annual gas savings} = \frac{5{,}430.5 \times 10^6 - \text{Btu/yr}}{1 \times 10^5 \text{ Btu/therm}}$$
$$= 54{,}305 \text{ therms per year}$$

Establish a Boiler Loading Schedule

Frequently, boiler plants have multiple boilers, each with its own auxiliary equipment. One or more boilers may be required to meet the load at any given time. Each boiler may be a different age, be in better or worse operating condition, and perhaps burn a different fuel.

As a result of different boiler characteristics, each one has a different efficiency. Each boiler will also have a different efficiency depending upon its load and firing rate. Most boilers do not have a straight line efficiency at all loads, but vary as shown in Figure 4-16. Boilers are typically less efficient at higher loads.

Since each boiler operates at a different efficiency, the boilers should be operated based upon a predetermined schedule, which will optimize the overall efficiency of the boiler plant. An optimized loading schedule will allow any heating demand to be met with the minimum energy input. When developing a boiler loading schedule, there are some general guidelines to consider, including the following:

- Boilers generally operate most efficiently in the 65–75% range of full load capacity as shown in Figure 4-16.
- It is usually more efficient to operate a small number of boilers at higher loads than a large number of boilers at lower loads.
- Boilers should be put in service in order of efficiency, with the most efficient boilers brought on line first; they should be taken off-line in reverse order of efficiency.
- Newer boilers with large capacities are generally more efficient than older, smaller boilers.
- Generally, load swings should be handled by the smallest and least efficient boiler

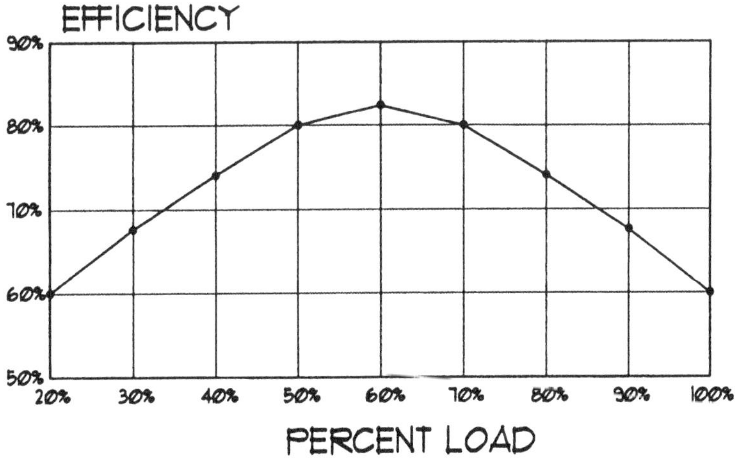

Figure 4-16. Boiler efficiency versus load.

In addition to a loading schedule for the boilers themselves, a loading schedule should be established for auxiliary equipment. Frequently, auxiliary equipment, such as deaerators, are piped such that they may serve multiple boilers. In this situation, the most efficient deareator should be paired with the most efficient boiler.

In addition to establishing a boiler loading schedule, procedures should be established for maintaining boilers in the standby mode for load fluctuations. As indicated above, it is generally advisable to operate fewer boilers at a higher firing rate. Standby boilers should be maintained off-line at the operating temperature and pressure. This may be accomplished by intermittent firing of either the ignitors or a main burner to replace ambient heat losses at no load. Guidelines for maintaining standby boilers include

- Shut all dampers and registers to minimize losses from the boiler.
- Establish and follow strict safety procedures for ignitor/burner light-off.

Clean Heat Transfer Surfaces

Both the fire side and water side of heat transfer surfaces can become fouled with deposits. These deposits, which form on the heat transfer surfaces, have a low thermal conductivity and actually become insulators that retard heat transfer. As a result, heat, which should have gone into the steam or hot water, is lost at the flue stack. Figure 4-17 shows a cross-section of a boiler tube with a thermal resistance analogy.

A primary indicator of a fouling problem is an increase in the flue gas temperature, although a higher temperature may be as a result of other conditions, such as excess or deficient air and fouled or eroded burners. Typically, flue gas temperatures range between 100°F and 150°F above the boiler water or steam temperature at full load. Higher temperatures may indicate fouling. The loss in boiler efficiency due to fouling can be estimated as a 1% efficiency loss for every 40°F increase in stack temperature.

Fire side deposits can include ash, soot and mineral deposits. If the deposits are allowed to build up on, or in, the tubes, heat transfer from the

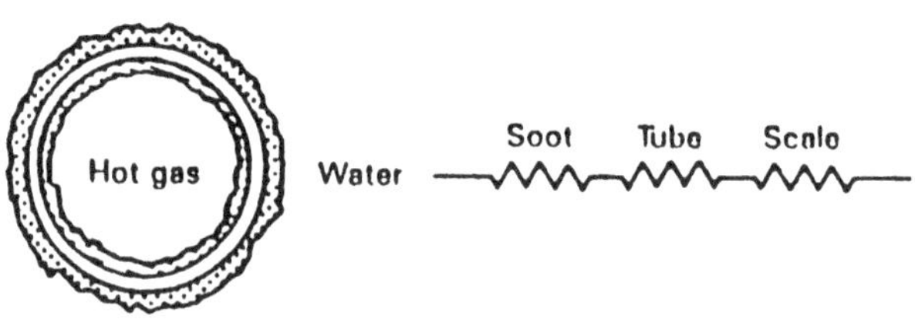

Figure 4-17. Cross-section of boiler tube with thermal resistance analogy.

hot combustion gases is retarded. Deposits that form in the convection section may also inhibit the draft in the combustion chamber. Since a full volume of air is not drawn into the firebox, the correct volume of fuel cannot be burned. This results in incomplete combustion, reduced heat transfer and wasted fuel. The degree of deposit accumulation in the combustion section can be estimated by measuring the draft loss between the firebox and flue stack. A high draft loss could be an indication of deposit buildup and restriction in the passageway.

Water side deposits are crystalline deposits that are a result of compounds in the water and whose solubility in water decreases with increases in temperature. These compounds include silica, carbonates of magnesium, calcium and sodium. A 1/32 inch layer of hard-iron silica scale can degrade boiler efficiency by about 7% as shown in Figure 4-18.

Water samples should be taken periodically to monitor the possibility of scale buildup. Samples should be taken of the blowdown water, raw feedwater, treated feedwater and condensate being returned to the boiler feedwater tank. These samples should be sent to a commercial water treatment laboratory for analyses and recommendations.

To reduce energy losses resulting from the buildup of deposits, heat transfer surfaces should be cleaned on an annual basis. They should be cleaned more frequently if there are indications of deposit buildup. Waterside deposits can be controlled by a proper water treatment program. Keeping the water clean and soft helps reduce scale formation.

Isolate Boilers that Are Out of Service

Light or variable heating loads on multiple boiler installations can frequently be met by a single boiler, with the remaining boilers idling or on standby. Idling boilers consume energy to meet standby losses.

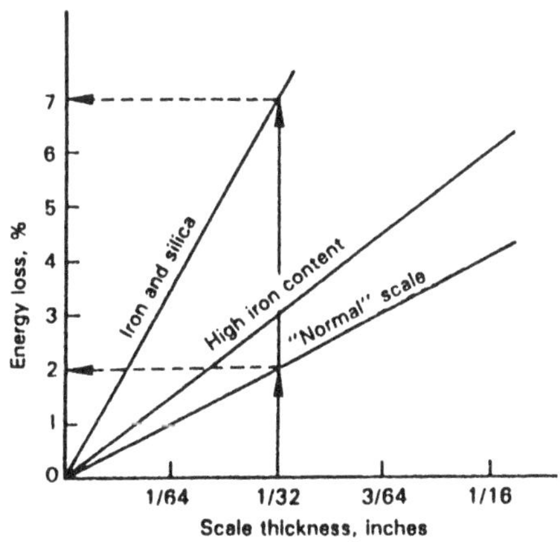

Figure 4-18. Boiler losses due to scale deposits.

Unless a boiler is about to be used to meet an expected increase in load, it should be isolated from the rest of the system by valving it off. Valves on the steam or hot water discharge and return to off-line boilers should be closed. Fuel line valves to off-line boilers should also be closed. Large boilers may be fitted with bypass valves and regulating orifices to allow the minimum flow to keep the boiler warm and reduce thermal stresses when it is brought on-line again. If the water side of a boiler is isolated, it is important to prevent the backflow of cold air, which could be brought in through the stack and could cause the boiler to freeze.

Off-line boilers should also have the air inlet dampers, or doors, to the burner closed. Closing the burner dampers, or doors, will reduce the losses resulting from a continuous induced flow of air through the off-line boilers. If the dampers are rusty, or difficult to close, they should be replaced or covered up to stop air leakage.

Reduce Boiler Blowdown

All boilers require blowdown to one degree or another. The amount of blowdown is directly related to the amount and quality of water brought into the system as make-up water.

The need for boiler blowdown was discussed earlier in this chapter. Blowing down a boiler has two main purposes: (1) to maintain a low concentration of dissolved and suspended solids in the boiler water, and (2) to remove sludge in the boiler to avoid priming, scaling, foaming and carryover.

As a result, blowdown can consume significant amounts of energy since heated water and sludge are discharged from the boiler. Although blowdown is necessary for the proper functioning of a boiler, the following items should be carefully considered to minimize energy consumption.

- Reduce blowdown by adjustment of the blowdown valve such that the controlling water impurity level is held at the maximum allowable level.
- Maintain continuous blowdown at the minimum acceptable level—this may be achieved by frequent manual adjustments.
- Reduce the amount of make-up water brought into the system by recovering more clean condensate. Recovery of steam condensate is discussed in "Chapter 5: Steam Systems."
- Maintain boiler feedwater treatment equipment to ensure that impurities are removed and the water is properly treated before being introduced into the system

Install and Maintain Proper Boiler Instrumentation:

In order for a boiler to operate at optimum energy efficiency, it is important that its performance of a boiler be monitored and recorded on a regular basis. A boiler may be producing the required steam or hot water output; however, unless proper instrumentation is available, it is difficult to measure the performance of the boiler and determine its operating efficiency.

To effectively monitor the performance of a boiler, the boiler plant, as a minimum, should have the following instrumentation for each boiler:

- stack thermometers
- steam supply pressure gauges
- heating water supply and return thermometers
- fuel meter
- run time recorder
- condensate return thermometer
- steam flow meter

Fuel Oil Additives

Fuel oil additives may be used in boilers to improve combustion efficiency, reduce high temperature combustion and minimize cold-end corrosion. Fuel oil additives may improve efficiency by allowing lower excess air levels and by reducing soot and fire side fouling of heat transfer surfaces. Energy savings from fuel oil additives is quite modest—up to a maximum savings of 2%.

There are a number of different additives available, and the type used depends upon the fuel oil characteristics. Following are some type of additives:

- Fuel handling additives, which include sludge and gum inhibitors, detergents, metal de-activators, color stabilizers, pour-point depressants, anti-static and anti-icing compounds, and corrosion inhibitors.
- Combustion additives, which reduce smoke and particulates, carbon monoxide, hydrocarbons, nitrogen oxides and sulfur oxides in the flue gases.
- Postflame treatment additives, which include soot removers, additives to control fire side corrosion and slag deposits, additives to enhance particulate collection and SO_x scavengers.
- Atomizers, which cause the oil to atomize into a finer mist that results in more complete combustion.
- Fuel oil emulsions, which produce an oil/water emulsion. The oil/water emulsion produces micro-sized droplets; as combustion takes place, micro-explosions of the water droplets occur, which produce very fine oil droplets. The very fine oil droplets have more exposed surface areas, resulting in more complete combustion.

Maintain Steam Traps

Steam traps are devices that allow liquid steam condensate to pass but do not allow steam to pass. Steam traps are used on practically all devices that consume steam and are also used to remove steam that has condensed in steam piping distribution systems. Traps prevent the wasteful loss of steam out of the system or into the condensate return.

If steam traps are not working properly, they can waste tremendous amounts of steam and energy. If steam traps fail in the closed position, the effectiveness and efficiency of the steam heating device will be significantly reduced. If steam traps fail in the open position, large quantities of steam will be lost out of the system or to the condensate return system.

Studies have shown that a high percentage of steam traps have failed in a steam distribution system at any given time. Traps should be inspected and repaired or replaced on a routine basis.

Steam traps are discussed in more detail in "Chapter 5: Steam Systems."

Repair or Replace Insulation

In order for water to boil and become steam, it must be heated to at least 212°F at standard pressure. Most steam boilers and steam distribution systems have surface temperatures from 350° to 450°F. To reduce heat losses from bare surfaces, insulation must be provided.

Often insulation is removed or damaged when repairs or modifications are made to the boiler or steam distribution system. For optimum energy efficiency, damaged or missing insulation must be replaced. Although repairing and replacing insulation can be rather labor-intensive, the payback period for repairing and replacing insulation is typically one to two years.

Provide Proper Water Treatment

Earlier, the need to maintain water side cleanliness on heat transfer surfaces and the need to reduce boiler blowdown were discussed. Both water side fouling and boiler blowdown can be minimized with proper treatment of boiler feedwater and boiler make-up water. Impurities in boiler water are typically removed both mechanically and with chemical treatments.

Dissolved oxygen and carbon dioxide in boiler water results in corrosion in the boiler and steam distribution system. Deaerators are used to thermo-mechanically remove oxygen and carbon dioxide from boiler feedwater; deaerators were discussed earlier in this chapter.

O_2 and CO_2 may also be removed chemically. Chemical degassifiers are used to remove dissolved gases in boiler water. Catalyzed sodium sulfate may be used to remove O_2. CO_2 in feedwater systems can be neutralized with a volatile amine.

Chelating agents may be added to the feedwater before it enters a boiler. A chelting agent is a chemical that combines with chemically unstable impurities (those more likely to form scale) to form stable soluble compounds. Chelants are used to keep calcium and magnesium in solution.

Precipitating agents may also be used to treat boiler feedwater. Precipitating agents combine with impurities to form insoluble compounds that settle out and may be removed as sludge. Precipitating agents are used to control salts of calcium and magnesium.

Boiler water, feedwater and boiler condensate should be sampled on a routine basis to ensure that the boiler water treatment equipment is functioning properly. If any changes in water quality are measured, the boiler water treatment equipment should be inspected to ensure that it is functioning properly before excessive scale builds up.

Reduce Steam Pressure

Often, boilers are operated to produce steam at an arbitrarily high pressure and produce steam at a pressure that is higher than necessary. Greater energy is consumed by operating steam systems at high pressures than at low pressures. Reducing the pressure of a steam system can result in energy savings of 1–2%. Energy savings result from lower steam temperatures and lower losses of steam.

For steam that is at saturation conditions (saturated steam), the temperature is directly related to the steam pressure, although the relationship is not linear. As a result, lowering the pressure of steam would also lower its temperature. Lower steam temperatures allow a boiler to operate more efficiently by lowering the temperature of the flue gases leaving the stack. The effect of flue gas temperatures on boiler efficiency was discussed earlier. Savings from reduced steam temperatures would also result from lower radiated heat losses from the boiler itself and from steam distribution piping, etc.

Energy savings also result from lower steam pressures due to reduced loss of steam. The mass flow of the steam through an opening or orifice is proportional to the magnitude of the pressure difference. Lower steam pressures would result in reduced losses through leaks in the system and result in lower losses due to flash steam.

When considering the possibility of reducing the steam pressure on a system, there are a number of issues that should be considered before reducing the system pressure. The pressure of the steam must be high enough to meet the process requirements of the devices the system is serving, plus the pressure losses through the distribution system at the required flow rate.

Lower steam pressures and temperatures will also result in steam with a lower enthalpy. This may result in higher steam flow rates for any given heating load. For example, the enthalpy of saturated steam at 135 psig is 1195 Btu/lb. If the pressure was reduced to 75 psig, the saturated steam enthalpy would be 1,186 Btu/lb. Reducing the pressure would result in a reduction in the steam enthalpy of 9 Btu/lb (1,195–1,186). Steam at lower pressures will also have a higher specific volume (ft^3/lb), which would increase the flow rate at a given load. Reducing steam pressure will also reduce the condensate handling capacities of steam traps since the capacity of a steam trap depends on the difference in pressure across the trap orifice.

Another issue to be considered is whether the boiler has a flue gas economizer. No savings will result from a reduction in flue gas temperature if the boiler is equipped with a flue gas economizer on the stack.

Energy savings should be carefully estimated when considering reducing steam pressure. The operational consequences must also be considered before reducing steam pressure to save energy.

Maintain Boiler Refractory Linings

All boilers loose heat by radiation and convection heat transfer through exterior surfaces. Typically, a boiler in good condition will loose about 1–2% of its heating capacity. For older boilers, the losses may be much greater if the refractory lining has deteriorated.

As a boiler is operated for many years, the periodic temperature swings can result in cracked refractory linings, which, in turn, can lead to loss of large amounts of heat. Linings that are cracked or deteriorated should be repaired or replaced.

The condition of refractory linings can be checked by measuring the outside surface temperature of a boiler. The outside surface temperature is a good indicator of a problem with the lining, and a technique such as infrared scanning with a surface thermometer is an effective method for inspecting the boiler surface for indication of deteriorated refractory linings.

4.7 ENERGY CONSERVATION MEASURES FOR COMBUSTION SYSTEMS AND BOILERS

The energy conservation measures for combustion systems and boilers discussed in the previous section were operation and maintenance procedures, which would improve the energy efficiency of existing combustion and boiler systems. Most of those measures were little, or no, cost measures that could be performed on an on-going basis to improve efficiency.

In addition to the low-cost Operation and Maintenance (O&M) procedures, there are also modifications that could be made to combustion systems and boilers to improve energy efficiency. These modifications, however, require a more significant cost investment and are permanent improvements. The modifications listed below are improvements that should be considered under any energy management program for existing, or proposed, combustion systems or boilers.

4.7.1 Install Flue Stack Economizer

A flue stack economizer is simply a heat exchanger placed in the exhaust stack of a boiler to recover heat from the flue gases leaving the boiler. Economizers are bent, finned tube heat exchangers that mount in the boiler flue. They are available in rectangular or cylindrical design. The temperature of flue gases leaving a boiler are typically on the order of 400°F or more for a properly functioning boiler. At an exit temperature of 400°F, there is considerable heat energy lost in the flue gas. Boilers that operate at pressures of 75 psig and 450°F, or more, are good candidates for economizers. Flue gas temperatures for boilers producing saturated steam at 75 psig are usually about 420–470°F. The overall efficiency of a system can be increased with an economizer by 2–7%, depending upon the fuel burned, boiler size and operating conditions. The payback period for a properly selected and installed economizer is in the range of 1–2 years.

For economizers to be economically feasible, their available heat must be matched with a suitable heat sink. Heat recovered from an economizer is typically used to preheat boiler feedwater; however, the recovered heat may also be used to preheat combustion air, to provide heat for a deaerator or for other process loads outside the boiler plant. Economizers used to preheat combustion air require more heating surface than feedwater heaters since the thermal resistance of air is much higher than that of water. As a result, economizers used to preheat air tend to be larger and more expensive than those used to preheat feedwater. The most common types of preheaters are tubular (recuperative) and the rotary regenerator (heat wheel).

A typical feedwater flue stack economizer is shown in Figure 4-19 and the installation of a feedwater economizer is shown in Figure 4-20. Boiler feedwater flows through the economizer tubes and recovers heat from the flue gas before the feedwater enters the boiler. For each 10°F increase in feedwater temperature, an increase in boiler efficiency can be expected.

Economizers systems are relatively expensive investments, relatively complex, and require special controls as well as careful attention to structural support considerations. Before selecting and installing an economizer, consideration should be given to the space available for it. The available draft must also be considered since the economizer will restrict the flow of flue gases to some extent. If a boiler is equipped with a forced draft fan or an induced draft fan, no change in fan equipment is normally required. The fin spacing on an economizer is affected by the type

Figure 4-19. Feedwater flue stack economizer (*Courtesy of Cleaver Brooks*).

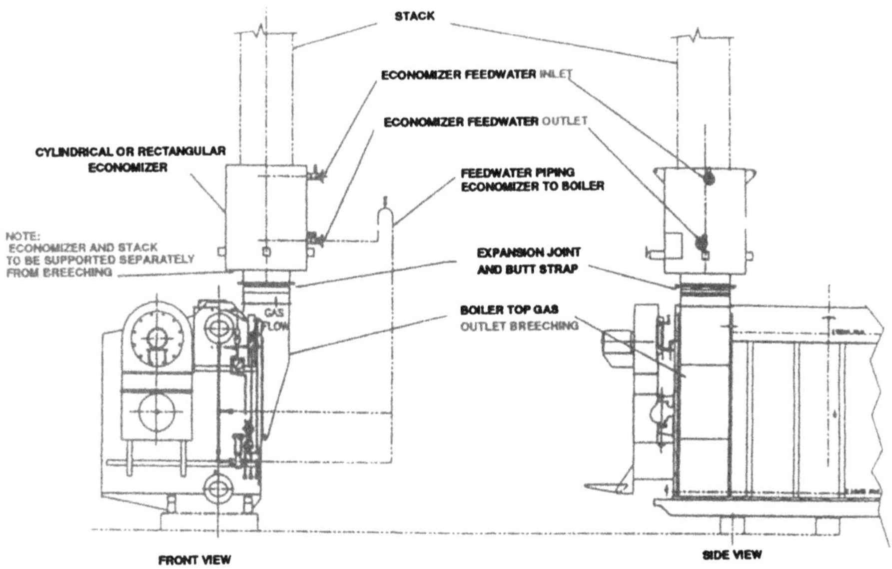

Figure 4-20. Typical installation of a feedwater economizer (*Courtesy of Cleaver Brooks*).

of fuel burned. Fuels, such as heavy fuel oil or coal, require wider fin spaces to reduce fouling on the heat transfer surfaces of the economizer. Soot blowers are also usually required to keep the fins clean.

The temperature of the water flowing out of the economizer must also be considered. The design flow rate must keep the temperature of the water well below the boiling point for the pressure of the water or steam will form in the economizer. Economizers are generally not designed to generate steam. If steam does form, it could result in water hammer, which may damage the economizer.

The outlet flue gas temperature is another important consideration; it should typically be a minimum of 300°F. Since most fuels contain sulfur, the flue gas will also contain sulfur. If the flue gases are cooled below their dew point temperature, they will begin to condense and sulfuric acid will condense on the colder metal surfaces. Sulfuric acid is highly corrosive.

Example 4.4. A 300 boiler horsepower firetube boiler without an economizer operates at 100% capacity with an efficiency of 80.5%. Flue gas leaves the boiler at 435°F. The feedwater enters the boiler at 220°F at the rate of 10,350 lb/hr.

It is proposed that an economizer be installed in the boiler flue to recover heat and preheat the feedwater. The data from the economizer manufacturer indicates that if heat were recovered from the flue gas such that the temperature leaving the economizer were 302°F, the feedwater would be heated to 260°F at the present flow rate.

Assuming the boiler operates at 100% capacity, how much heat will be recovered by the economizer? How much will the reduction in fuel consumption be? What will the percent reduction in fuel consumption be? What will the efficiency of the boiler be after the economizer is installed?

First, we will use the rule of thumb approximation to determine what we can expect for an increase in efficiency. As indicated above, a 1% increase in efficiency can be expected for each 10°F increase in feedwater temperature.

$$\text{Feedwater temperature rise} = 260 - 220 = 40^{\circ}\text{F}$$

$$\text{Expected increase in efficiency} = \frac{40}{10}(1\%) = 4\% \text{ approx}$$

The heat output of the boiler is

$$q_O = (300 \text{ Bhp})(33,472 \frac{\text{Btu/hr}}{\text{Bhp}}) = 10,041,600 \text{ Btu/hr}$$

The efficiency of the boiler is 80.5%; therefore, the heat input from the fuel is

$$q_{fi} = \frac{10,041,600}{0.805} = 12,474,040 \text{ Btu/hr input}$$

The heat recovered in the economizer is the difference between the entering and leaving enthalpies of the water times the flow rate. For water at saturated conditions:

$$h_{220^{\circ}\text{F}} = 188.22 \text{ Btu/lb} \qquad h_{260^{\circ}\text{F}} = 228.76 \text{ Btu/lb}$$

The heat recovered is

$$q_r = (10,350 \text{ lb/hr})(228.76 - 188.22) \text{ Btu/lb} = 419,560 \text{ Btu/hr}$$

The savings in fuel input is

$$q_{fs} = \frac{419,560}{0.805} = 521,195 \text{ Btu/hr}$$

The percent reduction in fuel consumption is

$$\frac{521,195}{12,474,040} = 4.1\%$$

The reduced fuel input rate is

$$q_{fi} = 12,474,040 - 521,195 = 11,952,845 \text{ Btu/hr}$$

The fuel rate is reduced with the economizer; however, the output of the boiler remains the same. Therefore, the efficiency of the boiler with the economizer is

$$\eta = \frac{10,041,600}{11,952,845} = 84\%, \text{ a } 3.5\% \text{ increase}$$

4.7.2 Recover Heat from the Boiler Blowdown

Earlier in this chapter, the need for blowdown from a boiler was discussed, as well as the two types of blowdown: bottom blowdown and surface blowdown. The operation of bottom blowdown is intermittent, while the operation of surface blowdown is continuous. In many applications, the bottom blowdown water must be cooled before it can be discharged into a sewer or drain.

Since the blowdown water is removed directly from the boiler drum after it has been heated, considerable amounts of energy are lost in the blowdown water. Significant amounts of heat can be reclaimed from extracting heat from the blowdown.

Heat recovery from bottom blowdown is usually not cost-effective since bottom blowdown is intermittent in nature. Surface blowdown is a constant, or continuous, low flow of boiler water. The blowdown flow is normally regulated by a flow control valve. Heat exchangers are used to transfer heat from the blowdown water to incoming make-up water. Heat exchangers used for blowdown heat recovery are typically of the counter-flow shell and tube type. A typical blowdown heat exchanger is shown in Figure 4-21. Blowdown heat recovery systems typically recover about 90% of the heat from the blowdown water.

Heat recovered from blowdown water is usually used to preheat incoming make-up water, although it can be used for other purposes. The make-up water quantity can be as much as 5–10% of the total boiler steam flow. Heat recovered from blowdown water can also be used to keep off-line boilers warm.

If a boiler is a high-pressure boiler and if the quantity of blowdown is significant, it is sometimes advisable to place a flash tank in the blowdown line prior to entering a blowdown heat recovery heat exchanger. The flash tank is used to recover low-pressure steam, which is generated by flashing when the high-pressure saturated water is throttled to a lower pressure. Flash tanks are discussed later in this section and in Chapter 5.

Potential savings for blowdown heat recovery systems are typically 2–5%. A blowdown heat recovery system usually returns the cost of the initial investment in one year, or less.

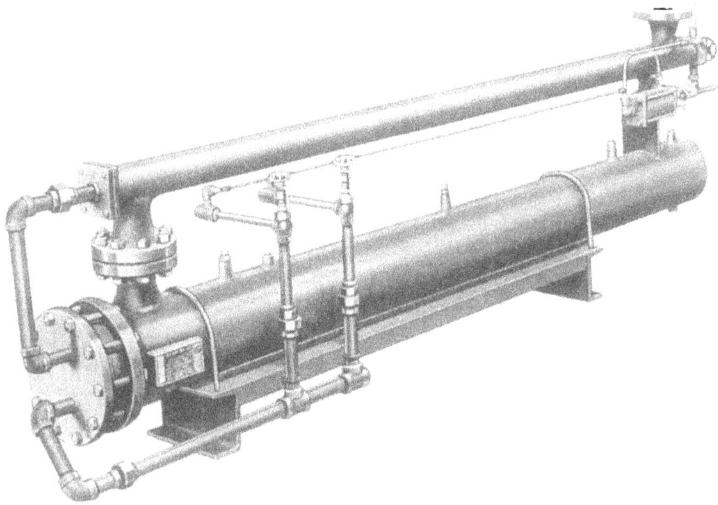

Figure 4-21. Blowdown heat recovery heat exchanger (*Courtesy of Cleaver Brooks*).

Example 4.5. A boiler produces saturated steam at 50 psig and operates for 5,000 hours per year. A continuous blowdown system removes water at the rate of 3 gal/min. The maximum temperature at which the blowdown water may be discharged into a sewer is 120°F. Make-up water at 80°F enters the boiler system at the rate of 15 gal/min.

If a blowdown heat recovery system with a heat transfer efficiency of 90% were installed, at what rate would heat be transferred to the make-up water, what would be the temperature of the make-up water exiting the heat recovery unit and what would the annual energy savings be?

The water in the boiler as it is blown down is at 50 psig and saturated. The enthalpy of saturated water at 50 psig is 250.2 Btu/lb. The water must be cooled to 120°F before being discharged. Assuming it is saturated, its enthalpy is 87.97 B tu/lb. The heat removed from the blowdown water, then, is

$$q_{BD} = (3 \text{ gal/min})(7.48 \text{ lb/gal})(60 \text{ min/hr})(250.2 - 87.97) \text{ Btu/lb}$$
$$- 218,426 \text{ Btu/hr}$$

The efficiency of the blowdown heat recovery unit is 90%; therefore, the heat transferred to the make-up water is

$$q_{MU} = 0.9(218,426) = 196,583 \text{ Btu/hr}$$

The temperature increase of the make-up water is

$$\Delta T = \frac{196{,}583}{15(7.48)(60)} = 29^{\circ}\mathrm{F}$$

The water exit temperature is

$$T_E = 80 + 29 = 109^{\circ}\mathrm{F}$$

The annual heat energy savings is

$$q = (196{,}583\ \text{Btu/hr})(5{,}000\ \text{hrs}) = 982.9 \times 10^{6}\ \text{Btu/yr}$$

4.7.3 Preheat Combustion Air

Earlier in this chapter, reducing excess combustion air was discussed as a possible energy conservation operating and maintenance measure. Since the combustion air that is brought in must be heated to the combustion temperature, energy is consumed as the air is heated in the combustion process.

Energy may be saved if the air is preheated prior to entering the combustion process. Combustion air can be heated up to 600°F for most fuels and up to 350°F for stoker-fired coal, oil and gas. It is estimated that for each 100°F increase in the incoming air, the result is an increase in boiler efficiency of about 2%.

In order for energy savings to result, the air must be heated with waste heat (heat that would otherwise be lost). There are several sources of waste heat for combustion air preheat, including flue stack economizers and blowdown heat recovery, as discussed above.

Waste heat may also be available from the boiler room itself. Ambient air in most boiler rooms is heated incidentally by hot boiler and pipe surfaces, which loose heat. As a result, warm air tends to accumulate near the ceiling of the boiler room. Heated air near the boiler room ceiling can be utilized directly by installing a fan and duct system to collect air from near the ceiling.

Some preheat systems employ a shroud mounted around the flue stack in addition to the ductwork and fan. These preheaters extract heat from the flue gases, through the wall of the flue, by passing the incoming combustion air around the flue in a shroud. Again, care must be taken not to cool the flue gases below their dew point temperature to prevent condensation.

4.7.4 Install Energy Management System to Monitor Flue Gases and Provide Trim Control

Combustion efficiency is a measure of the effectiveness of a boiler, or combustion system, to convert chemical energy with a fuel into heat energy. In order to achieve optimum efficiency, it is necessary that the fuel and air be supplied to the combustion process in the proper quantities. The purpose of the trim control is to provide air and fuel in sufficient quantities for complete combustion without providing more air than necessary. Air in excess of that required for stoichiometric combustion (excess air) is provided to ensure complete combustion. Air supplied above the required excess air amount results in wasted energy, as discussed earlier in this chapter.

The need to perform flue gas analyses on boilers and the importance of these analyses were also discussed earlier in this chapter. Although it is advisable to perform flue gas tests, such as an Orsat test, and adjust the boiler trim controls accordingly, there are limitations on the usefulness of this method in controlling boilers for optimum energy efficiency. First, these are manual tests that can be quite slow and time-consuming. Second, these are static tests that only measure the conditions at one point in time; they do not indicate conditions at other loads. Therefore, the boiler trim control will not necessarily provide the proper control at varying load conditions.

The control and efficiency of a combustion process can significantly be improved with a microcomputer-based energy management and control system. These systems measure the products of combustion in the flue gas and automatically adjust the trim controls to provide optimum combustion efficiency. The control panel of a microcomputer energy management system is shown in Figure 4-22. These systems continuously monitor the products of combustion in the flue gas, as well as other variables, and adjust the fuel or air flow to maintain the optimum set point. These systems also compensate for changes in ambient air temperature, barometric pressure and slight variations in fuel characteristics.

The flow of air and fuel to the boiler are metered by modulating fuel valves and air dampers as shown in Figure 4-23. In a typical microcomputer control system, input devices, such as sensors, switches and position encoders, are accessed at frequent intervals and requested to transmit data (steam pressure, on/off status, shaft position, O_2 level, etc.) to

Figure 4-22. Microcomputer boiler control panel (*Courtesy of Cleaver Brooks*).

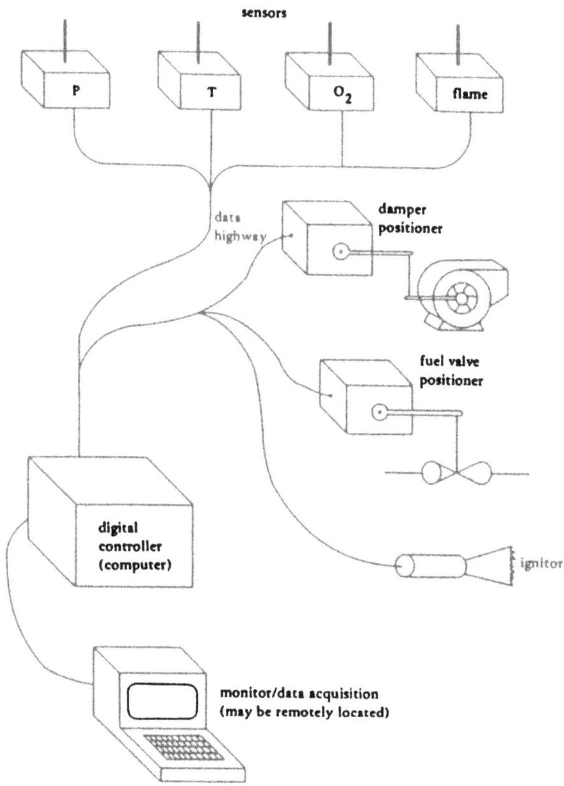

Figure 4-23. Microcomputer boiler control system.

the controller. The microcomputer then controls the combustion by means of a changeable program (software).

The advantages of a microcomputer control system include:

- Optimum combustion efficiency for varying loads and conditions.
- Control instructions and set points that can easily be changed.
- Systems that are usually easily expanded to include more features or additional boilers.
- Control of redundancy.
- Boiler operation diagnostics and performance data acquisition.
- Self-diagnostics of control system.
- Provision of lead-lag control for multiple boilers.

Fuel savings for boilers with microcomputer control systems can be between 4% and 6%. Disadvantages of microcomputer boiler control systems include the high cost and complexity of the systems.

The most common methods for measuring O_2 in the flue gas on a continuous basis are the paramagnetic sensor, the wet electrochemical cell and the zirconium oxide sensor. Solid state zirconium oxide sensors permanently installed in the flue stack are the most suitable for continuous monitoring.

Microcomputer-based combustion control systems tend to be rather costly. Typically, they are only cost-effective on systems with annual fuel costs of $60,000 or more. These control air most suitable for modulating boilers of 150 boiler horsepower and above. When selecting a microcomputer-based control system, performance and cost are the two main criteria. The most important performance parameters are the accuracy to which the steam pressure (or temperature) are controlled and the percent O_2 (present in the flue gas).

4.7.5 Replace Burner

Older boilers frequently have original burners that are inefficient. Older burners were typically designed to use more excess air than necessary to ensure complete combustion and safety. Many older boilers also have atmospheric burners, which usually cannot control excess air effectively. Finally, older burners may also be worn to a point where low excess air levels are not possible.

Newer, more efficient burners are designed for low-excess-air firing. Excess air and stack losses can be more easily controlled with a power burner that has electrically positioned control linkage, motor driven fans and automatic air dampers.

In some cases, energy conservation improvements may have significantly reduced the load on a boiler. Oversized boilers and burners can result in inefficient operation. In many cases, the efficiency of a boiler can be improved, without the cost of a new boiler, by installing a smaller burner that more closely matches the load. Newer burners are also available with larger turn-down ratios, which provide more efficient operation at light loads.

Newer fuel oil burners are available with flame retention heads. Flame retention head burners enable better mixing of the oil and combustion air, which allows less excess air. Newer fuel oil burners with steam or air atomization are also available. Air and steam atomization produces an aerosol in which fine oil droplets are supported by an expanding cone of air/steam, which promotes more complete combustion. Steam or air atomization burners are less sensitive to viscosity changes in the fuel oil, which allows the firing of a wide range of fuel oils.

4.7.6 Preheat Fuel Oil

Complete atomization of fuel oil in boiler/burners using heavier grades of fuel oil is often difficult to achieve. Heavier fuel oils, such as No. 4, No. 5 and No. 6, have much greater viscosities than lighter oils. The higher the viscosity of the oil, the more difficult it is to atomize, which can lead to incomplete combustion.

Preheating the fuel oil to control its viscosity will improve atomization, improve combustion and reduce energy consumption. Preheating fuel oil can increase efficiency by as much as 3% depending upon the particular fuel oil involved. Heat for preheating fuel oil should be waste oil, which has been reclaimed by one of the methods described in this chapter.

Care must be taken to not overheat the oil. Overheating could cause vapor locking and flame-out. Automatic viscosity controllers should be used: These feed back information on the viscosity of the oil to control the amount of preheating.

4.7.7 Install Soot Blower

When fuels such as fuel oil or coal are burned, the products of combustion frequently contain visible smoke. The smoke tends to accumulate on heat transfer surfaces as soot. There is more of a tendency for soot to form with heavier fuel oils and coal. The accumulation of soot on heat transfer surfaces retards the transfer of heat and reduces efficiency.

Soot blowers may be used to remove soot deposits from heat transfer surface. Soot blowers can remove fly ash and soot from the connective passes of a boiler. Wall blowers may be used to remove slag from the walls of the combustion chamber. The use of soot and wall blowers can improve efficiency by as much as 1%.

Soot blowers can be either fixed or retractable and use either air or steam as the blowing media. The availability of steam or compressed air would determine which media should be used. The choice of the blowing media may also depend on the chara cteristics of the deposits.

The introduction of large volumes of steam or air into the combustion process can upset flow patterns and upset the combustion process during soot blowing. The amount of steam or air blown into the furnace should be carefully controlled.

4.7.8 Install Flash Tanks

When blowdown is removed from high-pressure steam boilers, the blowdown water goes from high pressure to atmospheric pressure. As the pressure of the hot water is reduced to atmospheric pressure, part of the water flashes to low-pressure steam.

In many boiler plants, the flash steam is just vented to atmosphere and wasted. A flash tank is a device used to collect low-pressure flash steam and separate the steam from liquid blowdown water or steam condensate. A flash tank is shown in Figure 4-24.

Flash steam recovered from boiler blowdown can usually be used in the boiler plant itself. Typical uses for low-pressure flash steam include steam for deaerators, make-up water preheat, fuel oil preheat and combustion air preheat.

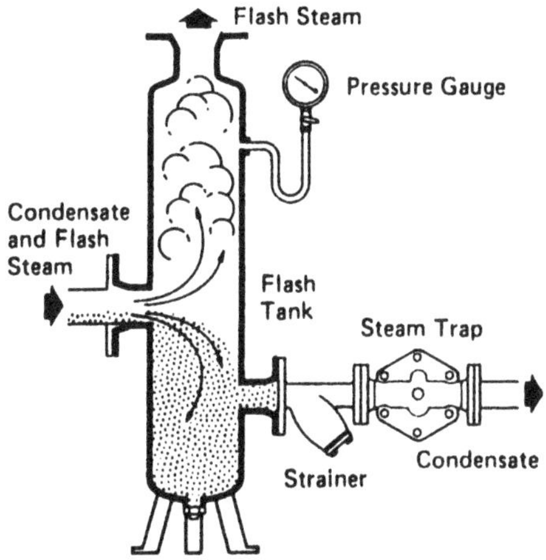

Figure 4-24. Flash tank.

Flash tanks are discussed in more detail in "Chapter 5: Steam Systems."

4.7.9 Install Waste Heat Boiler

Earlier, the advisability of installing a flue stack economizer to recover waste heat from boiler flue gases was discussed. Significant amounts of heat are lost up the flue stack. Boilers that produce high-pressure steam have flue gas temperatures of 400°F and more. An alternative method for recovering waste heat from flue gases are waste heat boilers.

Waste heat boilers are unfired boilers that produce low-pressure steam. They are mounted in the flue of a fuel fired boiler and recover heat from the flue gases to generate low-pressure steam.

Typically, flue gas temperatures should be in the 800–900°F range for a waste heat boiler to be feasible. There must also be a demand or need for the low-pressure steam generated by the waste heat boiler. Usually, they are economically feasible on large systems when there is a demand for steam that would otherwise require the purchase and installation of an additional boiler.

4.7.10 Upgrade Water Treatment

Earlier in this chapter, the importance of proper treatment of boiler make-up water was discussed. Unless make-up water is properly treated, significant scale buildup will occur in the boiler itself and in the steam distribution system. A buildup of scale will significantly reduce heat transfer within the boiler, which will also reduce efficiency.

Boiler make-up water is usually treated by filtering the water to remove solid impurities and by chemicals to soften the water. Alternate methods for water treatment to reduce scale buildup is magnetic or electrostatic field water treatment. Both of these methods reduce the chemicals that cause scale buildup without the use of chemicals and without the costs associated with chemicals. Savings also occur from reduced maintenance.

The general operating principle for the magnetic field water conditioning is based upon the interaction between a magnetic field and ions in the water. When the ions pass through the magnetic field, a force is exerted on each ion. The forces ions of opposite charges are in opposite directions. The redirection of the particles tends to increase the frequency with which the ions of opposite charge collide and combine to form a mineral precipitate, or insoluble compound. The ions precipitate out of the water before entering the boiler.

Electrostatic water conditioners operate on a different principle. Electrostatic conditioners impose a surface charge on the ions so that they repel each other. As a result, the two ions (positive and negative) of a kind needed to form scale are never able to come close enough together to initiate the scale-forming reaction. This reduces the scale formation on heat transfer surfaces.

Either of these methods may be used as a replacement for chemical treatment systems. Significant savings may result in saving the cost of chemicals for water treatment. Also, periodic cleaning and descaling of heat transfer surfaces is significantly reduced or eliminated.

4.7.11 Install Turbulators

Turbulators are thin strips of metal that have been bent and may be installed inside the tubes of a firetube boiler. The turbulators cause turbulent flow of the hot flue gases through the boiler tubes. Figure 4-25 shows a diagram of the flow of flue gases through turbulators. The turbulent flow of the flue gases inside the tube results in a higher rate of heat transfer between the flue gases and the water. As a result, the temperature of the flue gases leaving the boiler are lower, which increases the efficiency of the boiler. Energy savings up to 3% may be expected from the installation of turbulators, depending upon the application.

Turbulators are suitable only for some applications. They achieve the best effect in single- or double-pass boilers with tube diameters in excess of 2 inches. They are not recommended for boilers that burn high particulate fuels such as coal and residual fuel oil. Turbulators tend to increase both the collection of deposits and corrosion. Since turbulators reduce the leaving temperature of the flue gases, care must be taken to not cool the flue gases to a point where condensation will occur.

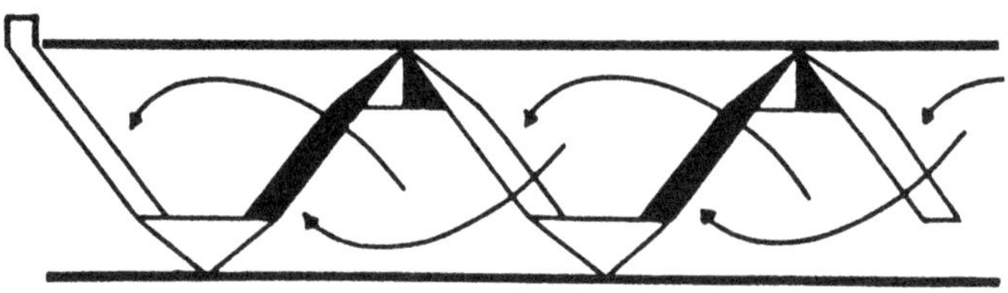

Figure 4-25. Flow of flue gases through a turbulator.

4.7.12 Install Cogeneration System

Occasionally, boiler plants are required to generate high-pressure steam for a particular process, as well as for other low-pressure applications. Generating steam at a pressure higher than necessary for an application and then reducing the pressure is quite wasteful and inefficient. If high-pressure steam is required for particular applications, cogeneration may be feasible.

Cogeneration is a process where high-pressure steam is expanded through a steam turbine to generate electricity. If steam is needed at a lower pressure, in addition to the high pressure, the steam could be expanded through a back-pressure turbine to the lower pressure rather than a pressure reducing station. As the steam is expanded through the back-pressure turbine, the turbine is used to turn a generator that generates electricity.

These systems are quite costly and are usually suitable only for plants with large rates of steam flow. The cost of electricity and electrical demand charges also affect the economic feasibility of cogeneration with back-pressure turbines.

Cogeneration is discussed in more detail in "Chapter 13: Cogeneration Systems."

4.8 ENERGY EFFICIENCY FOR NEW BOILERS AND COMBUSTION SYSTEMS

For facilities with boilers, or other combustion systems, these systems are one of the largest consumers of energy at the facilities. Since boilers, combustion systems and equipment consume significant amounts of energy, regardless of the type of facility, it is important that energy efficiency be a primary consideration when selecting new boilers and equipment for new, or retrofit, installations.

Earlier in this chapter, energy conservation for boilers and combustion systems were discussed. Several of the energy efficiency measures discussed as retrofit opportunities for boilers and boiler room equipment, such as economizers, are applicable to new and retrofit boiler installations. In addition to the previously discussed energy conservation measures for boilers and combustion systems, the following measures should be considered for new and replacement boilers and accessories.

The age of existing boilers and their expected remaining useful life should be considered. The estimated service life of most boilers is between 25 and 30 years. Older boilers frequently have low efficiencies, typically around 65%. Newer boilers typically have efficiencies between 80% and 90%, depending upon the type of boiler. Any boiler near the end of its useful life is a prime candidate for replacement.

When selecting new or replacement boilers, it is important to compare boiler efficiencies of prospective units. Even only a few tenths percent increase in efficiency can result in significant savings in fuel costs over the life of a boiler. The cost of the boiler itself is usually only a small part of the fuel costs over the life of a boiler.

4.8.1 Install Modular Boilers

Frequently, facilities have highly variable heating loads or have many heating loads at remote locations. Boilers tend to operate at their highest efficiency when operating near their peak capacity and when the heating load is nearly constant.

For variable heating loads, multiple modular boilers can operate more efficiently than a single boiler sized for the peak load. In most cases, full capacity is seldom required. A modular boiler system consisting of two or more smaller high efficiency boilers will increase energy efficiency. Fluctuations in heat load are met by firing one, or more, boilers as required. The smaller boilers operate closer to their peak operating efficiency when on-line. When not needed, they are off.

Losses in distribution systems can also be quite high. The flow of steam or hot water through a pipe results in a loss in pressure, as well as heat losses, even if the piping is well insulated. The larger the piping distribution system, the great er the flow and heat energy losses.

Many facilities have a central plant with one or more boilers. The steam, or hot water, is distributed to remote boilers through lengthy piping distribution systems. The installation of modular boilers near large heat loads can result in significant energy savings by reducing distribution system losses.

4.8.2 Install Condensing Combustion Boilers

Earlier in this chapter, the need to not cool flue gases below the dew point temperature was discussed. For most fuels, the leaving flue gas temperature must be at least 300°F to prevent condensation, even when economizers are used. High temperature flue gases waste significant amounts of heat energy and fuel.

Newer boilers are available with condensing combustion systems. Boilers with these combustion systems are capable of flue gas leaving temperatures below 100°F. These boilers are typically modular gas fired low-pressure steam or hot water type boilers. They are available with heating capacities up to 875,000 Btu/hr. Combustion efficiencies around 95% can be expected for boilers with condensing combustion.

4.8.3 Burners

When selecting new combustion equipment, such as boilers, the selection of the burner can have a significant impact on combustion efficiency. When selecting a burner for combustion equipment, there are several considerations that can affect combustion efficiency, including flame retention time, air/fuel mixing, excess air, turndown ratio and atomization.

Flame retention time has to do with how well the burner mixes the air and fuel, and how long the combustion gases spend in the furnace or combustion chamber. Complete mixing of the fuel and air increases combustion efficiency by reducing excess air. When selecting burners, one should keep in mind that the burners should have a long retention time to maximize heat release from the combustion gases. A good measure of retention time is the swirl number, which is a quantitative indication of the turbulence within the combustion chamber. Generally speaking, the higher the swirl number, the greater the flame retention and the higher the combustion efficiency.

New burners should be low excess air burners for most applications. Unless the burner is a direct fired burner and is being used to directly heat a process stream that will be fully utilized, more excess air than necessary will result in lower efficiencies. For most applications, high combustion efficiency is greatest when a low excess air burner is used for combustion.

The turndown ratio of a burner has a significant effect on its part load efficiency. When the turndown ratio exceeds 3 to 1, the performance of the burner at low loads will suffer. The flow of the air is a function of the square root of the air pressure drop. When the burner operation is at third of its full capacity, the pressure drop is one-ninth of the full capacity pressure drop. This significantly reduces swirl and turbulence in the combustion chamber, which reduces combustion efficiency. It is more efficient to use multiple burners than to have a single burner with a large turndown ratio.

Atomization burners that burn fuel oil were discussed earlier in this chapter. The ability of a burner to atomize the fuel oil, especially heavier fuel oils, is an important consideration when selecting new burners.

4.8.4 Fluidized Bed Combustion

When selecting boilers for coal fired applications, fluidized bed combustion should be considered. In fluidized bed combustion, crushed coal is burned in a bed of granulated limestone particles, which are maintained in suspension by a stream of air that is directed upward. The air velocity is maintained at a rate that just overcomes the gravitational forces on the particles, holding them suspended without carrying the particles out of the furnace.

High combustion efficiency can be achieved by using fluidized bed combustion. The process results in the complete combustion of the fines by enveloping the incoming coal in a circulation of hot solids, ensuring that all particles are burned. Advantages of fluidized bed combustion include:

- Much greater overall heat transfer than conventional coal fired units, which results in smaller, less expensive units.

- Lower combustion temperatures (1,500 to 1,600°F), which minimize the formation of NO_X and furnace slag.
- Enabling of combustion of high sulfur coals without flue gas treatment.
- Lower coal grinding costs since only crushed coal is required.
- Higher combustion efficiencies.

Since fluidized bed combustion boilers are smaller than conventional units, they are available as packaged boilers. Packaged fluidized bed boilers are available in sizes ranging from 2,500 to 50,000 lb/hr of steam.

4.8.5 Fuel Type

When selecting a new boiler, its fuel type is an important consideration. The various types of fuels have different heating values, costs, availability and handling expenses.

The characteristics of the fuel itself is a major consideration when planning for a new boiler. Natural gas requires little fuel preparation, mixes readily with combustion air and burns cleanly. Oil fuels require atomization prior to vaporization and mixing with the combustion air. Heavier grade fuels require preheating for proper atomization. Some burners are available as dual fuel burners, and can burn either light fuel oil or natural gas.

Combustion and handling of coal is quite complex. Typically, the coal must be crushed prior to burning. After burning, the ash must be disposed of. Many coal fired boilers require precipitators or dust collectors to remove pollutants from the flue gases.

The cost and availability of fuels is also an important factor. Sometimes gas is available at a lower cost of the utility and can shut the gas off during peak demand periods. In this situation, duel fuel burners should be considered.

The heating value of the fuel must also be considered. The lowest price fuel may not be the most efficient or best economical choice. When considering alternative fuels, the cost per Btu or the cost to produce a pound of steam should be considered.

Chapter 5

Steam Systems

Steam is one of the most common, and most important, fluids used in industrial, commercial and institutional facilities. Since it is generated from water, it is an abundant, low cost, effective heat transfer medium. A steam system uses the vapor phase of water to supply heat or kinetic energy through a piping distribution system. Steam is used for industrial process heating, space heating and in power cycles, such as the generation of electricity. Steam is also used for certain cooling processes, such as absorption refrigeration machines.

During the boiling and condensation processes, if the pressure is held constant, both water and steam can be present, and the temperature also remains constant. The temperature can uniquely be set by the pressure, which is relatively easy to control. As a result, it provides excellent temperature control of a process heating load. As steam changes phases (from a liquid to a vapor or from a vapor to a liquid), large amounts of heat are transferred to or from the steam. High rates of heat transfer can be obtained for relatively small heat transfer equipment, such a coils and heat exchangers. Thus, steam is widely used as a heating medium at many facilities.

Steam generation and usage typically accounts for about 50% of the energy costs at industrial, commercial and institutional facilities. Steam management is one of the most critical components of an overall energy management program. This chapter deals with energy management, the steam distribution system itself. The generation of steam is discussed in "Chapter 4: Combustion Systems and Boilers."

5.1 OVERVIEW OF STEAM PROPERTIES AND STEAM DISTRIBUTION SYSTEMS

When analyzing steam systems from an energy management standpoint, it is first necessary to have an understanding of the processes involved with steam systems. This section provides an introduction to steam processes and steam distribution systems.

5.1.1 Definitions and Terms Used in Steam Systems

When working with steam systems, there are a number of terms that are unique to these systems. See Glossary for definitions of useful terms that are commonly used when discussing steam systems.

5.1.2 Properties and Characteristics of Steam

A steam system uses the vapor phase of water to supply heat or kinetic energy through a piping distribution system. Steam carries relatively large amounts of energy in a small mass. Steam is pressure-temperature dependent, therefore, the system temperature can be controlled by varying the steam pressure or temperature. Steam systems that operate below 15 psig are classified as low pressure systems. Systems that operate at, or above, 15 psig are classified as high pressure systems.

There are two types of steam using equipment: thermal and mechanical. Thermal type equipment exploits the heat content of the steam. Steam temperature or the heat content of the steam are the most important parameters. Thermal type equipment include such devices as heat exchangers, heating coils, etc. Mechanical type equipment utilize the thermodynamic work available in it as it is expanded from a high pressure to a low pressure. Expansion usually takes place in a device such as a steam turbine, which drives pieces of rotating equipment, such as generators, compressors, pumps, etc.

Sensible heat (the enthalpy of the liquid) is the amount of heat, in Btu, required to raise the temperature of one pound of water from 32°F to the boiling point at a given pressure. It is registered as a change in temperature. Latent heat of vaporization (enthalpy of vaporization) is the amount of heat required to change one pound of water at the boiling point (saturated) at a given pressure to a pound of steam at the same pressure. The enthalpy of steam (total heat) is the combined enthalpy of the liquid and vapor. It is the total energy of flowing steam. The total enthalpy of steam is given by Equation (5.1).

$$h_g = h_f + h_{fg} \qquad \text{Eq. (5.1)}$$

where h_g = the enthalpy of the saturated vapor, Btu/lb
h_f = the enthalpy of saturated liquid, Btu/lb
h_{fg} = the enthalpy of vaporization, Btu/lb

Most of the heat content of steam is the latent heat (enthalpy of vaporization), which permits large quantities of heat to be transmitted with little change in temperature.

As long as both steam and water can exist at a given pressure, it is said to be at saturation conditions (saturated temperature and pressure). The steam vapor is said to be saturated steam and the water is said to be saturated liquid, or saturated water. If heat is added to water at a given pressure, the temperature of the water will increase. If the pressure is held constant and heat is added to the water, the temperature will increase until it can no longer exist as a liquid at that pressure and it will begin to boil.

These conditions are referred to as saturation temperature and pressure. The addition of any more heat, at that pressure, will cause the water to become steam. This vaporization of the water absorbs relatively large amounts of heat as it changes from the liquid phase to the vapor steam phase. This heat is referred to as the latent heat of vaporization, or just latent heat. Conversely, as steam condenses from a vapor to a liquid, it gives up large amounts of heat at a constant pressure.

As the phase change occurs, the temperature of the water and steam remains constant even though heat is added. After all the water has boiled into steam, additional amounts of heat (while the pressure is held constant) will cause the steam temperature to increase. Steam at a temperature above the saturation conditions is referred to as superheated steam.

Several terms frequently used to describe the conditions of steam are dry saturated steam, wet steam and quality. Dry saturated steam is steam at saturation conditions—it contains no liquid water. This is steam that has been completely vaporized so that it contains no droplets of liquid water. Wet steam, on the other hand, does contain some liquid water. It is a mixture of saturated steam and saturated water. The quality of wet steam is defined as the ratio of the mass of the vapor to the total mass of the vapor and water. The quality of wet steam can be calculated using Equation (5.2).

$$x = \frac{h - h_f}{h_{fg}} \qquad \text{Eq. (5.2)}$$

where h = the enthalpy of the mixture of steam and water, Btu/lb
h_f = the enthalpy of the saturated liquid, Btu/lb
h_{fg} = the enthalpy of vaporization, Btu/lb

Frequently, it is necessary to determine steam properties, such as enthalpy, at various conditions. Since this data is used so frequently, tables that list the properties of steam at various temperatures and pressures have been developed. These tables are referred to as steam tables. An abbreviated steam table is provided in Appendix B. Larger and more detailed steam tables are available in most thermodynamics textbooks.

5.2 STEAM AND CONDENSATE SYSTEMS

A typical steam and condensate system consists of a steam generator (boiler), steam piping distribution system, steam consuming equipment (steam loads) and a condensate return system. A typical steam and condensate system is shown in Figure 5-1. The steam generator produces steam at the highest pressure (and temperature). The pressure at which steam is produced is a function of the steam load, which requires the highest pressure plus the pressure losses associated with the steam

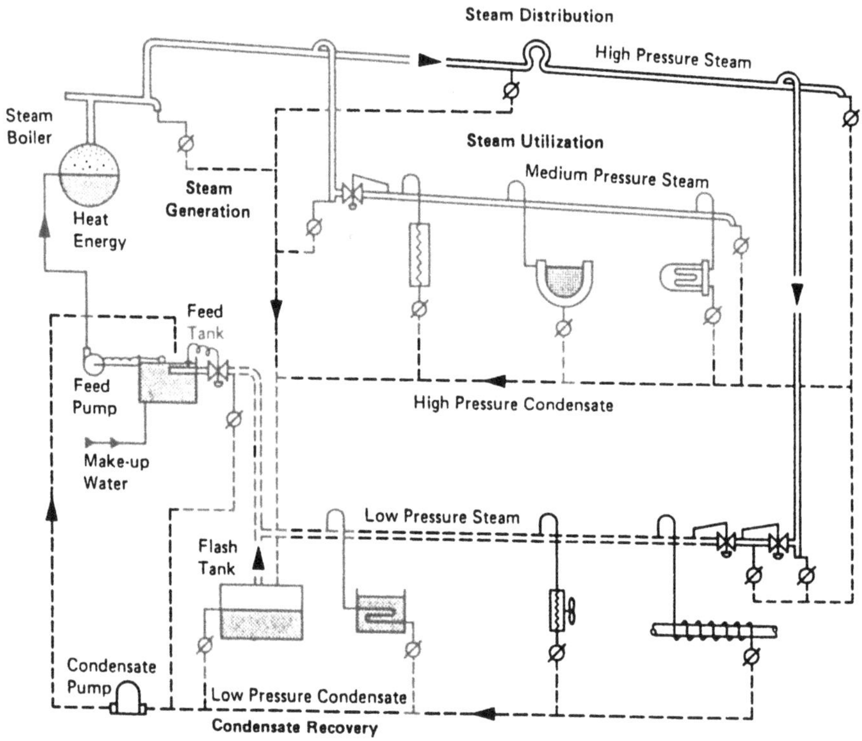

Figure 5-1. Typical steam distribution system.

distribution system. The steam produced by the steam generator may be either saturated or superheated steam.

Steam is carried from the boiler plant to the steam loads via a piping distribution system. Steam naturally flows from an area of high pressure to an area of low pressure. The piping distribution system includes steam manifolds (large pipes connecting multiple boilers together), steam mains (large pipes carrying steam from the boiler plant) and run-out piping to the individual steam loads. The steam distribution system may also contain shut-off valves, as well as steam.

As steam flows through the distribution system, heat is naturally lost to the surroundings, even if the system is well insulated. As the steam looses heat, some of the steam condenses to water to form condensate. When steam is consumed by a load, large amounts of heat are transferred from the steam, which results in most, if not all, of the steam condensing back to water.

After the steam has formed condensate, it must be removed from the system. If the condensate is not removed, the heat transfer properties of the steam are significantly reduced. Condensate in steam systems may also result in slugs of water, which can cause water hammer problems in the system. Water hammer can result in damage to steam system components. Once the condensate is removed from the steam side of the system, many systems return the condensate to the boiler to be reused. Occasionally, the condensate is not returned. Condensate may be returned to the boiler plant by a gravity return system, by a pumped condensate system or by a vacuum return system. Condensate return systems are discussed later in this chapter.

There are two methods used to determine the steam requirements for various pieces of steam consuming equipment (steam loads) and the total load on the system. One method is to consult tables, which list typical steam pressures and consumption rates for various types of equipment. A more accurate method is to calculate the design load or to consult the manufacturer of the steam consuming equipment to get the actual requirements.

5.3 STEAM AND CONDENSATE SYSTEM COMPONENTS

In addition to steam distribution piping and condensate piping, there are a number of components in a steam system used to control or condition the steam. This section provides an overview of steam system equipment and how it is used in a steam system.

Figure 5-2. Pressure reducing valve (*Courtesy of Spence Engineering Co*).

5.3.1 Pressure Reducing Valves

Steam produced by a steam generator must be at a pressure high enough for the steam load that requires the greatest steam pressure. Frequently, the steam system serves many loads, some of which require lower steam pressures. Where steam is supplied at pressures higher than required, one or more pressure reducing valves (pressure regulators), are required.

Pressure reducing valves (PRV) are used to reduce steam pressure to a safe point and regulate pressure to that required by the steam consuming equipment. Pressure reducing valves are available as direct operated, spring, weight or air-loaded. They may al so be air operated or pilot operated, using the steam as the operating medium. A typical pressure reducing valve is shown in Figure 5-2.

5.3.2 Temperature Control Valves

Steam is frequently used for comfort or process heating. Heating is accomplished by injecting steam directly into the medium to be heated or

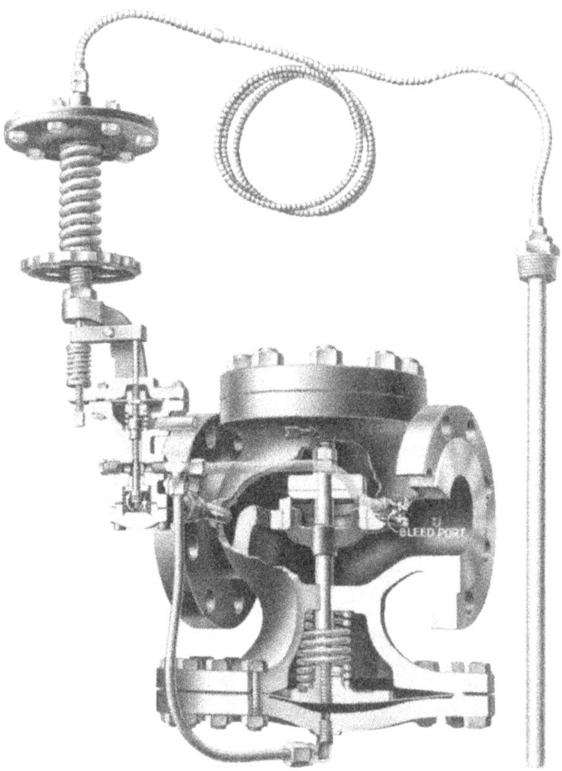

Figure 5-3. Temperature control valve (*Courtesy of Spence Engineering Co*).

by a heat exchanger, which transfers heat from the steam to the medium. When steam is used for heating, the temperature of the medium being heated is usually controlled. Temperature control is accomplished by one of two methods: permitting steam to enter a heat exchanger or medium intermittently, or by modulating the amount of steam that enters. Temperature control valves can be actuated electrically or pneumatically, or they can be self-contained. Self-contained temperature control valves have a wax or liquid filled temperature sensing element. Self-contained control valves require no external power source. A steam temperature control valve is shown in Figure 5-3.

5.3.3 Desuperheaters

Often, steam is generated at a temperature greater than its saturation temperature for a given pressure, usually for mechanical work applications such as operating steam turbines. When steam is at a temperature greater than its saturation temperature, it is referred to as superheated steam.

Most steam heat transfer equipment is designed to operate at saturation conditions. Certainly, the greatest heat is released from steam when it changes phase from saturated steam to a saturated liquid. For most steam systems that generate superheated steam, it is necessary to "desuperheat", or cool, the steam before it can be used for process or comfort heating applications.

The temperature of superheated steam may be reduced with a device called a desuperheater, which sprays controlled amounts of water into steam lines to control steam temperature at the desired level. A typical desuperheater is shown in Figure 5-4.

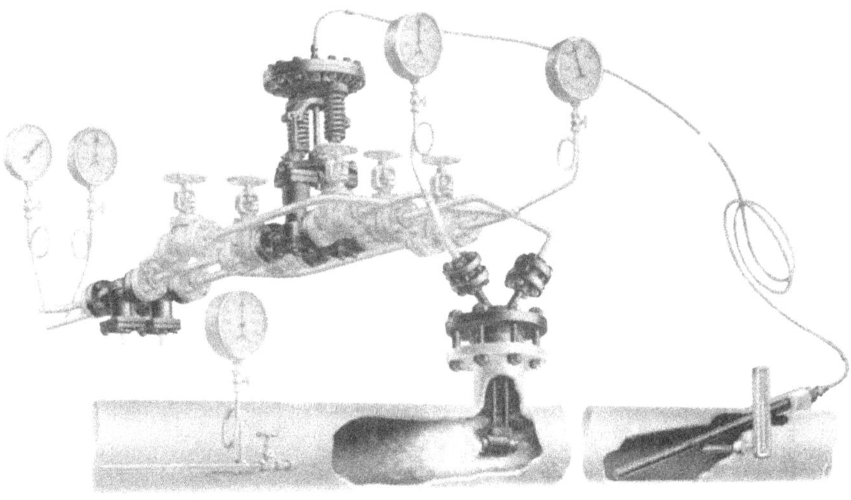

Figure 5-4. Steam desuperheater (*Courtesy of Spence Engineering Co*).

Although desuperheating steam (reducing its temperature) may appear wasteful and energy inefficient, it is not. When water is added to superheated steam to cool it, energy is not lost. As the water, not the superheated steam is sprayed, the total mass of the steam increases by the amount of water added, although it is at a lower temperature. The total energy content of the steam (enthalpy) remains constant. As a result, more steam is available at a lower temperature.

5.3.4 Heat Exchangers

In many applications where steam is used for heating, it is necessary for the steam to provide heating indirectly through a heat exchanger. Examples of steam heat exchangers include steam heating coils, which heat air and steam to liquid heat exchangers for heating domestic hot water, process hot water or process fluids.

Steam heating coils are frequently used in air handling units to heat air for comfort heating. Steam coils are steam to air heat exchangers. In most applications, the steam used in heating coils is of the low pressure variety (15 psig or less). Basic steam coils generally have tubes with fins on the in air side. The steam supply connection is at one end and the tubes are pitched toward the condensate return at the other end. A typical steam heating coil is shown in Figure 5-5.

Steam heat exchangers are also used to heat water or some other process liquid. Steam to liquid heat exchangers are also referred to as *converters*. Steam to liquid heat exchanges are usually shell and tube type exchangers. A typical shell and tube heat exchanger is shown in

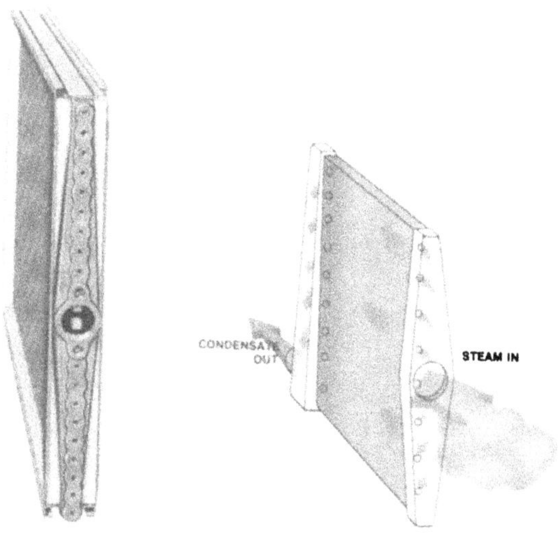

Figure 5-5. Steam heating coil (*Courtesy of the Trane Co*).

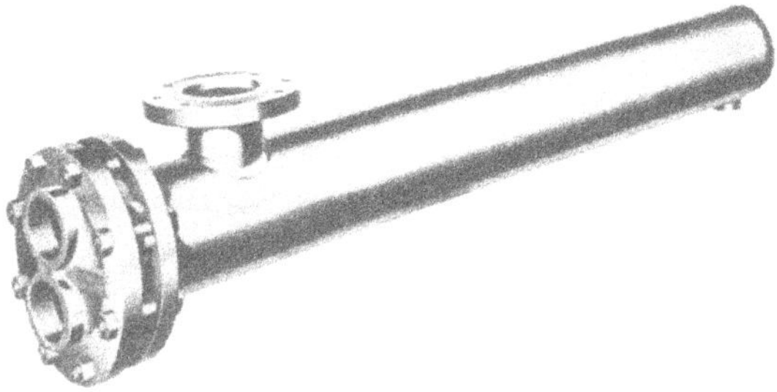

Figure 5-6. Shell and tube heat exchanger.

Figure 5-6. Usually, the medium to be heated (liquid) is circulated through the tubes while surrounded by steam in the shell of the heat exchanger. Steam usually enters the top of the shell while the condensate drains from the bottom. The capacity of the heat exchanger is a function of the number of tube passes and the length of the shell and tubes. Heat exchangers are discussed in more detail in "Chapter 14: Heat Recovery and Waste Heat Management."

5.3.5 Steam and Condensate Meters

In order to effectively manage the consumption of steam, it is necessary to measure the steam consumption or flow rates at various locations and at large consumers of steam. The consumption of steam can be directly measured by metering the flow of steam or condensate.

Condensate meters measure the flow of steam indirectly (by measuring the steam heating load or at some common point such as at a boiler plant.) Condensate meters are similar to water meters. They are usually positive displacement type meters, which entrap discrete quantities of liquid condensate and physically move it from the inlet to the outlet. Condensate meters are accurate only if no steam or condensate is consumed or lost in the heating process.

A more accurate method for measuring steam consumption is to meter the steam itself. All steam meters that directly measure the flow of steam make use of the kinetic energy of the steam as it flows.

There are two basic principles by which steam flow meters operate: measurement of pressure and direct measurement of fluid velocity. Orifice plates, venturi tubes and pitot tubes are referred to as "head" meters and measure steam flow by measuring pressure loss. Turbine meters and vortex shedding meters monitor the steam velocity directly.

Orifice plates are the most common type of steam flow meter. They measure flow by restricting the steam, by placing a plate with a hole in the steam line. As the steam flows through the hole (orifice), a pressure difference is created. The pressure difference is measured and converted to a flow rate. Orifice plates create a significant pressure loss, which results in an energy loss.

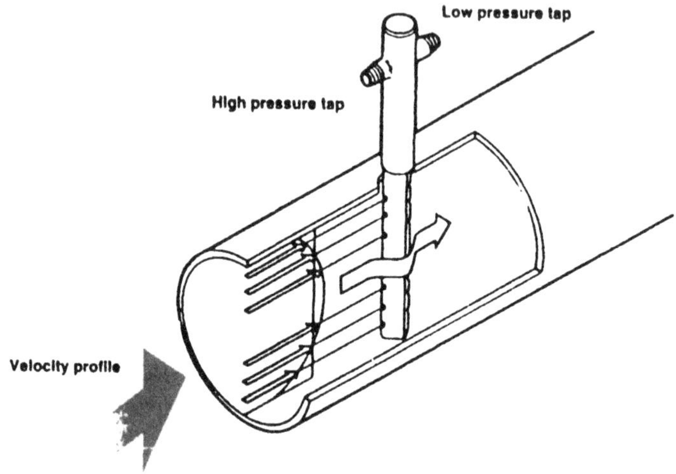

Figure 5-7. Averaging pitot tube flow meter.

Venturi meters have a venturi tube that creates a pressure drop as the steam flows through the meter. The venturi tube reduces the area through which the steam must flow, causing an increase in velocity and a resultant decrease in pressure. Pressure is measured at the inlet of the meter and at the throat of the ventrui. The difference in pressure between the two taps is then converted to a velocity, which may be converted to a flow rate.

A pitot tube meter measures flow by measuring the difference in total pressure (static pressure plus velocity pressure) and the static pressure in the steam flow. The difference between the total pressure and static pressure is the velocity pressure, which may be converted to a steam velocity and then to a flow rate. The venturi meter has a probe with holes on its upstream side to measure the velocity pressure of the steam, and a hole on its downstream side to measure static pressure. Separate pressure taps are provided for each set of holes to measure the differential pressure. A typical averaging pitot tube flow meter is shown in Figure 5-7.

Turbine meters have a turbine, which is rotated by the steam as it flows. The kinetic energy in the flowing steam causes the turbine to rotate. As the turbine rotates, its rotational speed is measured with a magnetic device that measures magnetic impulses. The rate of impulses is then converted to a flow rate since the rotational speed of the turbine is directly proportional to the steam flow rate.

Vortex shedder meters detect vortices in the steam flow as it flows; these vortices are created by an obstruction placed in the flow steam. The vortex formation is proportional to the velocity of the flow and therefore the rate of flow. The vortices are then detected by a very sensitive strain

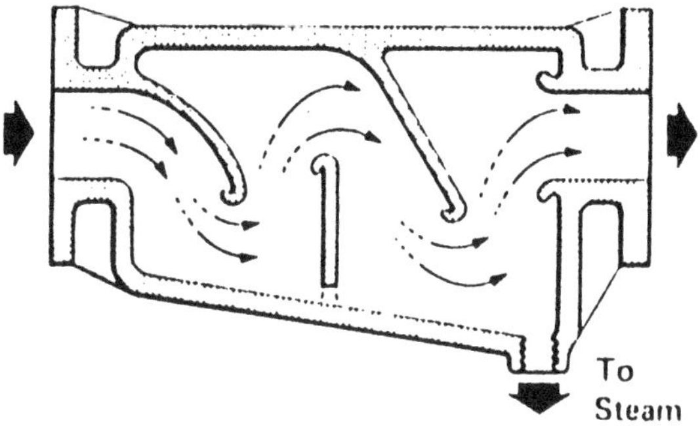

Figure 5-8. Steam separator.

gage imbedded in the obstruction, a magnetic pickup or an ultrasonic detector.

5.3.6 Steam Separators

Unless superheated steam is required, most steam systems use saturated steam and the boiler produces steam at near-saturation conditions. When steam is at saturated conditions, it can easily condense back to water as it passes through the system. Most steam heating applications require dry saturated steam, that is, saturated steam with no entrained condensate.

To deliver dry saturated steam at a steam load, steam separators are often used to remove the entrained condensate from the steam. A steam separator employs a set of baffles, through which the steam is passed, to separate the condensate from the steam. A steam separator is shown in Figure 5-8. As the steam passes through the baffles, it must change direction. As the steam changes direction, centrifugal force causes the water droplets to impinge on the baffles. The condensate is collected at the bottom of the separator, where it is drained from the system.

5.3.7 Steam Traps and Condensate Removal

Earlier in this chapter, the need to remove condensate from the steam system after the steam has condensed back to a liquid was discussed. It is important that condensate, as well as any non-condensable gases, be removed quickly and efficiently. Condensate must always be removed form the system as soon as it accumulates for the following reasons:

- Holding the condensate in the terminal heat exchange equipment reduces the heat transfer rate.
- The condensate can form a slug of water, which could cause serious damage from water hammer.

When steam is shut off to a steam heating device, the device cools, forming a vacuum inside the heating device. As a result, air is drawn into the device to take up all the space formerly occupied by the steam. When steam flows back into the heating device, the air must be removed from the system. If the air is not removed, the heat transfer capabilities of the system will be reduced and the air will cause corrosion within the system.

A steam trap may be defined as a self-contained valve, which automatically drains condensate and discharges air and non-condensable gases from a steam distribution system. The purpose of a steam trap is to remove condensate and non-condensable gases from a steam distribution system while not allowing steam to escape from the system. The steam trap is designed to differentiate between liquid or non-condensable gases from steam.

Steam traps distinguish condensate and non-condensable gases from steam by reacting to differences in temperature, density or thermodynamic properties of steam and condensate/non-condensable gases. As a result, there are three major classifications that are based upon the principle on which they operate:

- *Thermostatic traps*, which react to the difference in temperature of the steam and condensate.
- *Mechanical traps*, which react to the difference in densities of the steam and condensate.
- *Kinetic traps*, which operate on the difference in flow characteristics of the steam and condensate.

No single type of trap is best suited to all applications. Most steam systems require more than one type of steam trap.

Thermostatic traps are operated by changes in temperature. The two most common types of thermostatic steam traps are the bellows thermostatic and bimetallic thermostatic traps. The bellows thermostatic trap has an expandable bellows, which contains a fluid with a boiling point lower than that of water. A bellows trap is shown in Figure 5-9. The bellows containing the fluid expands and contracts in response to the temperature of the fluid surrounding the bellows inside the trap. The fluid inside the bellows closely parallels the temperature-pressure relationship of the steam saturation curve. When hot steam and condensate enters the trap, the fluid in the bellows cools and contracts, opening the valve allowing the liquid condensate to flow out of the trap. It is the difference in temperature between the steam and condensate that operates the trap. Bellows thermostatic traps are naturally self-venting.

The other common type of thermostatic steam trap is bimetallic. This type of trap has an element or disk made from metals with different coefficients of thermal expansion. Heat causes the disk to change shape, permitting the valve port to open or close. A bimetallic thermostatic steam trap is shown in Figure 5-10. The bending of the bimetal disk takes place at

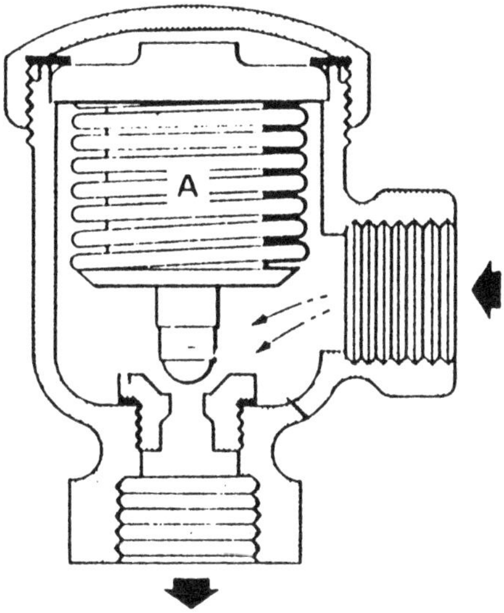

Figure 5-9. Bellows thermostatic trap.

a fixed temperature, so the trap tries to open and close at one particular temperature. The bimetallic trap, however, does not follow the steam saturation curve very well. When condensate enters the trap, it must cool considerably before the valve can open to release the condensate. While the condensate is cooling, more condensate is collecting, which may result in waterlogging. A bimetallic trap must be manually readjusted if the operating conditions, such as steam pressure or temperature, are changed from those originally set.

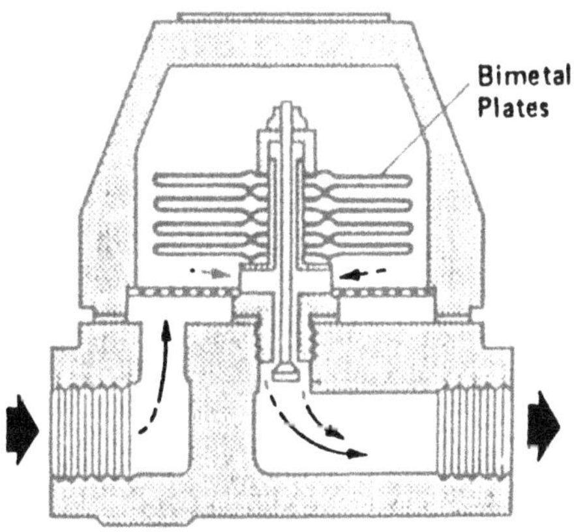

Figure 5-10. Bimetallic thermostatic trap.

There are three common types of mechanical steam traps: bucket traps, inverted bucket traps, and float and thermostatic traps. Bucket traps use a bucket float that is open at the top, as shown in Figure 5-11. Initially, the bucket is empty and rests on the bottom of the trap body, with the outlet valve open. As condensate enters the trap, the bucket floats upward and closes the discharge port. When additional condensate flows into the trap, the condensate overflows the top of the bucket, filling it with liquid condensate. As the bucket fills with condensate, it looses its buoyancy and drops down, opening the discharge valve allowing the steam pressure to push the condensate out of the bucket and trap. After the condensate flows out of the trap, the bucket again rests on the bottom of the trap body until the trap begins to fill with condensate again.

A variation on the bucket trap is the inverted bucket trap. Inverted bucket traps have a bucket mounted with the open end down, as shown in Figure 5-12. Initially, the bucket rests on the bottom of the trap body and the trap discharge valve is open. As liquid condensate, air and steam flow into the trap, the steam and air are trapped under the bucket, causing it to float. When the bucket rises, the trap discharge valve closes, trapping the steam and condensate in the trap. A small vent hole is usually provided at the top of the bucket to allow air and other non-condensable gases to escape from under the bucket. The steam trapped under the bucket is cooled by the liquid condensate causing the bucket to loose some of its buoyancy.

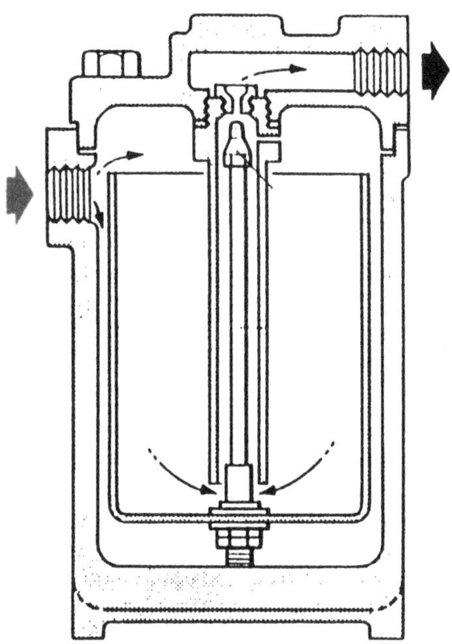

Figure 5-11. Bucket trap.

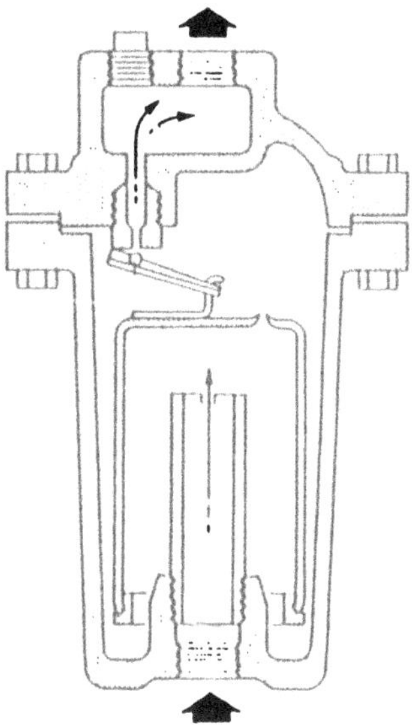

Figure 5-12. Inverted bucket trap.

Eventually, as the steam condenses, the weight of the bucket becomes greater than the buoyancy forces, the bucket drops, opening the discharge valve allowing the steam to force the condensate and air out of the trap. The bucket remains down until condensate again flows into the trap.

A float and thermostatic (F&T) trap has the ability to vent large amounts of air and non-condensable gases, as well as handle steam condensate. The F&T trap allows the continuous discharge of air, in addition to the condensate. The trap consists of a float ball, which opens and closes a discharge valve, as shown in Figure 5-13. The trap also includes a thermostatic vent near the top of the trap body. The thermostatic air vent is actuated by a thermostatic bellows.

Initially, the ball rests against the bottom of the body of the trap and the discharge valve is closed. As condensate flows into the trap, the float ball rises, opening the discharge valve. Any air that enters the trap during normal operation will collect at the top. The air cools the thermostatic bellows causing it to contract, opening the vent allowing the air to escape. If no air is present, the top of the trap will contain hot steam and the thermostatic vent will remain closed.

Kinetic traps operate on the difference between the flow characteristics of steam and condensate, and based on the fact that condensate discharging to a lower pressure contains more heat than necessary to maintain it in the liquid phase. The excess heat causes some of the condensate to flash back to steam at the lower pressure. Kinetic steam traps include thermodynamic traps, impulse traps and orifice traps.

Thermodynamic traps are also called disk traps. They contain only one moving part—a disk—as shown in Figure 5-14. A small disk covers

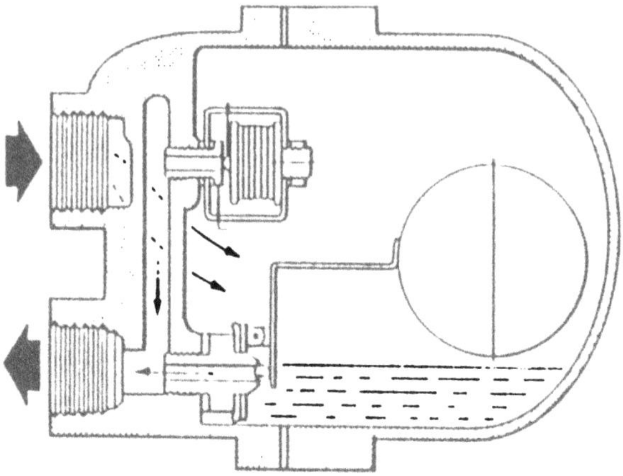

Figure 5-13. Float and thermostatic (F&T) trap.

the trap inlet orifice. Condensate or air, which move at relatively slow velocities, lift the disk off its seat, allowing the condensate and/or air to be discharged from the trap. When steam enters the trap, it passes through at a much higher velocity due to its volume. As the steam passes through the space between the disk and seat, it impacts the sides of the control chamber to produce an increase in pressure. The pressure imbalance between outside the disk and the side facing the seat causes the disk to snap shut, thereby preventing the passage of steam. When condensate again enters the trap inlet, the disk lifts off the seat, allowing the condensate to flow out.

Impulse traps are also known as piston traps. The trap consists of a main valve (piston) with a thin piston disk. The piston is free to move up and

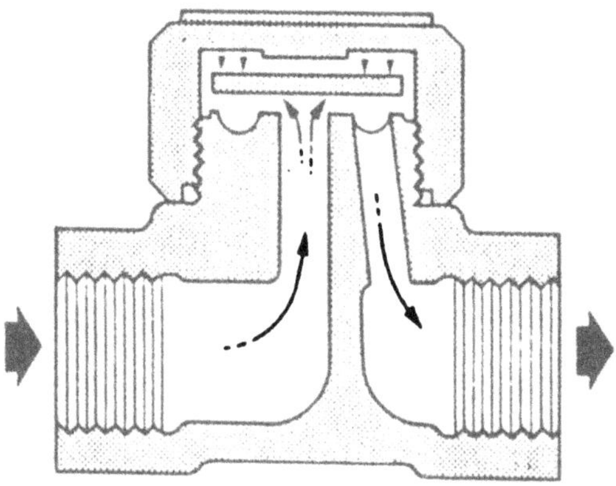

Figure 5-14. Thermodynamic (disk) trap.

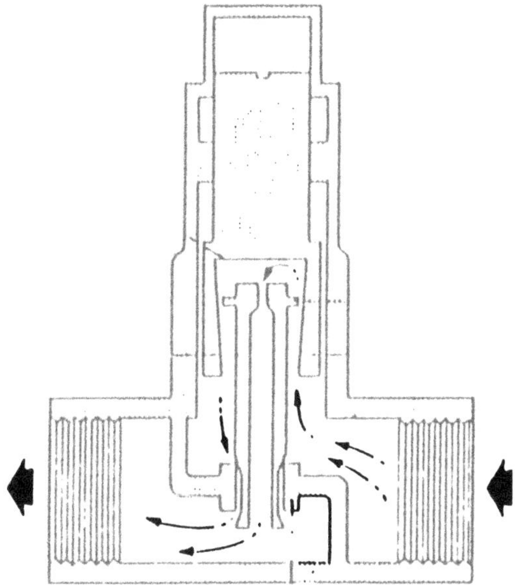

Figure 5-15. Impulse trap.

down within a tapered guide, as shown in Figure 5-15. Impulse traps are able to continuously pass small amounts of steam or condensate through a "bleed" orifice changing the pressure positions within the piston.

When cooler condensate enters the trap, it creates an unbalanced pressure on the underside of the piston disk decreasing the pressure. The reduced pressure lifts the piston and permits the valve port at the bottom of the piston to open, allowing the condensate to flow out. When live steam or very hot condensate that flashes to steam enters the trap control chamber, the increased pressure above the piston disk pushes the piston down, closing the valve port.

Orifice traps have no moving parts; they consist of a plate with an orifice in the center of the plate, as shown in Figure 5-16. The orifice trap is

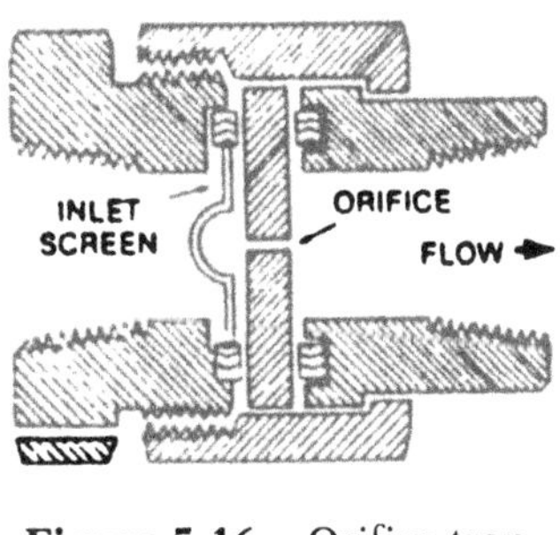

Figure 5-16. Orifice trap.

continuously open. Steam and condensate flow is controlled by two-phase flow through the orifice. The small orifice hole allows pressure in the steam to force condensate to drain continuously into the lower pressure return system.

This orifice has a much greater capacity to pass condensate than it does steam due to the significant difference between the density of liquid condensate and steam. As the condensate flows through the orifice, the condensate tends to flash into steam, which tends to choke the flow through the orifice.

The orifice trap continuously passes all condensate in addition to air, non-condensable gases plus a small controlled amount of steam. Orifice traps are selected such that the orifice limits the steam leakage rate to a small amount.

5.3.8 Condensate Return Systems

After steam traps remove condensate from steam consuming equipment and from steam distribution systems, the hot condensate is usually returned to the boiler plant to be reused. Since the condensate is at an elevated temperature, it still contains large amounts of thermal energy. Condensate return systems are classified into three types: gravity return, vacuum return and pumped return.

As the name implies, the flow of condensate in gravity return systems is caused by gravity. In gravity return systems, all steam consuming equipment must be located at an elevation higher than the steam generator. The condensate piping must be slopped down as it returns to the boiler plant to allow the condensate to flow by gravity. Gravity returns systems are typically suited for application in multistory buildings, where the boiler plants are at the lowest level.

Vacuum return systems use a vacuum pump to produce a negative pressure in the condensate return system. As steam traps discharge the condensate, as well as air and non-condensable gases, the vacuum pump draws the air and gases out of the system. As the air is drawn by the vacuum, the condensate is carried along with the air and gas. Vacuum return systems do not require that the condensate return piping continuously slopes back to the boiler. The vacuum can lift the condensate as the piping system returns to the boiler plant.

Pumped condensate return systems collect the condensate from steam consuming equipment and pump the condensate back to the boiler plant through a pressurized piping return system. Pumped systems are usually a combination of gravity return and pumped return. Condensate is usually drained by gravity to a collection point where it is pumped back to the boiler plant. Condensate pumps usually include a collection tank, as shown in Figure 5-17. Pump sets can be simplex (one pump) or duplex (two pumps). The pumps are usually energized by a float and switch inside the tank.

An alternative to electrically driven pumps is the pumping trap, which utilizes the pressure of the steam itself as the driving medium for the pump.

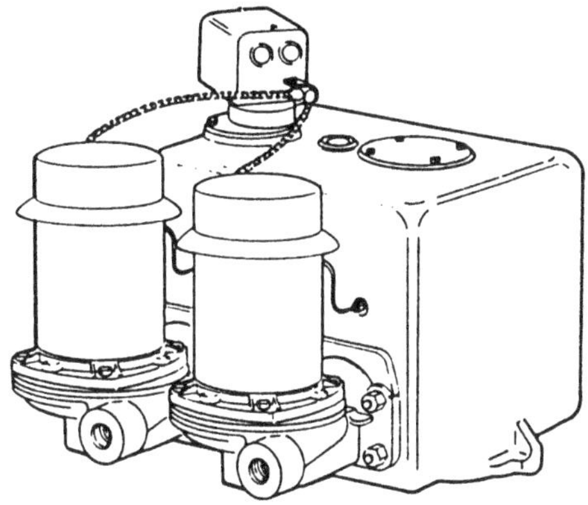

Figure 5-17. Steam condensate pump set.

A pumping trap is shown in Figure 5-18. Condensate enters the pump at the bottom, causing the float to rise to a point where it activates a float operated valve, which admits steam into the upper chamber. The steam forces the float down, discharging the liquid condensate out of the trap housing. A check valve is provided on the inlet pipe to prevent the condensate from flowing backward. After the condensate level drops, the float operated valve closes, preventing the steam from pressurizing the top of the float.

Condensate return systems are also classified as vented (open) return and nonvented (closed) return systems. Vented systems are open to the atmosphere and operate at atmospheric pressure. Vented return systems are applicable only if the condensate return is a gravity return. In high pressure steam systems, nonvented pressurized return systems are

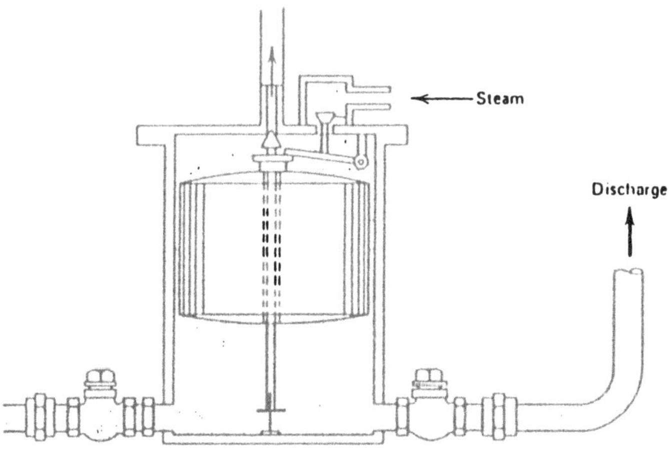

Figure 5-18. Pumping trap.

typically used to keep the return system above the saturation pressure. If the pressure were not maintained above the saturation pressure, the liquid condensate would have a tendency to flash into low pressure steam.

5.4 MAINTAINING AND OPERATING STEAM SYSTEM FOR OPTIMUM ENERGY EFFICIENCY

For steam and condensate systems to operate at peak efficiency and minimum energy consumption, it is necessary that they receive regular maintenance and service. Significant energy savings can be achieved by routine maintenance. Proper routine maintenance should be the first step in any energy management program for steam and condensate systems. The following are operations and maintenance procedures that will result in optimum energy efficiency for most steam and condensate systems.

Isolate Inactive Piping: Frequently, when a piece of steam consuming equipment is taken out of service, the steam run-out piping to the equipment is simply valved off at the equipment and the steam is left connected to the inactive piping. Inactive steam piping experiences the same heat losses as active sections of pipe.

Inactive piping must be positively isolated from the steam supply to reduce unnecessary losses. Instead of relying on shut-off valves, which may leak, it is better to isolate unused sections of steam piping by removing a short section of pipe and installing caps or blind flanges over the open ends of pipe.

Repair or Replace Insulation: A survey of the steam and condensate piping should be conducted to find missing or damaged insulation on piping and equipment. Any sections of pipe, valve bodies, unions, flanges and steam trap bodies should be insulated. Only thermostatic traps and their cooling legs should not be insulated.

A visual inspection of insulation should be conducted to find damaged insulation. Insulation materials are composed of millions of microscopic air cells, which form an effective barrier to heat transfer. If these cells become water logged or crushed, they will loose their ability to insulate and become ineffective.

Inspections should also be conducted to find excessive insulation surface temperatures, which would indicate damaged insulation. A survey of pipe and equipment surface temperatures should be made using infrared thermometry or thermography to locate areas of insulation deterioration.

Care should be exercised when working with older steam and condensate systems. Older insulation may contain asbestos, which must be properly handled for safety reasons.

Repair Steam Leaks: Steam leaks can represent a significant energy loss for steam systems. Leaks out of the system are relatively easy to spot by the plume of steam emanating from the leak. Leaks can also occur inside the steam system when steam leaks from the supply piping directly to the condensate return piping. Figure 5-19 is an estimate of the heat losses from leaks at various steam pressures.

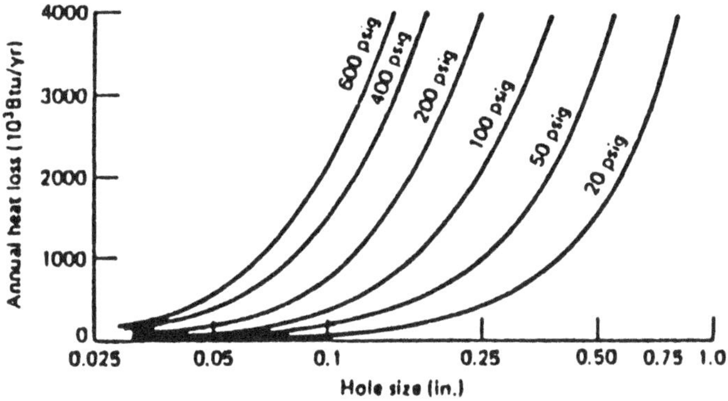

Figure 5-19. Heat losses from steam leaks.

Leaks inside the steam distribution system are invisible and hidden in the flash steam, making it very difficult to detect them. A systematic approach to locating and repairing steam leaks throughout a facility needs to be developed to ensure the success of an energy management plan.

A regular steam leak survey and repair program should be implemented. Upon detecting a leak, its location should be recorded on a piping diagram or schematic. The severity of the leak should also be recorded; it should be classified by type, line pressure, size, how often leaks occur and re-occur.

Bypass valves around steam traps are prime candidates for internal leakage. They are often opened to assist in system start-up or to vent excessive air, and inadvertently left fully or partially open. Other so urces of leaks include isolation valves and defective gaskets between pipe flanges.

Example 5.1. A maintenance crew conducting a steam leak survey finds steam leaking from a gasket between two pipe flanges in a 200 psig steam line. They estimate that the leak is about 0.02 inches wide and about 1/8 inches long. If the steam system is operated 8,000 hours per year, what would the annual savings be if the leak were repaired?

First the area of the leak is calculated

$$A_L = (0.02)(0.125) = 0.0025 \text{ in}^2$$

The diameter of an equivalent circle is calculated

$$D = \sqrt{(4)(0.0025)(\tfrac{1}{\pi})} = 0.056 \text{ in.}$$

From Figure 5-19, a hole in a 200 psig steam system would loose about 200×10^3 Btu/yr. Since the system operates only for 8,000 hrs. per year, the actual savings is determined by multiplying the value from Figure 5.19 by a ratio of the operating hours.

$$\text{Savings} = \frac{8{,}000}{8{,}760}(200 \times 10^3) = 182.6 \times 10^3 \ \text{Btu/yr.}$$

Note: This does not take into account the efficiency of the boiler.

Maintain Steam Traps: Losses due to inadequate maintenance of steam traps is the largest source of energy losses in a steam system. Steam traps are among the greatest source of energy waste in facilities with steam systems. Even when working properly, steam traps are quite inefficient. It has been estimated that, typically, 5 – 60% of the steam traps in a steam system have failed and are sticking open or closed. The cost of steam leakage through a malfunctioning trap can e xceed the cost of replacing the trap in one day. The cost of energy lost due to malfunctioning steam traps far outweighs the cost of proper maintenance and repair.

A typical steam trap may open and close over a million times a year. Some wear, malfunction and failure is inevitable. According to most manufacturers of steam traps, the life expectancy of a steam trap operating under normal conditions is only three to five years. The life expectancy of a steam trap is directly related to the pressure at which the trap must operate. In general, the higher the pressure, the shorter the life of the trap.

Steam traps can malfunction in two ways: They can stick in the closed position, causing condensate to back up into the steam system, or they can stick open, allowing live steam to discharge into the condensate system. When a trap sticks in the closed position, it is usually quickly detectable since flooding in a steam heating device will result in an obvious degradation in the performance of the heating device. Flooding can also damage the equipment as a result of water hammer, causing condensate to back up into other equipment. Since flooding could result in damage to the heat transfer equipment, traps are usually designed to fail in the open position.

Comprehensive maintenance programs for stream traps are among the most attractive energy management investments. Most traps can easily be repaired and have inexpensive replaceable parts. The basis for a steam trap maintenance program is a comprehensive trap inspection and testing program. As part of a trap inspection program, the traps should be reviewed—as to trap type and service application—to ensure that the most suitable and efficient trap is being used for each application. Steam trap applications and selection are discussed later in this chapter.

The main purpose of the inspection program, however, is to identify traps that are stuck open and blowing steam into the condensate return system. Since these leaks are internal to the system, they cannot readily be detected. When a trap is stuck open blowing steam can sometimes be detected by an increase in the condensate temperature entering a condensate receiver.

A visual inspection of traps can be made by placing a drain valve just downstream of the trap. When the drain valve is opened, if live steam blows out, the trap is stuck open. The key to this test is identifying live steam from flash steam. Live steam will tend to blow directly out of the valve at a high velocity and produce a rather narrow and straight steam cloud. Flash steam, on the other hand, will tend to produce a steam cloud, which is more billowy and will tend to slowly rise. A live steam cloud is shown in Figure 5-20 and a flash steam cloud is shown in Figure 5-21.

An alternative method for visually checking a steam trap is to install a sight glass immediately downstream of the trap. Sight glasses are shown in Figure 5-22. Sight glasses are relatively inexpensive and permit a quick visual inspection of trap operation without interfering in the normal operation of the steam system.

Another visual inspection method is to measure the temperature of the condensate pipe just downstream of a trap. A simple and easy way is to check the temperature with a temperature indicating stick or crayon. Theses ticks are available for various temperatures, depending upon the

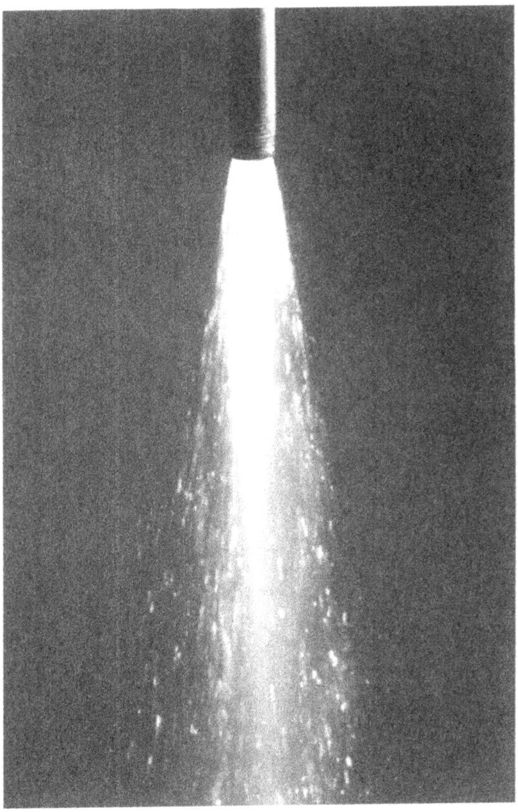

Figure 5-20. Live steam cloud.

Figure 5-21. Flash steam cloud.

steam system pressure and temperature. The test procedure involves simply striking the condensate pipe with the stick several inches downs team of the trap. The sick will leave a dry mark on the downstream pipe if

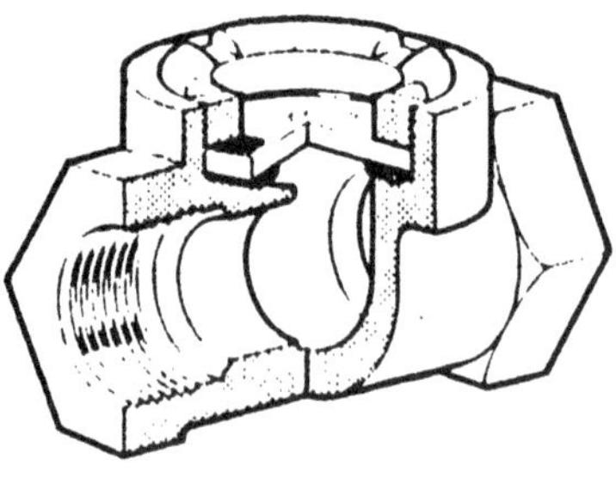

Figure 5-22. Sight glass.

the trap is working properly. If the trap is sticking open, the mark will melt or change color, indicating a higher temperature due to the live steam blowing through the trap.

The temperature of the condensate piping downstream of a trap may also be measured with a pyrometer or a direct temperature measuring device. Pyrometers are scanners that measure temperatures by measuring the infrared radiation emitted from a hot surface. Pyrometers tend to be fairly expensive and require some operator training to properly interpret the results of a scan. Direct temperature measuring devices have a sensor chamber mounted adjacent to the trap, which has an electric temperature sensor. The tip of the sensor is immersed in the condensate. The other end is electrically connected to an indicator, which indicates whether the trap is functioning properly or stuck open.

Checking the sound emitted from a trap as it operates is also a effective method for testing steam traps. Some traps can be heard opening and closing as they discharge condensate from the steam system. This method is more convenient when working with a closed condensate system.

Sound testing of steam traps can be done with a screwdriver or an industrial stethoscope. Ultrasonic detectors with earphones may also be

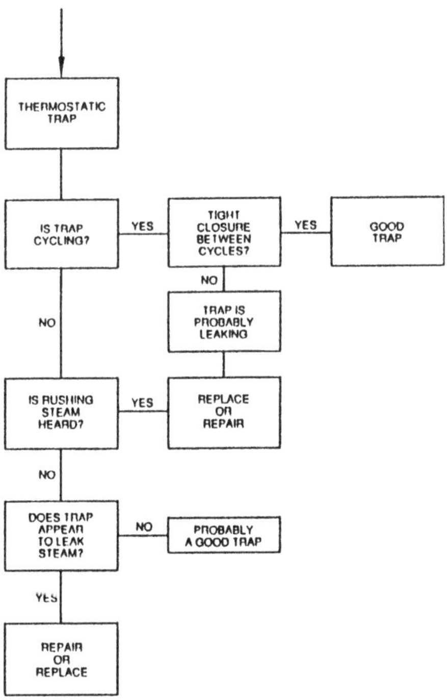

Figure 5-23. Thermostatic trap diagnosis tree.

used effectively. Ultrasonic detectors need careful tuning to match or suit the conditions of each trap. Steam blowing through a trap emits a very high pitched sound, which is produced by intense turbulence at the trap orifice. The flow of liquid condensate emits the lower-pitched and lower-intensity sound of a liquid flowing. However, the combination of condensate and flash steam immediately downstream of a trap orifice can sound very similar to condensate and live blowing steam at the orifice. Inspecting traps by the sound method is most effective with disk type traps and thermodynamic traps, although it can be used to some extent with other types of traps. The sound method is not as suitable for float and thermostatic traps.

Testing traps by the sound method requires someone who is familiar with the different sounds produced by a properly operating trap and a malfunctioning trap. He must be able to differentiate between a normally discharging trap and one that is discharging live steam with the condensate. The most successful trap inspection and diagnosis programs utilize a combination of testing methods.

When inspecting and diagnosing steam traps, there are some characteristics of each type of trap that should be kept in mind, including the following:

- Thermostatic traps modulate and discharge continuously. When testing by the sound method, listen for the rush of condensate or the

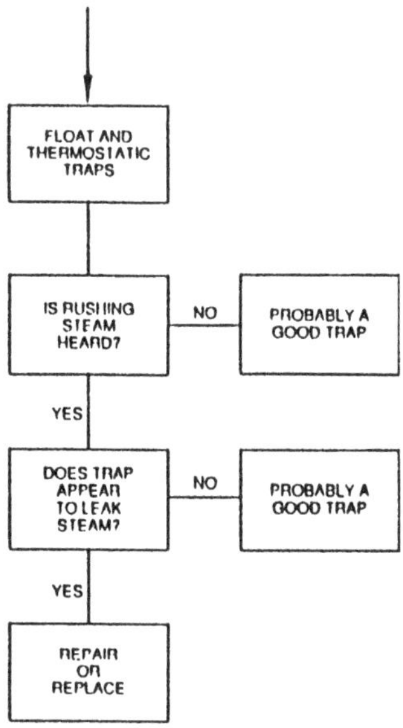

Figure 5-24. Float and thermostatic trap diagnosis tree.

hiss of live blowing stream. Visual tests must distinguish between flash steam and live steam.

- Float and thermostatic traps modulate as they discharge condensate, air and non-condensable gases. The sound is more intense for a failed element passing steam. Orifice may fail due to erosion. If the float is defective, trap will fail in the closed position. Visual tests must distinguish between flash steam and live steam.
- Bucket traps discharge at full capacity then close. The muffled sound of the bucket rattling against the body of the trap can be heard. If the trap looses its prime, the bucket will rattle violently and the rushing sound of blowing steam will be heard. The trap will fail in the closed position if the vent becomes clogged.
- Disk traps are the most suitable for the sound test method; the cycling of the trap is audible since the disk slams against the seat when it closes. If the seat is worn, the trap will be heard rapidly cycling. If the trap is passing live steam, a "machine gun" sound will be heard.

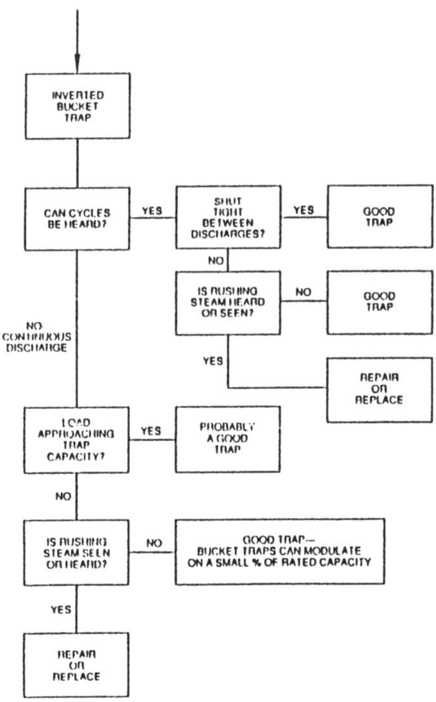

Figure 5-25. Bucket trap diagnosis tree.

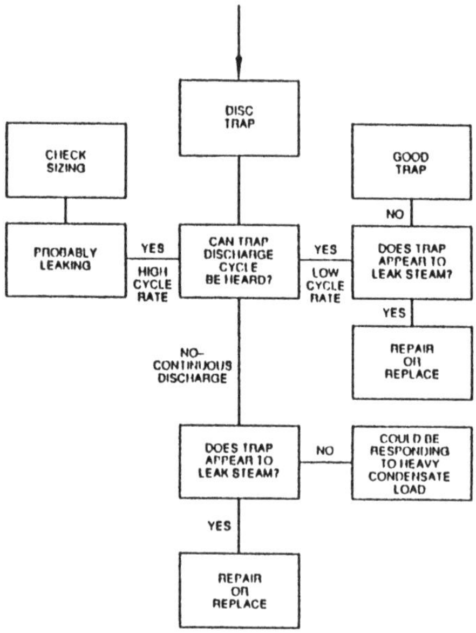

Figure 5-26. Disk trap diagnosis tree.

Figures 5-23, 5-24, 5-25 and 5-26 are diagnosis decision trees, which may be used when testing and diagnosing steam traps.

5.5 ENERGY CONSERVATION MEASURES FOR STEAM SYSTEMS

The energy conservation measures for steam systems discussed in the previous section were operation and maintenance procedures, which would improve the energy efficiency of existing steam systems. Most of those measures were little, or no, cost measures that could be performed on an ongoing basis to improve efficiency.

In addition to the low cost O&M procedures, there are also modifications that could be made to steam systems to improve energy efficiency. These modifications, however, require a more significant cost investment and are permanent. The modifications listed below are improvements that should be considered under any energy manage ment program for existing, or proposed, steam and condensate systems.

5.5.1 Insulate Steam and Condensate Piping and Steam Handling Equipment

Exposed uninsulated surfaces of steam piping, condensate piping and related equipment usually have surface temperatures in excess of 200°F. As

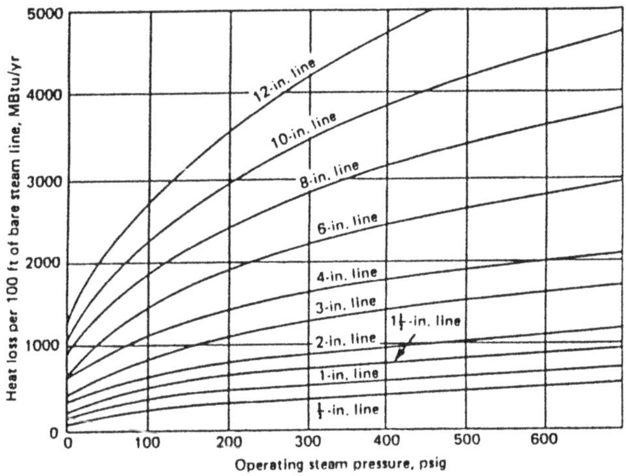

Figure 5-27. Heat loss from bare steam pipes.

a result, exposed uninsulated surfaces loose significant amounts of heat. Figure 5-27 shows the approximate heat loss from bare steam pipes at various pressures. Figure 5-28 shows the approximate heat loss from uninsulated surfaces at elevated temperatures.

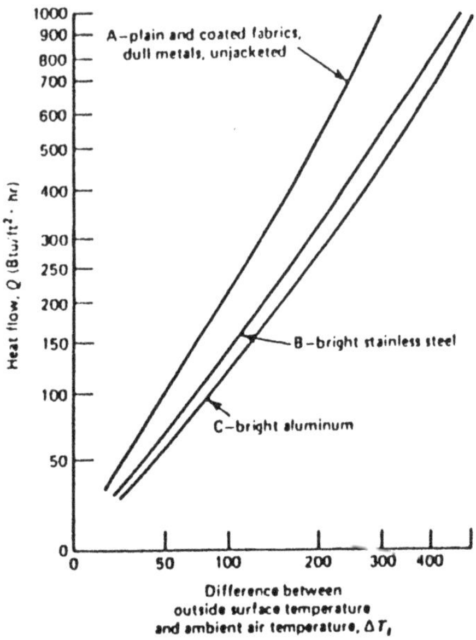

Figure 5-28. Heat losses from surfaces at elevated temperatures.

TABLE 5.1 Recommended Insulation Thickness' for Hot Piping

Fluid Design Operating Temperature Range, °F	Insulation Conductivity: Conductivity Range Btu in. (h ft³ °F)	Insulation Conductivity: Mean Rating Temperature °F	Nominal Pipe Diameter (in.): Runouts up to 2	1 and less	1-1¼ to 2	2-½ to 4	5 & 6	8 & up
Heating Systems (Steam. Steam Condensate, and Hot Water)								
Above 350	0.32-0.34	250	1.5	2.5	2.5	3.0	3.5	3.5
251-350	0.29-0.31	200	1.5	2.0	2.5	2.5	3.5	3.5
201-250	0.27-0.30	150	1.0	1.5	1.5	2.0	2.0	3.5
141-200	0.25-0.29	125	0.5	1.5	1.5	1.5	1.5	1.5
105-140	0.24-0.28	100	0.5	1.0	1.0	1.0	1.5	1.5

Insulation is often omitted, or removed during maintenance, on valves, piping and other hot surfaces. Whenever possible, steam piping and equipment should be insulated to reduce heat loss. Table 5.1 is excerpted from ASHRAE (1993) and gives recommended insulation thickness' for steam piping at various temperatures.

5.5.2 Steam and Condensate Metering

One of the difficulties in energy management for steam systems is that they are usually a totally unmeasured utility. Identifying steam consumption rates for major pieces of steam consuming equipment should be an integral part of any energy management program for steam systems.

Various types of steam and condensate meters were discussed earlier in this chapter. Steam meters should be placed at each major piece of steam consuming equipment and in the main branches of steam distribution piping.

5.5.3 Control Temperature of Processes

In many steam heating applications, the temperature of the heated medium is not controlled and the steam heater is allowed to "run wild." For optimum energy efficiency, all steam heating equipment should have some means of temperature control to prevent overheating. In process work, product quality is often dependent upon accurate temperature control.

Control may be achieved by (1) permitting steam to enter the heating unit intermittently, (2) varying the steam temperature, or (3) varying the amount of steam delivered to the heat exchange device. Most steam heating control applications involve the use of a control valve and control the temperature by methods 1 or 3. Temperature control valves for steam systems were discussed earlier in this chapter. A self-contained steam

Figure 5-29. Steam temperature control for hot water storage tank.

control valve for heating water in a storage tank is shown in Figure 5-29. Control systems are discussed in more detail in "Chapter 16: Energy Management and Control Systems."

5.5.4 Condensate Recovery

In some steam heating applications, the steam condensate is "wasted" and not returned to the boiler plant because it has become contaminated. Occasionally, the steam itself comes in contact with the process and the condensate becomes contaminated. Significant amounts of sensible heat is lost by not returning the condensate to the boiler plant to be reused. Condensate return systems and equipment were discussed earlier in this chapter.

In applications where steam condensate is wasted and not returned due to contamination, a heat exchanger may be installed to separate the steam system from the process that would contaminate it. A heat exchanger with a separate process hot water loop should be installed to separate the two systems. The process loop would contain the contaminated heating water or steam allowing the condensate in the main steam system to be returned to the boiler plant.

The energy savings would be a function of the temperature of the wasted condensate, the temperature of the make-up water and the steam flow rate to the process. Savings would also result from reduced chemical treatment required to treat the raw make-up water and from sewage treatment costs.

5.5.5 Install Flash Tanks

Condensate return systems usually operate at pressures (back-pressure) considerably below that of the steam supply system pressure. Many condensate return systems operate at atmospheric pressure and some condensate return systems operate at a vacuum.

Whenever a steam trap discharges liquid condensate into a low pressure condensate return system, much of the liquid condensate flashes to low pressure steam at the pressure of the condensate return system. The condensate that is discharged is at, or near, the saturation conditions of the steam. As the pressure is reduced when the condensate flows out of the trap, the temperature and reduced pressure causes the condensate to flash to steam. The greater the difference between the steam pressure and the condensate return pressure, the greater the percentage of flash steam. Equation (5.3) may be used to calculate the percent flash steam.

$$\%\ \text{flash steam} = \frac{100(h_{f1} - h_{f2})}{h_{fg2}} \qquad \text{Eq. (5.3)}$$

where h_{f1} = enthalpy of the liquid condensate at P_1
h_{f2} = enthalpy of the liquid condensate at P_2
h_{fg2} = latent heat of vaporization at P_2

Figure 5-30 may also be used to determine the percentage of flash steam.

Flash steam contains significant amounts of energy that can be recovered and reused. The energy value of the flash steam is essentially equal to the latent heat of vaporization of the flash steam at the pressure of the condensate retur n system. In order to collect and make use of flash steam, a flash tank is required. A typical flash tank is shown in Figure 5-31.

For flash tanks to be economically feasible, the low pressure flash steam must be available in sufficient quantities to justify the cost of the

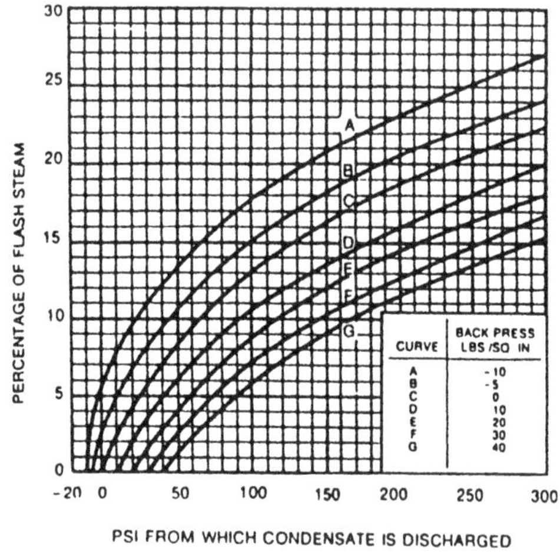

Figure 5-30. Flash steam.

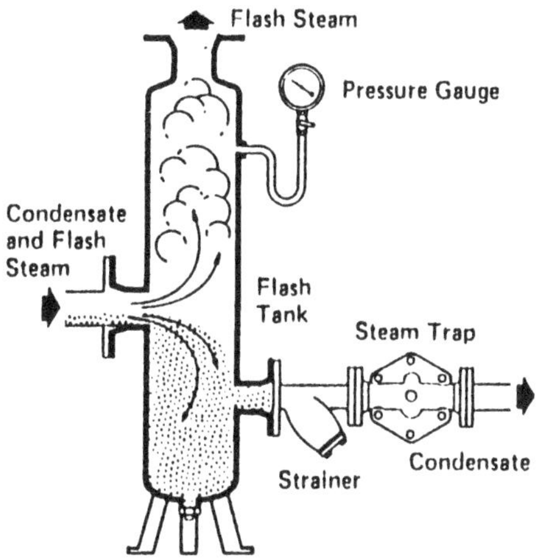

Figure 5-31. Flash tank.

flash steam recovery system. The demand for the low pressure flash steam must also be greater than the amount of flash steam available. If not all of the flash steam can be utilized, it must be vented to atmosphere and wasted. Typically, low pressure flash steam is used to supplement an existing low pressure steam system. After the low pressure flash steam is used for a low pressure application, the condensate must be returned to the boiler plant.

The cost of the piping for a flash steam recovery system could be significant if the source of flash steam is not close to the low pressure steam application. Typically, large piping is required to transport lower pressure steam to minimize pressure losses. Whenever possible, the flash tank and low pressure steam load should be in close proximity.

5.5.6 Cogenerate with Back-Pressure Turbines

Some steam generating plants are required to generate steam at a very high pressure in order to meet some process requirements that need high pressure steam. Usually, however, the high pressure steam system also serves other steam loads with requirements for much lower pressure steam. In such applications, steam pressure reducing stations are required to step the pressure down to a lower level. In some older systems, the steam was generated at a high pressure then indiscriminately reduced in pressure, with pressure reducing valves at the point of use.

An alternate method to pressure reducing valves for providing lower pressure steam from a high pressure system is a back-pressure turbine. It consists of a turbine in which the steam is exhausted at atmospheric pressure or higher. All of the condensing of the steam takes place downstream in the low pressure process. The back-pressure turbine

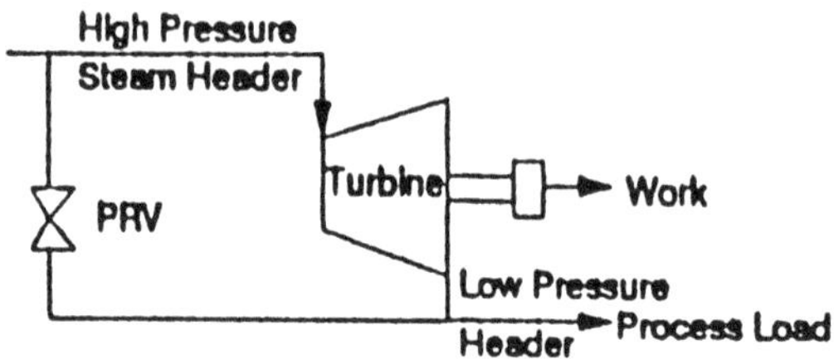

Figure 5-32. Back-pressure turbine schematic.

operates on the enthalpy difference between the steam inlet and outlet conditions.

In this application, the steam needed for the low pressure application is expanded (the pressure reduced) as it passes through the turbine. The expansion of the steam rotates the turbine generating work as a by-product. The turbine may be used to drive other mechanical equipment or to generate electricity. The low pressure steam discharged from the turbine may then be used for the low pressure steam application. A schematic diagram of a back-pressure turbine is shown in Figure 5-32.

There are a number of factors that affect the economic feasibility of cogenerating with a back-pressure turbine. The cost of the turbine and related equipment is quite high.

First, there must be a large enough demand for the reduced pressure steam to match the amount of steam required to drive the turbine. The

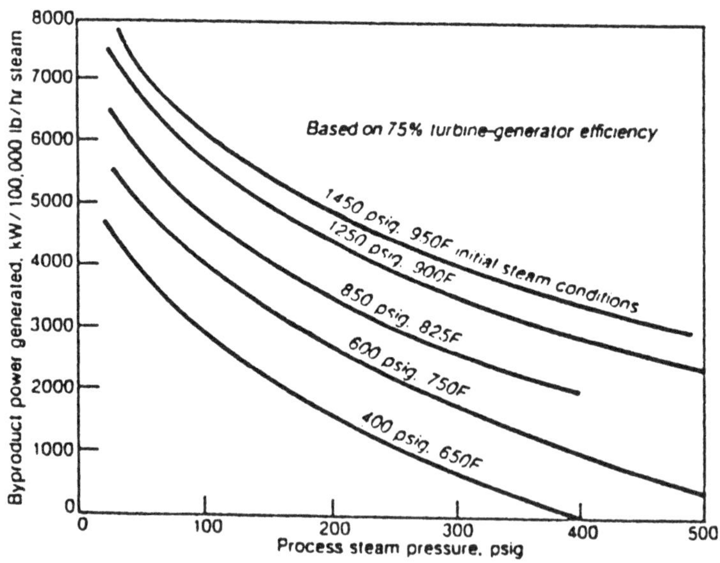

Figure 5-33. Power available from back-pressure steam turbines.

required steam flow rate through the turbine is usually quite high. The demand for the low pressure steam must be equal to, or greater than, the flow through the turbine.

The steam system must also operate at a high enough pressure for turbines to operate. Usually, the minimum pressure for a steam turbine to operate is at least 400 psig. The amount of power generated by a back-pressure turbine is a function of the inlet pressure. Figure 5-33 shows the power available from a back-pressure turbine at various pressures.

Deciding whether to cogenerate electric or mechanical power involves a number of economic trade-offs. In many cogeneration applications, the power produced cannot be used for mechanical work. As a result, back-pressure turbines are frequently used to generate electrical power. When electricity is generated, the electrical power generated is more than can be used on-site. Many cogeneration applications sell the excess electrical power to the local electrical utility. This is a complex and involved process. For additional discussion of cogeneration, refer to "Chapter 13: Cogeneration Systems."

5.6 ENERGY EFFICIENCY FOR NEW STEAM SYSTEMS

When designing and installing new steam distribution systems and condensate return systems, optimum energy efficiency should be taken into consideration.

Earlier in this chapter, energy conservation measures or improvements for existing systems were discussed. All of those measures should be considered for new steam systems. When installing new systems, however, the implementation of energy conservation measures is usually less costly than for existing systems. This section describes energy conservation measures for new systems, which may not be feasible for existing systems.

5.6.1 Steam Supply Piping

When designing and installing steam distribution systems, there are a number of guidelines that should be followed. This section provides general guidelines, which affect the energy efficiency of steam systems. More detailed design and installation guides are provided in ASHRAE (1996, 1997).

When designing and installing steam supply systems for energy efficiency, the following guidelines should be considered.

- Steam piping should be sized so that the pressure drops in all branches is nearly uniform.
- Pitch piping uniformly down in the direction of flow whenever possible.
- Insulate piping to avoid unnecessary heat loss.

- Condensate should be removed from steam systems as quickly as possible.
- Where horizontal pipe must be reduced in size, use eccentric reducers.
- Limit steam velocity in piping to 8,000 ft/min.
- Connect branch take-off lines to the top of steam mains.
- The total system pressure drop should not exceed half the initial pressure.
- Provide steam traps to remove condensate at all low points in the steam piping.

5.6.2 Condensate Return Systems

The design and installation of condensate return systems have a significant impact upon the performance of steam systems. Improperly installed condensate return systems can cause premature steam trap failure. This section provides general guidelines, which affect the energy efficiency of condensate return systems. More detailed design and installation guidelines are provided in ASHRAE (1996, 1997).

When designing and installing condensate return systems, the following guidelines should be considered.

- Condensate return piping must not be undersized as this will increase trap failure. Oversizing condensate piping rarely creates a problem.
- Pitch line downward in the direction of condensate flow.
- Insulate return piping and equipment well.
- Recover low pressure steam from condensate with flash tanks in high pressure systems.
- If condensate is contaminated and cannot be returned to the boiler plant, consider the installation of a heat exchanger to recover heat from the condensate.

5.6.3 Steam Trap Selection

The improper selection of steam traps is a significant cause of poor performance of steam systems. The proper selection of steam traps involves two main issues: selecting the proper type of trap for the application and correctly sizing the trap. No single type of trap is suitable for all applications.

For applications where there will be large amounts of air and non-condensable gases, a thermodynamic trap is most suitable. The mechanical type of steam traps are best suited for applications that require rapid condensate removal. Thermodynamic traps cannot release the condensate until it has cooled a set number of degrees below the steam temperature. This may result in a buildup of condensate in the steam-consuming equipment. In applications where there will be significant vibration to the system, thermodynamic traps are most suitable since they have only one moving part.

Major factors to consider when selecting and sizing traps include:

- The maximum condensate load.
- The operating pressure differential.
- Trap safety factor.
- Type of service, continuous or intermittent.
- Temperature and pressure of the steam and condensate.
- Reliability of the steam trap.
- Failure mode most likely to occur.
- Amount of air and non-condensable gases trap will handle.
- Corrosiveness of the condensate.
- If return is a gravity return, will the condensate have to be lifted?

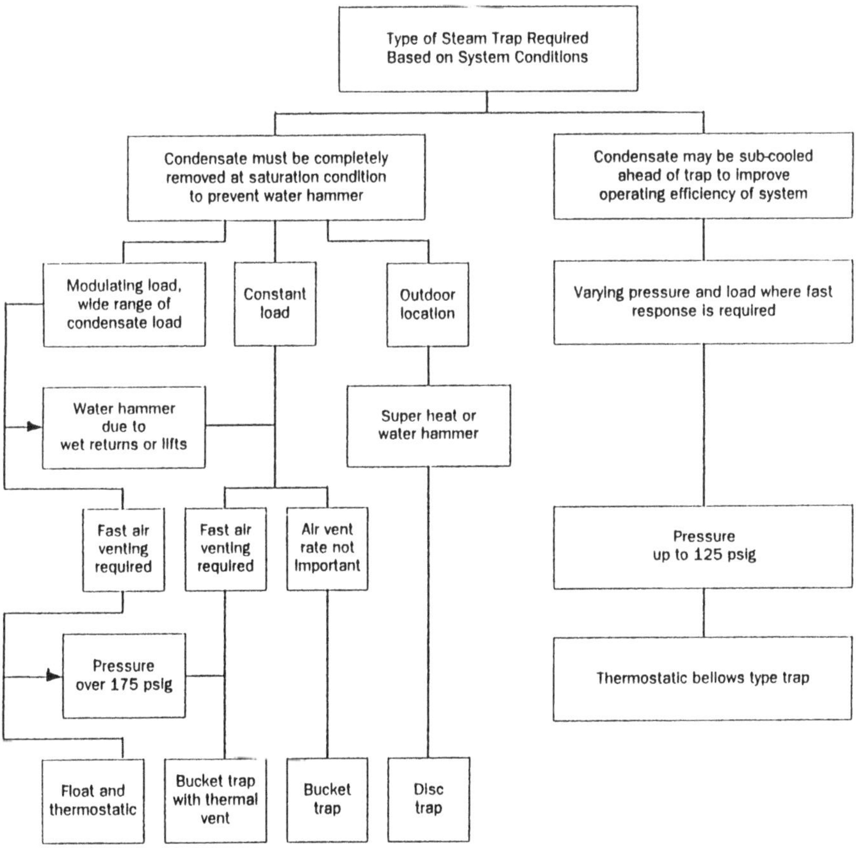

Figure 5-34. Steam trap selection tree.

- The extent to which condensate can build up (waterlogging) before being discharged.
- Expected service life of trap.
- Resistance to freeze damage if installed outdoors.
- Resistance to shock and waterhammer.

Figure 5-34 is a decision tree, which will aid in steam trap selection.

Steam traps should always be selected based upon the recommendations of the steam trap manufacturer.

Chapter 6

Hydronic and Pumping Systems

Pumps and hydronic systems are used in almost all central plants. Most facilities have at least one pumping system. A typical facility has many hydronic and pumping systems. With large numbers of pumps and systems in use, there is a significant opportunity for energy savings from these systems.

Frequently, pumps are selected and installed without much regard for energy efficiency. Many hydronic systems still use outdated methods of control, which are not very energy efficient, or have no controls at all.

This chapter primarily deals with systems that circulate water (hydronic) and with centrifugal pumps. These systems and pumps are the most commonly used pumping systems. Although most of the information in this chapter is for hydronic systems with centrifugal pumps, most of the energy saving opportunities are applicable to systems pumping other non-compressible fluids and to other types of pumps.

6.1 OVERVIEW OF HYDRONIC AND PUMPING SYSTEMS

When analyzing pumping systems from an energy management standpoint, it is first necessary to have an understanding of the processes involved with pumps and hydronic systems. This section provides an introduction to pumps and hydronic distribution systems.

6.1.1 Definitions and Terms Used in Pumping Systems

When working with pumping systems, there are a number of terms unique to these systems. See the Glossary for useful definitions of terms that are commonly used when discussing pumping systems.

6.1.2 Hydronic and Pumping System Types

Most facilities have systems that employ pumps to move some type of liquid. Pumping systems can be used to move any type of liquid. However, the most commonly pumped fluid is water, or a brine, which is primarily water.

Pumping and piping systems that pump liquids that are primarily water are referred to a hydronic systems. Specifically, a hydronic system is a pumping and piping system that circulates a fluid, which is primarily water and is used for heating and/or cooling. The heating or cooling application can be comfort heating and cooling or process heating and/or cooling. This section deals with hydronic systems used for heating and/or cooling applications. However, the principles may be used for other pumping applications.

The most basic classification for hydronic pumping systems is by open and closed systems. An open system is one that pumps the liquid to a point where it is discharged and not directly returned to the pump. An example of an open system would be a sump pump, which pumps water from a pit where it is openly discharged into a drain. In open systems, the pumped liquid is not directly recirculated to the pump. Closed systems have a complete piping circuit and the circulated liquid is returned to the pump. An example of a closed system would be a chilled water system where the water is circulated through a chiller to a cooling load, through a heat exchange device, and then recirculated back to the chiller. The pumped liquid travels in a complete circuit.

Closed circuit hydronic systems may also be classified according to their piping arrangement. The two most common arrangements are the reverse return and direct return. Direct return piping systems are the most common.

In a direct return piping system, the liquid is pumped to multiple loads and returned to the pump by the most direct route. A direct return piping circuit is shown in Figure 6-1. In direct return piping circuits, the flow direction of the liquid from the load is often opposite in direction to the supply flow. In direct return systems, the length of the supply piping and return piping through the subcircuits is not equal. This may result in unbalanced flow rates through the loads unless an artificial flow resistance is provided in the form of a flow balancing valve.

In reverse return piping circuits, the fluid does not flow directly back to the pump after leaving a load. The length of return pipe for each load is equal and the distance traveled by the fluid through the circuit is equal regardless of which load the fluid passes through. A reverse return piping circuit is shown in Figure 6-2. Reverse return piping systems are self-balancing and, if properly installed, require no balance valves to equalize flow.

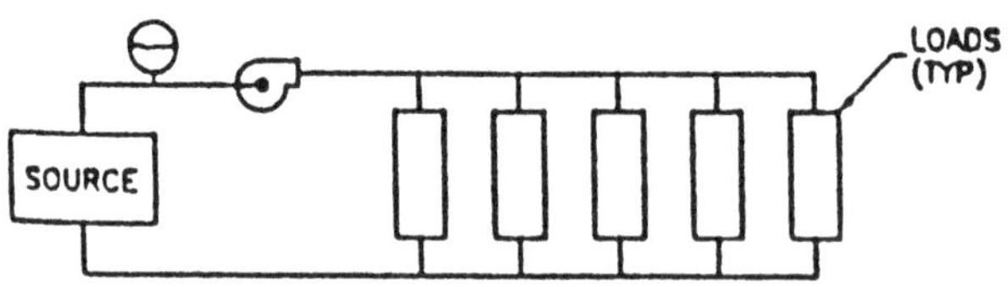

Figure 6-1. Direct return piping schematic.

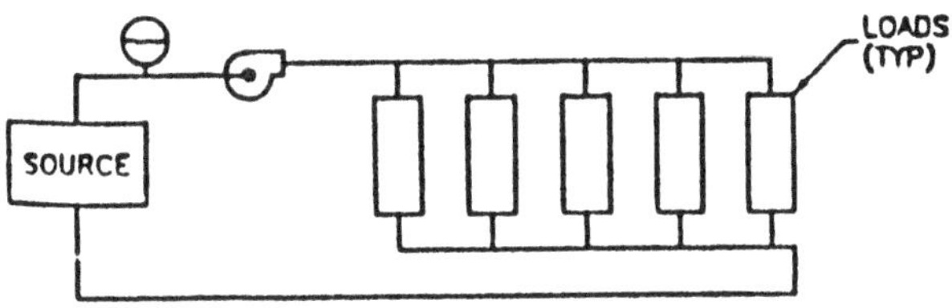

Figure 6-2. Reverse return piping schematic.

Hydronic systems are also frequently classified by temperature. The two broad classifications are chilled water and heating water. Chilled water systems use a chiller to provide refrigeration, which produces chilled water. See Chapter 7 for more information on chillers. The chilled water is then circulated by a pump and piping system to heat exchange devices or cooling loads. The temperature range of chilled water systems is primarily limited to a low of about 40°F and a high of about 60°F. Chilled water systems do not normally operate below 40°F due to the possibility of freezing the water. Temperatures higher than 60°F usually are too high for most cooling applications. Temperatures below 32°F can be produced if a brine solution is used in lieu of pure water. Brine solutions contain chemicals, such as glycol, which allow temperatures below the normal freezing temperature of water.

Heating water systems use a boiler or heat exchanger to generate heating hot water. See Chapter 4 for more information on boilers, and Chapter 5 for information on steam to hot water heat exchangers. The heating water is then pumped to heat exchange devices or process hot water loads. Since water at atmospheric pressure will boil and turn to steam at 212°F, it is necessary to pressurize the system to a pressure greater than the saturation pressure for water temperatures over 200°F. Heating water systems are classified by their maximum temperature.

- Low temperature water (LWT) systems have a maximum temperature limitation of 250°F and a maximum pressure of 160 psig.
- Medium temperature water (MWT) systems operate between 250 and 350°F, with pressures not exceeding 160 psig.
- High temperature water (HWT) systems operate at temperatures greater than 350°F and pressures up to 300 psig. The maximum supply water temperature is usually about 400°F.

The water flow rate for a heating load or a cooling load may be determined for any given water temperature change with Equation (6.1).

$$Q_W = \frac{q}{500\Delta T} \qquad \text{Eq. (6.1)}$$

where Q_W = water flow rate, gal/min
q = heating or cooling load, Btu/hr

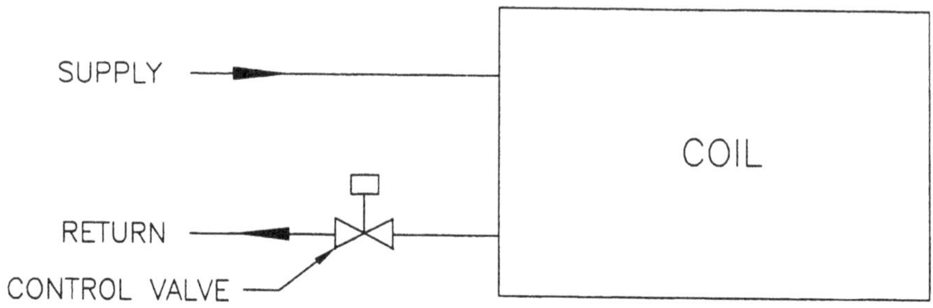

Figure 6-3. Two-way control valve.

500 = a constant for water
ΔT = water temperature change, °F

Example 6.1. A process heating load requires 100,000 Btu/hr. of heat. Heating water is supplied to a heat exchanger at 250°F and comes out at 230°F. What is the required flow rate for the heating water?

Using Equation (6.1), the water flow rate is

$$Q_W = \frac{100,000}{500\,(250 - 230)} = 10\,\text{gal/min}$$

6.1.3 Temperature Control for Hydronic Systems

When hydronic systems are used for either comfort or process heating/cooling applications, it is usually necessary to provide some type of temperature control. The simplest method for controlling the temperature is to just manually open or close a valve to admit or shut off water flow to

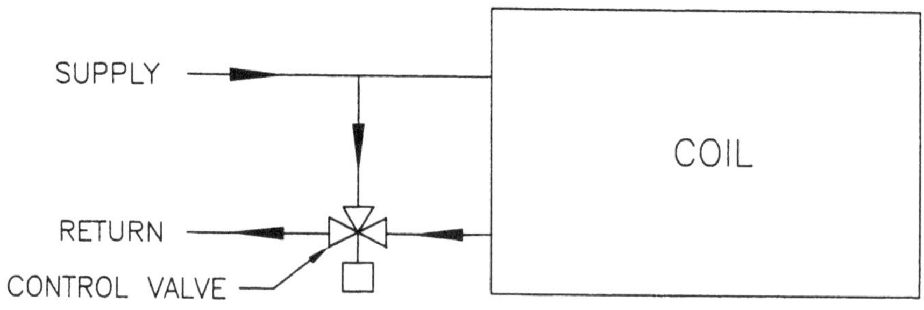

Figure 6-4. Three-way mixing valve.

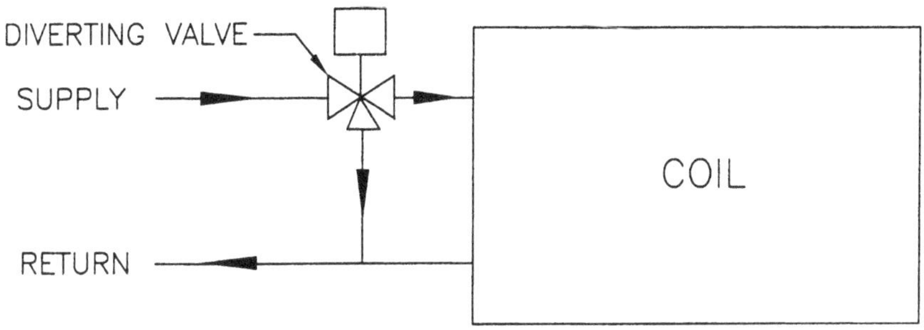

Figure 6-5. Three-way diverting valve.

the load. Open/closed control may also be provided by an automatically actuated two-position valve. When the temperature must be controlled more precisely, automatically actuated modulating control valves are used. Modulating control valves may be either two-way or three-way control valves.

Two-way control valves are straight through type valves with an inlet port and an outlet port. A schematic diagram of a two-way valve is presented in Figure 6-3. Two-way valves modulate open or closed to restrict the flow of heating water or chilled water based upon the demand by the heating/cooling load. The water flow through the load can be modulated or varied from full flow to no flow.

Three-way control valves allow a constant flow rate through the piping distribution system while controlling the temperature of the heating/cooling load. Three-way valves have three ports. They modulate the temperature of the load by either bypassing the water flow around the load or through the load. Three-way valves bypass flow by either mixing two entering streams into one leaving stream (mixing valve), as shown in Figure 6-4, or by splitting a single entering stream into two leaving streams (diverting valve), as shown in Figure 6-5.

6.1.4 Pumping Arrangements

Many hydronic systems have a single circulating pump to provide flow and pressure for the entire system. Frequently, though, hydronic systems have

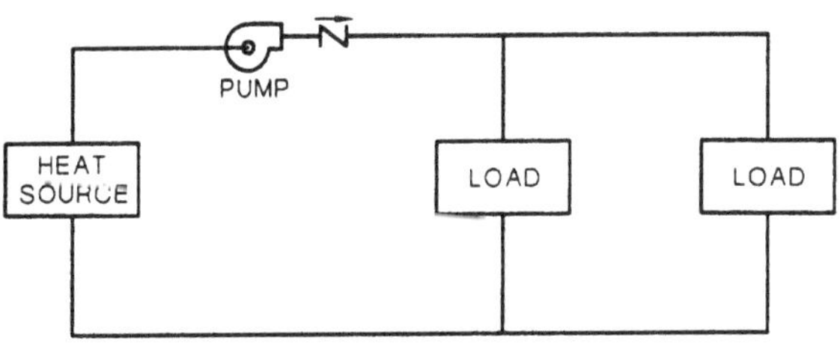

Figure 6-6. Single pumping arrangement.

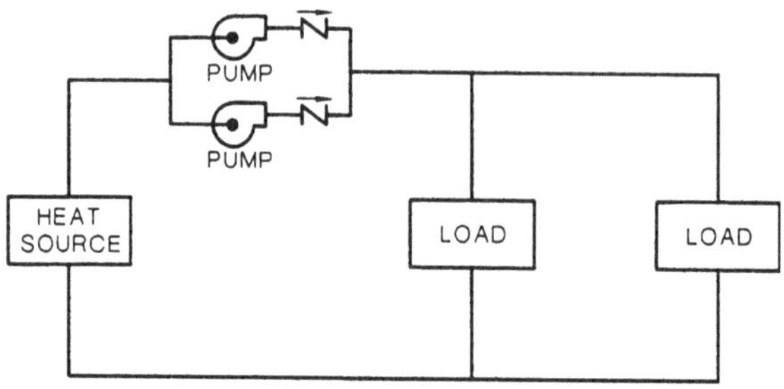

Figure 6-7. Parallel pumping arrangement.

multiple pumps. When multiple pumps are used, there are a number of arrangements available. For multiple pump systems, the most common arrangements are parallel, series, primary/secondary and primary/secondary/tertiary (PST).

Single pump systems are fairly simple. One pump is used to circulate the water through the loads and through a heating or cooling source. A schematic of a single pumping system is shown in Figure 6-6. The flow in the distribution circuit can be completely in series or in parallel.

When two, or more, main circulating pumps are used in a system, they are frequently piped in parallel. A parallel pumping arrangement is shown in Figure 6-7. Pumps may be piped in this arrangement with one pump operating and the other as a standby in case the first pump fails. Both pumps in a parallel arrangement may operate simultaneously. When both pumps operate in parallel, they must produce the same discharge head, although they may have different flow rates. Two pumps operating in parallel will produce a greater flow rate at the pump discharge head. The performance of pumps operating in parallel is discussed later in this chapter.

Multiple pumps may also be piped in series. A series piping arrangement is shown in Figure 6-8. For pumps to operate properly piped in series, they should have the same flow rate, although they may have different discharge head pressures. Pumps are typically piped in series to produce higher head

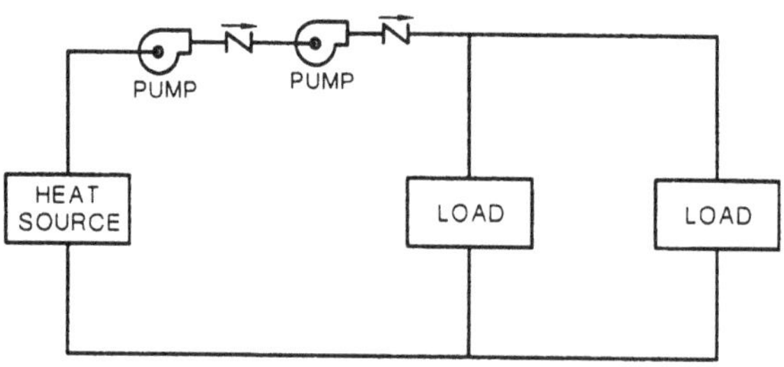

Figure 6-8. Series pumping arrangement.

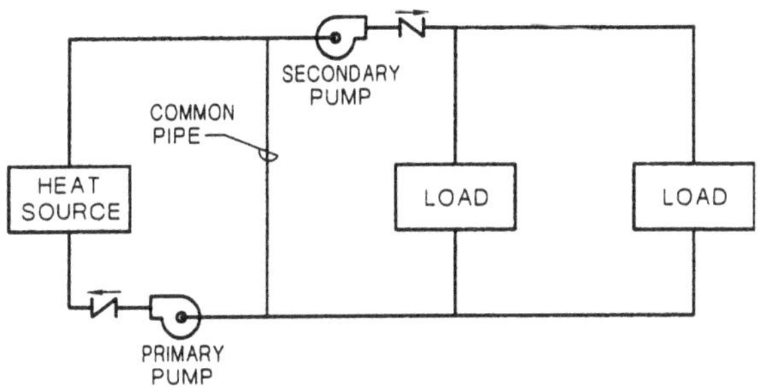

Figure 6-9. Primary/secondary pumping arrangement.

pressures for a given flow rate. Each pump boosts the system pressure by an amount equal to its discharge head. The performance of pumps operating in series is discussed later in this chapter.

Pumps may also be piped in a primary/secondary arrangement, as shown in Figure 6-9. Primary/secondary pumping arrangements are frequently used when it is advisable to decouple a primary loop from a secondary loop. Note that a common pipe is provided between the primary and secondary loops. The flow in the common pipe may be in either direction depending upon the flow rate of each pump. Primary/secondary pumping applications are discussed later in this chapter.

When a third pump is added to a primary/secondary pumping arrangement, it is called a primary/secondary/tertiary (PST) pumping arrangement. The third, or tertiary, pump is usually a small coil, or load, circulating pump. A PST arrangement is shown in Figure 6-10. Tertiary pumps are usually used to ensure a minimum flow rate through a load

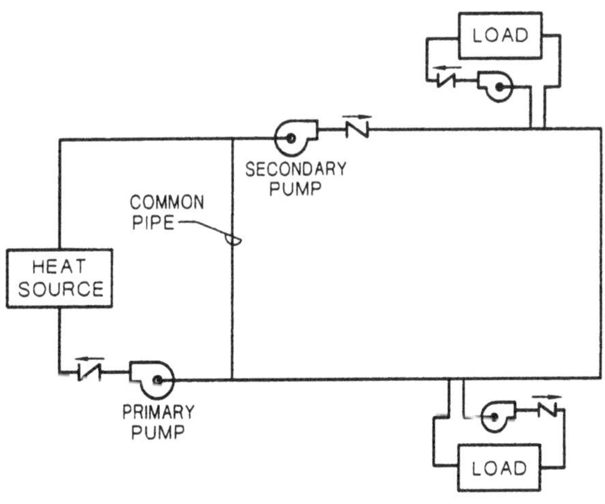

Figure 6-10. Primary/secondary/tertiary pumping arrangement.

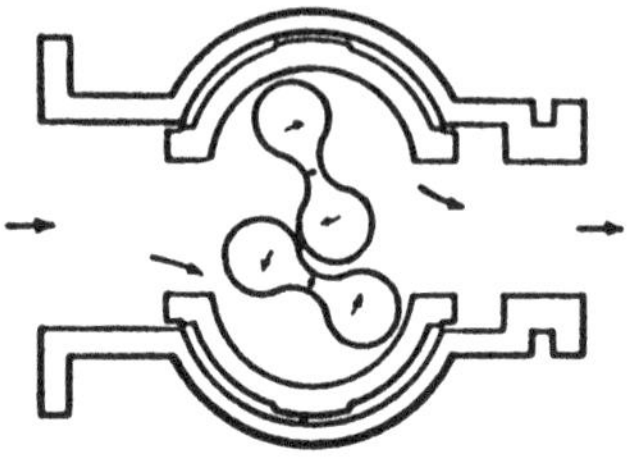

Figure 6-11. Lobe pump.

regardless of the actual demand for heating or chilled water. PST pumping systems are discussed later in this chapter.

6.2 HYDRONIC SYSTEM PUMPS AND ACCESSORIES

There are three basic classes or types of pumps in widespread use today. These pumps are rotary, reciprocating and centrifugal type pumps. These classifications are according to the mechanics by which the liquid is moved or pumped.

Rotary pumps are positive displacement type pumps used primarily to pump viscous liquids, such as light and heavy fuel oils, gasoline, tars, pitch, etc. They are used when a continuous flow of liquids at high pressures is required. Common rotary type pumps consist of gear pumps, lobular pumps, screw pumps and vane pumps. Figure 6-11 shows a lobe pump and Figure 6-12 shows a gear pump.

The displacement of a rotary pump varies directly as the rotational speed. Displacement can also be affected by slip, which is a loss in capacity due to leakage between the pump casing and rotating elements. For a constant liquid viscosity, the amount of slip varies as the discharge pressure increases.

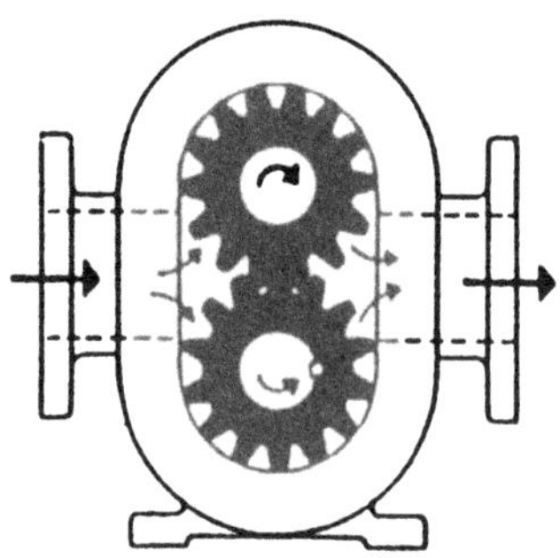

Figure 6-12. Gear pump.

Figure 6-13. Centrifugal pump with a volute.

The power input to a rotary pump increases with the viscosity of the pumped liquid. The efficiency of a rotary pump varies inversely with the viscosity of the pumped liquid.

Most rotary pumps are self-priming and can handle gas or air entrained in the pumped liquid.

Reciprocating pumps are also positive displacement pumps used for high head small capacity applications. Reciprocating pumps have a piston, which discharges a fixed quantity of liquid with each stroke. They are used for pumping clean, clear liquids. Typically, reciprocating pumps are used for services such as boiler feed, mud, and slush handling oil pumping and water pumping.

The flow of liquid from a reciprocating pump pulsates with the character of the pulsation's dependent upon the pump type and speed. The capacity of a reciprocating pump is directly related to the rotational speed of the pump and the volume displaced by the piston. Pump capacity is also affected by the viscosity and temperature of the pumped liquid.

Centrifugal pumps are, by far, the most common type of pumps in use. Centrifugal pumps employ a rotating impeller, which discharges the liquid into a progressively expanding spiral casing, or into a diffuser, to gradually reduce the liquid velocity and convert the velocity pressure into static

Figure 6-14. Centrifugal pump with a diffuser.

Figure 6-15. Vertical centrifugal pump (*Courtesy of Taco*).

pressure at the pump discharge. Unlike rotary and reciprocating pumps, centrifugal pumps are not positive displacement pumps; they are variable flow pumps. The flow rate of a centrifugal pump is a function of the pressure against which the pump discharges.

The casing of a centrifugal pump may be a volute type, as shown in Figure 6-13, or a diffuser type, as shown in Figure 6-14. The volute reduces the fluid velocity with a gradually expanding spiral casing while the diffuser casing changes the direction of the liquid flow.

Centrifugal pumps may have a vertical or horizontal pumping arrangement. Vertical pumps may simply be pumps mounted in a vertical position, or they may be pumps that pump the liquid in a vertical direction. A typical vertical centrifugal pump is shown in Figure 6-15. Horizontal pumps have the pump itself and the motor mounted in a horizontal position. They are frequently mounted on a common base. A typical horizontal centrifugal pump is shown in Figure 6-16.

Centrifugal pumps may be single or multi-stage pumps. Single stage pumps boost the pressure once before it is discharged. Multi-stage

Figure 6-16. Horizontal front end suction centrifugal pump (*Courtesy of Taco*).

Figure 6-17. Horizontal split case double suction centrifugal pump (*Courtesy of Taco*).

centrifugal pumps boost the fluid pressure more than once before it is discharged from the pump.

Centrifugal pumps may also be single of the suction (inlet) or double suction (inlet) type. Single suction pumps draw the liquid into the pump and use only one side of the impeller to pump the liquid. Double suction pumps draw liquid into both sides of the impeller and the impeller pumps from both sides. A typical double suction pump is shown in Figure 6-17. Double suction pumps are used where high flow rates are required from a single pump.

6.2.1 Pump Accessories

Pump installations usually include a number of accessories, along with the pump itself. Other than the accessories associated with providing electrical power to the pump motor, most accessories provide some sort of flow control for the pump and piping system. Flow control accessories commonly used with pumps include isolation valves, flow balance valves, check valves, etc. Other accessories include strainers, air eliminators and expansion tanks.

Whenever a pump is installed in a piping distribution system, it is advisable to install isolation valves on the pump inlet and discharge. The valves are necessary to isolate the pump from the piping distribution system for servicing. Installing the isolation valves allows the pump to be serviced without draining the pumped liquid from the system. Isolation valves are usually gate valves or butterfly valves.

Check valves are also commonly installed on the discharge side of the pump. These valves prevent the flow of the pumped liquid in the opposite direction through the pump when it is not operating. Check valves are particularly important in multiple pump installations, especially when the pumps are piped in parallel, as shown in Figure 6-7.

The flow rate from a centrifugal pump is a function of several variables, including the pressure against which the pump is pumping and also the size of the impeller. The impeller should be sized such that it matches the pressure of the distribution system at the required flow rate. Although it is frequently possible to trim an impeller to match the system requirements, it is difficult to exactly match the pressure and flow requirements of the system exactly by trimming the impeller alone. Since most pumps are selected such that they are oversized to ensure adequate capacity, it is usually necessary to install an artificial flow resistance in the system to offset the excess pump flow and head capacity. Flow control valves are frequently installed in the discharge piping of a pump to provide an artificial resistance and to balance the pumping system for the desired flow and pressure. Flow control valves are typically globe type valves, although butterfly valves may also be used. Gate valves are not suitable for this application.

Since most pump applications require an isolation valve, a check valve and a flow balancing valve in the discharge piping of the pump, a single multi-purpose valve may be installed instead of the three valves. A multi-purpose valve is a single valve, which serves all three purposes. A multi-purpose valve is shown in Figure 6-18. The valve can provide shut-off for serving or can be partially closed for balancing. Pressure gauge taps are provided for measuring the pressure loss across the valve, which can be used to indicate the flow rate. The valve also acts as a check valve to prevent backflow through the pump.

Although a pump may be pumping a clean liquid in a closed system, it is advisable that a strainer be installed in the suction piping to a pump. A strainer has a fine mesh screen, which removes solid particles from the liquid as it flows through the strainer. Strainers are particularly important for open systems where foreign materials may easily become entrained in the pumped liquid. Suction diffusers are often installed on the inlet of a

Figure 6-18. Multi-purpose valve (*Courtesy of Taco*).

front end suction pump. The suction diffuser contains a fine mesh screen to strain the flowing liquid, as well as straightening vanes to provide uniform flow into the pump suction.

Air occasionally becomes entrained in the pumped liquid as the result of leaks and from air entrained in make-up water. For hydronic systems to heat or cool effectively, it is important that the circulated water does not include entrained air. Air removal devices are frequently installed in the piping upstream of a pump to remove entrained air as it flows through the system.

Expansion/contraction tanks are also frequently required in hydronic heating or cooling systems. Although water is a non-compressible fluid, as it is heated or cooled, it does expand or contract to some extent. In closed hydronic systems, it is important that provisions be made for any expansion or contraction. Expansion/contraction tanks are used in closed systems to compensate for any expansion or contraction of the pumped liquid and to pressurize the system. Most expansion/contraction tanks have a bladder between the liquid and air, which is used to absorb expansion/contraction and to pressurize the system.

6.3 CENTRIFUGAL PUMP PERFORMANCE

6.3.1 Pump Head

The head developed by a centrifugal pump is a function of the impeller diameter and speed of rotation. Pump head may be expressed in pounds per square inch (psi), feet of water or any units of pressure. The total dynamic head (TDH) on a pump is the pressure against which a pump must discharge the pumped liquid at a given flow rate. The total dynamic head on a pump is a combination of two types of head in the distribution system: static head and friction or dynamic head.

Static head is applicable only to open distribution systems that do not recirculate the pumped liquid. Static head is also referred to as lift. Static head, or lift, is the pressure required to raise the elevation of the pumped liquid from one level to another. It is not a function of the flow rate. The static head consists of static suction lift and static discharge head. The static suction lift is the vertical distance, in feet, from the liquid supply level to the pump centerline. The static discharge head is the distance, in feet, to the point of free delivery of the pumped liquid.

Friction head or dynamic head, also measured in feet of the pumped liquid, is the equivalent head needed to overcome the flow resistance for the piping distribution system. Friction head exists on both the suction and discharge sides of the pump. The friction head varies with the flow rate of the pumped liquid, the pipe size, the type of pipe, the liquid being pumped and the flow resistance through any terminal equipment.

When centrifugal pumps have a suction lift, particular attention must be paid to the pressure and temperature of the pumped liquid as it enters the pump. As the liquid enters the eye of the pump impeller, the absolute pressure of the liquid is reduced due to the velocity of the liquid. As the pressure of the liquid approaches its vapor pressure, vapor bubbles begin to form in the liquid. This is then followed by the recollapse of the bubbles into liquid when the pressure increases to a point just above the liquid vapor pressure. This phenomenon is known as cavitation and should be avoided. As the bubbles collapse, a shock wave is set up by the implosion of the bubbles. The shock wave is propagated through the liquid until it is arrested by the surface of the impeller. Damage may result from prolonged cavitation. The minimum required pressure at the pump inlet, which will be great enough to prevent cavitation, is referred to as the Required Net Positive Suction Head (NPSHR).

Pump head, total dynamic head, static head and friction head in piping distribution systems are also discussed in the next section.

6.3.2 System Curves

The total dynamic head on a pump consists of both static head and dynamic (friction) head of the piping system. Although the static head is a constant for all flow rates, the dynamic head is a function of the flow fluid rate. As the flow rate increases, the dynamic head increases with the square of the fluid velocity.

The pressure drop caused by the friction of a fluid flowing through a pipe may be calculated by the Darcy-Weishbach Equation.

$$\Delta h = \frac{fLV^2}{2Dg} \qquad \text{Eq.(6.2)}$$

where Δh = friction head loss, ft.
f = friction factor, dimensionless
L = pipe length, ft.
D = inside pipe diameter, ft.
V = average fluid velocity, ft./sec
g = gravitational acceleration, ft./sec^2

The average fluid velocity in a pipe may be calculated by the Continuity Equation.

$$Q = A/V;\ V = Q/A \qquad \text{Eq. (6.3)}$$

where Q = volumetric fluid flow rate, ft.^{3}/sec
A = inside pipe area, ft.^{2}
V = average fluid velocity, ft./sec

Although the friction factor for various types of pipe are tabulated in many publications, such as Crane Company (1974), tabulated pressure losses for the flow of common liquids through various sizes of pipe are frequently used to calculate friction losses.

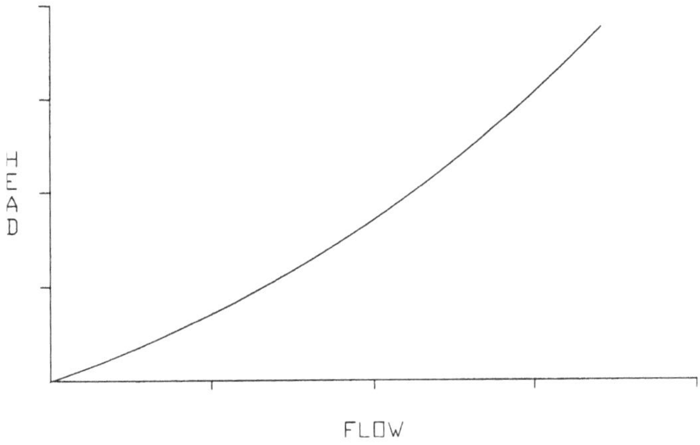

Figure 6-19. Typical system dynamic head curve.

If the dynamic head for a piping system is known for a given flow rate, the dynamic head for any other flow rate may easily be calculated by the following equation.

$$h_2 = h_1(Q_2/Q_1) \qquad \text{Eq. (6.4)}$$

where h_1 = dynamic head at a known flow rate, ft.
h_2 = dynamic head at a second flow rate, ft.
Q_1 = initial flow rate, ft.^{3}/sec or ft.^{3}/min
Q_2 = second flow rate, ft.^{3}/sec or ft.^{3}/min

Once the dynamic head is known at a single flow rate for a given piping system, Equation (6.4) may be used to determine the dynamic head for all flow rates. A plot of the dynamic head for a piping system versus flow rate is referred to as a system curve. A system curve is shown in Figure 6-19.

The system curve shown in Figure 6-19 is for a closed (recirculating) system. The curve passes through the point of zero flow and zero head. All of the head loss shown is dynamic head loss. Open systems, however, can include static head losses. In the case of a system with a static head loss, the total dynamic head curve does not pass through zero head on the vertical axis. The total dynamic head curve begins at the value of the static head, as shown in Figure 6-20.

6.3.3 Pump Performance Curves

Just as the relationship between flow and head for a piping system may be represented as a curve, the flow and head performance of a pump may be represented by a curve. A pump curve relates the total dynamic head (TDH) produced by a given pump to the flow rate. The curve is a graphical representation of the flow and head performance for a given pump and

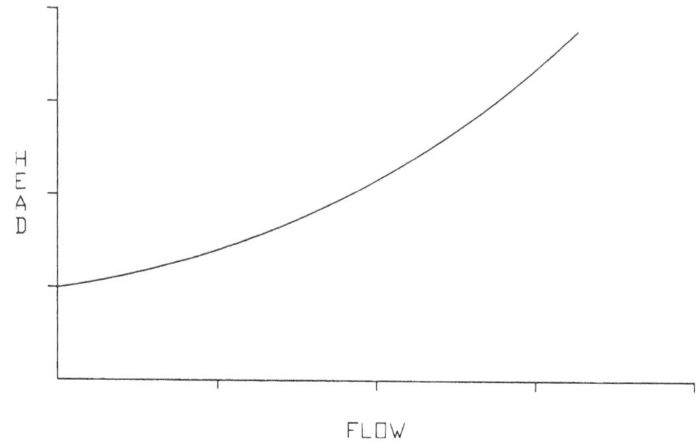

Figure 6-20. System curve with static head.

impeller. The performance will vary depending upon the impeller size, the rotational speed and the design of the pump. Pump curves are also referred to as head-capacity curves (H-Q). The total dynamic head shown on a pump curve represents the difference in pressure between the pump inlet (suction) and its outlet (discharge). The data from pump curves are determined by testing of the pump. Typically, pump performance curves are provided by the manufacturer of the pump. A typical pump curve is shown in Figure 6-21.

Pump characteristic curves may be described as being either flat or steep. Flat and steep performance curves are shown in Figure 6-22. Flat characteristic curves are typically installed in closed systems with modulating two-way control valves at the loads. Steep pump curves are

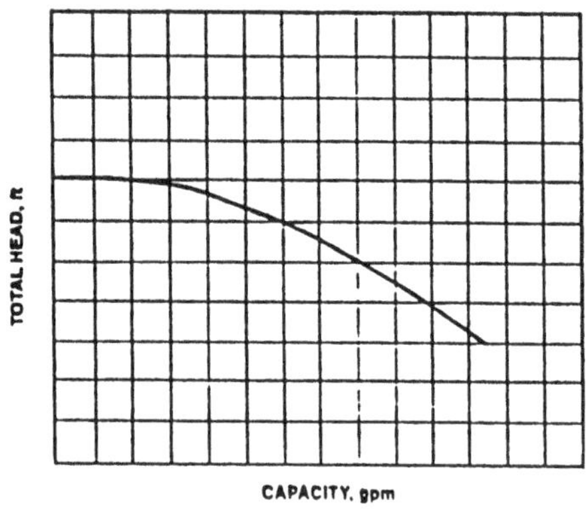

Figure 6-21. Typical pump curve.

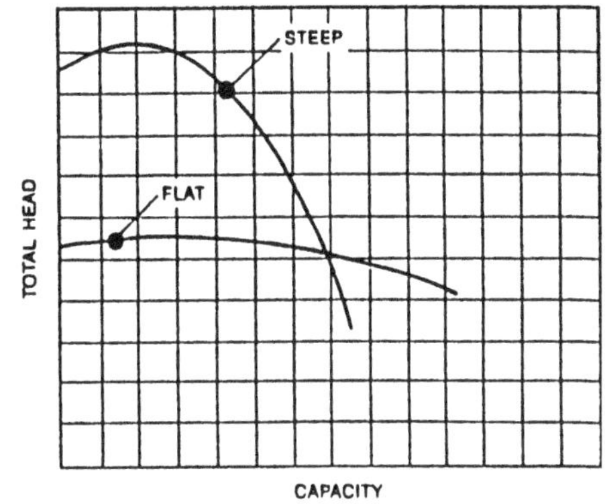

Figure 6-22. Flat versus steep performance curves.

usually installed in open pumping systems where high pressure and constant flow are desirable.

The actual energy or work input to the fluid is referred to as the water horsepower (Whp). Water horsepower is the theoretical power required to circulate the fluid through the piping system at a given flow rate and total dynamic head. The water horsepower for a pump circulating water may be calculated using Equation (6.5).

$$Whp = \frac{Q\Delta h}{3960} \qquad \text{Eq.(6.5)}$$

where Q = fluid flow rate, gal/min
Δh = total head, ft.
3,960 = units conversion, ft.-gal/min per hp

Fluid density does not affect the pump performance curve; however, it does affect the power requirements.

Brake horsepower (bhp) is the power required to operate the pump to produce water horsepower at a given flow and pressure. It takes into account the efficiency of the pump itself. Equation (6.6) may be used to calculate the brake horsepower for a pump; however, pump manufacturers usually provide this data as part of their pump curves.

$$Bhp = \frac{Whp}{\eta_p} \qquad \text{Eq. (6.6)}$$

where η_p - pump efficiency, %

In a manner similar to the head and flow relationship for piping distribution systems, the performance of a pump at one operating point may be used to predict the performance for other operating points. The performance of a pump at other operating points may be predicted using the affinity laws for centrifugal pumps. The affinity laws are a set of relationships that relate the dynamic characteristics of a pump at various hydrodynamic conditions. The laws can be described as similarity processes that follow these rules:

1. Flow (capacity) varies with the rotational speed of the impeller.
2. Head varies as the square of the rotational speed of the impeller.
3. Brake horsepower varies as the cube of the rotational speed of the impeller.

The affinity laws may be expressed mathematically by the following equations:

$$Q_2 = (D_2/D_1)\,Q_1 \qquad \text{Eq. (6.7)}$$

$$h_2 = (D_2/D_1)^2\,h_1 \qquad \text{Eq. (6.8)}$$

$$Bhp_2 = (D_2/D_1)^3\,Bhp_1 \qquad \text{Eq. (6.9)}$$

$$Q_2 = (N_2/N_1)\,Q_1 \qquad \text{Eq. (6.10)}$$

$$h_2 = (N_2/N_1)^2\,h_1 \qquad \text{Eq. (6.11)}$$

$$Bhp_2 = (N_2/N_1)^3\,Bhp_1 \qquad \text{Eq. (6.12)}$$

where Q = flow rate, gal/min
h = head, feet
D = impeller diameter, inches or feet
N = rotational speed, RPM
Bhp = brake horsepower, hp

Although the affinity laws may be used to calculate the performance of a pump at various operating conditions or with different impeller sizes, all of this data is typically included in the pump curves provided by a pump manufacturer. A typical manufacturer's pump performance curve is shown in Figure 6-23.

6.3.4 System and Pump Curves

Since the flow and head loss for a system can be represented by a flow/head curve and the performance of a pump can also be represented by a flow/head curve, the two curves can be combined to determine the operating point of a given pump and distribution system. The pump curve

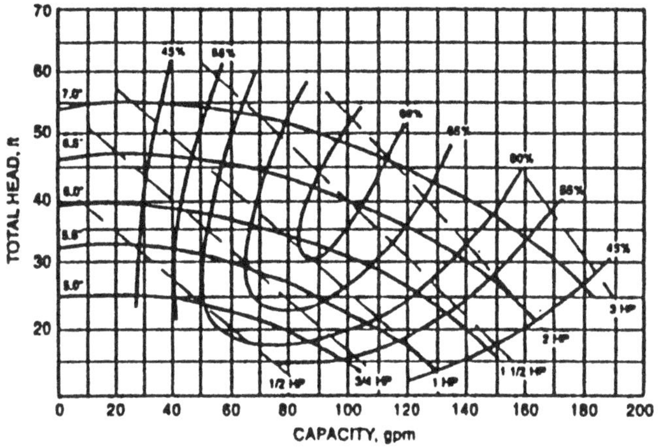

Figure 6-23. Typical manufacturer's pump curves.

represents energy put into the fluid by the pump. The system curve represents energy consumed by the distribution system to circulate the fluid. Since energy is not stored within the system, the energy (flow and pressure) put into the system must equal the energy lost by the system. This condition exists at the point of intersection of the pump curve and the system curve. A system/pump curve for a single pump system is presented in Figure 6-24.

If the pump is properly matched to the system, the operating point will be at the desired flow rate. If the system pressure loss at the desired flow rate is not properly calculated when selecting the pump, the system operating point will shift. If the pressure drop of the system is greater than estimated, the system curve and operating point will shift up and to the left.

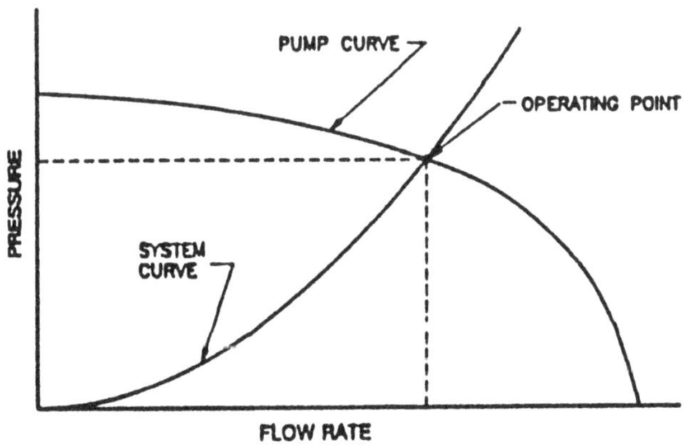

Figure 6-24. Pump curve and system curve.

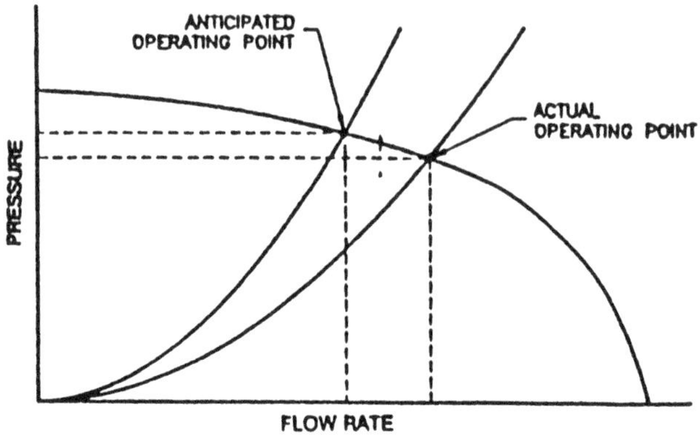

Figure 6-25. Shift of curve due to circuit unbalance.

As a result, the flow rate at the operating point will be less than the design flow rate. If the pressure loss for the piping system is overestimated, the system curve and operating point will shift down and to the right, as shown in Figure 6-25.

Typically, the piping distribution system pressure loss at the design flow rate is overestimated. This results in a higher fluid flow rate than design. To remedy the excess flow condition, balancing valves are typically placed in the pump discharge. The balancing valve adds an artificial pressure loss to the system that, when properly adjusted, produces the desired system flow and head.

Pumps with excess flow capacity may also be slightly modified to produce a lower flow rate. The flow rate for a centrifugal pump may be slightly reduced by trimming the impeller. Trimming the impeller slightly

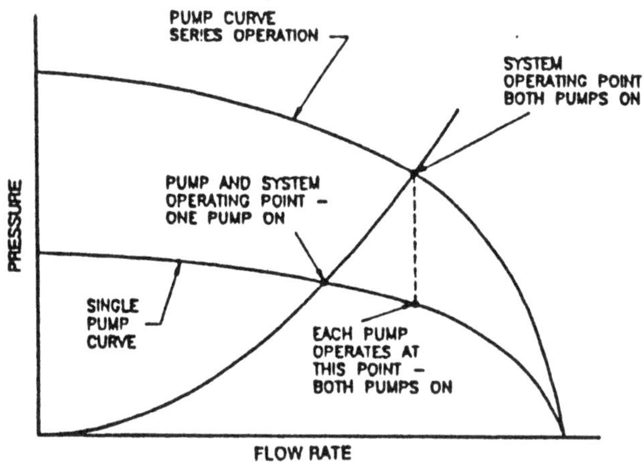

Figure 6-26. Curves for series pump operations.

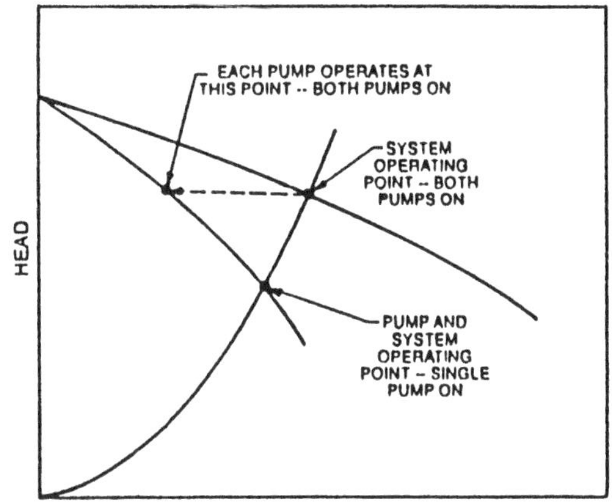

Figure 6-27. Curves for parallel pump operation.

reduces the impeller diameter and shifts the pump curve down and to the left.

All the system and pump curves discussed above are for systems with a single pump in operation. Frequently, however, pumping systems have multiple pumps. Various pump arrangements were discussed in Section 6.1.4. Depending upon the pumping arrangement and capacities of the pumps, multiple pumping arrangements produce modified pump and system curves for a given system.

When pumps are applied in series, each pump operates at the same flow rate and provides its share of the total flow. The flow rate through each pump is the same as the total flow through the system, however, each pump boosts the pressure by the amount of its head capacity for the flow rate. Figure 6-26 shows the operating curves for series pump operations.

Pumps are commonly piped in series for high head applications. For applications that require high flow rates, pumps are frequently piped in a parallel arrangement. When pumps are applied in parallel, each pump operates at the same pressure and provides its share of the system flow at that pressure. Figure 6-27 shows the operating curves for parallel pump operation. Note that single pump operation does not result in 50% of the total flow for both pumps. The system curve crosses the single pump curve considerably to the right of its operating point when both pumps are running.

6.4 MAINTAINING AND OPERATING HYDRONIC AND PUMPING SYSTEM FOR OPTIMUM ENERGY EFFICIENCY

Hydronic and pumping systems typically do not require extensive routine maintenance; however, significant energy savings can be achieved by proper routine maintenance and operation. Proper routine maintenance and operation should be the first step in any energy management program for hydronic and pumping systems. The following are operations and maintenance procedures that will result in optimum energy efficiency for most hydronic and pumping systems.

Run One Pump Instead of Two Whenever Possible: In the previous section, it was indicated that pumps are frequently installed in parallel for greater flow capacity. It was also indicated that whenever two pumps of equal capacity are installed in parallel and only one pump is operated, it produces more than 50% of the total flow for both pumps operating.

In variable flow systems, both pumps are typically kept on even when the demand drops to a point where one pump can carry the entire load. Whenever the required flow drops below 75% of the total flow for both pumps, one pump should be shut off and a single pump should provide the flow for the system.

If the demand is reduced to 50% of the total flow and two pumps are left operating in parallel, they will develop about 115% of the normal operating head and each pump will consume about 72% of its rated power. The combined power consumption of the two pumps would be about 145% of the rated power consumption for each pump. If one pump were stopped, the operating pump would consume 100% of its rated capacity, which would result in a 45% reduction in the power for two pumps.

Restore Internal Pump Clearances: As a pump is operated over a period of time, internal parts wear and clearances between internal parts increase. The rate at which parts wear and internal clearances increase depends upon the differential pressure across the clearances and the fluid pumped. The greater the pressure differential across the clearances, the greater the rate at which they wear. Corrosive or abrasive liquids also increase the rate of wear.

As running clearances increase with wear, a greater portion of the gross capacity of the pump is short-circuited through the clearances from the higher pressure discharge to the lower pressure suction side of the pump. The short-circuited fluid must be re-pumped, thereby wasting pumping energy. The effective, or net, capacity is reduced by an amount equal to the increase in internal leakage.

Pumps should be overhauled and internal clearances restored at least as often as recommended by the pump manufacturer. If a noticeable decrease in pump performance is observed, the pump should be overhauled as soon as possible.

Re-balance Hydronic Systems: When hydronic systems are installed, they are typically balanced to ensure that the terminal units receive the required flow rate. After a significant period of operation, parts wear and components, such as control valves, become out of adjustment.

After several years of operation, a hydronic system is probably no longer balanced and is no longer providing the required flow rates for optimum energy performance of the system. Hydronic systems should be periodically re-balanced so that the flow rates are within ±10% of the original design flow rates.

6.5 ENERGY CONSERVATION MEASURES FOR HYDRONIC AND PUMPING SYSTEMS

The energy conservation measures for hydronic and pumping systems discussed in the previous section were operation and maintenance procedures that would improve the energy efficiency of existing systems. Most of those measures were little, or no, cost measures, which could be performed on an ongoing basis to improve efficiency.

In addition to the low cost O&M procedures, there are also modifications that could be made to hydronic and pumping systems to improve energy efficiency. These modifications, however, require a more significant cost investment and are permanent improvements. The modifications listed below are improvements that should be considered under any energy management program for existing, or proposed, hydronic and pumping systems.

6.5.1 Trim Pump Impeller

In most pumping applications, pumps are almost always oversized. Oversizing occurs as the result of overly cautious safety factors, overestimation of pressure losses for various system components or because no analysis of the distribution system is performed when selecting a pump.

Whenever a pump is oversized, it produces flow at a rate greater than the design flow rate. As a result, it is necessary to reduce the flow by throttling it with a balance valve placed on the pump discharge. The balance valve reduces the flow by dissipating some of the energy put into the pumped fluid by the pump.

A more efficient way to balance oversized pumps is to trim the pump impeller. Impeller trimming is also referred to as proportional balancing. Trimming the impeller has the effect of reducing the diameter of the impeller so it pumps the fluid at a lower rate. Impeller trimming can eliminate unnecessary power consumption. Impellers should be trimmed such that the pump operates with the balance valve nearly fully open.

Example 6.1. A pump with a 9-inch-diameter impeller produces a flow of 500 gal/min through a system when it is not throttled. Since the required flow rate is only 400 gal/min, a balancing valve on the pump discharge is partially closed so only 400 gal/min results . If the pump impeller were trimmed to produce 400 gal/min with the balance valve fully open, what diameter impeller would be required and what would the savings in brake horsepower bc?

Referring to the pump affintiy laws, Equation (6.7) may be used to calculate the new impeller diameter.

$$Q_2 = (D_2/D_1)\, Q_1$$

$$D = D_1 (Q_2/Q_1) = 9\left(\frac{400}{500}\right) = 7.2 \text{ inches diameter}$$

Using Equation (6.9)

$$Bhp_2 = (D_2/D_1)^3\, Bhp_1$$

$$Bhp_2/Bhp_1 = (D_2/D_1)^3 = \left(\frac{7.2}{9}\right)^3 = 0.512 = 51.2\%$$

$$Bhp\ savings = 100\% - 51.2\% = 48.8\%$$

6.5.2 Primary/Secondary Pumping Arrangement

Earlier in this chapter, primary/secondary pumping arrangements were discussed. The function of the primary loop is to maintain a constant flow of supply water through the heat source. The heat source and primary pump may be cycled on and off based upon the (temperature of the water returning to the heat source.) The primary pumps need to have only enough total dynamic head to offset the dynamic losses in the primary loop at the required flow rate. These losses are typically relatively low and the flow rate is relatively high.

Flow in the secondary loop may be varied in response to the load or demand. The secondary loop may be a variable volume system. The total dynamic head in the secondary system is typically relatively high due to the piping distribution system losses, losses through loads and losses through control valves. The flow rate may be as high as the primary loop; however, the required flow is frequently less than the primary loop flow. A secondary loop can take advantage of the diversity, which usually occurs with multiple loads. As a result, the flow required to satisfy the loads in the secondary loop is usually less than the peak design flow rate. Secondary pumps are typically higher horsepower.

If an existing pumping system has a single pump with a heat source and multiple loads that are variable, consideration should be given to converting the system to a primary/secondary pumping system.

Example 6.2. A chiller water system with a single pump, as shown in Figure 6-6, operates for 8,760 hrs/yr. The pump supplies chilled water at 800 gal/min, at 140 ft head, and has a 40 Bhp motor operating at 1,750 rpm. It is determined that the loads need only 800 gal/min for 4,380 hrs/yr and 525 for 4,380 hrs/yr. The single pump is replaced with a primary/secondary pumping arrangement, as shown in Figure 6-9. The primary pump circulates 800 gal/min at 40 ft head with a 10 Bhp motor rotating at 1,750 rpm. The secondary pump has a two speed 30 Bhp motor and must provide 800 gal/min for 4,380 hrs/yr and 525 gal/min for 4,380 hrs/yr. Estimate the energy saved by installing the primary/secondary pumping system.

The energy consumed by the single pump arrangement is

$$E = (40\ Bhp)\,(870\ \text{hrs/yr})\,(1.43\ \text{Kw/hp}) = 501,072\,\text{Kwh/yr}.$$

For the primary/secondary system, the 10 Bhp primary pump operates for 8,760 hrs/yr, so the energy consumed is

$$E_{p1} = (10)(8,760)(1.43) = 125,268\ \text{Kwh/yr},$$

The secondary pump supplies 800 gal/min at 30 Bhp for 4,380 hrs/yr. The energy consumed by the pump at full flow is

$$E_{S1} = (30)(4,380)(1.43) = 187,902\ \text{Kwh/yr},$$

The secondary pump operates at 525 gal/min for 4,380 hrs/yr. Using Equation (6.10), the rotational speed of the pump at 525 gal/min is

$$N_2 = (\,Q_2/Q_1\,)\,N_1 = \left(\frac{525}{800}\right)(1,750) = 1148,\ \text{say 1150 RPM},$$

Using Equation (6.12), the brake horsepower is

$$Bhp_2 = (\,N_2/N_1\,)^3\,Bhp_1 = \left(\frac{1,150}{1,750}\right)^3(30) = 8.5\ \text{hp},$$

The secondary pump operates for 4,380 hrs/yr at 525 gal/min. The energy consumption is

$$E_{S2} = (8.5)(4,380)(1.43) = 53,239\ \text{Kwh/yr},$$

The annual energy consumed by the primary/secondary pumping arrangement is

$$E_{P/ST} = 125,268 + 187,902 + 53,239 = 366,409\ \text{Kwh/yr}$$

The annual energy savings for the primary/secondary pumping arrangement is

$$Savings = 501,072 - 366,409 = 134,633\ \text{Kwh/yr}.$$

Thus, there is a savings of about 27% of the system pumping energy.

6.5.3 Replace Three-Way Control Valves with Two-Way Control Valves

In the previous section, it was shown how a primary/secondary pumping arrangement could save energy if the secondary pumping loop had a variable flow rate. Although a secondary loop may have variable loads, which would allow variable volume pumping, the flow controls at the loads are frequently not suitable for variable flow or there are no controls at all. In order to take advantage of variable volume pumping, there must be flow controls in the secondary loop which allow variable volume.

Occasionally, hydronic heating or cooling applications have variable loads; however, the water flow to the load is constant with no control. In this situation, the water flow is allowed to "run wild" at a constant rate regardless of the heating or cooling load. In this situation, a modulating flow control valve, which modulates the flow in response to the temperature of the load, should always be installed.

Section 6.1.3 of this chapter discussed three-way and two-way modulating control valves. Three-way control valves modulate flow to a load by bypassing flow around the load when the load is satisfied. Two-way control valves directly modulate the flow of water through the load by varying the flow rate in response to the demand.

For a primary/secondary pumping system to be effective in saving pumping energy, the secondary loop must have variable volume. In order for the secondary loop to be variable volume, the loads must have two-way control valves that vary the flow in response to the demand. If a system has no flow control valves, two-way control valves should be installed. If the system has three-way control valves, they should be replaced by two-way control valves.

Two-way equal percentage control valves have better control characteristics than three-way valves. The modulation of a two-way valve is better correlated with changes in load-heat transfer and load-flow relationships.

6.5.4 Install Variable Speed Pump Drives

Earlier, the potential energy savings from using two-speed pumps in variable flow pumping systems was discussed. Even though the flow and head at the reduced speed may not closely match the actual flow requirements, the pump performance will "ride the curve" to meet the system flow and head requirements. The closer the pump flow and head capacity is to the actual system requirements, the more efficiently the pumping system will operate.

Typical load profiles for hydronic systems indicate that systems operate at 50% flow 15 times as much as they operate at 100% flow. Overheading of pumps by 50% or more is extremely common. Significant energy savings can be achieved by closely matching the pump performance to the actual system requirements.

TABLE 6.1. Pump and Drive Efficiency

	Efficiency-Variable Speed			Efficiency-Fixed Speed + Throttle			
Flow %	**Pump %**	**Drive* %**	**System %**	**Pump %**	**Pump & Valve %**	**Drive* %**	**System %**
25	59	84.0	49.6	47.0	16.3	85.0	13.9
50	66	85.1	56.2	53.5	22.8	88.0	20.1
75	74	85.6	63.3	63.0	37.2	89.5	33.3
100	71	86.0	61.1	71.0	71.0	90.0	63.9

*Drive efficiency typical for motor in 60-100 hp range.

Pumps can closely match system requirements when the pumps have variable speed drives on their motors. The variable speed drives cause the pump motor to rotate at various speeds in response to the system requirements. Table 6.1 shows the typical efficiencies of systems that are throttled and systems with variable speed drives for pumps with 60 to 100 hp motors. Speed of the pump is accomplished with a proportional integral derivative (PID) controller, which measures the differential pressure of the system across the supply and return piping. The PID controller then regulates the pump motor speed by controlling the output to a variable frequency drive supplying power to the pump motor. The controller maintains a constant head pressure in the system with varying loads by adjusting the variable frequency drive to the required speed. This reduces power input to the pump motor and provides a consistent differential pressure across control valves regardless of the load.

Frequently, pumps are installed in a parallel arrangement, as shown in Figure 6-7. When variable speed drives are used with pumps in a parallel arrangement, there are some precautions that should be taken. When two or more pumps operate in parallel, the pressures at the pump inlets and discharges are the same. When they operate at variable speeds, the same conditions must be met. Therefore, both pumps must operate at the same speed to produce the same discharge head. If a variable speed drive is installed on one pump in a parallel arrangement, it must also be used to control the other pump(s) at the same speed. All pumps in parallel must produce the same head and, therefore, operate at the same speed.

6.6 ENERGY EFFICIENCY FOR NEW PUMPS AND HYDRONIC SYSTEMS

Earlier in this chapter, energy conservation for pumps and hydronic systems were discussed. Several of the energy efficiency measures discussed as retrofit opportunities for pumps and hydronic systems such as variable speed drives and replacement of control valves. In addition to the previously discussed energy conservation measures for pumps and hydronic systems, the following measures should be considered for new and replacement pumping systems.

When considering the replacement of an existing pump with a new one, the service life of pumps should be kept in mind. The average service life of a base mounted pump is about 20 years while the service life for pipe-mounted pumps is about 10 years. The service life of control valves is about 15 to 20 years.

6.6.1 Selection of Centrifugal Pumps

The operating costs for a pump usually exceed the cost of the pump and motor in about six months. Since the operating cost for a pump and motor are the major costs associated with hydronic systems, it is important that new pumps be selected for optimum energy efficiency. Other than cost, the major issues to be considered when selecting a pump are the pump efficiency, the pump capacity and the pump motor type.

System curves, pump curves and the system/pump operating point were discussed earlier in this chapter. When selecting a pump, the system curve is typically plotted on the manufacturer's pump curve, such as the pump curves presented in Figure 6-23. The manufacturer's pump curves usually show the pump performance for various sizes of impellers. The pump and system will operate at the intersection of the pump impeller curve and the system curve. Ideally, the pump should be selected such that the operating point lies within the area of highest efficiency as indicated by the pump efficiency curves shown in Figure 6-23.

The ideal point to operate on a pump curve is known as the Best Efficiency Point (BEP). This is the point where the flow and head combine to provide maximum efficiency performance from the pump. Frequently, it is not possible to select a pump such that the operating point is within the BEP. In this case, a choice must be made between a pump with the operating point just to the left or just to the right of the BEP. A pump should not be selected with the operating point to the right of the BEP; it should be selected such that the operating point is just to the left of the BEP. Since the pump curve to the right of the BEP is steeper than the rest of the curve, slight changes in flow will cause large changes in the head. The curve to the left of the BEP is much flatter. In this region of the pump curve, changes in flow will result only in slight changes in head. Selection to the left of BEP allows for adjustments in flow and head conditions while providing a high level of efficient operation.

Although pumps should be selected such that the operating point is close to the BEP, the most important factor for energy efficiency when selecting a pump is proper sizing. In almost all cases, when a pump is selected, it is oversized in terms of both flow capacity and head. In some cases, it is due to the addition of a safety factor and, in many cases, because the pump is selected for the worst condition or maximum expected flow.

Pump power consumption will increase with increases in either head or flow as illustrated by the pump affinity laws discussed earlier. The speed of a pump varies directly with the flow (at a fixed head) and the head varies as the square of the speed (at a fixed flow). As a result, the power to drive a pump varies as the cube of the speed.

The head and flow capacity have a significant impact on power consumption. Oversizing pumps should be avoided. For optimum energy efficiency, the actual flow capacity and head requirements for a pump should be carefully determined and the pump should be selected such that it exactly meets the requirements of the system.

Whenever a new pump or replacement pump is selected, it should be specified with a high efficiency motor. The cost premium for a high efficiency motor over a standard efficiency motor is insignificant compared to the energy savings over the service life of the motor. High efficiency motors are discussed in more detail in "Chapter 10: Electrical and Lighting Systems."

Selection of Control Valves

Earlier in this chapter, the different types of control valves used in hydronic systems were discussed. For optimum energy efficiency, the use of three-way control valves should be avoided. Modulating two-way control valves should be used whenever possible.

Although the losses in piping distribution systems should be minimized whenever possible, low head loss control valves should not be used. Control valves should be selected for approximately 25% of the total pump head. The pressure drop of the valve directly affects the ability of the control valve to control the flow. A higher valve pressure drop results in a smaller valve pipe size and better control.

Chapter 7

Chillers and Chilled Water Systems

The function of a chiller is to remove heat from water (chilled water) as it is circulated throughout a chilled water distribution system. The circulated chilled water is frequently referred to as a secondary refrigerant. Chilled water systems typically consist of a chiller, pump(s), distribution piping and some sort of heat exchange device, such as a cooling coil, at the cooling load. Chilled water piping systems are discussed in "Chapter 6: Hydronic and Pumping Systems."

The major components that make up a vapor compression chiller include a liquid cooler (evaporator), a condenser, a compressor, a motor, a thermal expansion valve, internal refrigeration piping and chiller controls. All of the components are typically assembled as a single unit and arc referred to as a packaged chiller. Heat is removed from the secondary refrigerant in the evaporator and rejected by the condenser. The condenser may be air-cooled or it may reject heat to a water. If heat is rejected to water, a condenser water loop, including pump(s), cooling tower and piping is required. Cooling towers are discussed in Chapter 8. Refrigerant systems are discussed in greater detail in Chapter 12.

Chemicals are frequently added to the secondary refrigerant to prevent freezing of the water, allowing the systems to operate below the normal freezing temperature of water. Chemicals added to chilled water systems for freeze protection include calcium chloride, sodium chloride, ethylene glycol, propylene glycol, methanol and glycerin. When such chemicals are added to the water, the secondary refrigerant is referred to as a brine.

Chillers are frequently used for comfort air conditioning, as well as for process cooling. Process cooling is used extensively in many industries, including dairy processing, and for beverage industries, chemical processing, etc.

7.1 TYPES OF CHILLERS AND CHILLED WATER SYSTEMS

There are two general classifications for chillers: vapor compression and absorption. As the name implies, vapor compression machines produce a refrigerating effect by first compressing a vapor refrigerant, removing the heat of compression from the refrigerant to form liquid refrigerant, and then expanding the liquid to cool the circulated water. Absorption chillers use a physiochemical process to produce a refrigerating effect with heat.

Vapor compression chillers are the most common type of chillers used today. Vapor compression chillers are available with three types of refrigerant compressors: reciprocating, screw and centrifugal. The compressor serves as a refrigerant gas pump. Reciprocating chillers have positive displacement type compressors, which use pistons to compress the vapor refrigerant. Screw compressors are also positive displacement compressors. They employ two intermeshed helical grooved rotors (screws) to compress the refrigerant. Centrifugal compressors have a rotating impeller which uses centrifugal force to compress the refrigerant. The type of compressor is usually determined by the cooling capacity requirements of the chiller. Typical vapor compression chiller sizes are listed below.

Size	Compressor Type
Up to 25 tons	Reciprocating
25 to 80 tons	Reciprocating or screw
80 to 200 tons	Screw, reciprocating or centrifugal
200 to 800 tons	Screw or centrifugal
Above 800 tons	Centrifugal

Refrigerant compressors are discussed in greater detail in Chapter 12.

Absorption chillers produce a refrigerating effect with an absorption and evaporation cycle, which uses heat to produce the refrigerating effect. A secondary fluid (the absorbent) is used to absorb a primary fluid, a gaseous refrigerant that has been vaporized in the chiller evaporator. In the absorption system, the vapor refrigerant is first condensed and mixed into a solution with a liquid absorbent. Heat is removed in the evaporator section by absorption as the refrigerant is condensed. The solution is then transferred to another area in the chiller where heat is used to separate the refrigerant from the solution. Absorption chillers are available with capacities ranging from 100 to 800 tons.

As chilled water flows through the chiller evaporator, heat is transferred from the circulating water to the refrigerant. Chillers are typically designed to produce an 8°, 10° or 12°F temperature drop as the water flows through the evaporator section. For any given cooling load, a lower water temperature drop results in a higher chilled water flow rate. Conversely, a high temperature drop results in a lower water flow rate.

If the chilled water is expected to get near, or below, 32°F, chemicals should be added to the water to keep it from freezing. The solution is referred to as a brine. The amount of chemical added to the chilled water affects the heat transfer characteristics of the brine and the pumping head requirements. Greater brine concentrations usually result in lower heat transfer coefficients and greater specific gravity, which results in higher pumping heads. Hydronic and pumping systems are discussed in Chapter 6.

Chillers reject heat to ambient air through their condensing section (condenser). Condensers may be either air cooled or water cooled. Air cooled condensers reject heat to the ambient air directly through sensible heat transfer. Chillers with smaller cooling capacities tend to be air cooled. Larger capacity chillers tend to be water cooled. Heat is rejected to water as it flows through the chiller condenser. The water then rejects to ambient air via a condenser, fluid cooler (evaporative condenser) or a cooling tower. Heat rejection to ambient air can be just sensible, or it can be sensible and latent. Cooling towers are discussed in Chapter 8.

7.1.1 Reciprocating Chillers

Reciprocating chillers are chillers with reciprocating refrigerant compressors. Reciprocating compressors are constant volume type compressors. A reciprocating compressor employs one, or more, pistons operated from a rotating crankshaft. The pistons compress the gaseous refrigerant via a reciprocating action inside cylinders. Each cylinder compresses a fixed, or constant, volume of gaseous refrigerant. A typical reciprocating refrigerant compressor is shown in Figure 7-1.

The refrigerant vapor is drawn into the cylinder through an intake valve as the piston is on the downward, or intake, stroke. When the piston starts the upward, or compression, stroke, the valves are closed and the gaseous refrigerant is compressed. When the refrigerant reaches a predetermined pressure, a discharge valve opens, allowing the pressurized refrigerant to flow out of the cylinder. Reciprocating compressors are usually of the single acting piston type, although they may also be double acting.

Reciprocating compressors typically operate at a constant rotational speed. The capacity of a reciprocating compressor chiller is varied by loading and unloading of the cylinders as the load increases or decreases. Cylinders are unloaded by holding the suction valves open, which results in no compression for that cylinder. Multiple cylinder compressor chillers have staged cooling capacity by unloading cylinders to match the cooling load.

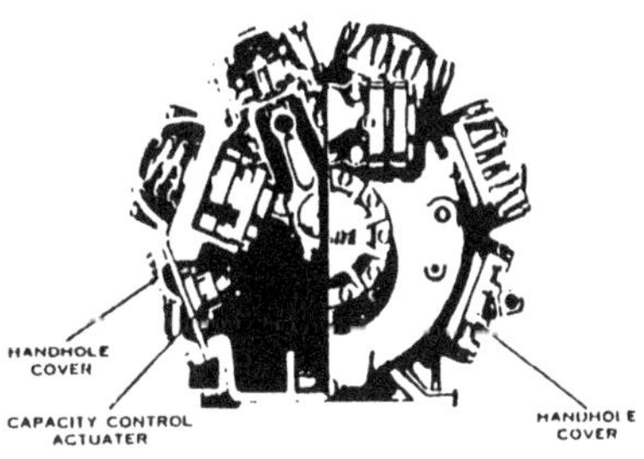

Figure 7-1. Reciprocating refrigerant compressor.

Compressors may have the motor mounted external to the compressor housing, in which case the compressor is referred to as an open compressor. Motors for open compressors are cooled by the ambient air surrounding the motor. Compressors may also have the motor mounted inside the sealed compressor housing. Then the compressor is referred to as a hermetic compressor and the motor is cooled by the refrigerant as it is circulated by the compressor. The selection of open or hermetic compressor is usually based upon the capacity of the chiller. Following are guidelines for compressor types.

Compressor Type	Chilling Capacity (tons)
Welded hermetic	25
Semi-hermetic	25 – 200
Direct drive open	over 200

A reciprocating hermetic compressor is shown in Figure 7-2 and a reciprocating open compressor is shown in Figure 7-3.

An important measure of a compressor's performance is the compression ratio, which is the ratio of the discharge (head) pressure to the inlet (suction) pressure of the refrigerant. The compression ratio may be expressed mathematically as:

$$r_c = \frac{P_h}{P_s} \qquad \text{Eq. (7.1)}$$

where P_h = head pressure, psia
P_s = suction pressure, psia

Higher compression ratios result in lower volumetric efficiency.

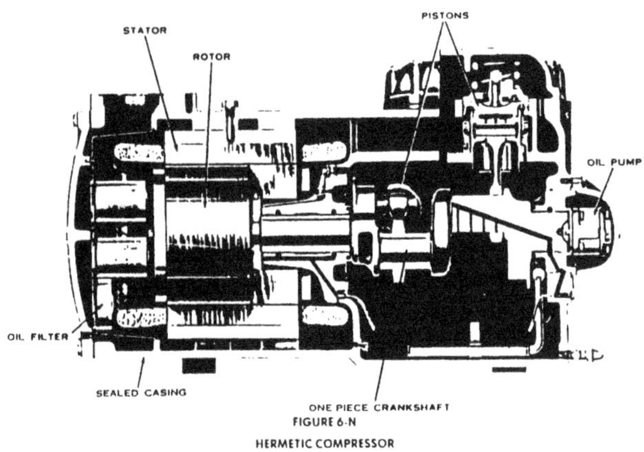

Figure 7-2. Reciprocating hermetic compressor.

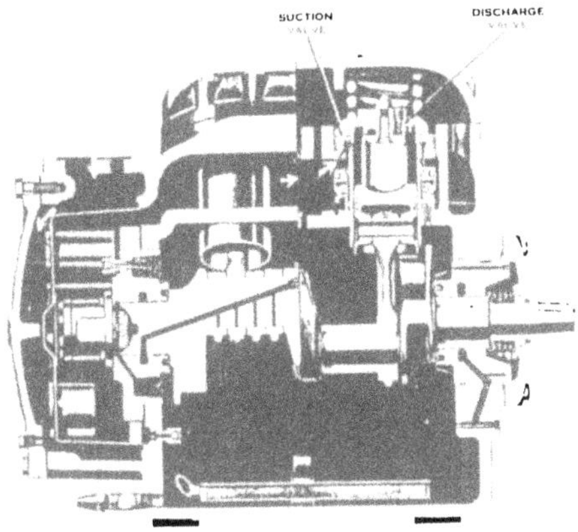

Figure 7-3. Reciprocating open compressor.

Reciprocating chillers typically us R-12, R-22 and R-134a as refrigerants. Since R-12 has been banned, it is being replaced by R-134a.

7.1.2 Screw Compressor Chillers

Screw compressors are also positive displacement compressors. Screw compressor chillers are typically available in factory manufactured units between 30 and 1,250 tons. Their best applications are in the 100 to 300 ton range. Screw compressors can provide stable operation over the entire range of the compressor. They can provide continuous capacity modulation from 100% down to 10%. Screw compressors are available as both hermetic and open compressors. A typical screw compressor chiller is shown in Figure 7-4.

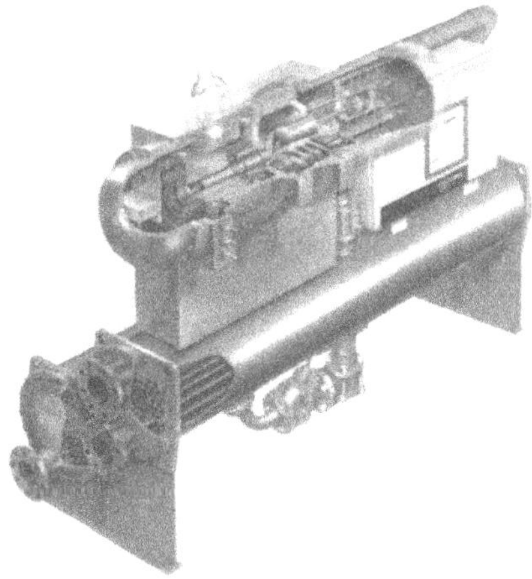

Figure 7-4. Screw compressor chiller (*Courtesy of The Trane Company*).

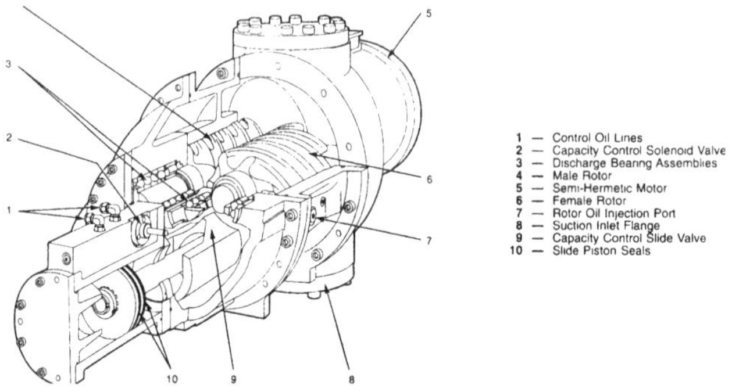

Figure 7-5. Screw compressor (*Courtesy of Carrier Corporation*).

A screw compressor consists of two mating helically grooved rotors, which rotate while enclosed in stationary housing. One rotor is male (with lobes) while the other is female (with flutes). A screw compressor is shown in Figure 7-5 The stationary housing has two ports: one is a refrigerant suction (inlet) port and the other is a discharge port. A motor drives the male rotor, which drives the intermeshing female rotor. When the rotors rotate, the refrigerant is trapped between the flutes of the female rotor and the lobes of the male as they intermesh. The operation of a screw compressor is shown in Figure 7-6. The female rotor may also be driven by synchronized timing gears. The compressor capacity is modulated by a sliding valve. Screw compressors can have single and twin rotors.

Since screw compressor chillers are positive displacement devices, they are well suited for high temperature lift applications such as thermal energy storage. Thermal storage is discussed in Chapter 15. Efficiencies of screw compressor chillers can be as good as 0.57 kW/ton.

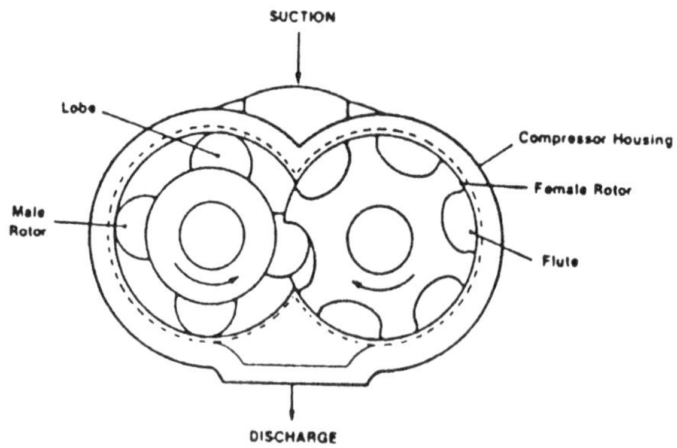

Figure 7-6. Screw compressor operation.

7.1.3 Centrifugal Compressor Chillers

Centrifugal compressors are not positive displacement machines, they are variable displacement machines. The displacement, or flow, of the refrigerant gas varies with the cooling load. Centrifugal compressors are large capacity units that handle large volumes of refrigerant. Centrifugal chillers usually have cooling capacities of 50 tons and more. A typical centrifugal chiller is shown in Figure 7-7.

Centrifugal compressors have one or more rotating impellers. As the impeller rotates, it imparts a centrifugal force (angular momentum) to the gaseous refrigerant in the form of kinetic energy (velocity). The kinetic energy is then converted in to pressure. A cross-sectional diagram of a centrifugal compressor is shown in Figure 7-8.

Centrifugal compressor chillers are available with both open and hermetic compressors. Open compressor chillers are available with cooling capacities up to 1,500 tons. The compressor may have one, or more, stages of compression. Hermetic compressor chillers are available in single chiller packages, with up to 2,000 tons cooling capacity. Hermetic compressor chillers are also available in one, or multiple, stages of compression. Electric motors for hermetic compressors may either be refrigerant cooled or water cooled.

Centrifugal chillers are most efficient when operated between 75% and 90% of their rated full load capacity. Operating a centrifugal chiller at very low load is inefficient and can result in unstable operation known as surge. Surge is a temporary reverse flow of the refrigerant gas that results from critically low refrigerant gas velocities. The operation of a centrifugal compressor at less than full load is limited by the surge point of the compressor. The surge point is the capacity at which surge will

Figure 7-7. Centrifugal chiller (*Courtesy of York International*).

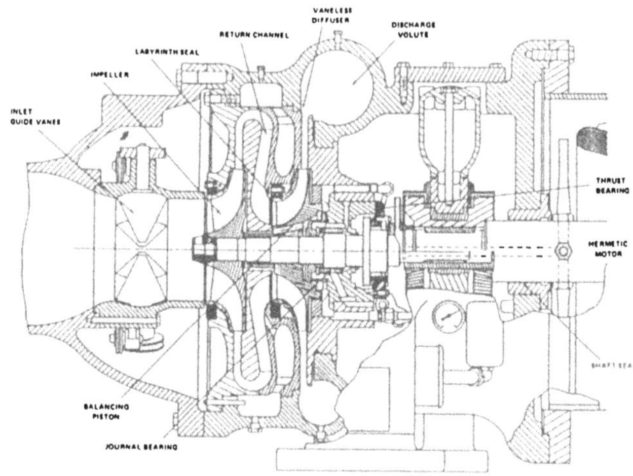

Figure 7-8. Cross-section of centrifugal compressor.

begin to occur. Operation of the compressor at capacities less than the surge point results in operation, which is said to be in the *unstable range*. Extended operation of a centrifugal compressor in the unstable range may result in physical damage to the compressor.

Direct drive reciprocating and screw compressors typically operate around 3,600 rpm. Centrifugal compressors can operate at speeds up to 30,000 rpm. Centrifugal compressor chillers can have multiple stages of compression. Multiple stage compressors with four stages are common. Chillers are available with as many as eight or nine stages. Multi-stage compressors are typically used for low temperature industrial applications with brine cooling.

A measure of the performance of a centrifugal compressor is the temperature lift of the refrigerant. The temperature lift is the condensing temperature of the refrigerant minus the evaporating temperature, and may be expressed mathematically as

$$\text{Lift} = T_C - T_E \qquad \text{Eq. (7.2)}$$

where T_C = refrigerant condensing temperature, °F
T_E = refrigerant evaporating temperature, °F

Chillers are usually sized based upon their capacity to meet the peak cooling requirements of a system. However, in most applications, the peak cooling load is rarely, if ever, reached. The chiller will usually operate at part load most of its operational lifetime. Therefore, some sort of capacity control for the chiller is required. For centrifugal chillers, there are several capacity control methods, including on-and-off control, hot gas bypass control, inlet throttling vanes, variable inlet guide vanes and speed control.

The on-and-off method is the most basic form of capacity control for a chiller. This method has the compressor operate at maximum cooling capacity until the load is satisfied, at which time the chiller is automatically shut off. When the cooling load again requires cooling, the compressor would automatically turn on and operate at full capacity. While the on-and-off method would appear to work in principle, there are problems associated with it. The most serious problem is that starting and stopping a large motor several times an hour would result in a short motor life. Starting the motor with the compressor fully loaded would also result in very large electrical demand charges.

Hot gas bypass control is a bypass installed between the compressor discharge and the compressor suction, or inlet. The cooling capacity is reduced by recirculating high pressure refrigerant vapor from the compressor discharge to the chiller evaporator as a hot gas. As the load is reduced further, the bypass opens wider to maintain a constant minimum flow rate through the compressor, which reduces the chance of surge due to low flow. Bypass control is not very efficient since there is no reduction in power input to drive the compressor when the bypass opens. Hot gas bypass is actually a false loading technique, which introduces a false load on the compressor. Hot gas bypass control is frequently used in conjunction with other control methods.

Inlet throttling vanes are butterfly type dampers that are located in the compressor inlet to control the flow of refrigerant into the compressor. They are effective in lowering the compressor surge point. As the refrigerant vapor flows through the partially closed vanes, the vanes produce a pressure drop in the refrigerant vapor, which reduces the mass flow of the refrigerant. Since the mass flow of the refrigerant is reduced, less refrigerant is compressed and the compressor drive operates at less load. Inlet throttling vanes are not a particularly efficient method for chiller capacity control. The reduction in drive input power to the compressor follows the reduction in cooling load fairly closely. However, an additional load is placed on the compressor drive by the pressure loss through the throttling vanes. As a result, the power input to the compressor drive does not fall off as quickly as the cooling load as it decreases.

Although the inlet throttling vanes are effective in reducing the chiller capacity by reducing the flow of refrigerant vapor to the compressor, an efficiency penalty is paid due to the turbulence created by the throttling vanes. Variable inlet guided vanes, or pre-rotation guide vanes, reduce the turbulence of the refrigerant vapor as it enters the compressor suction. Variable inlet guide vanes consist of multiple wedge-shaped vanes, which rotate open or closed in response to the cooling load on the chiller. The vanes are rotated by an operator and in unison by control linkages, which adjust the position of each vane in unison. As the vanes rotate, they impart a spin to the refrigerant gas, which is in the same direction as the rotating impeller. This greatly reduces the turbulence of the refrigerant, provides greater stability at low flow conditions and significant reductions in power consumption at low cooling loads. Variable inlet guide vanes lower the surge point, increase the stable operating range of the chiller and reduce compressor power consumption. The cooling capacity of the chiller can be reduced down to the 10–20% range. Variable inlet guide vanes also reduce motor overload on compressor start-up.

Variable speed control is based upon reducing the speed of the compressor impeller at reduced cooling loads. Reducing the speed of the impeller reduces the volume of the refrigerant vapor flow, which lowers the surge point and increases the stable operating range of the compressor. The head (discharge) pressure of the compressor is significantly reduced, with small changes in impeller speed. The capacity of the compressor is directly related to the change in speed while the pressure produced by the compressor is proportional to the square of the speed. The most common method for varying and controlling the speed of a centrifugal compressor is with adjustable speed motor drives. Variable speed drives are discussed later in this chapter and in Chapter 10.

7.1.4 Absorption Chillers

Absorption chillers do not use a mechanically driven compressor to compress the refrigerant. The absorption process uses certain chemical solutions, which have an attraction for the refrigerant, to create a pressure difference that compresses the refrigerant. Absorption refrigerant equipment operate on heat cycles during which a secondary fluid (an absorber) is used to absorb a primary fluid (a gaseous refrigerant) that is vaporized in an evaporator.

In the absorption refrigeration process, a gaseous refrigerant is vaporized by a heating source, condensed and then absorbed by a liquid absorbent. Like vapor compression chillers, the refrigerant is vaporized in an evaporator, which exchanges heat from the chilled water to the vapor refrigerant. The absorption process is rather complicated. An absorption chiller has few moving parts and requires little energy input in the form of work. The process does require significant amounts of energy input in the form of heat.

Volatile liquids, such as water or ammonia, are used as refrigerants. The refrigerant is evaporated at low pressure, which results in cooling at low temperatures. Chillers using water as the refrigerant utilize lithium bromide as the absorbent. Chillers using ammonia as the refrigerant employ water as the absorbent.

Absorption refrigeration is frequently used for industrial process refrigeration. Evaporator temperatures are possible from 50°F to as low as -75°F. Ammonia-water chillers are frequently used in the food and beverage industries.

In order to understand the absorption refrigeration process, it is first necessary to understand the terminology used. See the Glossary for definitions of terms frequently used in the absorption refrigeration process.

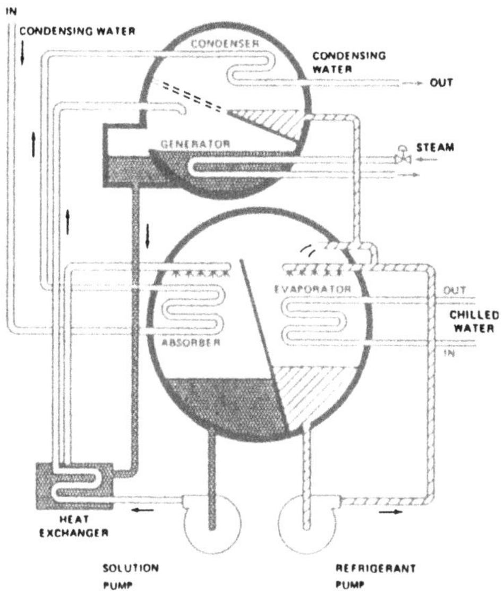

Figure 7-9. Schematic diagram of a two-shell absorption chiller.

Figure 7-9 shows a schematic diagram of a typical absorption chiller process. The absorption cycle begins in the *generator (concentrator)*. Generators are tube bundles submerged in the sorbent solution through which steam, hot water or other heating medium flows. Heat is supplied to the generator, which then causes the dilute (weak) solution to boil and vaporize. The solution never becomes hot enough for the absorbent to boil, only the refrigerant vaporizes. As the refrigerant boils away from the solution, the remaining liquid solution becomes stronger (more concentrated). This process is referred to a desorption.

The refrigerant vapor rises and flows through eliminators to the condenser, where it is cooled and condensed by water from a heat sink (usually a cooling tower). Both the evaporation (boiling) and condensing (cooling) of the refrigerant take place inside a vessel with a common vapor space maintained at a low pressure. The refrigerant vapor naturally flows upward due to the change in density of the vapor refrigerant. The condensed liquid refrigerant is collected in the condenser by a liquid trap where it flows by gravity to the evaporator.

The liquid flows into the evaporator, where it is collected in the bottom of the evaporator in a sump. A refrigerant pump draws the liquid refrigerant from the sump and pumps it to a distribution piping system with spray nozzles located over tube bundle through which the chilled water is circulated. As the refrigerant absorbs heat from the circulated chilled water, the chilled water is cooled and the refrigerant evaporates and becomes a vapor. The spray increases the rate of heat transfer between the water in the tubes and the refrigerant outside the tubes.

The absorber and evaporator share a common vapor space. The refrigerant vapor in the evaporator is drawn into the absorber due to the lower pressure in the absorber. The low pressure results from a strong attractive force between the absorbent solution and the refrigerant vapor. The molecules of refrigerant vapor condense into a liquid as they come in contact with the molecules of the absorbent solution. The greater the difference between the weak solution and strong solution, the greater the pressure difference. The absorption process releases heat of condensation and heat of dilution, which are removed by cooling tower water flowing through the absorber tubes. The condensed absorbent solution is collected at the bottom of the absorber where an absorber pump circulates the solution through a heat exchanger. The solution in the heat exchanger is cooled by solution on its way to the generator. This increases the efficiency of the absorption system by reducing the heat supplied to the generator section. After being cooled in the heat exchanger, part of the solution is sprayed over the absorber tubes to expose the greatest area of the solution to the molecules of refrigerant vapor and increase condensation. The rest of the dilute solution is pumped back to the generator, completing the cycle.

Chillers that use steam or hot water for the heating medium are referred to as indirect fired. Absorption chillers are also available; they use direct combustion of fossil fuels, such as natural gas or fuel oil, to produce the necessary heat for the process. Chillers using the combustion of fossil fuels are called direct fired. Machines that use combustion hot waste gas as the heating medium are referred to as heat recovery chillers. Direct fired absorption chillers are frequently used for both heating and cooling. Machines that perform both functions are referred to as a chiller/heater.

The absorption chilling process described earlier is a *single effect* system. The efficiency of the absorption refrigeration process can be improved by increasing the temperature of the generator. In single effect machines, the change in concentration of the solution and the difference between equilibrium temperatures of the boiling dilute and outgoing strong solutions for the generator are small. Absorption chillers are also available as multiple effect systems.

Absorption chillers are also made with a two stage generator. These chillers are called dual effect or double effect chillers. A dual effect chiller with a two stage generator is shown in Figure 7-10. In a double effect system, the first stage generator receives heat from an external source. The heat boils the refrigerant from the weak absorbent solution. The hot refrigerant vapor from the first, or primary, generator goes to a second generator, supplying heat for further refrigerant vaporization from the absorbent of intermediate concentration. The refrigerant has flowed from the first stage generator after being cooled by passing through a first stage economizer (heat exchanger). Other than the generator, all components of a dual effect system are the same as those for a single effect system.

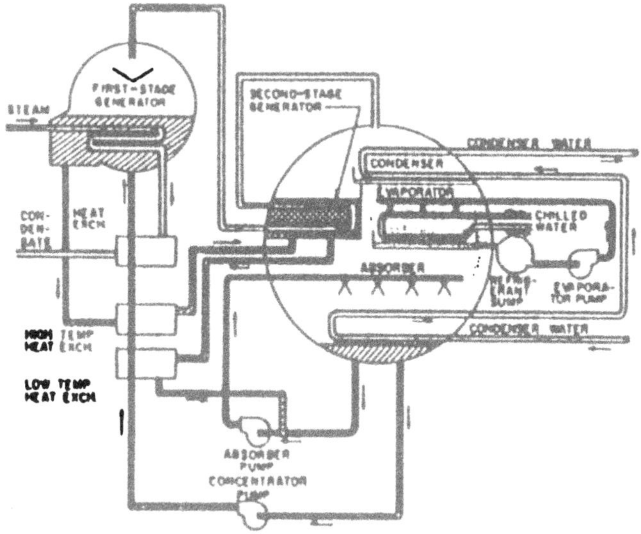

Figure 7-10. Dual effect absorption chiller.

Typical coefficients of performance (COP) for single effect chillers at the American Refrigeration Institute (ARI) rating are between 0.7 and 0.8; for small, indirect fired chillers, about 0.5; and as high as 1.0 for dual effect direct fired chillers.

7.2 CHILLER CONDENSERS AND EVAPORATORS

In addition to compressors, condensers and evaporators are major components of vapor compression chillers. Both condensers and evaporators are heat exchangers. The primary function of an evaporator is to cool the circulated chilled water as the liquid refrigerant vaporizes. The condenser, on the other hand, removes heat from the refrigerant and rejects the heat to the atmosphere, or some other heat sink.

7.2.1 Condensers

Condensers are heat exchangers, which remove heat from the refrigerant and reject heat to a heat sink, usually outdoor air. Condensers may be air cooled or water cooled. In air cooled condensers, the refrigerant is piped to the condenser located outdoors, or near a source of outdoor air, where a heat exchanger transfers heat from the refrigerant directly to the outdoor air. Air cooled condensers are usually suitable only for chillers with relatively low cooling capacity.

Condensers for large tonnage chillers are typically water cooled and reject heat to ambient air indirectly by circulating condenser cooling water through a cooling tower or fluid cooler. Cooling towers and fluid coolers are discussed in Chapter 8. Water cooled condensers for chillers are typically one of three types: (1) shell-and-tube, (2) shell-and-coil and (3) tube-in-tube.

Shell-and-tube condensers are used for chillers with capacities from 1 to 10,000 tons of cooling. In these condensers, the refrigerant flows through, and condenses, in tubes that are surrounded by the circulated cooling water. The tubes are mounted inside a large vessel (shell) through which the cooling water flows. The refrigerant tubes are usually copper in 3/4 and 1 inch outer diameter (OD) sizes. Sometimes the tubes have integral fins on the outside to enhance the transfer of heat from the refrigerant to the water. Since the water and refrigerant heat transfer film resistances are reduced with enhanced tubes, the effect of tube fouling becomes quite significant. Methods of dealing with tube fouling are discussed later in this chapter.

In shell-and-coil condensers, the containment of the refrigerant and cooling water is reversed from the shell-and-tube. Shell-and-coil condensers have the cooling water circulating through the tubes while the refrigerant surrounds the tubes. The condenser shell is a containment vessel for the refrigerant vapor as it is condensed into a liquid outside the tubes by the cooling water. In this arrangement, the tubes are not mechanically cleanable or replaceable. Chillers with shell-and-coil condensers typically have cooling capacities up to 15 tons.

Tube-in-tube condensers have two coaxial tubes, one with the other. The refrigerant may be condensed in either the inner tube or the annular space between the tubes. Chillers with tube-in-tube condensers are typically available with cooling capacities up to 50 tons.

7.2.2 Evaporators

Evaporators (liquid coolers) remove heat from the circulated chilled water as the liquid refrigerant evaporates or vaporizes. As the liquid refrigerant vaporizes, it produces a cooling effect which cools the chilled water. Typically, there are two types of evaporators used in chillers: direct

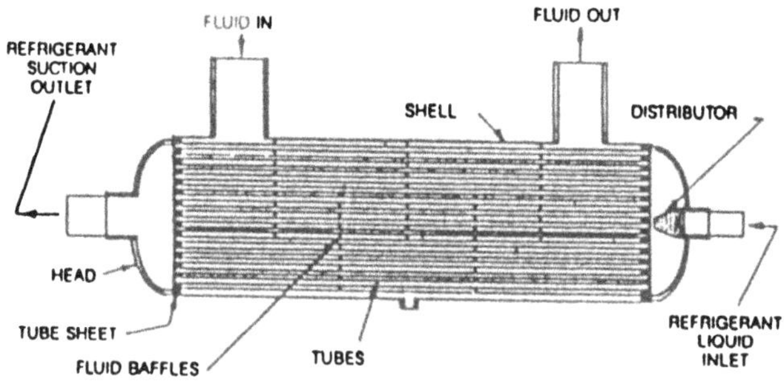

Figure 7-11. Direct expansion shell-and-tube evaporator.

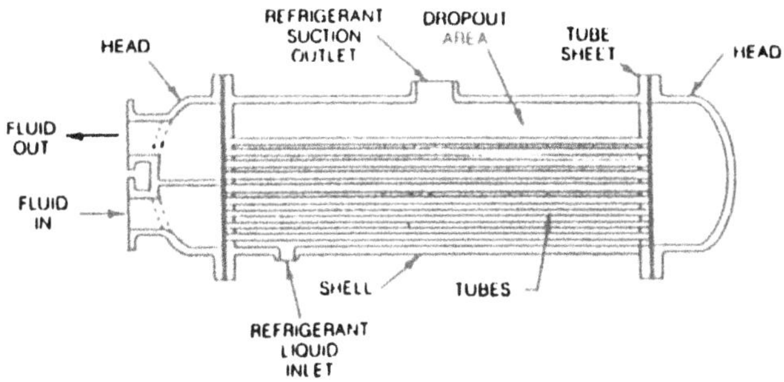

Figure 7-12. Flooded shell-and-tube evaporator.

expansion (DX) and flooded shell-and-tube. In DX evaporators, the refrigerant vaporizes inside the tubes, which are surrounded by the water to be cooled. A diagram of a direct expansion evaporator is shown in Figure 7-11. DX evaporators are suitable for use on chillers with positive displacement compressors, such as reciprocating or screw compressors. The cooling capacity of the chiller is controlled by modulating the flow of refrigerant through the evaporator with a thermal expansion valve (TXV).

Flooded shell-and-tube evaporators have the chilled water flowing through the tubes, which are surrounded by liquid refrigerant. A diagram of a flooded shell-and-tube evaporator is presented in Figure 7-12. As the water flows through the tubes, it is cooled as the refrigerant vaporizes. Flooded shell-and-tube evaporators are used on chillers with rotary screw or centrifugal compressors. Liquid refrigerant is fed into the bottom of the evaporator shell. As the refrigerant boils and becomes vapor, the density change causes it to rise to the top of the shell where it flows to the compressor suction.

7.3 CHILLER PERFORMANCE

Before evaluating the energy consumption of various types of chillers and energy improvements for chillers, it is first necessary to understand the various energy performance measurements and standards. The most common energy performance measurements for chillers are the coefficient of performance (COP), efficiency ratio (kW/ton), Integrated Part Load Value (IPLV) and Application Part Load Value (APLV).

To compare the performance of various chillers and chiller modifications, it is first necessary to set standard operating conditions under which the performance of a chiller is to be evaluated. The Air-Conditioning and Refrigeration Institute (ARI) publishes standard operating conditions for chiller performance evaluation. The applicable standards are ARI Standard 550 Centrifugal and Rotary Screw Water-Chilling Packages, ARI Standard 560 Absorption Water Chilling Packages and ARI Standard 590 Positive Displacement Compressor Water-Chilling

Packages. For centrifugal chillers, ARI Standard 550 sets the following operating conditions:

- 44°F Leaving chilled water temperature
- 54°F Return chilled water temperature
- 2.4 gal/min/ton (gpm/ton) chilled water flow rate
- 3.0 gal/min/ton (gpm/ton) condenser water flow rate
- 85°F Entering condenser water temperature
- 95°F Leaving condenser water temperature
- 0.00025 hr.-ft.^{2}-°F/Btu chilled water fouling factor
- 0.00025 hr.-ft.^{2}-°F/Btu condenser water fouling factor

Fouling results from the fact that fluids will foul the fluid side heat transfer surface, which reduces the overall heat transfer coefficient of the chiller heat exchangers (evaporator and condenser).

A heat pump is any device that removes heat from a cool source and discharges the heat of a warmer heat sink. A refrigeration system, such as a chiller, is a heat pump since it removes heat from the chilled water (a cool source) and discharges the heat to ambient air (a warmer heat sink). A heat pump always transfers more heat energy than work energy consumed, or input, because a net gain of total energy is transferred from the low temperature to the high temperature environment. Heat pumps, such as chillers, are reverse heat engines and, therefore, are limited by the Carnot Cycle Coefficient of Performance (COP), which is given by Equation (7.3).

$$\text{COP}_{\text{Carnot}} = \frac{T_H}{T_H - T_L} \qquad \text{Eq. (7.3)}$$

where T_L = the low absolute temperature in the cycle, °R
T_H = the high absolute temperature in the cycle, °R

However, due to the losses in efficiency, an actual heat chiller has a lower COP than would be predicted by Equation (7.3).

The actual COP of a chiller is a measure of the heat absorbed by the system and is divided by the energy input to the system. Therefore, the actual COP of a chiller is the ratio of the heat transferred by the refrigeration system for a low temperature sink (the chilled water) to the motivating power input (work). The COP for a chiller is given by Equation (7.4).

$$\text{COP} = \frac{q_c}{q_i} \qquad \text{Eq. (7.4)}$$

where q_c = cooling capacity, Btu/hr.
q_i = energy input, Btu/hr.

A measure of chiller performance that is easier to understand and apply is the efficiency ratio; it uses units that are commonly available from chiller operational logs and are essentially the inverse of the COP with different units. The energy efficiency ratio is given by Equation (7.5) and has units of kW/ton

$$\eta = \frac{W_i}{q_c} \qquad \text{Eq. (7.5)}$$

where W_i = the work input to the chiller, kW
q_c = cooling capacity, tons of refrigeration

Note: 1 ton refrigeration = 12,000 Btu/hr.

Both the chiller COP and efficiency ratio are based upon the steady-state performance of the chiller. These ratings provide the performance of the chiller, assuming that it is operating at optimum load conditions and that the load is constant. For most chillers, the optimum load is that at which the best efficiency is obtained near 100% of full load cooling capacity. Rarely, however, do chillers operate near optimum load continuously. Chillers usually operate at reduced loads for the majority of their operational life.

The Integrated Part Load Value (IPLV) is a method for rating chiller efficiency, which takes into account the typical load profile of most chillers. The IPLV is a representative measure of seasonal chiller efficiency (in kW/ton). It provides a measure of chiller efficiency over a specified range of load and operating conditions, using a formula contained in ARI Standard 550, and is based on operation under ARI design conditions. The IPLV assumes the chiller operates at 75 – 100% capacity for 10% of the operating hours; 50 – 75% for 50% of the operating hours; 25 – 50% for 30% of the operating hours; and below 25% capacity for 10% of the operating hours. The IPLV is given by Equation (7.6).

$$\text{IPLV} = \frac{0.1(A+B)}{2} + \frac{0.5(B+C)}{2} + \frac{0.3(C+D)}{2} + 0.1D \qquad \text{Eq. (7.6)}$$

where A = chiller efficiency ratio at 100% capacity, kW/ton
B = chiller efficiency ratio at 75% capacity, kW/ton
C = chiller efficiency ratio at 50% capacity, kW/ton
D = chiller efficiency ratio at 25% capacity, kW/ton

Note: Chiller efficiency ratios are at ARI standard conditions.

Example 7.1. A chiller operating under ARI standard conditions has the following efficiency ratios at the listed percent of full load capacity.

- 100% capacity, η = 0.610 kW/ton

- 75% capacity, η = 0.572 kW/ton
- 50% capacity, η = 0.590 kW/ton
- 25% capacity, η = 0.702 kW/ton

What is the Integrated Part Load Value for this chiller?

Using Equation (7.6),

$$\text{IPLV} = \frac{0.1(0.610 + 0.572)}{2} + \frac{0.5(0.572 + 0.590)}{2} + \frac{0.3(0.590 + 0.702)}{2} + 0.1(0.702)$$

$$\text{IPLV} = 0.614\,\text{kW/ton}$$

Although the IPLV can be useful in comparing chiller performance, the IPLV comparison is valid only if the chillers are operating at the ARI standard conditions listed earlier. Frequently, the use of IPLV is not appropriate because the operating conditions assumed by ARI Standard 550 do not describe the actual operating conditions for most chiller installations. To account for actual operating conditions, the time-weighting values of Equation (7.6) may be adjusted to fit the actual operating conditions of the chiller, which results in a similar performance value called the Application Part Load Value (APLV). The APLV is based upon selected application conditions and is an attempt to normalize the chiller manufacture's part load rating for a particular application. APLV is given by Equation (7.7) and is calculated in the same way as for IPLV, except at actual operating conditions.

$$\text{APLV} = \frac{0.1(A + B)}{2} + \frac{0.5(B + C)}{2} + \frac{0.3(C + D)}{2} + 0.1D \qquad \text{Eq. (7.7)}$$

where A = chiller efficiency ratio at 100% capacity, kW/ton
B = chiller efficiency ratio at 75% capacity, kW/ton
C = chiller efficiency ratio at 50% capacity, kW/ton
D = chiller efficiency ratio at 25% capacity, kW/ton

Note: Chiller efficiency ratios are at actual operating conditions.

To obtain the part load performance figures for use in Equation (7.7), it is usually necessary to obtain the data from the manufacturer of the chiller. It is necessary to provide entering water temperatures, leaving water temperature's water flow rates, and fouling factors for both the chiller evaporator and condenser. The chiller manufacturer may then calculate the customized part load performance figures for use in Equation (7.7).

Although the APLV is useful in comparing the energy consumption for different chillers, it must be recognized that it has limitations. One major shortcoming of the APLV is that it does not include the impact of electrical demand charges on actual operating costs. Another factor to consider is that the actual chiller operation may very well have a different time-load profile than that assumed by Equation (7.7). Finally, the APLV does not account for any auxiliary equipment, such as chilled water pumps, condenser water pumps and cooling tower fans.

7.4 MAINTAINING CHILLED WATER SYSTEMS FOR OPTIMUM ENERGY EFFICIENCY

For chillers and chilled water systems to operate at peak efficiency and minimum energy consumption, it is necessary that they receive regular maintenance and service. Proper routine maintenance should be the first step in any energy management program for chilled water systems. The following are maintenance procedures that will result in optimum energy efficiency for most chillers.

Check Operational Controls: The operational controls of chillers and chilled water systems should be checked periodically to ensure that they are functioning properly and are calibrated.

Clean Evaporator and Condenser Tubes: Chiller evaporators and condensers provide the heat exchange surfaces for the transfer of heat from the chilled water to the refrigerant and from the refrigerant to the cooling tower water, respectively. As the evaporator and condenser surfaces become fouled by water deposits, scales and dirt, etc., the heat transfer capability is reduced. When the heat transfer capability of the heat transfer surfaces is reduced, chiller efficiency is reduced and energy consumption increases. The tubes through which chilled water and cooling tower water flows must be cleaned on a regular basis to remove builtup scale and deposits. In addition to cleaning the tubes on a regular basis, fouling can be reduced by proper chemical water treatment of the chilled water and, especially, the condenser water if an open cooling tower is used. The buildup of scale and other deposits may also be reduced by maintaining the flow of water through the tubes at a velocity between 3 and 11 feet per second.

Maintaining Proper Refrigerant Charge: Too much, or too little, refrigerant reduces the chiller's heat transfer capacity, which results in higher head pressures and increased power consumption. In addition to increasing head pressure, improper refrigerant levels can also decrease the evaporator temperature. For every 1°F that the evaporator temperature can be increased, a 1–1/2% reduction of the full load energy input can be obtained.

A sight glass should be provided in the evaporator shell of centrifugal chillers to monitor refrigerant levels. In the case of reciprocating chillers, bubbles in the refrigerant liquid sight glass is an indication of low refrigerant charge, while high discharge pressure and/or low refrigerant temperature is an indication of excessive refrigerant charge.

Eliminate Refrigerant and Air Leaks: In high pressure chillers, leaks will result in loss of refrigerant from the chiller. In low pressure chillers, leaks will result in air being drawn into the refrigerant. The result of low refrigerant was discussed above. Air that leaks into a chiller will collect in the condenser and displace the refrigerant vapor, which will result in higher condenser pressures. This condition increases head pressure, which results in higher energy consumption. The following procedure may be used to check for excess air in the condenser of a chiller:

1. Measure the refrigerant pressure in the condenser.
2. Convert the pressure to an equivalent temperature for a temperature/pressure table for the refrigerant in use.
3. Measure the temperature of the liquid refrigerant.
4. Subtract the temperature measured in step 3 from the temperature obtained from the table in step 2.

If the difference in the two temperatures is more than 2°F, the condenser probably has excess air in the refrigerant. Typically, for each 2°F difference in the temperatures, the chiller power consumption will increase by 3% of the full load power input.

Since it is common for air to leak into most low pressure chillers, it is advisable to install a refrigerant purge unit that automatically removes excess air without loosing significant amounts of refrigerant.

Maintain Motor Efficiency: The chiller compressor motor consumes the greatest amount of energy used by a chiller. The most common cause of motor performance degradation is lack of cooling. For chillers with hermetic motors, periodic checks should be made to ensure that the motor has adequate flow of refrigerant and that there are no restrictions to the refrigerant flow, such as a clogged refrigerant filter. Open motors should be checked to ensure that there is adequate ventilation and that the flow of ventilation air is not restricted by clogged intake and exhaust openings, or by clogged air intake filters.

Maintain Equipment Logs: A log of chiller performance should be kept on a daily basis. Normal operating conditions should be established and recorded at initial start-up. Changes in chiller performance will quickly become evident if complete logs are kept on a daily basis. An analysis of chiller operating logs can reveal potential problems and allow corrective maintenance to be performed before significant amounts of energy are wasted.

Sample and Analyze Oil Annually: An analysis of oil will reveal the presence of destructive acids and detect metal deposits, which would indicate abnormal wear.

Maintenance procedures listed above are applicable to all types of chillers. Absorption chillers have some additional maintenance procedures which are will help ensure peak energy performance. Maintenance procedures which will improve the energy performance of absorption chillers include:

1. Start and operate equipment to ensure proper timing sequence and draft conditions for direct fired absorption chillers.

2. Check and adjust solution and refrigerant levels.
3. Check operation and performance of purge pump and purging system
4. Adjust burner and perform combustion analysis to ensure optimum performance on direct fired chillers. See Chapter 4 for combustion analyses.
5. Adjust control valves on steam fired and hot water fired units.
6. Inspect steam traps on steam fired units. See Chapter 5 for inspections of steam traps.

7.5 OPERATING CHILLERS AND CHILLED WATER SYSTEMS FOR OPTIMUM ENERGY EFFICIENCY

Optimum energy consumption for chiller operation is a complex function of several factors, including chilled water temperatures, condensing water temperatures and cooling loads. Chilled water system energy consumption is a function of chiller operation, chilled water pump operation and cooling tower operation. When considering operational changes that may improve energy efficiency, the effect of changes on all components must be considered.

This section covers some of the low cost operation and maintenance to keep chilled water systems operating at peak energy efficiency.

7.5.1 Isolate Off-Line Chillers

Frequently, chiller installations have multiple chillers serving a common chilled water system. Usually, each chiller has enough cooling capacity to provide cooling for part of the total load, but not all of the load. Multiple chillers provide the flexibility to serve variable cooling loads. However, chilled water systems are frequently piped with chillers in a parallel flow arrangement.

During light load periods, or as the result of load reduction due to the implementation of energy conservation measures, the load is significantly reduced and not all chillers are required to meet the cooling load. As a result, chillers are shut off when the load does not require all chillers to be on-line. Since the chillers are usually piped in parallel, chilled water is circulated through all of the chillers, including the off-line chillers. This mode results in pump energy being wasted and the on-line chillers must operate a higher load to offset the chilled water mixing effect. Pump and fan energy may also be wasted by circulating condenser water through the off-line chillers.

Off-line chillers should be isolated by closing the isolation valves in the chilled water piping and condenser water piping, and shutting off unnecessary pumps.

7.5.2 Operate Chillers in Proper Sequence

When multiple chillers of different sizes or types are connected to a common cooling load, each chiller should be evaluated to determine which chiller, or combination of chillers, will meet the cooling load with least energy consumption. Each chiller, and combination of chillers, should be evaluated at part load and full load operation to determine which combination of chillers consumes the least amount of energy at various system load conditions. As a general rule, chillers should not be operated at less than 40% capacity unless absolutely necessary. Below 40% capacity, the efficiency of most chillers drops off significantly. Operation of a chiller below 40 % may also result in surge conditions. Chillers typically operate most efficiently between 70% and 40% of the full load capacity. Above 70%, the condenser pressure rises and the compressor must work harder, which results in higher power consumption.

During system part load operation, the smallest chiller, or combination of chillers, that will handle the load should be in operation. A single chiller operating near full capacity is usually more energy efficient than multiple chillers operating at part load.

7.5.3 Operate Condenser and Chilled Water Pumps in Proper Sequence

Frequently, chilled water systems have multiple chilled water and condenser water pumps piped in a parallel piping configuration and the flow from all (usually two) pumps are required during peak cooling loads. During light load periods, pumps may be cycled off and the flow reduced to more closely match the cooling loads.

Reducing the condenser water flow should be done with caution. Reducing the condenser water flow rate to an operating chiller may increase the chiller head pressure and energy consumption of the chiller.

7.5.4 Control Electrical Demand Charges

Electrically powered chillers usually have large electric motors to provide power for the chiller compressor. When the compressor is started, the motor requires a large in-rush, or demand, of current to start it. Most commercial and industrial facilities pay a demand charge, in addition to the charge for electricity consumed (see "Chapter 10: Electrical and Lighting Systems"). Peak demand, as well as the peak cost for demand, typically occurs in the summer months during the day. If several chillers are started at once, as well as any other large electrical load, the demand charge can be quite large and quite costly.

When starting chillers, only one should be started at a time and allowed to become fully operational before another chiller or any other large electrical loads are started.

7.5.5 Maintain the Proper Flow Rate of Chilled and Condenser Water

A reduced flow of condenser water increases the compressor head pressure, thereby increasing compressor power consumption. A 20% reduction in condenser water flow rate can result in a 3% increase in full load energy consumption by the compressor. Common causes of reduced water flow include partially closed valves, clogged cooling tower nozzles, dirty strainers and filters, scale buildup in the condenser and air in the water piping.

An increase in chilled water flow rate through the evaporator will result in increased water pressure drop through the evaporator and an increase in pumping energy consumption. Pumping systems are discussed in Chapter 6. Common causes of improper chilled water flow rates include faulty chilled water control valves, modification to the chilled water distribution system, which results in short-circuiting, and improper hydronic balancing of the chilled water system.

7.5.6 Reset Chilled Water Temperatures

Most chilled water systems operate at less than peak cooling capacity for the majority of their operational life. Many cooling loads are not constant and, in many cases, are a function of outdoor air temperatures. At reduced loads, the chilled water system can usually meet the reduced cooling load at slightly higher chilled water temperatures.

Raising the chilled water temperature reduces the compressor head, or discharge pressure, which results in reduced power consumption. Reducing the head on the compressor reduces the pumping work required by the compressor. For each 1°F increase in chilled water discharge temperature, an average energy savings of 2% of the compressor full load power input can be expected. In addition to increasing the leaving chilled water temperature, increasing the chilled water temperature range can result in energy savings. The chilled water temperature range is the difference between the chilled water return temperature entering the chiller and the supply temperature of the water coming out of the chiller. Chilled water systems are typically designed to operate with a temperature range between 8°F and 12°F at peak cooling load. Increasing the temperature range can save energy by reducing the chilled water flow rate, which would reduce the energy consumption by the chilled water pumps. Increasing the temperature range may actually decrease the chiller efficiency; however, this increase is more than offset by pump energy savings. For every 1°F increase in the chilled water temperature range, chilled water pump power consumption will decrease by about 10%. This would be approximately 3% of the total system energy consumption. See Chapter 6 for additional discussion of pumping systems.

The optimum operating conditions for the chilled water temperatures would be the highest chilled water supply temperature and the largest temperature range, yet would still satisfy all cooling loads served by the chilled water system. To determine the optimum chilled water temperatures at off-peak cooling loads, the following procedure is suggested:

1. Raise the leaving chilled water temperature 1°F at a time and allow the system operation to achieve a steady-state condition.
2. Observe the position of each chilled water control valve supplying major cooling loads.
3. Continue to raise the chilled water temperature 1°F at a time, allowing the system to stabilize until one control valve (the one with the greatest cooling load) is wide open. If one or more than one control valve is in the full open position, reduce the chilled water supply temperature until only one valve is in the full open position.

7.5.7 Reset Condenser Water Temperature

Chiller efficiency is directly related to the refrigerant condensing temperature. As the condensing temperature is reduced, the refrigerant pressure in the condenser is also reduced. Lower condensing pressures result in lower power input to the compressor. For each 1°F reduction in condenser water temperature, approximately 1.5% reduction in full load input power consumption can be expected.

Typically, condenser water piping systems have a cooling tower bypass valve, which bypasses water around the cooling tower, or evaporative cooler, to control the chiller temperature of the water to the chiller condenser. To reset the condenser water temperature, set the local controls operating the cooling tower bypass valve and tower fans to the lowest temperature allowed by the chiller (consult the chiller manufacturer for recommendations). At full load, this is usually around 70°F.

The lowest condensing water temperature for the chiller may not result in optimum energy consumption for the entire system. A lower condensing water temperature may require that the cooling tower fans operate for longer time periods, increasing the energy consumption of the cooling towers. The operation of the entire system should be evaluated to determine the optimum operating conditions for peak energy efficiency. See Chapter 8 for additional considerations regarding the performance of cooling towers.

7.6 ENERGY CONSERVATION MEASURES FOR CHILLED WATER SYSTEMS

The energy conservation measures for chilled water systems discussed in the previous section were operation and maintenance procedures that would improve the energy efficiency of existing chilled water systems. Most of those measures were little, or no, cost measures that could be performed on an ongoing basis to improve efficiency.

In addition to the low cost O&M procedures, there are also modifications that could be made to chilled water systems to improve their energy efficiency. These modifications, however, require a more significant cost investment and are permanent improvements. The modifications listed below are improvements that should be considered under any energy management program for existing, or proposed, chilled water systems.

7.6.1 Reduce Cooling Loads on the System

Whenever the opportunity to reduce the cooling loads on a system is available, it should seriously be considered and implemented if at all feasible. Reducing cooling loads will directly reduce the energy consumption of the chilled water system.

Occasionally, chilled water systems serve some cooling loads that are highly variable and some that are constant. Whenever possible, consideration should be given to providing two separate systems, or multiple systems, for the two types of cooling loads. Some cooling loads vary with time of day, some with production schedules and some with the season. If variable cooling loads are served by dedicated systems, such as spot coolers, there is opportunity to shut off the cooling system when there is no cooling load. Consideration should be given to grouping variable loads with similar cooling load schedules, while constant cooling loads should be served by another. Smaller constant cooling loads are usually better served by installing a small packaged chiller or air conditioning unit.

7.6.2 Interconnect Chilled Water Systems

For systems with variable cooling loads, the energy efficiency of the chilled water system drops off and the kW/ton of cooling increases during light load periods. Most chillers operate most efficiently between 40% and 80% of full capacity. Operating below 40% capacity results in a higher kW/ton of cooling.

When there are multiple systems serving variable cooling loads at the same facility, there is opportunity for interconnecting the two systems during light load periods. This would allow one of the chillers, the associated pumps and the associated cooling tower to be shut off during light load periods. The on-line chiller would then operate closer to optimum load conditions.

To interconnect two chilled water systems, it is necessary to provide interconnecting piping between the two chilled water distribution systems and the two condensing water systems. Two position automatic or manual shut-off valves should be provided in the interconnecting piping of each system. The valves would be open to interconnect the systems and closed to isolate the systems.

7.6.3 Retrofit Chiller Compressor Motor-Driveline

When chillers are originally sized and selected, they are sized for the expected peak cooling loads. Often conservative estimates of cooling loads are made and chillers are selected on a conservative basis, which results in oversized chillers. One of the previous energy saving measures recommended reducing the cooling loads on chilled water systems. As a result of conservative chiller selection and/or energy conservation measurers, chillers frequently have cooling capacities that are greater than the actual peak cooling load for the system.

Since chillers operate near peak efficiency when they are operating near full capacity, most chiller systems never operate near maximum efficiency. If a chiller cooling capacity is greater than the maximum cooling load, the chiller can be downsized to more closely match the actual cooling loads. Downsizing a chiller involves replacing the compressor driveline and motor. Downsizing a chiller can produce significant energy savings for the system. For older chillers, the replacement compressor-driveline motor is frequently more energy efficient since the motor can be replaced with a high efficiency motor.

7.6.4 Modify Chilled Water Piping

A common practice in designing and installing chilled water piping systems has been to use either a three-way control valve to control cooling at a cooling load or to allow the chilled water to "run wild" (have no chilled water flow control). The three-way control valve either allows the water to flow through the cooling load, or around the load. A typical chilled water piping schematic with three-way control valves is shown in Figure 7-13. These piping arrangements are installed so that the flow of chilled water through the chiller could be held constant. Typically, the cooling output of the chillers is controlled by providing a constant flow of water while varying the chilled water temperature drop through the chiller to meet the system cooling load.

This arrangement tends to increase system energy consumption due to increased pumping energy. The use of three-way control valves, as well as allowing cooling loads to run wild, requires that the chilled water flow rate be constant, despite load variations, during all operating hours of the chilled water system. The flow rate is at maximum even during light load periods.

A modification to the chilled water piping system, which will result in better chilled water pumping efficiency, is to decouple the chillers from the primary chilled water loop by providing primary/secondary chilled water loops. A primary/secondary chilled water piping schematic is shown in Figure 7-14. In this arrangement, the primary chilled water pumps provide a constant flow of water to the chiller and have to provide only enough pumping head pressure for the flow through the chiller and associated primary loop piping. A secondary pump provides variable flow through the

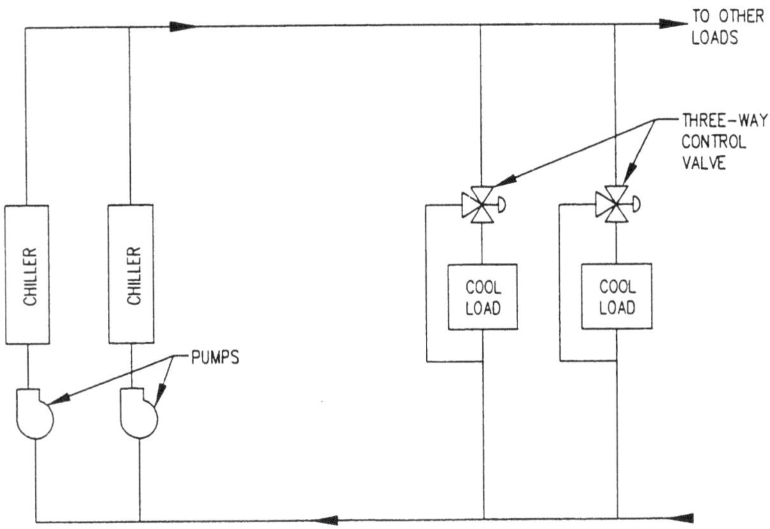

Figure 7-13. Chilled water piping schematic with three-way control valves.

chilled water distribution system to the cooling loads. Flow to the cooling loads is varied by modulating two-way control valves at the cooling loads. Variable flow pumping is provided by staging pumps with different flow capacities or by providing variable flow pumps. For additional information on hydronic and pumping systems, see Chapter 6.

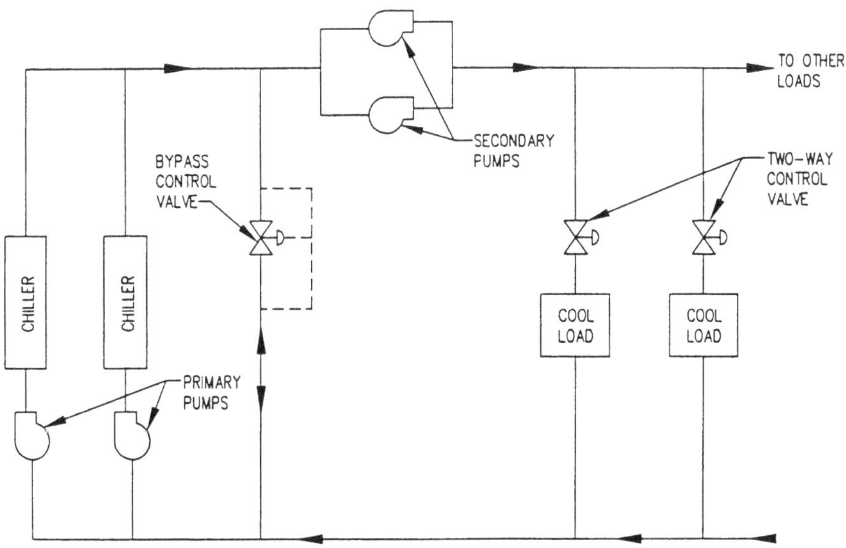

Figure 7-14. Chilled water primary/secondary piping schematic.

7.6.5 Free Cooling

Free cooling is providing chilled water cooling without operating the chiller compressor. Actually, free cooling is a misnomer since the cooling is not actually free. Some energy is consumed by operating the pumps, cooling tower fans and other equipment in the chilled water system, which may consume energy. Significant savings can be achieved, however, whenever it is possible to provide suitable cooling while the chiller compressor is off.

Although many cooling loads vary with seasonal changes, and with changes in outdoor air temperature, some cooling loads are fairly constant throughout the year. Process cooling loads are fairly constant regardless of the season or outdoor air temperature. Whenever a cooling load requires chilled water throughout the year, there is a potential opportunity for free cooling. The opportunity exists whenever the ambient outdoor air wet bulb temperature is less than the chilled water supply temperature. If condenser water is available in the 35°F to 50°F range, equipment such as heat exchangers, pumps, valves and piping can be added to a chilled water system, which will provide cooling without running the chiller compressor. Whenever it is possible to provide chilled water without operating the chiller compressor, significant energy savings are possible for the chilled water system.

Since the condenser water loop is an "open" loop exposed to ambient air, a heat exchanger is required to separate the chilled water loop from the condenser water loop. When an open cooling tower is used, the condenser water may be too contaminated or dirty to be used directly in the chilled water loop. To use the condenser water for cooling, a heat exchanger is usually required to transfer heat from the chilled water to the condenser water loop. For additional information on free cooling by this method, see "Chapter 8: Cooling Towers and Fluid Coolers."

Vapor migration of the refrigerant within the chiller is another method of free cooling for chilled water systems. In order to provide free cooling by vapor migration, the chiller itself must be suitable for this application and the option must be included when the chiller is manufactured. Vapor migration free cooling is discussed in Section 7.7, "Selecting Chillers for Optimum Energy Performance."

When installing a free cooling system for chillers, it is important that the chillers be able to automatically start and operate during periods when the condenser water temperature is low enough for free cooling. Centrifugal, reciprocating and screw chillers typically require condenser water between 55°F and 65°F for smooth start-up and operation with full condenser water flow.

Absorption chillers typically require 60°F to 65°F condenser water. For free cooling applications, tower water is usually required at temperatures below 55°F for even partial free cooling. Care must be taken so that there is adequate cooling to satisfy the cooling load during the changeover period from free cooling to regular cooling provided by the chiller with the compressor operating. Consult the chiller manufacturer regarding operation of the chiller with below-normal condenser water temperatures.

7.6.6 Thermal Storage

In Section 7.5.4, the effect of chillers on electrical demand and demand charges resulting from chiller operation was discussed. The rate at which electric utility companies charge for electrical demand usually varies with the season and with the time of day. Most utility companies usually charge a higher demand during the daytime hours and a lower demand charge during nighttime hours. Whenever there is an opportunity to shift cooling loads, so they occur during low demand charge times instead of periods of higher demand charges, there is an opportunity for cost savings.

Usually, it is not possible to shift the cooling loads time from a period of high demand charges to a period of lower demand charges. It is possible, however, to shift the time at which chilled water is generated from a high cost demand charge period to a lower demand cost period with thermal storage. Essentially, chilled water is generated during low cost demand periods (usually at night) for later use during high cost demand periods (usually during the day).

There are two common methods for storing chilled water off-peak for on-peak cooling loads. The simplest method involves generating and storing chilled water in large water storage tanks for later usage. The second method involves the generation of ice during off-peak demand periods. The ice then melts and provides cooling during no-peak demand periods. These thermal storage methods, as well as other methods, are discussed in greater detail in "Chapter 15: Thermal Energy Storage Systems."

Thermal storage usually saves little, if any, energy. The advantage of thermal storage is shifting the consumption of energy generating chilled water from a period of high cost demand charges to a period of lower cost demand charges. Although there may be little energy savings with thermal storage, there can be significant cost savings.

7.6.7 Install Liquid Pressure Amplification on Refrigeration System

Reciprocating compressors for chillers are designed to maintain a constant discharge or head pressure. The head pressure is a pressure that will allow condensation of the refrigerant at high ambient temperatures. However, when lower ambient temperatures are available, which is the majority of the time, energy is wasted since the condenser temperature and pressure are higher than necessary. Energy could be saved if the head pressure were allowed to "float" with the ambient temperature. In most refrigeration systems, however, this is not feasible due to the pressure differential required by the thermal expansion valve (TXV). A typical direct expansion refrigeration system schematic is presented in Figure 7-15. The pressure difference is necessary to maintain a liquid seal at the TXV and to provide

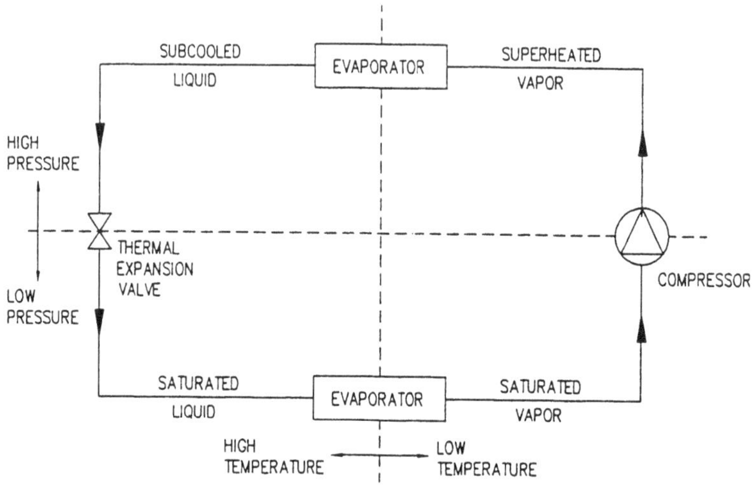

Figure 7-15. Direct expansion refrigeration systems schematic.

stable valve operation without excessive evaporator superheat. Excessive refrigerant superheat at the evaporator discharge may result in compressor overheating and failure.

Liquid pressure amplification (LPA) systems allow the compressor head to float with the ambient air temperature. The addition of LPA to a refrigeration system involves the addition of a refrigerant pump between the liquid receiver to the condenser and the TXV. A refrigeration system schematic with liquid pressure amplification is shown in Figure 7-16. The

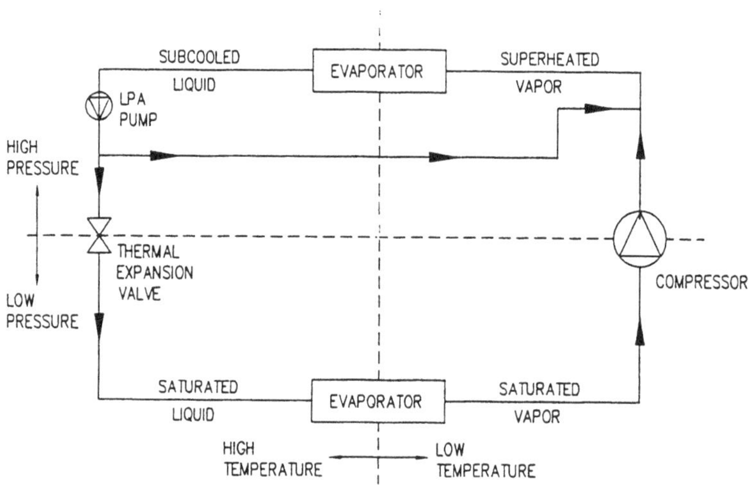

Figure 7-16. Refrigeration system schematic with liquid pressure amplification.

purpose of the refrigerant pump is to increase the refrigerant pressure of the saturated liquid, which increases the pressure of the saturated liquid from the liquid receiver. As a result, the liquid refrigerant entering the TXV is less likely to contain flash gas, which is vapor refrigerant in the liquid line that cannot be used to absorb heat as it vaporizes. The LPA pump also provides additional pressure to ensure adequate pressure differential across the TXV, which provides for proper operation of the TXV. By maintaining proper TXV operation, the evaporator capacity is maintained while the compressor head pressure is allowed to float.

LPA systems are suitable only for chillers with reciprocating compressors. Since the installation of LPA systems are application specific, it is difficult to provide an indication of anticipated savings. The performance of an LPA system is a function of several factors, including:

- Ambient air temperature: A low condensing temperature is required to decrease the compressor head pressure. The ambient temperature must be low enough to not only remove sufficient heat from the condenser coils, but also offset the increased energy consumption resulting from the installation of the refrigerant pump.
- System load profile: If the cooling load varies directly with the ambient temperature, the load will be minimum when the opportunity for energy savings by the LPA is the greatest. A cooling load that is constant, or not related to ambient temperatures, would result in the greatest savings.
- Operating time period: If suitable conditions exist for operation of the LPA system during peak electrical demand cost, savings in addition to the energy cost savings can be achieved.

Presently, liquid pressure amplification is a patented proprietary technology and very little independent third-party testing of the systems have been done. Before installing an LPA system, the manufacturer of the chiller should be contacted for recommendations.

7.6.8 Install Variable Frequency Drives on Chillers

Earlier in this chapter various means of capacity control were discussed for the different types of vapor compression chillers. Most chillers spend the majority of their operational life at less than full capacity. Capacity control for part load operation included unloading cylinders, hot gas bypass, opening and closing inlet vanes, etc. All the methods for chillers were control methods for compressors rotating at constant rotational speed.

Although the above capacity control methods are effective, the energy losses inherent in these methods directly affect the efficiency of the chiller. Another effective capacity control method for chillers is to control the rotational speed of the compressor. Higher loads require the compressor to rotate faster, while lower loads require slower rotational speeds.

Adjustable speed drives (ASDs) are available on new chillers, and as a retrofit for existing chillers, to control the speed of the compressor. Adjustable speed drives are also known as variable frequency drives. These drives vary the rotational speed of a compressor by varying the frequency at which electrical power is supplied to the compressor motor. The efficiency of the variable speed drives themselves is over 90%.

The application of ASDs to control chiller capacity alone does not produce much, if any, energy savings. The best application of ASDs is to use them in conjunction with other capacity control methods, such as inlet vanes. Typically, the greatest energy savings are achieved by allowing the ASD to provide capacity control with the inlet vanes wide open at higher loads while the inlet vanes are used at lower loads.

Since the application of ASDs to chillers for capacity control and energy savings requires a sophisticated control logic to control operation of the ASD as well as other control methods, variable speed drives must be specifically designed for chiller applications. For additional information on variable speed drives, see "Chapter 10: Electrical and Lighting Systems."

7.6.9 Chiller Automation and Control Systems

Although many of the operational energy conservation measures and modifications to chilled water systems can be successfully implemented with only manual or local, control it is difficult to control various operational energy saving measures for chillers s imultaneously to achieve optimum savings without some sort of system automation. Chiller automation and control systems can control multiple variables in a chilled water system to simultaneously achieve optimum system performance and maximize the energy savings.

Chiller automation systems are typically direct digital control (DDC) systems, which are electronic control systems. The automation system uses a microprocessor and software for control logic. Software generated output signals provide full control flexibility, which allows sophisticated control sequences.

Control loop parameters, such as set point, can be adjusted from a central control system terminal through software selections and modifications. Changes in control point settings can be performed more accurately because the software performs proportional-integral-derivative (PID) control. This reduces offset and overshoot, which saves energy and allows the chilled water system to reach the optimum operating conditions quickly. For additional information on automation and control systems, see "Chapter 16: Energy Management and Control Systems."

The following are operational energy conservation measures that typically could be optimized for maximum energy savings.

Optimum Chiller Selection: When multiple chillers of different types and/or different sizes serve a common chilled water system, the automation system selects the best combination of chillers and accessories (cooling towers and pumps) to meet the cooling load at minimum energy consumption.

Optimum Start/Stop Control: The automation system controls the starting and stopping of chillers so that unnecessary chillers and accessories are shut off when not needed. The automation system also controls the timing of chiller start-up so large electrical loads are not started simultaneously, resulting in a large electrical demand.

Chilled Water Reset: The automation systems automatically reset chilled water temperatures to the highest temperature that meets the actual cooling load.

Condenser Water Reset: The automation system provides the lowest permissible temperature water to the chiller condenser without increasing the cooling tower fan energy consumption beyond the chiller energy savings.

Free Cooling: When conditions are favorable for free cooling, the automation system automatically switches the chilled water system over to the free cooling mode.

Records Chilled Water System Energy Consumption: If a problem develops with a component in the chilled water system, which wastes energy, records kept by the automation system alert the system operator if the energy consumption is greater than normal.

Periodic Maintenance: The automation system keeps track of maintenance requirements for the chilled water system and alerts the system operator when maintenance is required.

7.7 SELECTING CHILLERS FOR OPTIMUM ENERGY PERFORMANCE

Chillers and chilled water systems are one of the largest consumers of energy at many facilities. Since chillers consume significant amounts of energy, regardless of the type of facility, it is important that energy efficiency be a primary consideration when selecting new chillers for new or retrofit installations.

7.7.1 General Considerations When Selecting Chillers

Earlier in this chapter, energy conservation for chilled water systems was discussed. Several of the energy efficiency measures discussed as retrofit opportunities for chillers, such as variable frequency drives, are applicable to new and retrofit chiller installations. In addition to the previously discussed energy conservation measures for chillers, the following measures should be considered for new and replacement chillers.

Avoid Oversizing Chillers: When estimating cooling loads for chilled water systems, it is important that an accurate estimate be made of the required cooling capacity. Frequently, cooling loads are estimated on the high side to ensure adequate cooling capacity in the event of any errors due to unknown conditions. Often, safety factors are also added to the estimate, which results in chillers that are grossly oversized. Since chillers operate at peak efficiency when they are between 40% and 90% of full load capacity, oversized chillers will result in wasted energy due to the inefficiency of the chiller at part load conditions.

Select Chiller Type Based Upon Cooling Load: Section 7.1 of this chapter discussed the various types of chillers and the typical cooling capacities for each type. The capacity sizes listed should be used as a guide for selecting the type of chiller most suited for a particular application. When considering chiller types, the refrigerant type must also be considered. Since the production of R-11 and R-12 has been banned, these refrigerants are not available in new chillers. Refrigerants R-22 and R-123 are still in production and available; however, they are scheduled for phaseout in the years 2020 and 2030, respectively. R-134a, which is used in medium pressure chillers, has no phaseout schedule and should be available indefinitely.

Consider Multiple Chillers: For variable cooling loads, multiple chillers of different sizes and different types should be considered. As the cooling load varies, the chiller, or combination of chillers, with the best energy performance should be selected to meet the cooling load.

Select a Suitable Heat Rejection Method: Chillers reject heat to the atmosphere with one of three heat rejection components: air cooled condensers, evaporative fluid coolers and cooling towers. Air cooled condensers are typically used with chillers having smaller cooling capacities. The performance of an air cooled condenser is a function of the ambient air dry bulb temperature and all of the cooling is by sensible heat transfer. Larger chillers typically use evaporative fluid coolers or open cooling towers. The performance of these devices is a function of the ambient air wet bulb temperature, which is usually lower than the dry bulb temperature. These heat rejection devices also take advantage of the evaporative process, which significantly increases the heat rejection capacity. Cooling towers are discussed in Chapter 8.

Consider Part Load Efficiency of Chiller: Studies have shown that most chillers spend only about 2% of their total operating hours at full load capacity. Therefore, it is important to consider the part load efficiency of a new or replacement chiller. Section 7.3 of this chapter discussed various methods for evaluating the part load performance of chillers.

Select Chillers with High Efficiency Motors: The chiller motor is the largest energy consuming component of a chiller. Whenever possible, chillers with high efficiency motors should be selected for new and retrofit installations.

Consult Energy Codes and Standards: Many local and state governments have adopted energy codes, which provide minimum standards for buildings and energy consuming equipment, such as chillers. Many energy codes are based upon ASHRAE Standard 90.1

TABLE 7.1 Minimum Performance of Water-Cooled Chillers

	ARI Efficiency		IPLV	
Capacity(tons)	**COP**	**kW/ton**	**COP**	**kW/ton**
Over 300*	5.2	0.68	5.3	0.66
150-300	4.2	0.83	4.5	0.78
Under 150	3.8	0.93	3.9	0.90

* For chillers using refrigerants with ozone depletion factors less than or equal to those for HCFC-22,the requirements are reduced to COP=4.7 (0.75 kW/ton) and IPLV =4.8 (0.73 kW/ton).

(1993), Energy Efficient Design of New Buildings Except Low-Rise Residential Buildings. Table 7.1 lists the standards for water cooled chillers.

7.7.2 Recover Chiller Waste Heat

As part of the normal cooling process, chillers remove heat from a cooling load (heat source) and discharge the waste heat to a heat sink, usually outdoor ambient air. In addition to the heat removed from the cooling load, the chiller must also reject the heat generated as the result of compressing the refrigerant in the chiller.

Since chillers reject significant amounts of heat to the atmosphere, there is opportunity to recover the heat, which would otherwise be rejected for heating purposes. Recovered heat can be used for a number of purposes, such as space heating, domestic hot water, process hot water and for preheating. Typically, heat is available from chillers in the 95°F to 120°F range. To recover waste heat from a chiller, it must be designed for heat recovery applications.

Heat recovery chillers have two separate refrigerant circuits and two condensers. One condenser is for operation in the heat recovery mode and heat is rejected to water circulated through the condenser, which absorbs the rejected heat for recovery purposes. Since all of the heat that must be rejected by the chiller may not be used by the heat recovery system, a second condenser is necessary for normal operation of the chiller. The second condenser rejects heat to circulated water, which rejects heat to the atmosphere with a heat rejection device like a cooling tower.

When considering a heat recovery chiller, an analysis must be made about the potential use for the rejected heat and the cost of heat from other sources. Since heat recovery chillers must have higher condenser temperatures for heat recovery, the efficiency of the chiller itself will probably be lower than for a conventional cooling only chiller. The energy consumption of a heat recovery chiller is typically 3 to 5% higher than a conventional chiller. Heat recovery systems are also discussed in Chapter 14.

7.7.3 Vapor Migration Free Cooling

In Section 7.6.5 of this chapter, the application of free cooling to chilled water piping systems was discussed. That method of free cooling described a method whereby chilled water could be made while the chiller compressor was off by adding a heat exchanger and modifying the chilled water and condenser water piping systems.

Another method of free cooling is the vapor migration method. This is a factory installed option that is available on some centrifugal chillers. It also allows the production of chilled water with the chiller compressor off. Chilled water can be produced by vapor migration whenever condenser water is available at temperatures lower than the required chilled water supply temperature. Up to 46% of the chiller's total cooling capacity can be produced by vapor migration.

Vapor migration free cooling occurs as a result of a free flow of refrigerant through a compressor bypass connection between the chiller evaporator and condenser. The circulation of cold water from a cooling tower through the condenser, which is mounted vertically above the evaporator, cools and condenses the refrigerant into a liquid. The liquid refrigerant flows by gravity to the evaporator where the chilled water warms the refrigerant and vaporizes it. The warmed refrigerant vapor flows up to the condenser, where the cycle is repeated. The circulation of the refrigerant is a result of the difference in density between the refrigerant in the evaporator and in the condenser. The circulation of the refrigerant will continue as long as there is a temperature difference between the cooling tower water and the chilled water.

When specifying chillers with vapor migration free cooling, they should be specified with automatic changeover. The automatic changeover feature will allow the chiller to switch from free cooling to compressorized cooling when cooling water temperatures are not low enough, or when cooling loads become too great for vapor migration free cooling.

7.7.4 Consider Absorption Chillers

Indirect fired absorption chillers are considerably less efficient than equivalently sized electrically driven chillers. As indicated earlier in this chapter, the coefficient of performance (COP) for indirect fired single effect absorption chillers is around 0.7 to 0.8. There are two situations, however, in which indirect fired absorption chillers may be cost-effective: These two are very high electrical demand charge situations or when there is a suitable supply of waste heat to operate the absorption chiller.

Typically, electric utilities have the largest demand charges during the daytime hours of the summer months. Absorption chillers can be used as "peak shaving" chillers during periods of peak electrical demand. Chilled water from the absorption chillers can either replace, or supplement, the chilled water from electrically driven chillers.

The operation of indirect chillers can also be economically attractive when there is a suitable source of waste heat (heat which would otherwise be rejected to the atmosphere or other heat sink). Single effect, indirect fired chillers can operate on steam with a pressure as low as 5 psig or hot water with a temperature as low as 180°F. Obviously, the waste heat must be available at the same time as cooling is required.

7.7.5 Consider Direct Fired Chiller/Heaters

Chiller/heaters serve a dual purpose: They operate as a chiller when cooling is required and also as a heating source, such as a boiler, when heating is required. They can also provide heating and cooling simultaneously, In applications that require both a chiller and a boiler, a chiller/heater is a single piece of equipment that can perform both functions. Since a single piece of equipment serves two purposes, the cost of a separate boiler can be saved, which reduces the installed first cost of a complete central energy plant. Direct fired chillers are available as double effect and can have a COP of 1.0. They are about 40% more efficient than indirect single effect absorption machines.

Most chiller/heaters operate with either natural gas or fuel oil as a source of heat for the absorption process. However, some chiller/heaters are available as dual fuel units, which can operate on either natural gas or fuel oil. Gas and fuel oil prices are usually at their lowest (during the summer months) when electrical demand charges are the highest. Some gas utilities offer financial incentives for installing gas fired chiller/heaters. Additional savings are possible if interruptable gas is used with light fuel oil as a backup if the gas supply is off.

Chapter 8

Cooling Towers and Fluid Coolers

Cooling towers and fluid coolers are heat exchangers that exchange heat between a liquid (usually water or a refrigerant) and ambient air. They are used to reject heat from a liquid to the environment, usually ambient air. Most air conditioning systems, as well as many industrial processes, generate waste heat that must be removed from the system. Unless the heat is stored for some other use, it must be dissipated. This is the function of a cooling tower. Although cooling towers are frequently utilized in many industrial facilities and are used to transfer large amounts of energy, they are frequently neglected and are prime candidates for improvements in energy conservation.

8.1 BASIC PRINCIPLES OF COOLING TOWERS AND FLUID COOLERS

8.1.1 Definition of Terms Used for Cooling Towers and Fluid Coolers

For an explanation of terms used for cooling towers and fluid coolers, see the Glossary.

8.1.2 Basics of Cooling Tower Operation

Atmospheric air is a mixture of a number of gases (primarily nitrogen and oxygen), as well as various quantities of water vapor. Atmospheric air also contains impurities such as airborne dust and pollen in addition to many gaseous pollutants. The thermal condition of air can be defined by its dry bulb and wet bulb temperatures. The dry bulb temperature of an air-vapor mixture is simply the temperature that would be read from an ordinary thermometer. Changes in dry bulb temperature indicate changes in the sensible heat of the mixture. The wet bulb temperature of an air-vapor mixture is a measure of the moisture content of the air-vapor mixture at a given dry bulb temperature. The dry bulb and wet bulb temperatures of a mixture define its state point. The enthalpy of the mixture is a measure of its total heat content.

Evaporation is a thermodynamic process whereby a liquid, such as water, changes from the liquid phase to the vapor phase. In order for a liquid to undergo evaporation, heat is required to cause the change in phase. This heat is referred to as the latent heat of vaporization, and represents a significant part of the heat rejection capacity of a cooling tower.

A cooling tower cools the circulated process hot water primarily through heat transfer to the ambient air as it passes through the tower and, to a lesser extent, by mass transfer. Ambient, or atmospheric, air is typically circulated through a cooling tower by one or more methods, which may include fans, natural convective currents, natural wind currents or the induction effect from the hot water spray inside the tower. Cooling of the process hot water is accomplished by spraying the water into the ambient air stream as it passes through the tower. Cooling primarily occurs by the transfer of both sensible and latent heat to the ambient air flow.

Sensible heat transfer occurs between the hot water and the air stream primarily by direct contact between the water and air. As the water is sprayed into the air, latent heat is transferred via evaporation of the water into the air stream. As the water evaporates, mass transfer also occurs as part of the water vapor is carried away by the air stream. Approximately 1/4 to 1/3 of the cooling effect in a cooling tower occurs as the result of conduction and convection heat transfer between the water and air. This heat transfer is a function of the surface area of the air/water contact and the temperature difference between the hot process water and the dry bulb temperature of the air. The rest of the cooling effect of a cooling tower occurs by evaporation and by the loss of water that is carried away by the air stream. The mass of the water leaving the tower is less than the mass of the water entering the tower. Sensible heat transfer diminishes as the water temperature and air dry bulb temperature become nearly equal to each other. Latent heat transfer decreases as the air becomes saturated with water vapor.

The heat transfer, or cooling effect, can be increased by increasing the area and the time of contact between the water and air, up to a point. Heat transfer between the water and air can be improved by the addition of splash bars or filming-type fill, which increases, the surface area of contact between the air and water, and expose a very large water surface to atmospheric air.

There are several variables that directly affect the thermal performance for any given cooling tower, including the water flow rate, the entering and leaving water temperatures, the ambient air flow rate, and the entering and leaving air wet bulb temperatures. The temperature difference between the entering air wet bulb temperature and the leaving water temperature is referred to as the approach, and is shown graphically in Figure 8-1 for a counterflow cooling tower (a tower in which the water flow direction is opposite to the direction of air flow). The difference between the entering and leaving water temperatures is referred to as the range and is also shown graphically in Figure 8-1.

The amount of heat transferred from the hot water to the air is proportional to the difference in enthalpy between the air entering and

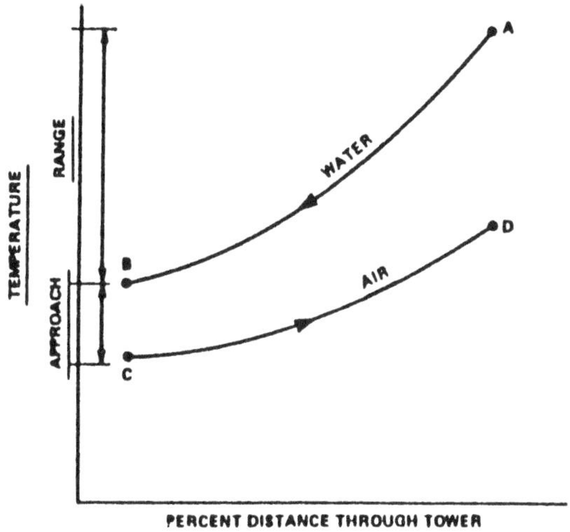

Figure 8-1. Temperature relationship between water and air in a counterflow cooling tower.

leaving conditions, as shown on the psychrometric chart of Figure 8-2. Referring to a psychrometric chart, it may be observed that the lines of constant enthalpy are very nearly parallel to the line of constant wet bulb temperature for an air-vapor mixture. Since this is true, it may be concluded that for any given tower and thermal load, the thermal performance of a tower is primarily dependent upon the entering air wet bulb temperature.

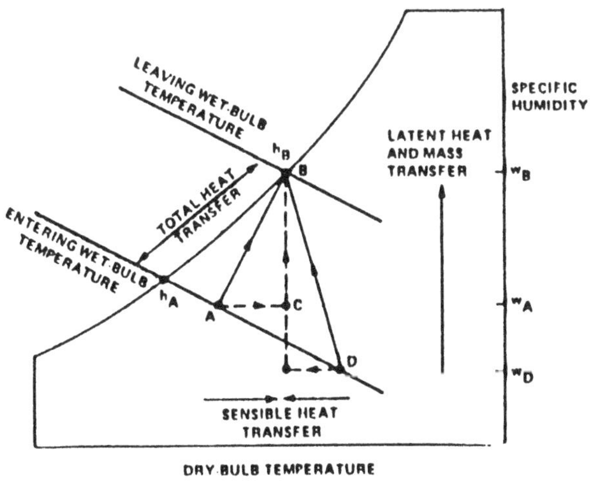

Figure 8-2. Psychrometric analysis of air passing through a cooling tower.

8.2 TYPES OF COOLING TOWERS

There are two basic types of cooling towers: the direct-contact type and the indirect-contact type. In a direct-contact tower, the hot water comes in direct contact with the cooling air as it flows through the tower. Heat from the water is transferred directly to the air via sensible heat transfer and latent heat transfer (evaporative cooling). Direct-contact towers are also referred to as "open loop" towers.

Indirect-contact cooling towers are frequently referred to as fluid coolers. They usually contain two separate water circuits. The hot process water flows through a heat exchanger, such as a coil, through which sensible heat is transferred to the cooling air. The heat transfer can be increased with evaporative cooling, which is accomplished by a second water circuit that sprays water on the exterior of the heat exchanger. Water in the exterior loop evaporates as the air passes through, thereby increasing the heat transfer to the cooling air through latent cooling. As a general rule, indirect-contact towers are not as efficient as the direct-contact type since the heat exchanger is not 100% efficient in the transfer of heat to the cooling air. Indirect towers are frequently used when the primary water circuit may not become contaminated by the ambient air and, therefore, is contained in a "closed loop." An indirect-contact cooling tower is shown in Figure 8-3.

Another basic type or classification for cooling towers is the method by which ambient air is circulated through the tower as it cools the water. Ambient air circulation may be accomplished by natural air draft (flow) or

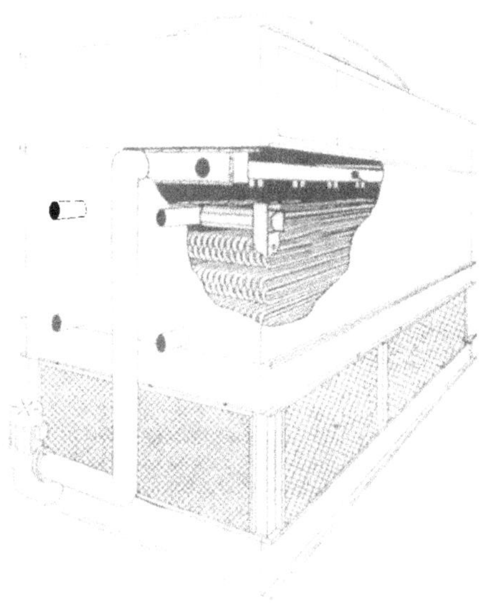

Figure 8-3. Indirect-contact cooling tower (*Courtesy of Evpaco*).

mechanical draft. Natural draft towers have no mechanical device, such as a fan, to circulate air through the tower. Air flow results from natural induction (aspiration), which is caused by the hot water spray inside the tower. Natural draft towers operate as the result of a density differential caused by the hot water spray. As the hot water is sprayed into the air, the air is heated, becomes less dense and rises due to the buoyancy of the lower air density. The hot water is sprayed in a parallel flow pattern, which induces air flow through the tower from bottom to top. Since natural draft towers are aspirated by a hot water spray, they are typically open on the inside and do not contain fill to obstruct the air or water flow.

One type of natural draft tower is the atmospheric spray tower, as presented in Figure 8-4. Atmospheric towers are usually small towers that utilize natural draft air flow for relatively small thermal load and low water flow applications. Since atmospheric towers depend upon natural air flow, their performance is susceptible to adverse wind conditions. They are best suited for small heat load applications when precise control of the cold water temperature is not required.

Another type of natural draft tower is the hyperbolic natural draft tower. A hyperbolic natural draft tower is shown in Figure 8-5. These types of towers are use for processes that have large heat loads, such as electrical generating plants or chemical plants. These towers can handle water flow requirements of 250,000 gal/min and can be over 50 feet in height.

Mechanical draft towers use one, or more, fans to provide forced air circulation through a tower. The fans circulate air at a controlled rate and, therefore, the thermal performance of mechanical draft towers is more

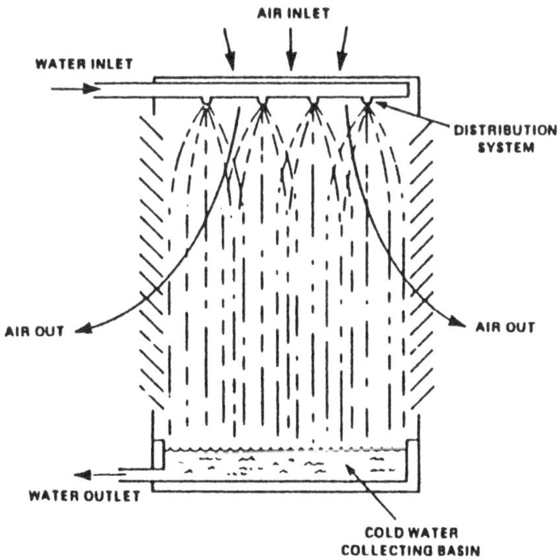

Figure 8-4. Atmospheric spray tower.

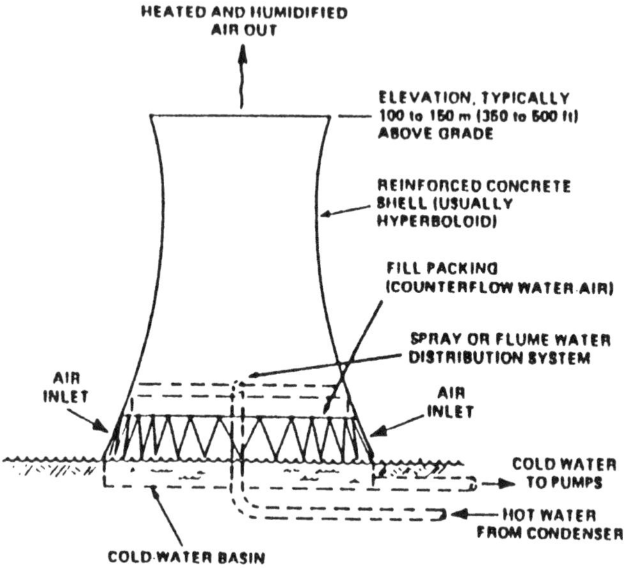

Figure 8-5. Hyperbolic cooling tower.

predictable and controllable. The use of fans also regulates the air flow through the tower in response to changing thermal loads and changing ambient conditions.

Mechanical draft towers may be either of the forced draft or induced draft type. Forced draft towers have the circulation fan located where the

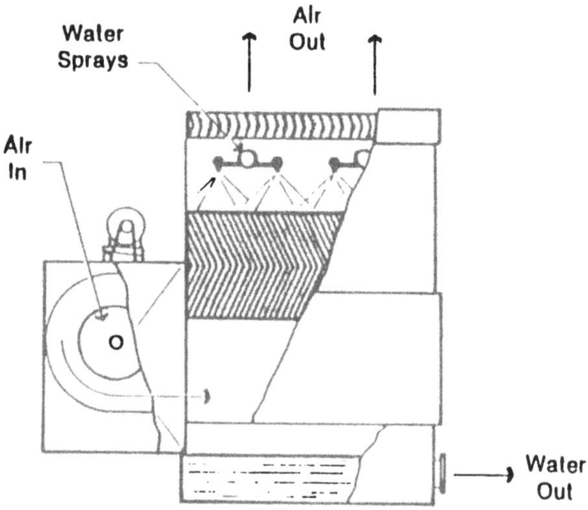

Figure 8-6. Forced draft counterflow tower.

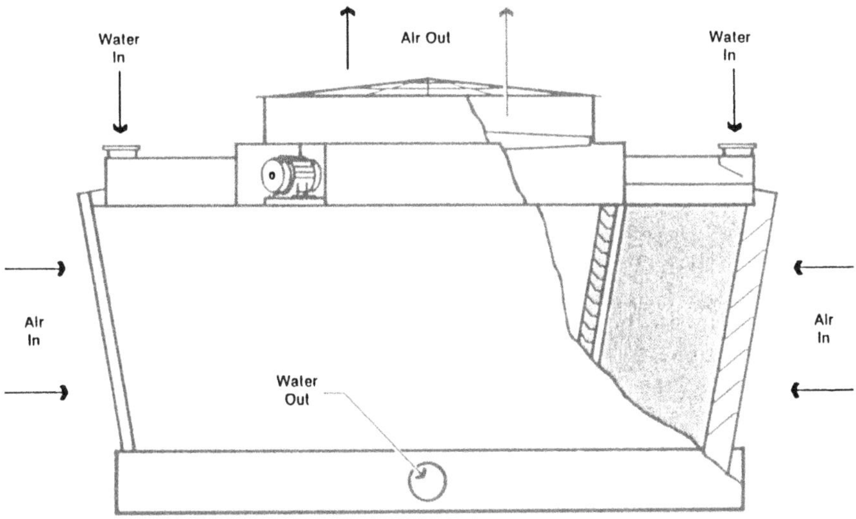

Figure 8-7. Induced draft crossflow cooling tower.

ambient air enters the tower. The fan then pressurizes the air, forcing it to flow through the tower. Forced draft fans typically have high air velocities at the entrance and low velocities at the tower discharge. Forced draft towers may be located indoors, with the intake air and discharge air ducted outdoors. Forced draft towers usually use centrifugal fans for air circulation. A forced draft tower is shown in Figure 8-6.

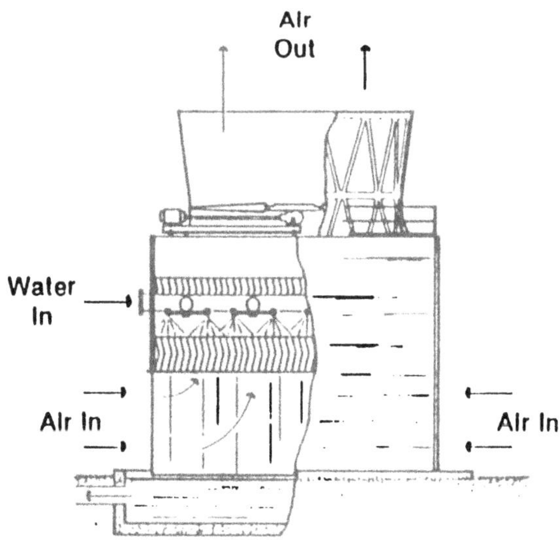

Figure 8-8. Induced draft counterflow tower.

Induced draft towers have the fan at the tower discharge and draw air into the tower. These towers typically have relatively low inlet air velocities and relatively high outlet velocities. Induced draft towers may discharge the air either horizontally or vertically. Vertical discharge cooling towers tend to have less of a problem with recirculation of hot discharge air than forced draft towers. Induced draft towers typically have one, or more, propeller type fan. A typical induced draft cooling tower is shown in Figure 8-7.

Induced draft towers may also be classified according to their air flow pattern. In a counterflow tower, the air moves opposite to the direction of the water flow. Typically, the hot water spray moves vertically from top to bottom, while the induced air flows from the bottom and is discharged from the top of the tower. A typical induced draft counterflow tower is presented in Figure 8-8.

The air flow pattern may also be a crossflow arrangement. In a crossflow tower, the air flow is perpendicular to the water flow for part, or all, of the

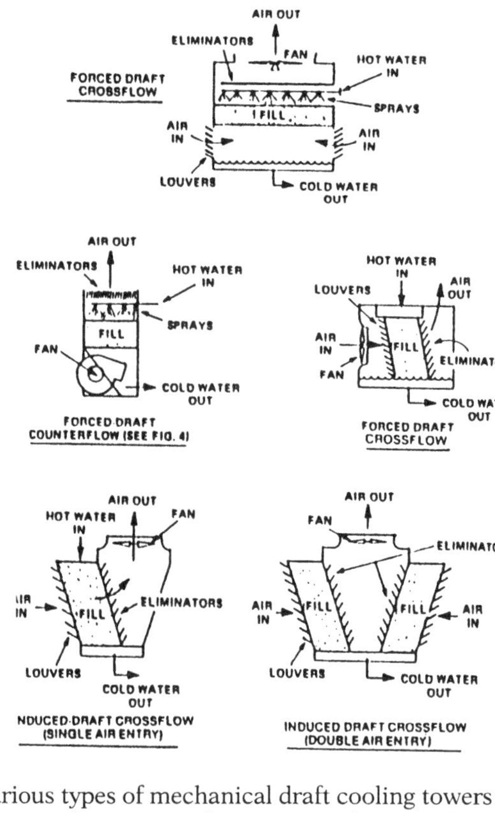

Figure 8-9. Various types of mechanical draft cooling towers (*Copyright 1992 by the American Society of Heating, Refrigerating and Air-Conditioning Engineers, Inc. from the ASHRAE HVAC Systems and Equipment Handbook. Used by permission*).

way through the tower. The tower shown in Figure 8-6 is a crossflow arrangement. Crossflow towers may have a single or double air inlet configuration.

Figure 8-9 summarizes the various types of mechanical draft cooling towers discussed above.

8.3 FACTORS AFFECTING THE PERFORMANCE OF COOLING TOWERS

As indicated earlier, for a given tower, the thermal performance is primarily a function of the ambient air wet bulb temperature. The lower the wet bulb temperature of the ambient air, the colder the process water can be, with the freezing temperature of the water being the lowest allowable temperature. The ambient air conditions are a function of the prevailing weather conditions at the facility. Appendix A provides weather design conditions for various locations in the United States. Obviously, a facilities engineer/manager has no control over ambient weather conditions; however, tower selections must be based upon the design ambient conditions at the site.

The process heat load also has a significant effect on tower performance. The thermal load of the process usually determines the water flow rate and the water temperatures. The heat load determines the tower-entering temperature. Usually, the facilities engineer cannot control these temperatures. For any given set of ambient conditions, the size and cost of the cooling tower is a function of the process load.

The location of the tower at the facility site can have a significant impact on the thermal performance of a cooling tower. Interference from other structures, especially other cooling towers, can adversely affect a tower's performance. If another cooling tower is located on the windward (upwind) side of a given tower, the heat and moisture discharged from the leeward tower frequently will reduce the cooling capacity of a tower. This can occur as the result of heat and moisture being drawn into the downwind tower, as shown in Figure 8-10. The interference caused by the

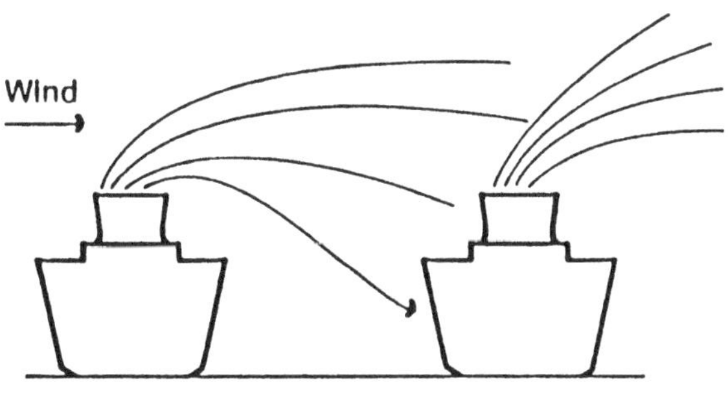

Figure 8-10. Tower interference.

upwind tower increases the dry bulb and wet bulb temperatures of the ambient air entering the downwind tower.

Obstructions that restrict the free flow of air into and out of a tower can also affect tower performance. A tower should be located such that any obstructions, such as walls, buildings and other large structures, do not restrict the free flow of air. The tower manufacturer should be consulted regarding minimum distances between a tower and any obstructions.

The volume and velocity at which ambient air flows through a tower has a significant impact on tower performance. Generally speaking, for any given set of ambient conditions for a given tower, the greater the volume of ambient air circulated through a tower, the colder the return water. As will be discussed later, the rate of air flow is frequently used to control the temperature of the return water from the tower. When the ambient air wet bulb temperature becomes low enough, the rate of air flow through forced draft towers is reduced to maintain a minimum return water temperature.

The prevailing wind at the tower location can influence tower performance. The effect that wind has on tower performance is a function of the wind velocity and direction. As the wind flows past a tower, the wind can cause a relative high pressure on the windward (upwind) side of the tower and a relative low pressure on the leeward (downwind) side. If the high pressure area is such that it is at the tower air inlet, increased air flow can result in an improvement of tower performance. Conversely, if the relative low pressure is at the tower inlet, air flow through the tower may be reduced. In addition to disturbing the normal air flow pattern through a tower, wind may cause discharge air from the tower to be recirculated to the tower inlet. If there is a significant amount of recirculation, there can be a significant adverse effect on the performance of the cooling tower.

Recirculation of the tower air discharge (the tower plume) will cause a significant reduction in the cooling performance of a tower. A typical recirculation situation is shown graphically in Figure 8-11.

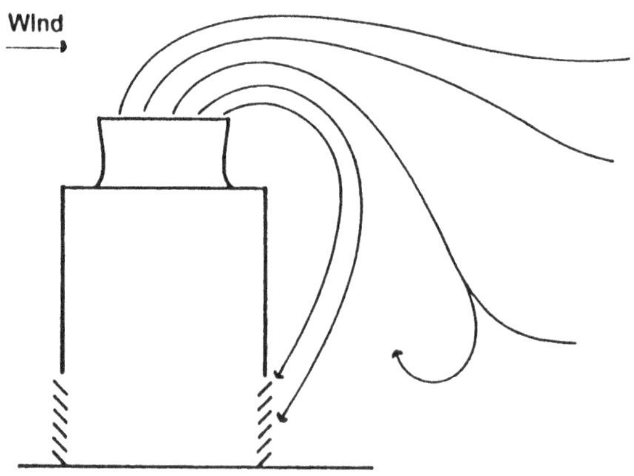

Figure 8-11. Recirculation from a cooling tower.

The air in a tower plume always has greater dry bulb and wet bulb temperatures than the surrounding ambient air. This is due to the fact that the air in the tower plume has absorbed heat and moisture from the process water. The air from a tower plume is usually saturated with moisture. If part of this air is recirculated, the air entering the tower is at a higher dry bulb and wet bulb temperature than it would be if only ambient air were entering the tower. The potential for recirculation is primarily a function of wind direction and velocities; however, tower construction can have an effect. Higher wind velocities usually increase the potential for recirculation. The plume discharge velocity also has a significant influence on the potential for recirculation. Greater tower discharge velocities tend to reduce the tendency for recirculation. The length of time and the surface area of the process water that is exposed to the circulated air also affects tower performance and is usually a function of tower design. If the height of a tower is increased, the length of time for heat and mass transfer to occur is increased. There are a number of limitations on tower height, however. There may be structural, as well as other physical limitations on tower height. The tower height also has an effect on pumping horsepower for the process water, which could increase overall energy consumption for the system.

The heat and mass transfer between the circulated air and the process water can be improved by methods other than increasing tower height. Providing an even distribution of water inside the tower improves its thermal performance. Providing a uniform water spray of small droplets significantly increases the surface area of the water in contact with the air. A uniform spray of small droplets is usually accomplished by spraying the water through nozzles. The effectiveness of spray nozzles is a function of

Figure 8-12. Cooling tower spray nozzle.

Figure 8-13. Splash type fill.

the nozzle design, water pressure, nozzle installation pattern, spray chamber height and the effect of any obstructions within the tower spray chamber. A typical spray nozzle is shown in Figure 8-12.

Heat and mass transfer between the water and air can also be improved by the addition of fill within the tower spray chamber. Fill is placed inside the tower such that the water flows over the fill as it falls within the tower. As the water splashes and cascades over the fill, the contact between the process water and the circulated air is increased. There are two basic types

Figure 8-14. Film type fill.

of cooling tower fill: the splash type, as shown in Figure 8-13, and the film type, as shown in Figure 8-14. The splash type fill breaks the water up and changes its vertical flow path causing it to cascade through successive offset levels of parallel splash bars. The fill causes the water to splash into smaller droplets as well as wets the surface of the individual splash bars.

Film type fill causes the water to spread its water flow in a thin, even film as it flows down through the tower. The film fill provides large surface areas for the water to flow over, thereby significantly increasing the surface area of contact between the water and air. Tower fill can be made of any material that can be formed or molded into shaped sheets. Polyvinyl chloride (PVC) plastic is popular due to its resistance to corrosion and decomposition.

Finally, the cleanliness of the circulated process water can affect the thermal performance of a cooling tower. A tower is essentially an air washer—a washer that washes unfiltered outdoor air. Since ambient air typically contains dust, insects, leaves, etc., tower water can quickly become contaminated with debris. Tower water is also subject to microbial growth within the water. The warm water with debris can facilitate the growth of microorganisms. Unless steps are taken to periodically clean and chemically treat tower water, the thermal performance of the tower will quickly deteriorate.

8.4 SELECTING COOLING TOWERS FOR OPTIMUM ENERGY EFFICIENCY

When a new cooling tower is being selected as part of a new system, or as a replacement, the facilities engineer has the most control over tower energy efficiency. This is the time when the engineer has the most options. Submittals from tower manufacturers should be carefully reviewed for energy efficiency as well as installed cost. A lower cost inefficient tower could result in a much higher life cycle cost.

8.4.1 Tower Fans and Fan Motors

Forced draft cooling towers typically have centrifugal fans while induced draft towers typically have propeller type fans. Forced draft towers have higher static pressure requirements than induced draft towers in order for the ambient air to flow through the tower. This is due to the fact that the fan must force the air flow through the tower at a higher static pressure. For any given air flow rate, the horsepower required for air to flow through a resistance is proportional to the static pressure loss through the resistance. Equation (8-1) expresses this relationship mathematically.

$$Bhp = \frac{QP_s}{6{,}356\eta_f} \qquad \text{Eq. (8.1)}$$

where Bhp = the brake horsepower supplied to the fan, hp
Q = the air flow rate, ft^3/min

P_s = the static pressure, wg
6,356 = a unit conversion factor
η_f = the fan static efficiency, %

It may be observed from Equation (8.1) that the required fan brake horsepower is directly proportional to the static pressure for any given air flow and fan efficiency. Propeller fans can have a lower efficiency than a centrifugal fan; however, the static pressure loss through an induced draft tower is usually much less than through a forced draft tower. Therefore, the peak fan horsepower for towers should be compared when selecting and specifying a cooling tower. In general, induced draft fans require a lower fan brake horsepower than centrifugal fans.

Induced draft cooling towers may also have improved energy efficiency over forced draft towers due to a reduced tendency for recirculation of the heat- and moisture-laden discharge air. For forced draft towers, the fan is on the tower air inlet. This results in a relatively low pressure, low velocity area at the tower air inlet while a relatively high pressure exists at the tower outlet. With a relative low pressure at the outlet, there is a natural tendency for recirculation. For induced draft towers, the same pressure relationship exists; however, the air discharge is at a much higher velocity. The higher velocity tends to force the discharge air away from the tower, reducing the tendency for recirculation.

Fan motors are a major consumer of energy by a cooling tower system. Whenever possible, high efficiency motors should be specified for a new or replacement tower. Depending upon the size of the tower and its temperature control method, multi-speed motors may also be suitable for a cooling tower. Electric motors are discussed in more detail elsewhere in this book.

The fan laws provide relationships between fan speed, air flow, pressure and power consumption for a given fan and operating conditions. The fan laws are given in "Chapter 9: Air Distribution and HVAC Systems." It may be observed from the fan laws that the air flow rate varies directly as the rotational speed of the fan. It may also be observed that the power required to drive the fan varies as the cube of the fan speed. These relationships show that significant power savings can be achieved for a given fan if it is operated at a lower air flow and rotational speed.

Velocity (regain venturi) stacks installed on the fan discharge may also reduce the fan motor power consumption. Venturi stacks can increase the air flow through a tower by as much as 7% at the same fan motor power consumption.

8.4.2 Cooling Tower Capacity Control

Earlier it was discussed that the capacity control of a given tower was primarily a function of the ambient wet bulb temperature. For any given process, water flow rate and air flow rate, the capacity of a tower is inversely proportional to the ambient wet bulb temperature. As the ambient wet bulb temperature increases, the capacity of the tower decreases. Conversely, as the wet bulb temperature decreases, the tower cooling capacity decreases.

When selecting and sizing a tower, it is usually necessary to size it for a summer design wet bulb temperature. As the seasons change and the ambient wet bulb temperature decreases, the process cooling load frequently remains constant, or declines only slightly. As a result, cooling capacity control is usually required for cooling towers. Typically, there are two general methods to control tower capacity: modulation of the water flow and modulation of the air flow.

The capacity of a tower can be reduced when the ambient wet bulb temperature decreases by reducing the rate at which water flows through the tower. Tower water flow can be modulated by either variable flow pumping or a bypass piping arrangement. Variable flow pumping may utilize multiple pumps with different flow capacities or variable speed pumps. In either variable flow arrangement, the flow of water through the tower is modulated to maintain the return water temperature at a desired set point temperature. A bypass piping arrangement is shown in Figure 8-15. In this piping arrangement, the circulated process water is either passed through the cooling tower or bypassed around it. The relative proportion of water flow through the tower or around is a function of the return water set point temperature and the ambient wet bulb temperature.

Although the variable pumping method and the bypass piping arrangement method are effective capacity control methods and save on system pumping energy consumption, they are not best methods for capacity control. In order for a cooling tower to function properly, it is necessary that the water spray pattern inside the tower be maintained. Reduced water flow to the tower will result in reduced water flow spray to the nozzles. Since the nozzles are designed for full flow operation, reducing the water flow rate or pressure would have a negative impact on the spray

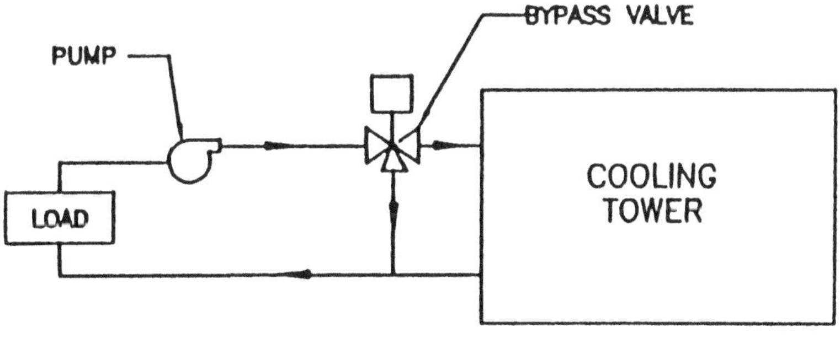

Figure 8-15. Cooling tower bypass control.

pattern. The reduced water flow will cause the spray pattern to deteriorate to a point where the water does not spray; rather, it just falls straight down to the tower basin. A lower water flow rate through a tower may also promote freezing within the tower when ambient conditions are low.

The energy demand by the cooling tower fans can be four to five times that of the pumps circulating the process fluid. Therefore, from an energy management standpoint, as well as an operational standpoint, control of the air flow through a tower is a much more effective method for controlling the capacity of a tower. Controlling a tower cooling capacity by modulating the air flow rate is energy effective and will result in fewer operational problems.

The simplest and easiest method for controlling air flow through a mechanical draft tower is to cycle the fan on or off in response to the required cooling capacity. This method of control may be suitable for small towers; however, it is not suitable for large towers. Frequently, cycling of the fan motor can lead to premature motor burnout. If this type of control is employed, it is recommended that some type of control be included, which limits the frequency of start/stop cycles. As the ambient air temperature decreases, the frequency of motor starts increases until the maximum number of start/stop cycles per hour is reached, at which time the control will become ineffective.

An improvement to the fan cycling method is multi-speed motors, which usually have two speeds with variable torque design. Variable torque motors vary the torque directly with the motor rpm. Typically, the low speed is half of the fast speed. This allows for three control settings: off, half speed and full speed. Since the torque and motor speed are reduced at low speed, significant energy savings can be achieved with two speed motors at a relatively low cost. Forced draft cooling towers usually use double inlet centrifugal fans to circulate ambient air through a tower. Centrifugal fans are required due to the higher static pressure losses as the air flows through a forced draft tower. Since they circulate large quantities of air at relatively high static pressures, the motor horsepower for these fans is usually quite large. The fans and motors are usually suitable for on/off control with the limitations listed above. Forced draft cooling towers also use multi-speed motors frequently. Centrifugal fans may also be controlled by two other methods: discharge dampers and variable frequency motor drives.

Discharge dampers are modulating dampers that are mounted at the discharge of a centrifugal fan. The fan is operated at full speed, while the dampers modulate the air flow rate by modulating the static pressure against which the fan must operate. As the dampers are closed, the static pressure increases and the fan "rides up the fan curve." As the fan rides the fan curve, the air flow rate varies with the additional static pressure loss imposed by the discharge dampers. Although discharge dampers are an effective means of capacity control, they are not as effective from an energy management standpoint. The fan and fan motor operate at constant speed as the discharge dampers provide an artificial pressure loss to the system.

The discharge dampers merely dissipate the excess energy provided by the fan motor as a pressure loss.

Variable frequency motor drives are an effective means to control a tower capacity from both a control standpoint and an energy management standpoint. Variable frequency drives vary the frequency of electrical power supplied to the tower fan motors, which, in turn, varies the rotational speed of the fan. The cost of variable frequency motor drives is becoming more attractive as they gain widespread use and can provide significant electrical energy savings for tower fan motors. When considering variable speed drives for tower fans, caution should be exercised during selection of the control system for variable frequency drives. Drive controls must ensure that the fan not be allowed to operate in an overspeed condition, nor should the fan be allowed to operate at its critical speed, which could result in excessive fan vibration.

Induced draft towers use propeller type fans almost exclusively. These fan are suitable for moving large volumes of air with a relatively low static pressure. The air flow capacity of propeller fans may also be controlled by on/off cycling, multi-speed fan motors and variable frequency motor drives as discussed above. Propeller fans may also be controlled with gear speed reducers and by varying the pitch of the fan blades.

Gear reduction units use differential gears of positive engagement to vary the rotational speed of a propeller fan in response to the cooling load, while the fan motor operates at a constant speed. They typically have spiral bevel and helical gear sets in a gear box, although worm gears may also be used. Gear reduction units are used on towers with 5 hp and larger motors. Although there is some loss in efficiency through the gear train, gear reducers are able to transmit power efficiently and can provide effective capacity control.

Variable pitch propeller fans are fans on which the pitch angle of the fan blades is varied as required by the load on the cooling tower. The fan and motor rotate at a constant speed. The blade pitch angle is increased or decreased, which changes the angle of attack as the air passes over the fan blade. Propeller fan blades are airfoil-shaped and produce motion of air in an axial direction as they rotate. For any given rotational speed, the rate at which a given propeller fan moves air is a function of the blade pitch angle. As the blade pitch angle is increased, the fan moves more air. Conversely, as the blade pitch angle is reduced, the air flow rate through the fan is reduced. Variation of the blade angle can result in a nearly uniform modulation of air flow.

The fan motor power consumption is reduced as the blade pitch angle is reduced and the fan is unloaded. The load on the fan motor is proportional to the fan blade pitch angle for any given propeller fan. A 10°F reduction in the ambient wet bulb temperature from 65°F to 55°F can result in a savings in fan power of up to 65%.

The pitch of the blades on a variable pitch propeller fan may be changed manually; however, most variable pitch fan blades change pitch automatically in response to the required return water temperature. Since the

TABLE 8.1 Power Consumption Comparison for Typical 400 Ton Cooling Towers (Hensley 1981)

Colling Tower Type	Operating Fan Motor HP	Fan Motor kW	Tower Pump Head (ft)	Additional Pump Motor kW	Total Operating kW
Counterflow with centrifugal fan	40	32.4	23	6.9	39.3
Induced draft with propeller fan	20	16.2	10	3.0	19.2

fans rotate at a constant speed, it is not necessary to be concerned with overspeed conditions or fan critical speeds.

Typical figures for energy consumption for 400-ton cooling towers are presented in Tables 8.1 and 8.2 for comparison purposes. Table 8.1 compares the total operating power consumption for a system with a forced draft tower with centrifugal fans to an induced draft tower with propeller fans.

The systems for both of the towers in Table 8.1 are assumed to have cooling capacities of 400-tons; the fan and pump motors have efficiencies of 92%; and the pump efficiency was assumed to be 82%. It may be observed from Table 8.1 that the total operating power (kW) of the induced tower with propeller fans is about 49% that of the forced draft tower with centrifugal fans.

Table 8.2 compares the total fan operating power draw of a 400-ton induced draft tower with that of a 400-ton forced draft tower for the

TABLE 8.2 Typical Power Consumption for Various Tower Control Methods (Hensley 1981)

Operating Situation	Hours of Operation	Average Power Draw (kW)	Propeller Fan Energy Consumption (kWH)	Centrifugal Fan Energy Consumption (kWH)
Constant operation at full capacity	1202.2	Propeller=16.2 Centrifugal=32.4	19475.6	38951.2
Single speed fan cycling	Propeller=765.3 Centrifugal=852.7	Propeller=16.2 Centrifugal=32.4	12397.9	27627.5
Two speed fan cycling	Propeller=1132 Centrifugal=1146	Propeller=4.3 Centrifugal=8.55	4867.6	9798.3
Variable pitch fan at constant speed	1202.2	Propeller=2.72 Centrifugal=5.44	3270	6540
variable speed control	1202.2	Propeller=1.99 Centrifugal=3.98	2392.4	4784.8

various control methods discussed above. The forced draft tower is assumed to have centrifugal fans while the induced draft tower is assumed to have propeller fans. It may be observed from Table 8.2 that the induced draft tower with a propeller fan and variable speed control had the lowest operating power requirement, followed by the induced draft tower with a propeller fan and variable pitch control. Table 8.2 is based on the assumption that the induced draft tower with the propeller fans operate slightly fewer hours due to the fact that induced draft towers can produce some cooling with the fans off as a natural draft tower.

When controlling the capacity of a tower by varying the air flow rate, consideration should be given to the air discharge velocity at part load operating conditions. A low tower exit velocity relative to the ambient wind can increase the possibility of recirculation of the discharge air into the tower air inlet. It is recommended that the tower discharge velocity be at least 1.4 times that of the prevailing ambient wind. For a 10 mph (880 fpm) wind, the air discharge velocity should be at least 1,200 fpm.

8.4.3 Drift Reduction

Drift is water droplets that have become entrained in the air stream as it passes through the tower and are carried outside the tower as the air is discharged. The amount of drift may be reduced by the use of drift eliminators in the tower. The amount of water that is lost through drift is a function of the tower configuration, water flow rate, air velocity as it flows through the tower and the design of the drift eliminator. Water loss, or drift, is undesirable since it results in an increase in water costs, as well as increased chemical costs. A tower plume, which contains a large amount of drift, may also result in a chemical buildup on objects downwind of the tower.

Drift eliminators are inertial impact separators that are used to reduce the amount of entrained water droplets in a cooling tower discharge air plume. Water droplets are removed by causing the discharge air to make sudden changes in direction as it flows through the tower. As a result, centrifugal force separates the droplets of water from the air, returning them to the tower basin. Drift eliminators are usually made of a corrosion-resistant material, such as wood or PVC plastic. Typical drift eliminators are shown in Figure 8-16.

Generally, well designed drift eliminators can reduce the amount of drift to as low as 0.2% of the water flow rate. The performance of drift eliminators is measured in two ways: the water collection efficiency and the air pressure loss caused by the drift eliminators. Drift eliminators are classified by the number of directional changes, or passes, the air must make. Generally, the greater the number of passes, the better the efficiency and the greater the air pressure loss. The selection of drift eliminators involves a trade-off between the eliminator collection

Figure 8-16. Drift eliminators (*Courtesy of Evapco*).

efficiency and the power requirements imposed on the fan by the increased pressure loss.

8.5 OPERATION AND MAINTENANCE OF COOLING TOWERS FOR ENERGY EFFICIENCY

In order for a cooling tower to operate at peak efficiency, it is important that it be well maintained. There are some operation and maintenance procedures that can have a significant effect on a cooling tower energy consumption, and these should be the first step in any energy management program for cooling towers.

The cleanliness and quality of the water circulated in the cooling tower loop can have a significant impact on tower efficiency and energy consumption. Since a cooling tower essentially operates as an air washer, the water can collect large amounts of dissolved solids and other impurities in a very short period of time. These deposits of dissolved solids can cause scaling, corrosion and/or silt deposits. The buildup of deposits on heat transfer surfaces can reduce heat transfer efficiency. In addition to a buildup of dissolved solids, tower water can also have a buildup of algae, slime and other biological growth. The growth of biological microorganisms is usually controlled by the addition of chemicals and scale inhibitors.

One method of controlling the buildup of contaminants in the tower water is through a process called *"blow down."* Blow down involves bleeding off a small percentage of the circulated water and discharging it to a suitable sewer or collection system. Mechanical filtration may also be used to remove dissolved solids entrained in the tower water. Deposits of silt and mud can partially be controlled with periodic blow down of the tower water; however, the deposits may have to be removed by physical cleaning occasionally.

The buildup of biological organisms can reduce tower efficiency by (1) forming an insulating coating on heat transfer surfaces, (2) restricting fluid flow, (3) promoting corrosion and (4) attacking tower components. The growth of algae can result in clogged water spray nozzles. Fungi, which includes molds and yeast, can accelerate the decay of wooden tower components. The most common method for controlling biological organisms in tower water is by chemical treatment of the water. A chemical treatment company should be consulted for advice on proper chemical treatment. An alternative method for treating tower water is ozone treatment, which is discussed later in this chapter.

In addition to maintaining a cooling tower for maximum efficiency, the manner in which the tower is operated may result in energy savings. Frequently, the process or equipment served by a cooling tower may save energy if the cold water supplied by the tower is lower than the normal set point temperature. Colder tower return water usually results in energy savings at little or no additional cost for tower operation. For cooling towers serving chillers, studies have shown that there would be approximately 3.5% savings on electrical costs for the compressors and condensers for each 1°F reduction in water supplied to the condenser (Berger 1987a).

8.6 ENERGY CONSERVATION MEASURES FOR COOLING TOWERS

In addition to operation and maintenance procedures that improve the efficiency of cooling towers, there are some energy conservation measures that can result in energy savings. Unlike operation and maintenance procedures, these improve ments involve some capital cost and should be evaluated on a life cycle cost basis. The measures may be considered as improvements to existing tower systems, as well as for new tower installations. Some of the energy conservation measures suggested for new towers may be suitable as a retrofit when renovating an existing tower. Any tower installation that is over 15 years old is a likely candidate for either replacement or renovation. The typical service life for cooling towers ranges from 20 to 35 years.

8.6.1 Free Cooling

Free cooling may be suitable for chilled water systems that employ a chiller to produce chilled water for process or comfort cooling. Mechanical refrigeration (a chiller) is used to produce chilled water typically at a temperature of 40–55°F. Since the ambient summer wet bulb temperature is frequently much higher than the required chilled water temperature, a chiller is required to produce the lower water temperature. However, during the winter months, the ambient wet bulb temperature is much lower and a cooling tower is capable of producing water at temperatures in the same range as that produced by a chiller. Free cooling is also referred to as water-side economizer operation. Free cooling is a method whereby a cooling tower is used to produce lower temperature water in the process water loop while the chiller remains off. Typically, free cooling is available whenever the ambient wet bulb temperature is 50°F, or lower. This can result in significant electrical energy savings when outdoor conditions are suitable.

Actually, free cooling is not really free. Whenever the cooling tower is operating, energy is expended by the pumping system and occasionally by the cooling tower fans. However, significant energy savings can occur whenever the chiller can remain off. In addition to the savings resulting from the chiller being off, the load on a cooling tower is reduced whenever the chiller in the system is off, since chillers are not 100% efficient as they produce chilled water. For a refrigerant compressor machine, power is typically consumed at a rate of 0.80–1.1 kW/ton of cooling, which means that the cooling tower must reject heat at a rate of about 1.25 times the process load. Absorption chillers typically consume power at a rate of about 2.5 kW/ton, which requires that a tower reject heat at a rate approximately 2.5 times the process load. Therefore, free cooling offers an attractive opportunity for energy savings in systems with chillers. There are three methods by which free cooling may be achieved: direct free cooling, indirect free cooling and vapor migration cycle cooling.

Direct free cooling is the simplest and easiest free cooling method; however, it also has the greatest opportunity for problems. Direct free cooling is also referred to as free cooling by interconnection of water circuits, and is shown schematically in Figure 8-17.

Direct free cooling involves the interconnection of the cooling tower water loop with the chilled water loop using bypass valves and interconnecting piping. In this arrangement, the cooling tower water flows through the tower and directly into the chilled water loop, bypassing the chiller. The bypass valves may be open/closed either manually or automatically.

This arrangement makes maximum use of the opportunity for free cooling and probably involves the least capital cost. It is, however, the least recommended method for free cooling. The primary disadvantage of this arrangement is that the relatively dirty tower water may contaminate the relatively clean chilled water. The use of direct free cooling usually requires that the facility engineer devote considerable attention to water treatment. This application usually results in higher water treatment costs. The use of some sort of filtration of the water is strongly recommended since the tower water is not as clean as the chilled water loop.

Since the two water circuits are combined into one, one of the circulating pumps must be bypassed—usually the chilled water pump. In

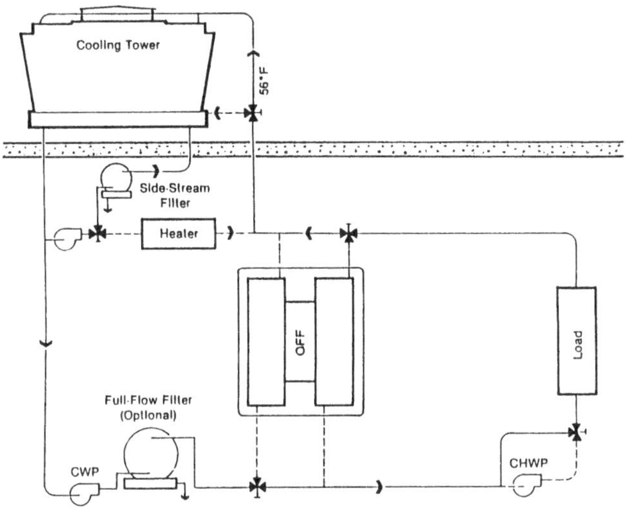

Figure 8-17. Direct free cooling.

some applications, the cooling tower water flow rate may be different from the required flow rate for the chilled water loop, which may result in this method being unsuitable.

Indirect free cooling is similar to direct free cooling except that a heat exchanger is provided between the cooling tower water loop and the chilled water loop. The heat exchanger provides total separation of the two water loops. An indirect free cooling schematic is presented in Figure 8-18.

Separation of the tower water loop and the chilled water loop provides several distinct advantages over direct cooling. The greatest advantage is that the two water streams do not intermix. The relatively dirty tower water does not mix with the relatively clean chilled water. This reduces the need for water treatment since the chilled water loop remains a closed loop. Another advantage is that the flow rate for the tower water need not be equal to the flow rate for the chilled water. The heat exchange between the two loops can be controlled with automatic flow control valves.

There are two main disadvantages to indirect free cooling: the added cost of the heat exchanger and some loss in efficiency. The primary cause for the loss in efficiency is due to the fact that the efficiency of heat exchangers are not 100%. In order for heat to transfer through the heat exchanger, there must be a temperature difference between the fluids in the two loops. Plate and frame heat exchangers are frequently used for indirect free cooling applications. They have relatively large heat transfer surfaces and can effectively transfer heat between two fluids with as little as a 2°F temperature difference between the two fluid streams. Loss of

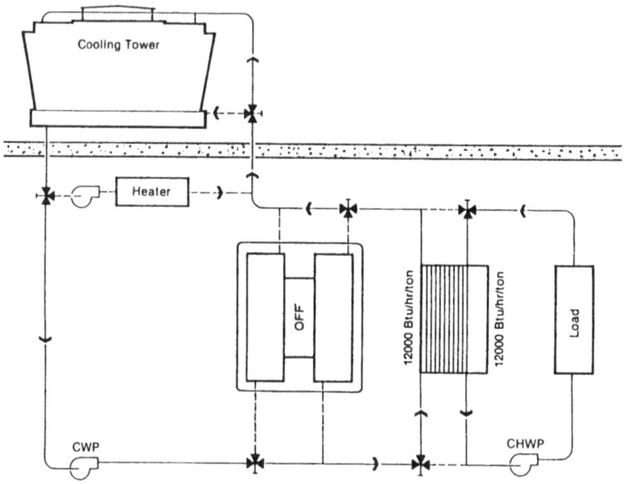

Figure 8-18. Indirect free cooling.

efficiency in indirect free cooling systems results from the additional pressure loss as the fluids flow through the heat exchanger as well as from the fact that a set of pumps must operate in each water loop during free cooling.

Plate and frame heat exchangers provide a highly efficient means of transferring heat between the tower water loop and chilled water loop. To accomplish efficient transfer of heat, plate and frame, heat exchangers have large surface areas through which heat is transferred. Heat transfer plates are used to separate the two fluid streams. These plates are frequently made of stainless steel or some other corrosion-resistant material, which transfers heat efficiently. As a result, these heat exchangers can be costly. When considering indirect free cooling, it is advisable to perform a life cycle analysis using the estimated installed cost for the indirect free cooling system and the projected energy savings.

Vapor migration is a form of indirect free cooling. In vapor migration, the chiller compressor remains off while the refrigerant in the chiller flows between the condenser and evaporator as a result of buoyancy due to changes in the density of the refrigerant. Vapor migration free cooling is shown schematically in Figure 8-19. It is suitable for a limited number of applications where the chiller is suitably equipped for vapor migration operation.

In the vapor migration process, the chiller compressor remains off while the refrigerant is cooled in the condenser by the cooling tower water. The cooling results in a migratory flow of refrigerant between the condenser and evaporator. As the chilled water is cooled by the refrigerant in the evaporator, it evaporates and rises to the

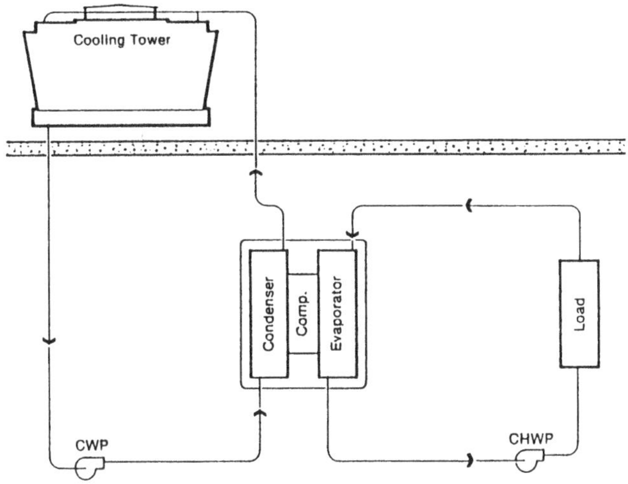

Figure 8-19. Vapor migration free cooling.

condenser. The tower water cools the refrigerant in the condenser, where it condenses into a liquid and flows back to the evaporator by gravity.

Since the heat transfer between the tower water and chilled water is primarily limited to the latent heat resulting from the phase change of the refrigerant, the cooling capacity of vapor migration free cooling is limited to about 25–35% of the full load cooling capacity. In addition, the tower water usually must be 45°F or lower to accomplish adequate heat transfer. This limits the time period when ambient conditions are suitable for free cooling.

8.6.2 Water Filtration Systems

Earlier, the need for cleanliness of the cooling tower water was discussed. Although chemical treatment and tower water blow down provide effective means to maintain clean cooling tower water, water quality can also be improved through water filtration. Filtration is particularly important if direct free cooling is employed. Debris in tower water can clog tower spray nozzles and condenser coils, as well as damage pump impellers.

To improve the cleanliness of tower water, filtration systems can be installed in addition to the regular strainers in the tower water piping. Typically, these filters are side filtration systems, as shown in Figure 8-20. They are designed to continuously filter the tower water by drawing water from the tower basin. These filtration systems normally circulate about 5% of the total circulation rate; however, the filtration rate may be as high as 10%.

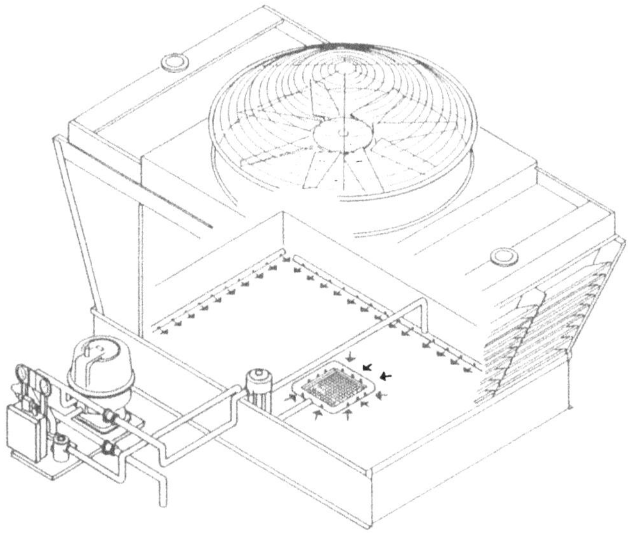

Figure 8-20. Cooling tower filtration system.

8.6.3 Improved Water Distribution

For a cooling tower to operate efficiently, it is important that the circulated water be in contact with the circulated ambient air as much as possible. This is a function of the water distribution system, which consists of tower water piping, spray nozzles, tower fill and water collection basin.

The most important component in the tower water distribution system is the fill. The performance of a tower is dependent upon the ability of the fill to promote maximum contact between the water and air for the maximum amount of time, while providing the least possible restriction to the flow of the ambient air. The various types of fill were discussed earlier and shown in Figures 8-13 and 8-14.

Splash type tower fill tends to be less effective compared to cellular fill, which has a lower resistance to air flow. In splash bar fill, the droplets of water bounce from one layer to the next lower layer, which results in spherical shaped water droplets. Cellular, or film, fill spreads the water out in an even thin film, which increases the surface contact area between the water and air. The film contains much more surface than splash bars, resulting in a greater area of water/air contact. Since the design of cellular fill permits air to flow with less of a static pressure loss, it is much more efficient.

Generally speaking, significant improvements in performance can be made by replacing the fill in the tower. A significant improvement in efficiency may be obtained by replacing splash type fill with cellular fill, with appropriate redistribution sections at various levels, and adjusting the water flow rate. For towers with the splash type fill, consideration should be given to replacing flat splash bars with V-shaped splash bars. V-shaped splash bars improve performance by partially turning the water droplets into film as they flow down the angular surface of the splash bar.

When selecting tower fill, the fill should have the following characteristics and features:

- be chemically inert
- not support scale or be susceptible to attack by algae, bacteria or fungi
- be lightweight and strong
- have a large surface area per unit volume
- have low resistance to air flow

In order for the tower fill to maximize the contact between the circulated water and air, it is imperative that the water be evenly distributed within the tower. The purpose of the spray nozzles are to evenly distribute the water to the tower fill. A typical nozzle is shown in Figure 8-12. When a nozzle is partially or completely clogged, the result is a dry spot over and through the fill. Since air will flow in the path of least resistance, a large portion of the air flows up through a dry spot, wasting energy and reducing the capacity of the tower. It is important to check all spray nozzles to ensure that they are not clogged and are providing an even water spray pattern. Any clogged nozzles should be removed, cleaned, and reinstalled or replaced.

For crossflow or counterflow towers, providing a uniform flow is the key to efficient operation. For maximum performance, the water distribution system must provide a uniform water spray pattern over the fill. The performance of spray nozzles in older towers may not be as efficient as that of newer nozzles. A more uniform water spray pattern can be obtained by installing older nozzles with metered orifice nozzles.

8.6.4 Ozone Treatment of Tower Water

Earlier, the problems associated with algae and biological growth in tower water were discussed. For many years, the primary method for treating tower water to prevent biological growth has been chemical treatment. Since towers loose water by drift, by blow down and by evaporation in the tower exhaust plume, water must continuously be replaced. The loss of tower water by drift, blow down, and as part of the plume along with the continuous supply of make-up water results in a dilution of the chemicals in a tower water system. A typical tower operation is shown in Figure 8-21. Treatment chemicals must periodically be replaced in order to control biological growth, which can be costly. If tower water is not properly treated to control biological growth, a *"bio-film"* will tend to accumulate on the sides and components of the cooling tower, retarding heat transfer and reducing the capacity of the cooling tower. In addition to the direct

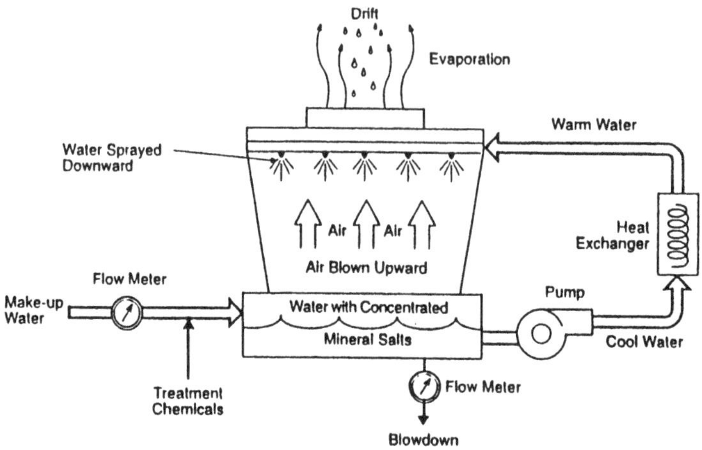

Figure 8-21. Typical cooling tower operation.

problems caused by the bio-film itself, the bio-film promotes the buildup of scale, which adheres to the bio-film.

For many years, ozone, which is a form of oxygen, has been used as a disinfectant. A small amount of ozone can control bio-film by acting as a biocide, which kills the microorganisms in the water. Ozone is a powerful biocide and virus deactivant. It will oxidize many organic and inorganic substances. Under the conditions that typically exist in a cooling tower, ozone is very unstable and has a half-life of less than 10 minutes. The ozone dissipates quickly and will not build up in the tower water. Since ozone is so effective, the need for blow down is reduced and the need for chemical treatment is significantly reduced. This results in reduced energy costs, reduced chemical costs and reduced water costs. There may also be a reduction in wastewater treatment costs. Manufacturers of ozone treatment systems claim an average gain in efficiency of about 10%, and up to a 20% improvement in chiller performance for systems with chillers. Case studies have shown that the savings in chemical costs are also a significant contributor to the cost-effectiveness of ozone treatment systems.

Typically, ozone cooling tower treatment systems consist of an air compressor, an air dryer, water and oil coalescing filters, a particle filter, ozone injectors, an ozone generator and a monitoring/control system. In the ozone generation process, ambient air is compressed, dried and then ionized to produce ozone. A typically ozone tower treatment process is shown in Figure 8-22. Ozone generation is accomplished by passing the compressed air through a high voltage alternating current (6–20 kV) field. As the air is exposed to the electrical field, the oxygen disassociates and forms single oxygen molecules to form ozone (O_3). The ozone is then injected into the water with a ventrui in a recirculation line from the tower water basin. Since the solubility of ozone is a function of temperature, the

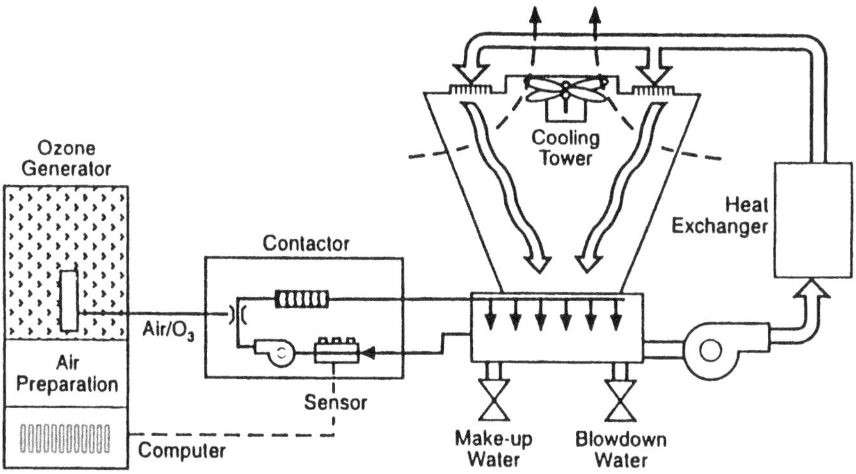

Figure 8-22. Process for ozone treatment of cooling tower water.

basin, which has the coldest water temperature, is the optimum injection point.

Ozone systems available for tower water treatment are modular units that are fully self-contained with an independent water circulation system. The size and cost of such systems are dependent upon the operating water temperature and the ambient conditions for the tower. An estimate of the installed cost of the system, the expected savings and the life cycle costs of the improvement should be determined prior to installing an ozone treatment system.

Chapter 9

Air Distribution and HVAC Systems

Almost all occupied buildings require some type of heating and ventilation; many also require cooling. Frequently, a manufacturing process needs a controlled environment. Occasionally, a process requires special ventilation, heating or cooling. A comfortable environment inside a building can increase worker productivity and efficiency by reducing worker fatigue and reducing environmental conditions that may be harmful to worker health.

Many facilities have processes that require specific environmental conditions, including proper temperature, humidity, air motion, air quality and air cleanliness. Frequently, processes require that airborne contaminants generated by the process be collected and treated before being discharged from the building or returned to the area. Exhaust systems are used to collect and remove airborne contaminants—such as dust, fumes, smoke, vapors and heat—which can create unhealthy or undesirable indoor environmental conditions.

Since almost all buildings, and many processes, require heating, ventilating and air conditioning (HVAC), these systems consume significant amounts of energy. Maintaining environmental conditions inside a building, or for a process, can be a large portion of the energy budget for a facility. Energy can be saved by the proper operation of environmental control systems, such as air distribution and HVAC systems.

9.1 BASIC PRINCIPLES OF HVAC SYSTEMS

9.1.1 Definition of Terms Used for HVAC Systems

See Glossary for definitions of terms related to HVAC systems.

9.1.2 Heating Loads

Almost all occupied buildings and facilities located in cold climates require some type of heating. Even unoccupied buildings frequently require heating to prevent freezing of the building's contents and any water piping inside the building. The capacity of heating systems is determined by the amount of heat loss from the building. The amount of heat loss is a function of the inside temperature, the outside air temperature, the insulating properties of the exterior surfaces of the building, and the amount of air that enters the building by ventilation or natural air leakage, called infiltration.

The purpose of a building heating system is to maintain a desired indoor air temperature and indoor comfort conditions despite the loss of heat to the outdoors. The amount of heat loss from a building is directly related to the difference between the indoor and outdoor temperatures. A heating load for a building occurs as the result of heat loss, or transfer, from inside the building to outside.

Heat loss from a building, or space, occurs by two methods.

1. The transmission of heat through the building's exterior surfaces such as walls, roofs, etc.
2. Through infiltration and/or ventilation, which exchanges warm indoor air with colder outside air.

The purpose of the heating system is to maintain the desired indoor temperature, say 70°F. In order to do this, it must provide the amount of heat equal to the heat loss from the space at the desired indoor temperature. Figure 9-1 shows a typical heat balance for a space requiring heat.

For the temperature of the heated space to remain constant at the desired set point, the heat supplied by the system must be equal to the combined transmission, ventilation and infiltration heat losses. The heat loss (combined heat losses) for a space are determined based upon an assumed constant outdoor winter design temperature. The winter design temperature is based on the climatic conditions for the location of the building and on the intended usage of the building. Methods for determining the heat loss for a building are beyond the scope of this book. The reader may wish to refer to ASHRAE (1997) and Mull (1997) for heat loss calculation methods.

The heating load for a space is usually determined assuming that there is no heat contribution from internal heat sources, such as lights, people, equipment, etc. The peak heating load usually occurs at night when the outside temperatures are the lowest. At night the building usage is usually quite low and internal heat sources may not be present. If a credit for some internal heat source is taken, the heat source (heat gain) must be relatively constant. That is, the heat source must occur 24 hours per day.

The credit for the thermal storage of the building itself is also usually not considered, although it can have an effect. In buildings with large thermal

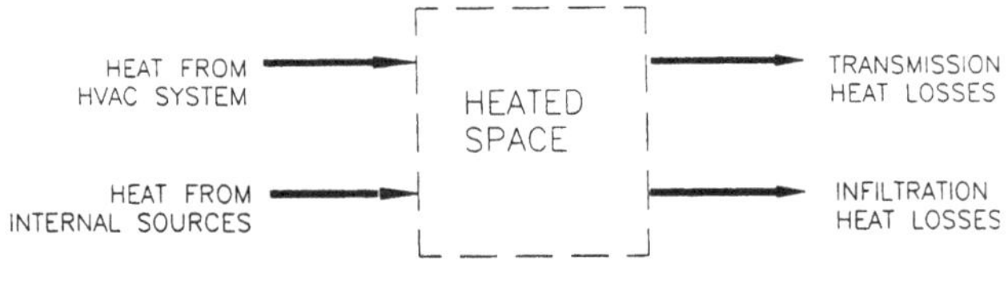

Figure 9-1. Heated space heat balance.

mass and large daytime internal heat gains, the heating load may be mitigated to some extent.

9.1.3 Heating Systems and Equipment

Heating for buildings may be provided by a number of different systems and equipment. The most common heating methods include air handling units with hydronic or steam coils, unit heaters and gas fired furnaces.

Hydronic systems are systems that provide heating to a space by circulating hot water. Heated water is usually generated by a boiler and circulated through a piping distribution system with a pump. Hydronic systems are generally very energy efficient; however, the need for a central plant to generate heating water is usually a large part of the cost of an HVAC system. Boilers are discussed in Chapter 4.

In addition to the main central plant equipment, a hydronic system consists of one, or more, circulating pumps, piping and some sort of heat exchange device. Heat exchange devices transfer heat between the circulated water and the air in the space, or the air supplied to the space by an air handling unit. Heat exchange devices include hydronic heating coils in air handling units, unit heaters and radiators.

Hydronic systems can have several piping arrangements. Usually there is supply piping that distributes the circulated water from the central plant to the heat exchange devices, and return piping that returns the water to the primary heating device in the central plant. Such an arrangement is called a two-pipe system since there is a supply piping system and a return piping system.

Hydronic systems have a number of advantages. First, since water has a much higher specific heat than air, the cross-sectional area of a pipe required to carry water is much less than the cross-sectional area for a duct carrying air for any given thermal load. Therefore, a hydronic system requires much less space in occupied areas of the building than does an air system.

Hydronic systems also require less pumping horsepower than that required to circulate air for any given thermal load. This can result in lower operating costs. Third, since the heat transfer device is frequently located in the conditioned space, hydronic systems are usually easier to zone and provide individual temperature control. Hydronic and pumping systems are discussed in Chapter 6.

Steam systems are similar to hydronic systems except that steam is distributed as the heating medium in lieu of heated water. Circulating pumps are not required for steam heating systems since the pressure of the steam itself is adequate for steam flow. Steam condensate (steam that has condensed back to water) is usually returned via a gravity drainage system. Occasionally the condensate is pumped if gravity drainage is not possible.

Steam systems are somewhat different from hydronic systems in that the steam undergoes a phase change. When water is heated, it undergoes a phase change as it is converted to steam, usually in a boiler. Steam is converted back to water (condensate) in a heat exchange device, usually a steam heating coil or a steam radiator. Again, it undergoes a phase change. It is the phase change that releases the majority of the heat from steam. For example, steam at 15 psig and 250°F will release about 945 Btu per pound of steam as it condenses from steam to condensate. The phase change occurs at a nearly constant temperature.

In order for the steam to condense in a heat exchange device, the steam itself must be kept inside the device until it condenses. After the steam becomes condensate, it must be removed from the heat exchange device. This is usually accomplished with a steam trap. The trap will not allow steam to pass; however, it will allow condensate to pass. A steam trap is placed on the discharge piping of the heat exchange devices to hold the steam in while allowing the condensate to flow out. Steam systems are discussed in Chapter 5.

A heating coil comprises many fins, between which are air flows, coil tubes and headers. The fins increase the heat transfer area in contract with the passing air. Although much of the heat transfer from a coil is through the fins, the primary source of heat transfer is round tubes, or pipes, that run between the coil headers at each end. The tubes are installed perpendicular to the fins and air flow. If the fins run vertically, the tubes run horizontally. The number of tubes in the direction parallel to the air flow designates the number of rows of the coil. A typical hot water heating coil is shown in Figure 9-2, and a typical steam heating coil is shown in Figure 9-3.

The tubes are usually interconnected by return bends to form a serpentine arrangement. The fins are bonded to the tubes to provide good conduction heat transfer between the fluid in the tubes, the tubes themselves and the fins. Water coils and steam coils usually have aluminum fins and copper tubes. A header is usually provided at one or both ends of the tubes to connect and distribute the fluid to and from

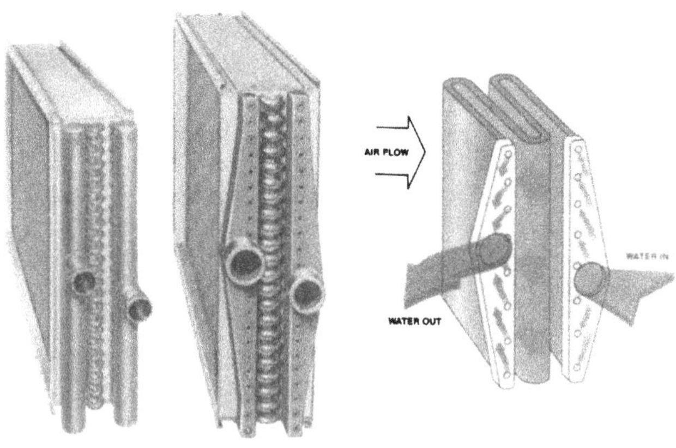

Figure 9-2. Water coil.

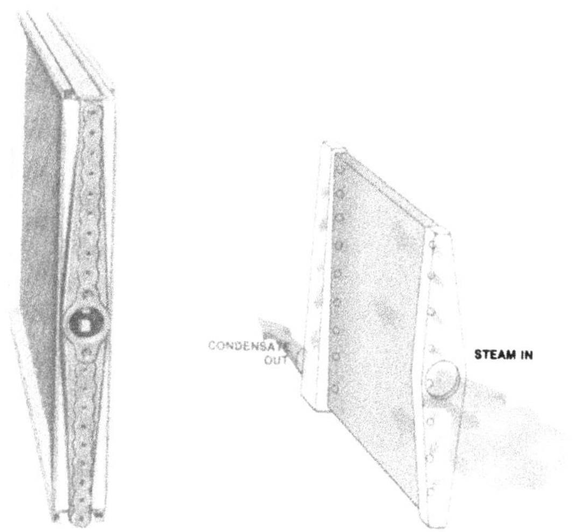

Figure 9-3. Steam heating coil.

the tubes. Air vents and drains are usually provided on the headers of water coils.

A unit heater consists of a heating element (a coil or furnace) and a small fan enclosed in a single cabinet. Their discharge can be horizontal or vertical. Unit heaters are usually installed to provide heating in unfinished spaces such as warehouses, manufacturing areas, garages, etc. The heating medium for these heaters is usually heating water or steam. Unit heaters can also have electric heating coils or be gas-fired with gas furnaces. A typical unit heater is shown in Figure 9-4.

Figure 9-4. Unit heater (*Courtesy of The Trane Company*).

Radiators are frequently used for heating in both finished and unfinished areas of buildings; they are usually located against exterior walls and especially under windows. Radiators can use steam, heating water or electricity as their heating medium.

Older buildings still utilize vertical sectional cast iron radiators. Although cast iron radiators are still in use today, they have been replaced by finned tube radiators in new buildings. The finned tube radiator has no moving parts. Steam and heat ing water radiators consist of a pipe through which the heating medium flows, and metal fins are bonded to the pipe. A finned tube element is shown in Figure 9-5.

The fins increase the heat transfer area of the heating element. Air flows through the fins by natural convection heat transfer. As the air is heated, it becomes less dense and rises. Air that rises is then replaced by cold air from below, which, in turn, is heated.

Warm air furnaces heat air by burning a fossil fuel in a combustion process. The fuels most widely used for furnaces are natural gas, liquefied petroleum (LP) gas and fuel oil.

A warm air furnace is a combustion and heating device in which gas or oil is directly fired to heat the air through a heat exchanger—or air is directly heated—in order to supply warm air to the conditioned space. A furnace may consist of multiple burners, a heat exchanger, combustion air blower an ignition device and controls. The furnaces discussed here are commercial furnaces used in commercial HVAC systems, however, warm air furnaces are widely used for residential heating applications. The main difference between residential furnaces and commercial furnaces are the size and heating capacity of the equipment. The heating capacity of commercial systems typically range from 150 MBh to over 1,000 MBh.

Gas furnaces may be direct-fired or indirect-fired. Direct-fired furnaces have burners located in the conditioned air stream. Figure 9-6 shows a direct-fired burner. The fuel is burned directly in the air

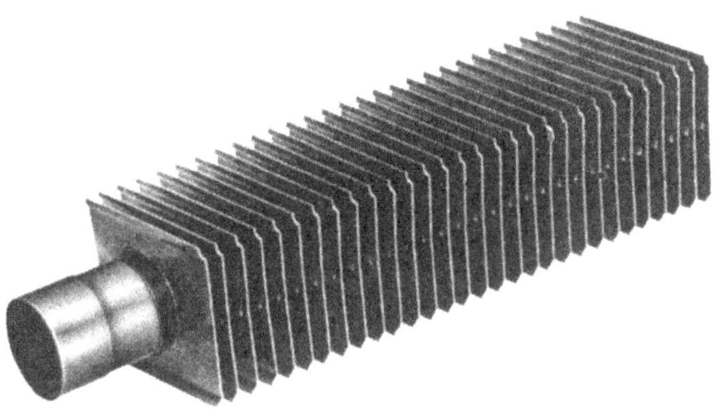

Figure 9-5. Finned tube element (Courtesy of the Trane Company).

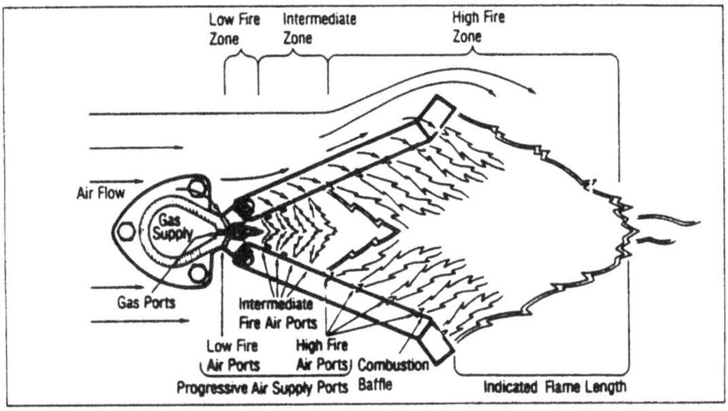

Figure 9-6. Direct fired burner (*Courtesy of Reznor*).

stream and the combustion products mix with the air passing through the burner. These types of burners are most commonly used on 100% outdoor make-up air units.

Indirect-fired furnaces do not allow the products of combustion to mix with the circulated air. A heat exchanger is employed to transfer heat from the burning fuel to the air supplied to the conditioned space. A heat exchanger is shown schematically in Figure 9-7.

Heat exchangers are usually of a clamshell design and completely surround the burner. A heat exchanger typically has a two pass arrangement.

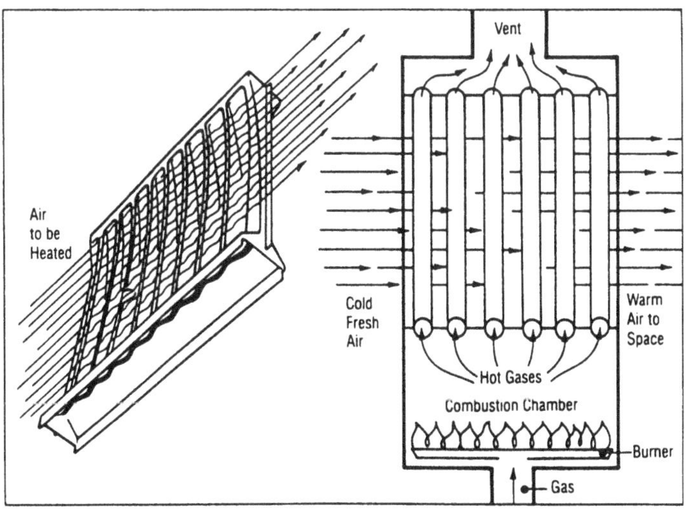

Figure 9-7. Heat exchanger (*Courtesy of Reznor*).

Burners may be classified as atmospheric or power burners. Atmospheric burners draw combustion air in at atmospheric pressure; power burners employ a fan to provide combustion air at a higher pressure. The power burner may be forced draft vented (blower on the upstream side of the burner) or induced draft vented (blower on the downstream side of the burner). Power burners are employed to provide better combustion and higher efficiency than atmospheric burners. A power draft blower is used to extract the combustion products and discharge them to a vent pipe or chimney.

The heating capacity of gas furnaces is controlled by a gas valve and an ignition device. The gas supply to a burner is controlled by a pressure regulator and a gas valve, which controls the firing rate. Smaller furnaces are controlled by a room or duct thermostat, which opens the gas valve on a call for heat and closes it when the thermostat is satisfied. Larger furnaces usually have a two-stage gas valve controlled by a two-stage thermostat. The two-stage gas valve has two open positions (part open or fully open) to control the flow of gas, depending on the heating load.

There are three methods used to ignite the fuel/air mixture when the gas valve opens: The first is a standing pilot light, which is a small continuous flame. Ignition may also be an intermittent pilot flame, which is off when the burner is off and lights prior to the opening of the main gas valve. The third method is by direct spark or hot surface ignition, which involves igniting the main burners directly.

Furnace safety controls include a temperature-sensitive fan switch, which is exposed to the circulating air stream. The fan switch will close the gas valve if excessive temperatures are detected. Other safety controls include a pilot light and gas pressure sensors.

Furnace efficiency is an important factor to consider when selecting a warm air furnace. The performance of warm air furnaces is usually measured by one of the following parameters:

1. Steady State Efficiency is the efficiency of a furnace when operated under steady state, equilibrium conditions. It is calculated by measuring the energy input, subtracting the losses for exhaust gases and dividing by the full input.

$$\eta = \frac{q_l - q_L}{q_l} \times 100\% \qquad \text{Eq. (9.1)}$$

where q_I = fuel input, Btu/hr
q_L = flue losses, Btu/hr

2. Utilization Efficiency is obtained by assuming the furnace is 100% efficient and subtracting losses for exhaust sensible and latent heat, cyclic effects, infiltration of combustion air and pilot-burner effect.

3. Annual Fuel Utilization Efficiency (AFUE) is the same as Utilization Efficiency, except that losses from a standing pilot light during non-heating season time are deducted.

$$\eta = \frac{\text{annual energy output (Btu/hr)}}{\text{annual energy input (Btu/hr)}} \times 100\% \qquad \text{Eq. (9.2)}$$

Gas furnaces presently have steady state efficiencies that vary from 75–95%. Atmospheric and power venting furnaces typically have efficiencies from 75–85%. Condensing gas furnaces can have steady state efficiencies as high as 95%. Furnaces are usually rated in terms of fuel, or heat, input and heat output.

LP gas furnaces are very similar to natural gas furnaces. The major difference between the two furnaces is the pressure at which the gas is injected into the burners. For natural gas, the pressure is usually controlled at 3 to 4 inches wg. For LP furnaces, the pressure is usually 10 to 11 inches wg.

Fuel oil furnaces are also similar to natural gas furnaces. The major difference between oil and gas furnaces are the combustion systems, heat exchanger and a barometric draft regulator, which is used in lieu of a draft hood. Heat exchangers for oil furnaces are usually heavy gauge steel formed into a welded assembly.

9.1.4 Cooling Loads

The purpose of a building cooling system is to maintain a desired indoor temperature, and sometimes relative humidity level, despite the addition of heat to the space from both outdoor conditions and indoor sources. Unlike heating loads, the space cooling load is not just a function of the difference between the indoor and outdoor air dry bulb temperatures. The cooling load for a space or a building is the result of many complex variables. Heat gain to the space occurs from many sources, both inside and outside of the space.

Heat may take one of two forms—sensible heat or latent heat. Sensible heat is heat that is associated with a temperature change of a substance, such as air. Latent heat is heat that is associated with a phase change of a substance, such as condensation or evaporation. In the case of cooling loads, it is heat that is associated with the evaporation or condensation of moisture or water. For a cooling system, it is heat that is added to the space by evaporation of moisture (such as perspiration) and must be removed as the moisture, or latent heat, is condensed out of the air.

As previously stated, the total heat gain for a space, or a building, is a result of many sources. Basically, the sources of heat gains may be divided into two major categories: internal heat gains and external heat gains. Internal heat gains result from sources inside the space itself, such as from lights, people, equipment, etc. External heat gains are from sources outside

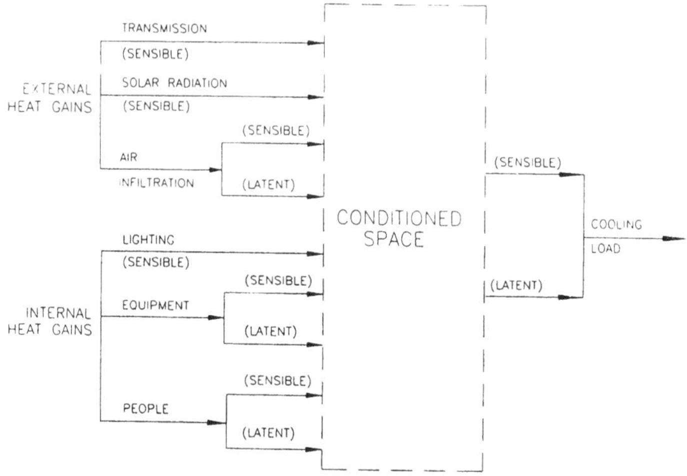

Figure 9-8. A typical heat balance for a space requiring cooling.

the space, or building, and are a result of weather conditions. External heat gains include heat transmission through exterior walls, solar heat gains through windows, and introduction of outside air into the space in the form of outdoor ventilation and infiltrated warm air. Figure 9-8 depicts a typical heat balance for a space requiring cooling.

Before discussing heat gains, some definitions of terminology are in order. The distinction between the instantaneous heat gain for a space, the cooling load for a space and the heat extraction rate from the space by the cooling system should be understood.

The instantaneous space heat gain is the rate at which heat is being generated in the space, or is entering the space, at any given instant in time. It does not take any thermal storage characteristics of the space into account.

The cooling load is the actual heat transfer to the air in the space at any given time. The space cooling load is different from the space instantaneous heat gain due to the thermal storage characteristics of the space itself. The cooling load is the rate at which heat must be removed from the space to maintain the room dry bulb and wet bulb temperatures at set values.

The heat extraction rate is the rate at which heat is actually removed from the space by the HVAC system. Due to the nature of HVAC systems and temperature control systems, the rate of heat removal is intermittent, causing "swings" in space air temperatures. Most thermostats have a "dead band" temperature range. When the space air temperature is in this range (usually 70°–78°F), neither heating nor cooling is required. Only when the space air temperature rises above the cooling set point temperature (78°F) does the space receive cooling, or have heat removed.

Heat is removed until the space air temperature drops below the set point temperature. This results in a non-constant or varying heat extraction rate, which is frequently different from the actual cooling load at any given time.

The external heat gain (external gain) to a space or building is heat gain from outdoors primarily through external surfaces. It includes heat transmission through walls, roofs, doors, windows, etc. External gains also include solar heat gain through glass windows and glass doors (fenestration); as well as the addition of heat to a space by ventilation, or the infiltration of outside air into the space.

Unlike heat losses, heat gains through exterior surfaces may not be assumed to be steady state. Solar radiation on exterior surfaces and the thermal storage affect of the surfaces are the major reasons why external heat gains must be treated as non–steady state heat transfer.

The calculation of cooling loads for HVAC systems is fairly involved and outside the scope of this book. An overview of cooling loads is presented here since they result in considerable energy consumption by HVAC systems. The reader may wish to refer to ASHRAE (1997) and Mull (1997) for additional information on cooling load calculation methods.

The American Society of Heating, Refrigerating and Air Conditioning Engineers (ASHRAE) has developed two calculation methods that take thermal storage and transient heat conduction into account for sunlit exterior surfaces. One method, introduced by ASHRAE in 1972, is the Total Equivalent Temperature Differential (TETD) method. In the TETD method, various components of space heat gain are added together to obtain an instantaneous rate of space heat gain. The space heat gain is converted to an instantaneous cooling load by the Time-Averaging (TA) technique of averaging the radiant heat gain with related values for previous hours. Space heat gains are separated into radiant and convective gains. The convection portion is assumed to be an instantaneous cooling load, while the radiant gain is time-averaged to get a cooling load. The number of hours the radiant gain is averaged is based upon the thermal mass of the space and exterior surfaces.

The second calculation method is the Transfer Function Method (TFM), which utilizes a series of "weighing factors" (coefficients of room transfer functions). The TFM weighing factors use time-averaging to approximate the heat gain for the present hour and for several previous hours, in order to account for thermal storage of the space and determine the resultant cooling load. Since the actual calculation procedure for the TFM is so complicated that a computer would be necessary to perform the calculations, a simplified method suitable for manual calculation was developed by ASHRAE. The simplified method uses Cooling Load Temperature Differences (CLTD) for heat gains through sunlit walls and roofs and for heat conduction through glass (fenestration). The method also uses Cooling Load Factors (CLF) for solar gains through glass and for internal heat sources. Both CLTD and CLF take into account non–steady state heat transfer caused by thermal storage of the building mass.

CLTDs were calculated and tabulated for various exterior surfaces for three types of room space construction; light, medium and heavy thermal characteristics. It is assumed that the heat flow through a similar wall or roof can be determined by multiplying the surface area and overall heat transfer coefficient (U Value) by a tabulated CLTD.

Internal heat gains are gains that are generated within the conditioned space itself. Typical examples of internal heat gains include gains from people, lights and energy-consuming equipment located within the space.

ASHRAE has developed CLF for internal heat gains to account for the thermal storage of the building mass. Again, buildings have been classified into Light Mass, Medium Mass and Heavy Mass construction groups. Cooling loads and CLF for interior heat gains are primarily a function of the length of time since the heat gain started, or entered the space, and the thermal mass of the space. As sensible heat is absorbed by the building mass, the temperature of the mass increases. As the temperature of the mass increases, it is able to absorb less heat, causing the cooling load to approach the instantaneous heat gain. After the heat gain stops, or leaves the space, the thermal mass continues to reject the stored heat to the space, resulting in a cooling load.

When calculating cooling loads that result from internal heat gains, a diversity factor is occasionally applied to the loads. Diversity of cooling loads result from not all of the cooling load being on or used at a particular time. As an example, all of the lights in a large building are rarely on at the same time. Some rooms may be occupied while others are not. For large spaces or buildings, some of the lights may be on while others are off. Diversity factors account for variable loads and can range from 70–100 % depending on the size and use of the building. The application of diversity factors requires some knowledge of the operation of a building and should be applied with considerable judgment. For small spaces, it is likely that all internal gains will be on and, therefore, a diversity factor of 100% should be used when calculating cooling loads for small spaces.

In addition to exterior and interior heat gains, outdoor ventilation air and infiltration of outdoor air can be a significant part of the total cooling load on an HVAC system. In addition to being a significant sensible cooling load, outdoor air is frequently the greatest source of latent cooling loads.

Air conditioning involves the control of both the temperature and moisture content of the air in the conditioned space. The space temperature and moisture content are controlled by various thermodynamic processes provided by the HVAC system.

Figure 9-9 graphically presents the typical processes used in HVAC systems on a skeleton psychrometric chart. The lines show the variation in properties that the air-vapor mixture would pass through in going from one state point to another. State point 0 is assumed to be the beginning point of

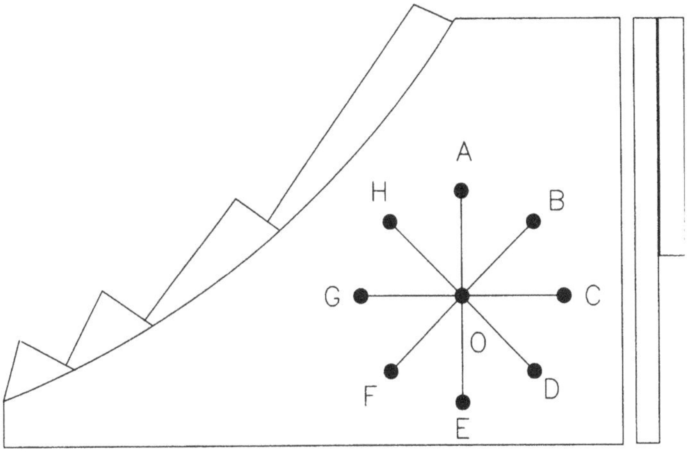

Figure 9-9. Basic HVAC processes.

the process. The processes in actual HVAC systems are frequently more complex and involved than those shown in Figure 9-9; however, most HVAC processes can be analyzed by combinations of the basic processes. Any change in the state point of the air-vapor mixture can be broken down into sensible and latent heat changes. These changes can be plotted on a psychrometric chart.

From this point on, the air-vapor mixture will just be referred to as air. Referring to Figure 9-9, the basic processes may be observed.

Humidification Only (state point O to A) and Dehumidification (state point O to E): These processes involve an increase (O to A) or a decrease (O to E) in the humidity ratio with no change in the dry bulb temperature of the air. These are purely changes in latent heat, with no change in sensible heat content.

Sensible Heating Only (state point O to C) and Sensible Cooling Only (state point O to G): These processes involve an increase (O to C) and a decrease (O to G) in the dry bulb temperature of the air. These are purely changes in sensible heat, with no change in the moisture content. The humidity ratio (W) remains constant.

Combined Cooling and Dehumidification (state point O to F): A result of the reduction in both dry bulb and wet bulb temperatures. Cooling coils normally perform a process like this where both sensible and latent heat of the air decreases.

Combination Heating and Humidification (state point O to B): The result of an increase in both dry bulb and wet bulb temperatures of the air. In this process, both sensible and latent heat are added to the air.

Evaporative Cooling Only (state point O to H): A process in which cooling and humidification occur. The wet bulb temperature of the air remains constant as the dry bulb temperature is reduced. This is an adiabatic (no net heat transfer) process, whereby sensible heat is removed, with the heat going to vaporize an additional amount of water added (added latent heat).

Chemical Dehumidification (state point O to D): A process in which moisture is chemically removed from the air. This is a constant enthalpy process, with the latent cooling nearly equal to the sensible heat increase.

9.1.5 Cooling Systems and Equipment

Most cooling and refrigeration systems operate as direct expansion refrigeration systems, which are discussed in this section. Some chillers operate on the absorption refrigeration principle. Absorption chillers are discussed later in this section and in Chapter 7.

Direct expansion (DX) systems are classified as direct systems by ASHRAE Standard 15: Safety Code for Mechanical Refrigeration. A direct system is one in which the evaporator or condenser of the refrigerating system is in direct contact with the air to be cooled. Chillers are classified as indirect systems. An indirect system is one in which a secondary coolant (such as water) is cooled by the refrigerating system and then circulated to the air to be cooled.

The most common system for producing a refrigerating affect is the DX system. DX systems operate on a thermodynamic cycle that transfers heat from a low temperature region to a higher temperature region. Since heat only flows naturally from a high temperature region to a low temperature region, it is necessary to apply energy, in the form of work, to a refrigeration system. Work is provided by a compressor that compresses the refrigerant vapor. A simple DX refrigeration system is shown in Figure 9-10.

In the DX system a refrigerant is circulated by the compressor. The refrigerant is a substance that will undergo a phase change (change from a gas to a liquid and back to a gas) as it goes through the cycle. The refrigerant enters the compressor as a gas and is compressed. Increasing the pressure of the gas causes its temperature to increase and it becomes a superheated vapor. The superheated refrigerant vapor then passes through a condenser, where it rejects much of its heat. The heat is usually rejected directly to the atmosphere or to a circulated cooling fluid, such as water. When the refrigerant comes out of the condenser, it is a saturated vapor at high pressure. (A saturated substance is one in which vapor or liquid phases can be present at a given temperature and pressure). The saturated vapor refrigerant then passes through an expansion valve. The expansion valve reduces the pressure of the refrigerant, which also reduces its temperature. The refrigerant then becomes almost 100% saturated liquid at a much lower temperature. The saturated liquid refrigerant then passes through the evaporator. As the name implies, the liquid refrigerant evaporates and absorbs heat from the circulated air that is passed over the evaporator. The evaporator in an air conditioning system is the cooling

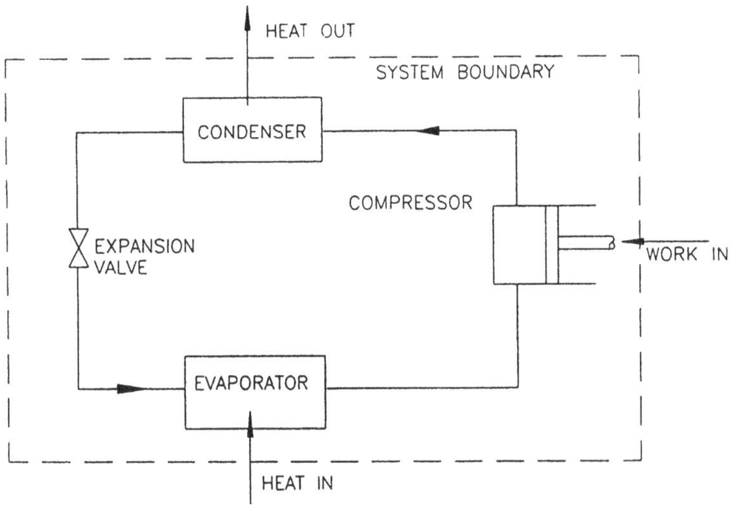

Figure 9-10. A simple refrigeration system schematic.

coil. When the refrigerant comes out of the evaporator, it is again a vapor and the cycle is repeated.

The most common type of cooling system is called a split system, where the evaporator coil and expansion valve are located in the furnace or air handling unit, while the compressor and condenser coil are in an outdoor unit. Refrigerant piping connects the two units and carries the refrigerant between them. The outdoor unit is usually called a condensing unit. Large commercial condensing units can have cooling capacities of more than 150 tons. A large commercial condensing unit is shown in Figure 9-11.

When the outdoor unit contains the compressors, the condenser coils and fans, it is referred to as a condensing unit. If the outdoor unit has just the condenser coils and fans, with the compressors located elsewhere, the unit is referred to as a condenser.

A variation on the DX cooling system is the heat pump. A heat pump is used for heating a space instead of cooling it. It is almost identical to the cooling system except that the direction of heat exchange is reversed. In winter, heat is transferred from the outdoors to indoors by the refrigeration process. For a heat pump, the evaporator coil is outdoors and the condenser coil is indoors. Most heat pump systems function as a cooling system in summer and a heating system in winter. This is accomplished by a switching valve in the refrigeration piping that changes the function of the evaporator and condenser coils by changing the flow of the refrigerant.

Frequently, large buildings and large facilities are cooled from a central plant. Central plants usually include one or more chillers to

Figure 9-11. Large commercial condensing unit (*Courtesy of Carrier Corporation*).

produce cooling, which is distributed to individual air handling and fan coil units by chilled water. Chilled water systems typically supply chilled water between 42°F and 50°F. If the water were any colder, it would require an anti-freezing solution and be called a brine. Brine systems are generally used only for process cooling where lower temperatures are required. Typically the water in HVAC systems is cooled by 8°, 10° or 12°F as it passes through the chiller. As the chilled water passes through the chilled water coils in air handling units, the water temperature rises by an equivalent amount before it is returned to the chiller. The chilled water is circulated throughout the system by one or more chilled water pumps.

Most chillers operate on the same principle as the DX systems. The main difference is that the evaporator removes heat from the circulated water instead of air. Electrically driven chillers have a DX refrigeration system and are usually reciprocating, rotary screw or centrifugal compressor chillers. Reciprocating chillers use pistons inside of cylinders to compress the refrigerant. Reciprocating chillers typically have cooling capacities from 20 to as high as 200 tons. Rotary screw chillers use a rotating helical screw to compress the refrigerant. Screw chillers have cooling capacities from 50–750 tons. Centrifugal chillers use a rotating wheel, or impeller,

Figure 9-12. Centrifugal chiller (*Courtesy of York International*).

and centrifugal force to compress the refrigerant. A centrifugal chiller is shown in Figure 9-12.

Centrifugal chillers typically have cooling capacities ranging from 100–2,000 tons in a single machine.

Chilled water may also be produced by absorption chillers, which do not operate on the DX cycle. Absorption chillers use heat instead of work from an electrical motor to produce a refrigerating effect. Most absorption chillers use a lithium bromide and water solution as the circulated refrigerant. The cooling cycle is based on the absorption of heat from the chilled water when the refrigerant (water) evaporates. The refrigerant solution in the chiller is maintained at a near-perfect vacuum. The absorption cycle utilizes the affinity that lithium bromide has for water while under a vacuum.

The refrigerant (water) is cooled by latent cooling as it evaporates and is absorbed by a strong concentration of lithium bromide solution. As the lithium bromide solution becomes diluted by the water, it is transferred to another section where it is heated. The heat causes the water to be driven off. The lithium bromide is then returned to the absorber to again absorb the water refrigerant. The water vapor that was driven out of the lithium bromide is condensed by cooling water and returned to th e evaporator to repeat the cycle. Water is chilled as it is circulated through the tubes of the evaporator.

Smaller chillers, like reciprocating chillers, have condensers that are air cooled. Larger chillers, like centrifugal and absorption chillers, have condensers that are water cooled. Water cooled chillers reject heat to the atmosphere by a cooling tower, which is usually located outdoors. A typical cooling tower is shown in Figure 9-13.

Figure 9-13. Cooling tower (*Courtesy of Evapco, Inc.*).

The amount of heat rejected by a cooling tower is increased if the circulated water comes in contact with the outdoor air. Most cooling towers spray water into the outside air as it is circulated through the cooling tower by fans. Spraying the water increases the evaporation of the water, which increases the heat rejection rate. As the water evaporates into the air, latent heat is transferred from the water to the air. Cooling towers are discussed in Chapter 8.

Air cooling coils in HVAC systems cool the circulated air by circulating either chilled water or a refrigerant through the coil as a heat transfer medium. Both chilled water coils and refrigerant coils are similar in construction to heating water coils. The coils have tubes through which the heat transfer fluid is circulated. They also have fins to serve as extended heat transfer surfaces. The construction of chilled water coils and refrigerant coils are very similar to heating water coils, with the main difference being that cooling coils usually have more rows than other types of coils. Coil row offerings may vary from 2 to 12 rows, with 4-, 6- and 8-row coils being most widely used for HVAC applications.

Cooling coils that have many rows and close fin spacing have larger surface areas for heat transfer from the flowing air to the cooling coil. These types of coils tend to have low bypass factors. Coils with comparatively wide fin spacing and few rows tend to have higher bypass factors. The velocity of the air passing through the coil also affects the bypass factor. Typical air face velocities through cooling coils range from 300–800 fpm. Higher face velocities tend to result in higher bypass factors while lower velocities result in lower bypass factors.

The actual surface temperature of a cooling coil with chilled water or with refrigerant is not uniform, but varies with the location on the cooling coil. On a multi-row cooling coil, the surface nearest the entering air is in contact with the warmest air and, therefore, is warmer than the average or mean surface temperature. The coil surface near the discharge side of the coil is in contact with air that has already been cooled to near the discharge temperature. The coil surface temperatures near the discharge side of the coil tend to have lower surface temperatures.

In order to analyze the effect of cooling coil on the air passed through the coil, it is necessary to assume an effective coil surface temperature that would produce the same leaving conditions as the actual nonuniform surface temperatures. This point is commonly called the apparatus dew point (ADP) temperature. The ADP of a coil is a function of the log mean temperature difference between the fluids and the fluid heat transfer properties of the air and cooling fluid. See Chapter 2 for a discussion of LMTD. The psychrometrics of cooling coil processes were discussed earlier in this chapter. A more detailed discussion of cooling coil processes is available in ASHRAE (1997) and Mull (1997).

Chilled water coils in HVAC systems use water that has been chilled or cooled by a chiller. The chiller cools the water through a refrigeration process. Chilled water is typically supplied to the cooling coil in an air handling unit AHU at a temperature between 40°F and 55°F. Lower water temperatures may be used; however, there is a risk of freezing if the water temperature drops to 32°F somewhere in the system and the water does not contain an anti-freeze additive, such as glycol. Higher water temperatures will result in higher apparatus dew point temperatures, which will result in less dehumidification of the air. Most chilled water systems are designed to operate with a 10°F or 12°F increase in water temperature as the water flows through the coil and absorbs heat.

The main difference between heating water coils and chilled water coils are the number of rows and the fin spacing. Chilled water coils require more rows and closer fin spacing since the LMTD for chilled water coils is much less than that for heating water coils, and because of the dehumidification requirements for cooling coils. In order to provide the required high rate of heat transfer with a smaller LMTD, chilled water coils usually are in a counterflow arrangement. That is, the coldest air meets the coldest water and the warmest air meets the warmest water. The coil tubes are usually in a staggered arrangement with many rows. This allows lower leaving air temperatures, frequently lower than the leaving water temperatures; lower leaving air temperatures provide better dehumidification.

Although the construction of water coils and refrigerant coils are similar, the operation of refrigerant coils are more complex than water coils. DX refrigerant systems operate on a thermodynamic cycle that transfers heat from a low temperature.

Since the refrigerant flows into the coil as a liquid and evaporates to a vapor, it absorbs heat from the air. In order to accomplish this, the liquid refrigerant must evenly flow into the coil and be evenly distributed to various tube circuits within the coil. For even coolant distribution and to reduce pressure drop, the flow paths of the refrigerant in a DX coil are always divided into a number of refrigerant circuits. DX coils use a distribution nozzle and multiple feeder tubes to provide a flow of refrigerant to all circuits in the coil.

In a DX coil, the refrigerant is expanded and evaporated directly inside the tubes of the coil. The DX coil acts as the evaporator in the refrigerating system. In comfort HVAC systems, the evaporating temperature of the refrigerant inside the coil tubes i s usually between 37°F and 52°F. At these temperatures, the surface temperature of the coil is usually lower than the dew point temperature of the air entering the coil. The refrigerant flow rate into the coil must be properly metered and distributed to the coil tubes. DX coils used in comfort air conditioning applications have either a capillary tube or a thermostatic expansion valve (TXV) to regulate the flow of refrigerant. The TXV is located at the refrigerant inlet to the evaporator coil. A typical TXV and evaporator coil arrangement is shown in Figure 9-14. The TXV provides smooth evaporator cooing capacity modulation by maintaining a fixed temperature of the refrigerant as it leaves the coil. The majority of the heat transfer from the air to the refrigerant occurs as the refrigerant changes phase from a liquid to a vapor. However, the refrigerant in most DX coils is heated to a temperature higher than that required to completely vaporize the refrigerant. This additional heating of the vapor is called superheating. The number of degrees that the refrigerant is heated above that required for vaporization is referred to as "the degrees of superheat." Superheat represents only a small percent of the coil's heat transfer capacity; however, it is important for refrigeration system control and for compressor protection.

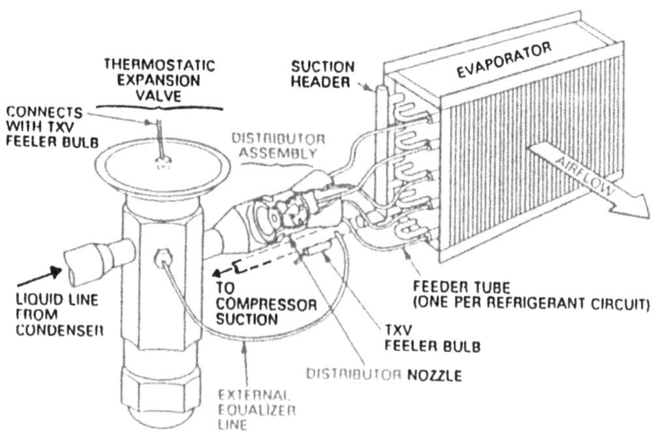

Figure 9-14. Thermostatic expansion valve and evaporator coil.

Part load operation and control for DX coils is more critical than for water coils. At part load operation, the evaporating temperature is lowered because less refrigerant is evaporated in the coil. This may result in temperatures below freezing and frost formation on the coil, which blocks the flow of air and severely reduces the rate of heat transfer. A TXV cannot modulate down to zero refrigerant flow. Below a certain point, the clearance between the valve needle and seat alternate between too far open and closed (overcapacity with too little superheat and undercapacity with too much superheat). The stable operating range for most TXVs used in HVAC systems is from around 100% to about 50% of design capacity.

DX systems use fluorinated hydrocarbon refrigerants (also called fluorocarbon refrigerants). The most commonly used refrigerants are HCFC-22 (R-22) and HFC-134a (R-134a).

Cooling for spaces is frequently accomplished with unitary air conditioners. Unitary or packaged HVAC units are, just as their name implies, a complete HVAC unit in a single package or cabinet. Packaged units usually have a complete direct expansion cooling system along with fans and, sometimes, heating. The cooling system includes a refrigerant compressor, evaporator coil, condenser coil and condenser fan. Heating is usually through a gas furnace or electric coil, although heating water and steam coils are also available. Packaged units with a heat pump cycle are also available.

Packaged units are, quite popular since they usually have a lower first cost. Since everything is factory built and installed in a single cabinet, equipment costs are comparatively low. Installation cost is usually low since it is usually necessary only to set the unit, connect the utilities and connect the distribution ductwork (if any).

Packaged units are, however, not capable of providing very precise control of temperature and humidity. These units are primarily for basic comfort heating and cooling, and are not suitable for process cooling or a situation with high latent cooling loads. Since these are packaged units, there are a limited number of options with the components. It is usually not possible to "build up" a custom unit for a special application.

The most common type of unitary air conditioner is a package rooftop air conditioner. As the name implies, rooftop HVAC units are packaged units mounted on the roof of a building and ducted down through the roof to the spaces below. Packaged rooftop units are also available with a horizontal duct discharge for grade mounted units. Rooftop units are usually installed on a roof curb, which supports the unit and provides a weatherproof seal around the duct openings through the roof. Rooftop units are completely factory assembled units that have a direct expansion refrigeration system, heating system, fans and controls in a single weatherproof cabinet. A typical rooftop unit is shown in Figure 9-15.

The most common forms of heating are gas furnaces and resistance electric heating coils. Units with heating water coils and steam coils are also available. Small- and medium-size units are available as heat pumps.

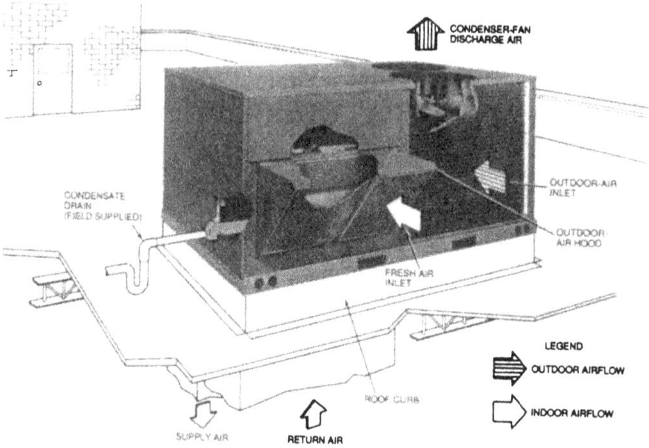

Figure 9-15. Rooftop HVAC unit (*Courtesy of Carrier Corporation*).

Rooftop units are usually used for commercial application—in office buildings, retail spaces and other spaces requiring purely comfort cooling. They are quite popular because of their low first cost and the fact that they don't use any space inside the building. Usually, the only required connections to the unit are the supply/return ductwork, electrical power and gas (if applicable). One of the major disadvantages to rooftop units is that they can cause considerable noise and vibration in the spaces near the unit, particularly under it.

Rooftop units are available with cooling capacities ranging from as low as 2 1/2 tons to as high as 125 tons in a single unit. Smaller units tend to have constant air volume while larger units can have variable air volume. Small- and medium-size units can use an air bypass arrangement to achieve variable airflow to the spaces.

9.2 BASIC PRINCIPLES OF VENTILATION AND AIR DISTRIBUTION SYSTEMS

9.2.1 Ventilation

Most occupied buildings and indoor facilities require some type of ventilation. Ventilation may be required by codes in order to maintain reasonably healthy indoor air conditions for the occupants; it may also be required to limit indoor air temperatures to reasonable levels. Finally, many processes require ventilation as a part of the process itself, or to control the level of indoor air pollutants.

Minimum ventilation requirements for industrial buildings are frequently controlled by local, state and national codes. These include safety and health codes and standards set forth by the National Institute of Safety and Health (NIOSH), the American Conference of Governmental Hygienists (ACGIH), Occupational Safety and Health Administration (OSHA), the Environmental Protection Agency (EPA) and the National Fire Protection Association (NFPA). Standard industry practice may also set minimum ventilation requirements. Minimum ventilation requirements are set by code for the health and safety of workers and other building occupants.

Ventilation is also frequently required to provide maximum temperature control in a building or facility that is not mechanically cooled. In addition to general ventilation, heat relief is often required for workers who work near heat generating equipment or hot processes. Heat relief ventilation removes sensible heat via a temperature increase of the air flowing through a space.

When processes give off noxious or toxic fumes, ventilation must be provided to prevent the buildup of toxic chemicals in the indoor environment. There are two types of exhaust systems used to ventilated indoor processes that emit pollutants, general exhaust and local exhaust. General exhaust systems remove air from an entire workspace without considering specific processes or operations. Local exhaust is typically for processes that release airborne chemicals and pollutants. Indoor air contaminants can include dust, fumes and mists. Local exhaust controls contaminants at the source (the process or operation) before it enters the general indoor environment. Local exhaust provides better contaminant control and lower exhaust air flow rates. Newer process equipment that produces air pollutants frequently has integral exhaust connections and is designed for a specific air flow.

9.2.2 Supply Air Distribution Systems

Supply air distribution systems usually include at least a supply air fan and some type of ductwork to distribute the circulated air to the desired location. This section provides an overview of the most common types of air distribution systems commonly used for HVAC systems. Central air distribution systems can basically be divided into three major categories, which describe how heating, ventilating and cooling capacity control is achieved. The major types of HVAC air systems are constant volume/ single path, constant volume multi-path and variable air volume systems.

The most basic type of central air HVAC system is the constant volume single path system. As the name implies, the airflow rate is constant. There is only one path for the air to follow as it leaves the AHU and is distributed to the conditioned space. This type of system is best suited for a single zone since the system is either in the heating or cooling mode at any given time. The system may be in the "dead band" mode, in which it is neither heating nor cooling, just circulating air. It is not possible to simultaneously heat one zone and cool another with a single system. A

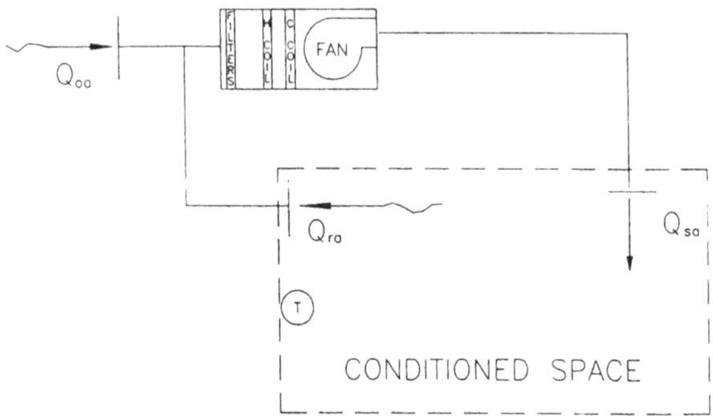

Figure 9-16. Constant volume single path system schematic.

schematic diagram of a constant volume single path system is presented in Figure 9-16.

Constant volume single path systems normally have an AHU with heating and/or cooling coils and ductwork in a series flow path. Sometimes the heating coil may be replaced with a gas or oil fired furnace. The heating or cooling is cycled on or off in response to a load in the conditioned space sensed by a thermostat.

A variation on the constant volume single path system is the constant volume reheat system. This system can provide either heating or cooling to multiple zones simultaneously with a constant air volume system. A schematic diagram of a constant volume reheat system is shown in Figure 9-17.

In this type of system, air is circulated at a constant rate and leaves the AHU at a set temperature, usually about 55°F. Just before the air enters the conditioned zone, a duct-mounted reheat coil increases the temperature of the air entering the zone in accordance with the zone heating or cooling

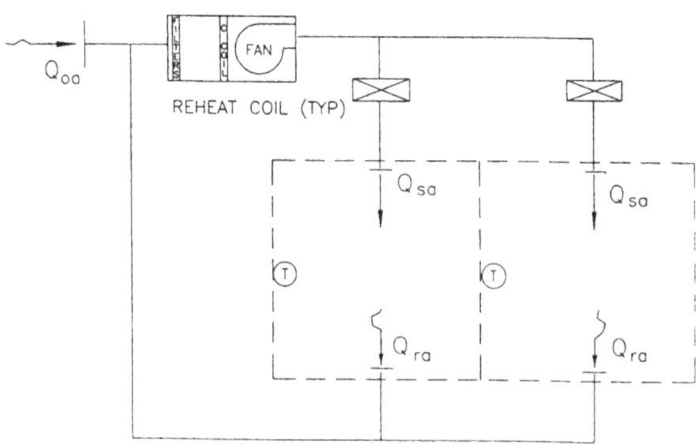

Figure 9-17. Constant volume reheat system schematic.

load sensed by the zone thermostat. Reheat coils can be electric, steam or heating water. A separate preheat coil may also be provided in the AHU to preheat the incoming outdoor air as necessary.

Constant volume reheat systems are rarely used on new installations since they are very energy-inefficient. Air is first cooled to a set supply air temperature and then heated to meet the zone requirements. Some energy codes no longer allow constant volume reheat systems.

As the name implies, multi-path systems allow the conditioned air to follow multiple parallel paths to each zone. The temperature of the air supplied to each zone is a function of the heating or cooling load in that zone. Typically, the AHU has both heating and cooling coils in order to provide either heated or cooled air to a zone as required. The two most common multi-path air systems are the multi-zone system and the dual duct system.

Multi-zone systems usually have an AHU with a cooling coil and multiple heating coils. A typical multi-zone air handling unit is shown in Figure 9-18.

Each zone has its own supply air duct from the AHU allowing the air to flow through multiple parallel supply air ducts to the zones. Each zone has its own reheat coil inside the AHU. The zone reheat coil is controlled by a zone thermostat. Air is passed through the cooling coil in the AHU and is cooled to a predetermined temperature, usually about 55°F. The airflow is then divided inside the AHU into separate paths for each zone. Still inside the AHU, the air for each zone is then either passed through a reheat coil or bypassed around the reheat coil according to the zone heating or cooling load. The proportion of air

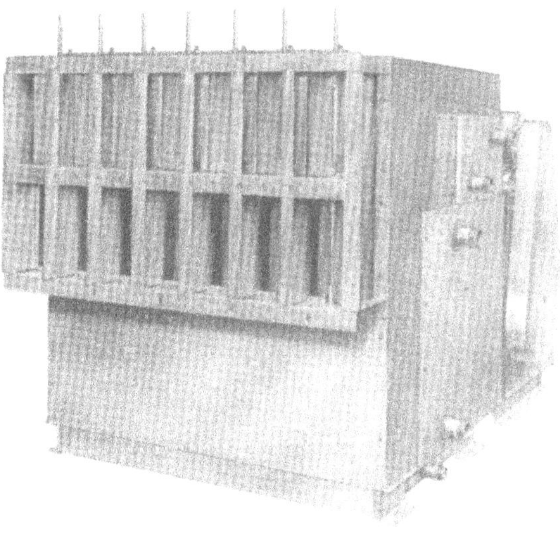

Figure 9-18. Multi-zone air handling unit (*Courtesy of York International*).

passed through, or around, the reheat coil is controlled by the zone thermostat.

Since the supply air is first cooled then reheated or bypassed to provide the desired "mixed" air temperature, these systems are rather energy-inefficient and energy-wasteful. Very few multi-zone systems are installed in new facilities today; most multi-zone systems are found in older facilities.

The second multi-path system in use today is the dual duct system. As the name implies, there are dual (two) ducts to each space or zone. One duct carries heated air while the other carries cooled air. There can be a single or dual air handling units for the system. Either cooled air or heated air, or both, are supplied to the space according to the thermal load in the space. In order to provide air to the space at the required temperature, a dual duct terminal is provided for each zone to mix the heated and cooled air in the required proportion. The dual duct terminal mixes the two air streams in accordance with the requirements of the zone thermostat.

Other than the constant volume single path system, the variable air volume (VAV) system is the most common type of air system installed in nonresidential buildings today. As the name implies, the rate of airflow in the system is not constant, but varies in accordance with the total sensible cooling load for the space. The supply air is usually cooled and supplied to the distribution ductwork at a constant temperature, although the air temperature may be reset in accordance with the cooling load. A schematic of a typical VAV system is presented in Figure 9-19.

As may be observed from Figure 9-19, the VAV system is also a single path system; however, it is not constant air volume. The amount of cooled air admitted to the conditioned space is a function of the sensible cooling load in the space, which is sensed by a thermostat. As the temperature of

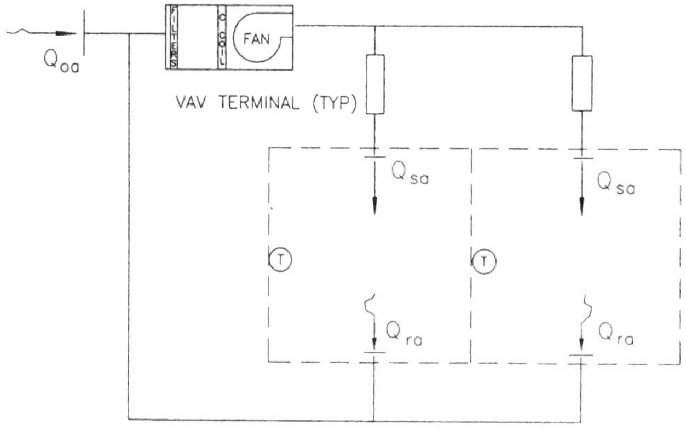

Figure 9-19. VAV system schematic.

the space increases, an air valve in a VAV terminal closes, reducing the airflow as the cooling load is satisfied.

Heat for the space is usually not provided by heating the supply air in the AHU. Frequently heat is provided by baseboard radiators around the exterior perimeter of the conditioned space. Heat may also be provided in the VAV terminal with a reheat coil in the VAV terminal itself.

Since the quantity of air supplied to the spaces varies, it is good design practice to provide some method to control the airflow of the AHU fan. This is normally accomplished by three methods; fan discharge dampers, fan inlet vanes and variable frequency motor drives. Each method varies the flow of the AHU fan in response to changes in static air pressure in the main supply air duct.

The first, and least desirable, are discharge dampers, which are automatically actuated dampers that are placed on the outlet, or discharge, of the AHU. The dampers open and close in response to the pressure in the main supply air duct. As the VAV terminals open and the pressure in the main duct drops, the discharge dampers open to allow more air from the fan into the ductwork. As the VAV terminals close and the duct pressure increases, the discharge dampers close. Discharge dampers are so mewhat energy-inefficient since the fan is rotating at full speed while the dampers block off the airflow. Discharge dampers may also result in undesirable high air pressures in the AHU.

A variation on discharge dampers is the bypass system. Air is simply bypassed around the conditioned space and returned to the AHU. Air is

Figure 9-20. Fan with inlet vanes (*Courtesy of York International*).

normally "dumped" into a return air plenum or duct by a motorized relief damper. The fan in the AHU is still rotating at full speed fully loaded. Bypass systems are typically found on smaller commercial air systems.

A second, and more energy-efficient, way to control the AHU fan air volume is with inlet vanes on the fan. These vanes are located at the air inlet on the fan and open or close in response to the pressure in the main supply air duct. Figure 9-20 shows a centrifugal fan with inlet vanes.

Since the inlet vanes restrict the amount of air entering the fan, the fan rotates in an unloaded condition, even though it is rotating at full speed. This reduces the load on the electric motor driving the fan and reduces the power consumed by the motor. Inlet vanes are more costly than discharge dampers, but they are also more effective.

The third, and most energy-efficient, way to control the airflow of a fan is a variable frequency motor drive. These drives vary the frequency of the electric power to the fan motor in response to the pressure in the main supply air duct. The varying electrical frequency in turn causes the motor and fan to rotate at different speeds. When the air pressure in the duct drops, the fan speed increases to supply more air. When the air pressure increases, the fan speed decreases accordingly. Although variable frequency drives tend to be more expensive than the other methods to control air volume, they are also much more efficient. Frequently, the energy cost savings of the variable frequency drives justifies the additional first cost. Variable frequency drives are discussed in more detail in Chapter 10.

9.2.3 Exhaust Air Systems

Ventilation is normally required in occupied buildings to prevent the buildup of excessive indoor air temperatures and to control the buildup of indoor air pollutants. Heat and pollutants are normally controlled by exhausting high temperature air from a building; they are removed after they have entered the indoor air.

Outdoor air must be brought into a building to replenish air that is exhausted. Make-up air may be brought into a building indirectly by leakage around doors, windows, etc., or it may be brought in by a mechanical make-up air system, which may heat and/or cool the air as it is brought into the building.

There are two types of exhaust systems: (1) general exhaust and (2) local exhaust. General exhaust systems provide exhaust for an entire workspace without considering specific operations. General ventilation dilutes the indoor air to reduce the space temperature and/or pollution level. General exhaust systems are generally sized based upon volumetric air changes in the space. Frequently, codes set the minimum requirements for general ventilation. General ventilation can be by natural draft; however, it is usually provided by mechanical systems, such as fans.

Local exhaust systems remove heat and pollutants generated at the source. Local exhaust is preferable since it provides better control of the contaminant or heat, with less air than would be required by general exhaust. Close-capture exhaust systems utilize less air than general exhaust systems since they depend on the capture velocity of the air to remove heat or airborne pollutants rather than volumetric airflow rates. In close-capture local exhaust systems, it is the capture velocity that determines the required airflow rate rather than volumetric dilution.

Local exhaust systems typically may have five basic components: (1) a hood or point of entry, (2) a duct system, which transports the exhaust air, (3) an air cleaning device, which removes contaminants from the airstream, (4) an air moving device, which provides motive power for overcoming the system flow resistance and (5) an exhaust stack, which discharges exhaust air to the atmosphere. The size of the hood is determined by the size of the source producing the contaminant. The exhaust airflow rate is a function of the size of the hood and the required capture velocity for the contaminant. Additional information about local exhaust systems may be found in ACGIH (1988).

9.2.4 Air Handling Units

An AHU is a set of modular components consisting, at the minimum, of a fan in a cabinet with a motor. Other components that can make up a complete AHU include cooling coils, heating coils, filters and dampers. Usually each component is an individual module, such as a coil section, which can be added to the unit as required for a particular application. The cabinets for AHUs are usually constructed of sheet metal panels. The sheet metal is usually either painted or galvanized steel. ARI Standard 430 defines AHUs as "a factory made, encased assembly consisting of a fan or fans and other necessary equipment to perform one or more of the functions of circulating, cleaning, heating, cooling, humidifying, dehumidifying and mixing of air, but which does not include a source of heating or cooling. This device is capable of use with ductwork having a total static resistance of at least 1/4 inch wg."

Air handling units are selected and sized based on the system heating and/or cooling loads, air circulation rate and type of air distribution system. Factory assembled AHUs are available from 900 cfm to 63,000 cfm. The most commonly used air distribution systems include constant volume, multi-path (multi-zone and dual duct) and VAV. A complete AHU should perform some or all of the following functions:

- provide air movement and circulation
- cool and dehumidify thc air
- heat or temper the air
- humidify the air
- filter and clean the air
- provide adequate outdoor ventilation
- provide a means of mixing outdoor and recirculated air

Air handling units can be divided into two basic configurations, which describe the location of the fan with respect to the coil section; blow-thru and draw-thru. The terminology for central station AHUs is shown in Figure 9-21, along with the various components and their typical location on a modular AHU. A blow-thru unit has the fan located upstream of the coil section. The fan blows or pushes the air through the coils. In the case of the blow-thru unit, the cooling coil removes the heat added to the air stream by the fan. The unit configuration does not affect the total system cooling load; however, the fan heat for a blow-thru AHU does not become part of the space cooling load. Since the air is blown through the coils, it can take more than one parallel path after leaving the fan. This permits

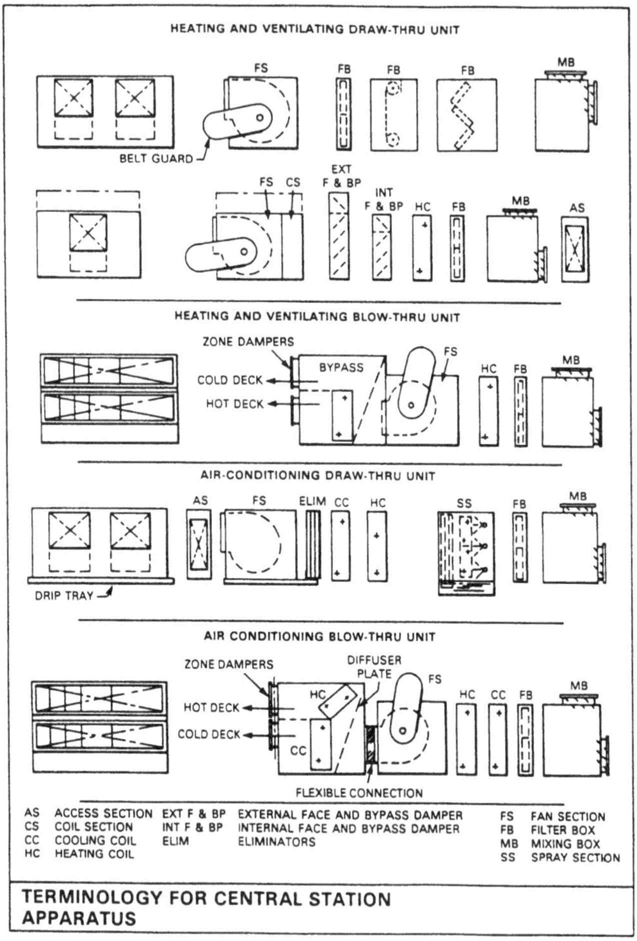

Figure 9-21. Terminology for central station air handling units.

what are called multi-zone systems and AHUs. Multi-zone AHUs are units with several separate temperature controlled zones to serve different areas of a building with different heating and/or cooling load characteristics. Blow-thru units can also be used for dual duct units. A more detailed discussion of the different types of air distribution systems can be found earlier in this chapter.

In the majority of single fan applications for central station AHUs, the fan is located downstream of the coil section and draws or pulls the air through the coils. Since the draw-thru unit has the fan located downstream of the cooling coil, the fan heat enters the cooled air stream and becomes part of the room sensible cooling load by increasing the supply air temperature to the conditioned space. This results in a higher required airflow rate for a given space sensible cooling load.

Air handling units can also be classified according to the path of the air as it travels through the unit. In a horizontal unit, the flow of air is completely horizontal. The fan and coils are in a horizontal line. In a vertical AHU, the fan is mounted above the coil section and the air flows vertically to the fan after passing through the coil section. Horizontal and vertical arrangements for AHUs are shown in Figure 9-22.

The AHUs presented in Figure 9-22 are both horizontal discharge units. The discharge for AHUs may also be vertically through the top or bottom (for horizontal units).

Most AHUs include some method of cooling, dehumidifying, and sometimes heating the air as it is circulated through the unit. These cooling, dehumidifying and heating processes are usually performed by cooling and heating coils.

Mechanical cooling and dehumidification of air is accomplished by one of two types of cooling coils: chilled water and DX. Air is cooled with a chilled water cooling coil by circulating cooled or chilled water through a coil as the heat transfer fluid or heat transfer medium. Air is cooled by DX coils as the result of circulating a refrigerant through the coil as the heat transfer medium.

Cooling and dehumidification of air is accomplished by passing the circulated air through a coil with a heat transfer medium that is cooler

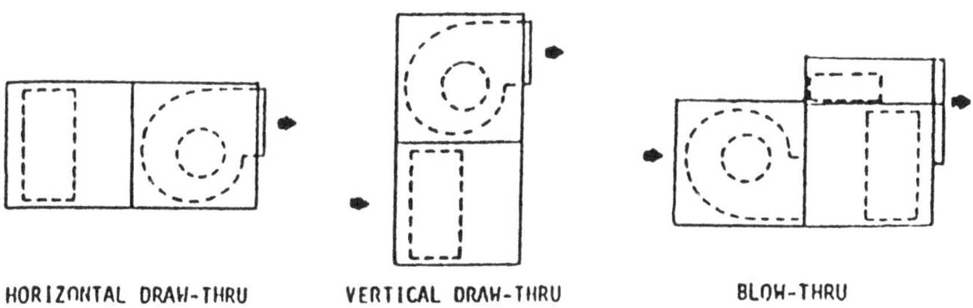

Figure 9-22. Air handling unit arrangements.

than the air. The coil serves as an extended surface heat exchanger between the heat transfer fluid and the air. Whenever air is cooled below its dew point temperature, moisture will condense out of the air.

Heating of air may also be accomplished by circulating a heat transfer fluid through a coil with air passing through it, although heat may also be produced in a coil electrically. The heating fluids most commonly used in HVAC systems are heated water and steam.

Air filters are essential in AHUs since they perform a necessary function that provides two benefits in HVAC systems.

- They remove airborne particulate matter before the circulated air enters the conditional space, thereby reducing the level of contaminants in the space.
- They remove airborne particulate matter before it adheres to the inside of AHU components and the air distribution ductwork.

Mixing sections, or mixing boxes as they are commonly called, are used to mix two converging air streams in the AHU, usually system return air and outdoor air. A mixing box is a plenum with two air inlets, each of which has a set of volume control dampers. Mixing boxes provide the proper mixing of return air and outdoor air in the required quantities. The sets of dampers in the mixing box inlets can be permanently adjusted to provide set quantities of return and outdoor air, or can be actuated to modulate the relative quantities of the two air streams. Modulating dampers are usually automatically adjusted by a pneumatic or electric actuator. The modulation range can be from 100% outdoor air (economizer) to the minimum outdoor air required by code. Occasionally, filters are included in the mixing box to both mix and filter the air as it enters the AHU. These components are referred to as combination filter mix boxes.

AHU accessories include such items as dampers, moisture eliminators and humidifiers. Dampers are used to control the flow of air. In addition to their use in mixing boxes, dampers may also be used for temperature control purposes. Dampers may be employed to pass air either through the face of a heating or cooling coil or to bypass the air around the coil. The air passing through the coil is either heated or cooled, depending on the type of coil. Air passing around the coil is unaffected, i.e., neither heated nor cooled and dehumidified. The two air streams are then mixed inside the AHU after passing the coil to achieve the desired temperature. Dampers of this type are called face and bypass dampers.

Humidifiers are occasionally needed to add moisture to the air stream in order to maintain minimum relative humidity conditions in the conditioned space. Humidification is frequently required in winter when low temperature, low moisture air is brought into a building for outdoor ventilation purposes. Low humidity levels in the conditioned space can also be a problem during economizer operation, when 100% outdoor air is brought into the building by the air system for cooling purposes during mild weather.

9.2.5 Fans and Fan Performance

There are many types of fans used for HVAC applications. The type of fan used for a particular system depends on the application or required characteristics. For example, some types of fans are best suited for high airflow volumes and low static pressures. Others are better suited for high pressure low flow situations; some fan types are better suited for VAV applications than others. Other considerations include space available, noise levels and the system in which they are installed.

The most common types of fans used for HVAC systems, which will be discussed in this chapter, include axial flow fans and centrifugal fans. Axial flow includes propeller, tubeaxial flow and vaneaxial flow. Centrifugal fans include forward curved, backward inclined, backward curved and airfoil fans. Special applications of fans include inline centrifugal, plug fans and power exhaust/ventilator fans. The special fans are usually an adaptation of the two basic types of fans.

Both centrifugal and axial flow fans increase the total pressure, of the air by producing velocity pressure, which is converted to static pressure in the air distribution system (ductwork). In an axial flow fan, the airflow direction is parallel to the rotational axis of the fan. In centrifugal fans, air is radially discharged from an impeller. The air turns a net 90° from the fan inlet to its outlet.

A typical propeller fan is shown in Figure 9-23. Propeller fans are used for moving air against a relatively low static pressure, generally less than

Figure 9-23. Propeller fan (*Courtesy of the Trane Company*).

1 inch wg. The performance of these fans is very sensitive to changes in airflow resistance. Propeller fans are well suited for moving high volumes of air at little or no static pressure differentials. The propeller fan can be selected over its entire operating range as long as there are no high noise level restrictions. These fans tend to be noisy. Propeller fans may be either belt driven or direct drive fans.

Tubeaxial (duct fans) have propeller type blades in a short cylindrical housing, as shown in Figure 9-24. There are no straightening vanes. These fans are suitable for relatively moderate static pressures, less than 2 inches wg.

Vaneaxial fans are very similar to tubeaxial fans. The main difference is the addition of airflow straightening (guide) vanes on the fan discharge and occasionally on the fan inlet. Vaneaxial fans have a propeller type configuration, with hub and airfoil type blades mounted in a cylindrical housing. The housing normally includes straightening vanes on the discharge side of the impeller. Vaneaxial fans are more efficient than other axial flow fans and can develop higher static pressures. A vaneaxial fan is typically mounted on the centerline of the duct and produces an axial flow of air.

Vaneaxial and tubeaxial fans are generally used for handling large volumes of air at relatively low static pressures. The maximum efficiency of these fans is about 65%. The operating range for tubeaxial and vaneaxial fans is between 65% and 90% of the airflow range. Disadvantages of these fans include relatively high noise levels and lower efficiency.

Figure 9-26 shows a typical centrifugal fan. There are three commonly used types of centrifugal fans based on the design of the impeller: forward

Figure 9-24. Tubeaxial fan (*Courtesy of the Trane Company*).

Figure 9-25. Vaneaxial fan (*Courtesy of American Coolair*).

curved (FC), backward inclined/backward curved (BI) and radial. Flow within a centrifugal fan is primarily radial through the fan impeller. In a

Figure 9-26. Centrifugal fan (*Courtesy of Twin City Fan and Blower*).

centrifugal fan, air enters the impeller and is accelerated out through the blade passages by centrifugal force as the impeller rotates. The blast area is the cross-sectional area just above the cutoff; it prevents the air from recirculating inside the fan and discharges the air out of the fan housing. The blast area is always smaller than the outlet area of the centrifugal fan. The blast area and the airflow rate determine the maximum discharge air velocity from the fan. Single width centrifugal fans draw air into the impeller from only one side. Double width fans draw air in from both sides of the impeller.

A double width fan is basically two single width fans, side by side, with two inlets and a single outlet with no partition in the scroll.

Forward curved (FC) centrifugal fans have impeller blades that curve toward the direction of rotation. They are also referred to as "squirrel cage" fans. An FC fan impeller is shown in Figure 9-27. These fans have relatively low tip speeds, require relatively little space and are relatively quiet in operation. They are generally used in low static pressure applications. Although FC fans rotate at relatively slow speeds, they are used for producing high volumes of airflow. The typical operating range of FC fans is from 30–80% of the wide open volume of the fan. The maximum static efficiency of these fans is around 60–80%, and generally the peak static efficiency is slightly to the right of the peak static pressure on a fan performance curve. The horsepower curve has an increasing slope and is said to have an overloading characteristic. Overloading of the fan motor occurs if the system resistance is overestimated and the fan operates at a higher airflow rate, thus overloading the motor.

Air leaves an FC fan at a greater velocity than the tip speed of the impeller. For this reason, the FC fan operates at a lower rotational speed

Figure 9-27. Forward curved centrifugal fan (*Courtesy of Twin City Fan and Blower*).

than other types of fans. Since the fan impeller rotates at a relatively low speed, the fan can have a lightweight construction for low pressure applications.

Backward inclined (BI) centrifugal fans have impeller blades that are inclined opposite to the direction of the impeller rotation, which causes the air to leave the fan at a velocity less than the blade tip speed. A typical BI fan impeller is shown in Figure 9-28. BI fans typically rotate at relatively higher speeds, resulting in higher tip speeds and higher efficiency. They have relatively low noise levels. These fans have non-overloading motor horsepower characteristics, i.e., the point of maximum horsepower occurs near the optimum operating point.

BI fans typically rotate faster than an equivalent FC fan. The normal selection range for a BI fan is approximately 40–85% of the wide open airflow. Maximum static efficiency is about 80% and generally occurs near the high airflow end of its normal operating range.

Advantages of the BI fan include higher efficiencies, quiet operation and a non-overloading motor horsepower curve. Since BI fans rotate at relatively high speeds, their construction is much stronger than that of FC fans, which makes them more suitable for higher pressures. Disadvantages include the higher rotational speeds, larger physical sizes and unstable operation at block tight (shut-off) static pressure.

Airfoil fans (AF) are a refinement of the BI fan. AF fans are similar to BI fans but have airfoil shaped blades. The airfoil shaped blades improve the static efficiency of the fan up to about 86% and slightly reduce the noise generated by the fan.

Figure 9-28. Backward inclined fan impeller (*Courtesy of Twin City Fan and Blower Co.*).

TABLE 9.1 Comparison of Centrifugal Fan Types

Feature	Forward Curved Blades	Radial Blades	Backward Inclined Blades
Efficiency	Medium	Medium	High
Space required	Small	Medium	Medium
Speed for press rise	Low	Medium	High
Noise	Poor	Fair	Good

Centrifugal fans with radial impellers have blades that are straight, or are mounted in a radial direction from the hub. Radial bladed fans are generally narrower than other types of centrifugal fans. Consequently, they require a larger diameter impeller for a given capacity. This increases the cost of the fan, which is a major reason why they are not commonly used in HVAC applications. Radial fans are frequently used for material handling systems; they are well suited for handling low volumes of air at relatively high static pressures.

Table 9.1 presents a synopsis of the advantages and disadvantages of the most widely used centrifugal fans.

Centrifugal fans are also available in an in-line configuration, as a tubular centrifugal fan, as shown in Figure 9-29. Tubular centrifugal fans generally consist of a single width airfoil impeller in a cylindrical housing. The impeller discharges the air against the inside of the cylinder to produce static pressure at the fan outlet. These fans typically have a static efficiency of about 72% and have a slightly higher noise level than other types of centrifugal fans.

Figure 9-29. Tubular centrifugal fan (*Courtesy of Twin City Fan and Blower*).

At any fixed volume flow rate (cfm) through a given system, a corresponding pressure loss, or resistance to the airflow, would exist. The system resistance is the sum of all pressure losses through filters, coils, dampers and ductwork. If the flow rate is changed, the resulting pressure loss, or resistance to airflow, will also change. For a fixed system (a system with no changes in damper settings), the system resistance varies as the square of the airflow rate. The relationship governing this change for such a system is

$$P_{s2}/P_{s1} = (Q_2/Q_1)^2 \qquad \text{Eq. (9.3)}$$

where

P_{s1} = the static pressure at the initial conditions
P_{s2} = the static pressure at the final conditions
Q_1 = the airflow at the initial conditions
Q_2 = the airflow at the final conditions

If the airflow rate is increased from 100–120% of the initial design, the system resistance will increase to 144% of the initial design. A decrease to 50% flow would result in a decrease in resistance to 25% of design.

Using Equation (9.3), the system static pressure can be determined for an airflow rate if the static pressure for the system has been determined at a given flow. If the system resistance were calculated for a number of rates of flow and plotted on a graph, with the airflow on the horizontal axis and the system resistance on the vertical axis, a system resistance curve would result. A system curve is a graphical representation of the relationship between the system airflow and the static pressure loss for the entire system at various airflow rates. A system curve is a graphical representation of Equation (9-3). A typical system curve is shown in Figure 9-30.

Example 9.1. An air distribution system has a static pressure loss of 2 inches wg at an air flow rate of 5,000 cfm. If the airflow rate was increased to 7,000 cfm, what would be the static pressure loss for the system?

Using Equation (9.3),

$$P_{s2} - P_{s1}(Q_2/\ Q_1) = (2 \text{ in wg})\left(\frac{7{,}000 \text{ cfm}}{5{,}000 \text{ cfm}}\right) = 3.92 \text{ in wg}$$

The theoretical horsepower (hp) required to drive a fan is the power that would be required if there were no losses in the fan itself (such as friction losses, drive losses, etc.). That is, it is the power that would be required if

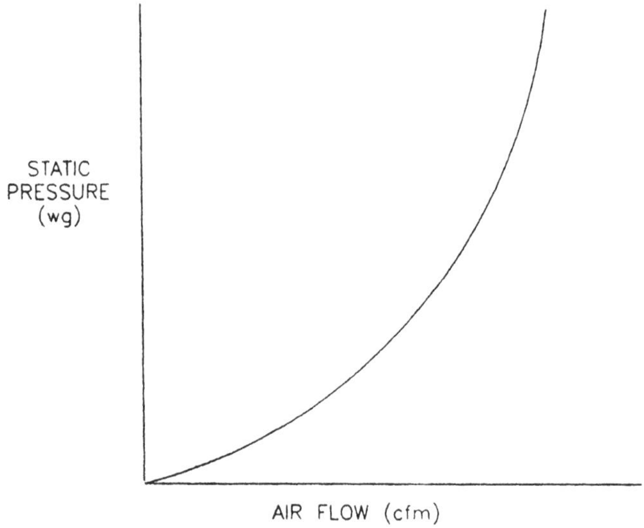

Figure 9-30. System curve.

the fan efficiency was 100%. The theoretical power required by a given fan is given by Equation (9.4).

$$W = QP_T/6,356 \qquad \text{Eq. (9.4)}$$

where W = the power to drive the fan in horsepower
Q = the airflow in cfm
6,356 = a conversion factor to make the units consistent

Brake horsepower (Bhp) is the actual power a fan requires for a given set of operating conditions, i.e., airflow, static pressure and fan efficiency. It is the power to input to the fan shaft. The brake horse-power required to drive a fan is always larger than the theoretical horsepower due to energy losses inside the fan itself. The brake horse-power required for a fan can be determined only through an actual test of the fan.

After testing a fan to determine the required brake horsepower, both the total efficiency and the static efficiency of the fan can be calculated. The total efficiency of the fan is based on the total pressure produced by the fan, whereas the static efficiency is based on the static pressure. The total fan efficiency may be calculated using Equation (9.5).

$$\eta_{FT} = [(QP_T)/(6,356 \text{ Bhp})] \times 100\% \qquad \text{Eq. (9.5)}$$

where η_{FT} = the total efficiency of the fan, %
Bhp = the brake horsepower required to drive the fan
P_T = the total pressure of the fan, inches wg

The static efficiency may be calculated using Equation (9.6).

$$\eta_{FS} = \left[(QP_s)/(6,356\ \text{Bhp})\right] \times 100\% \qquad \text{Eq. (9.6)}$$

where η_{FS} = the static efficiency of the fan in percent
P_s = the static pressure of the fan in inches wg

The fan efficiency is the flow work output of the fan divided by the work input.

A fan performance curve (fan curve) is a graphical representation of the relationship between the fan airflow and the static pressure produced by the fan. Just like the system curve, a fan has the airflow rate plotted along the horizontal axis and the static pressure along the vertical axis. Fan curves are obtained from a series of laboratory tests on a fan with various levels of restriction at the end of the test duct. Lines connecting the test points form the fan performance curve.

Fan performance curves typically show the static pressure versus airflow. Some fan curves also include total pressure, static efficiency and total efficiency. A typical fan curve for a centrifugal forward curved fan is shown in Figure 9-31 and a typical fan curve for a backward inclined and airfoil fan are shown in Figure 9-32.

Usually, fan curves show the entire range of fan performance, from free delivery (no obstruction or resistance to airflow) to no delivery (airtight system with no airflow).

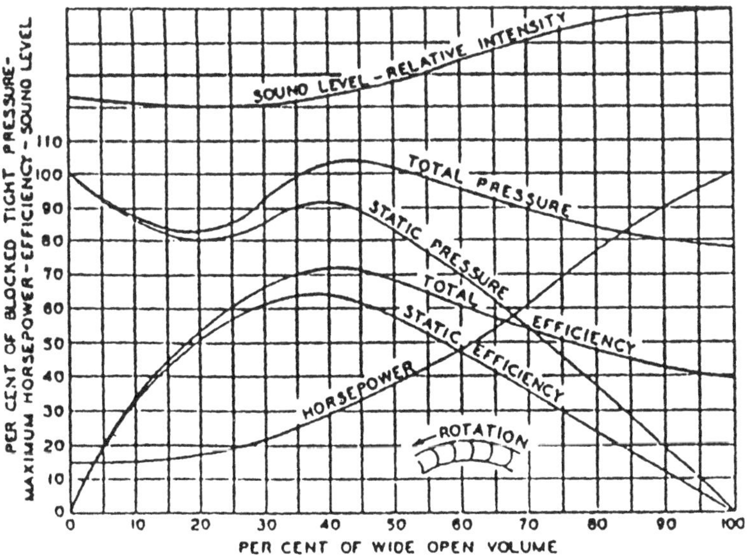

Figure 9-31. Typical fan curve for a forward curved centrifugal fan.

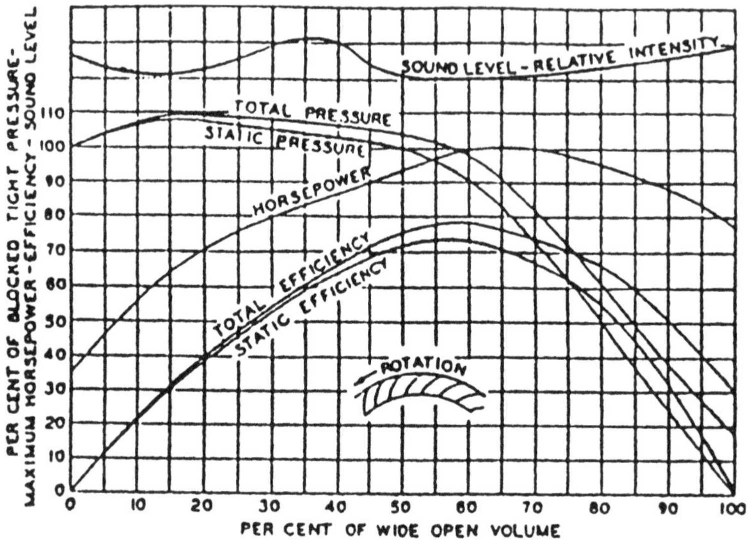

Figure 9-32. Typical fan curve for a backward inclined and airfoil centrifugal fan.

The performance of a fan at various flow and pressure conditions may also be presented in a tabular form called a multi-rating table. In addition to publishing performance curves for their fans, most manufacturers also publish multi-rating tables for their fans. A separate fan curve and multi-rating table is usually provided for each type of fan and each diameter impeller manufactured. Multi-rating tables show the fan rotational speed (rpm), airflow rate, static pressure and horsepower for each size impeller. Interpolation between values is frequently necessary to obtain specific values such as speed. A typical fan curve and multi-rating table for a BI fan is shown in Figure 9-33.

Most multi-rating tables do not cover the entire range of performance from no delivery to free delivery. They usually cover only the typical operating range.

The operating point at which the fan and system will perform is determined by the intersection of the system curve and the fan curve. A typical fan and system curve is presented in Figure 9-34. Note that a fan will operate along its performance curve. If the actual air system resistance is different from the design resistance, the operating point will be different from the design point and the volume of air delivered will be different from the design airflow. Figure 9-35 shows the variation from system design that would occur if the actual static pressure for a system were greater than the calculated or design static pressure.

Notice that an increase in system static pressure would cause the system curve to shift to the left, resulting in a reduction in airflow. The operating point for the fan and air system will shift up or down along the fan curve depending on the static pressure for the system.

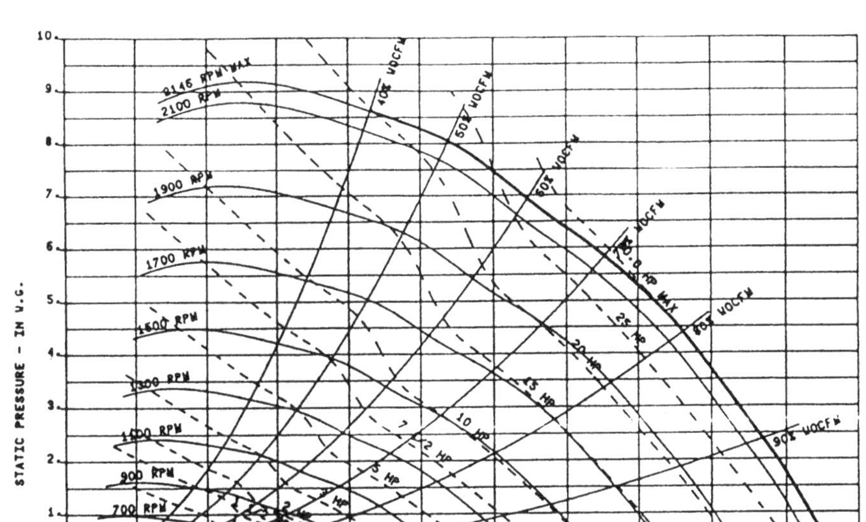

SIZE 30 22.25" BI VDT-PIGV

CFM Std Air	Outlet Vel	Total Static Pressure 0.25 RPM	0.25 BHP	0.50 RPM	0.50 BHP	0.75 RPM	0.75 BHP	1.00 RPM	1.00 BHP	1.25 RPM	1.25 BHP	1.50 RPM	1.50 BHP	1.75 RPM	1.75 BHP	2.00 RPM	2.00 BHP	2.25 RPM	2.25 BHP	2.50 RPM	2.50 BHP	2.75 RPM	2.75 BHP	3.00 RPM	3.00 BHP	3.25 RPM	3.25 BHP
10000	1587	1042	3.37	1084	3.84	1128	4.34	1170	4.86	1210	5.39	1250	5.93	1290	6.49	1331	7.08	1372	7.69	1413	8.32	1454	8.97	1494	9.66	1533	10.36
11000	1746	1138	4.39	1176	4.89	1216	5.42	1255	5.98	1292	6.56	1329	7.14	1365	7.73	1401	8.34	1438	8.97	1476	9.63	1513	10.31	1551	11.00	1587	11.71
12000	1905	1234	5.59	1270	6.14	1305	6.70	1342	7.30	1377	7.91	1411	8.54	1445	9.18	1478	9.82	1511	10.47	1545	11.15	1579	11.85	1613	12.57	1648	13.31
13000	2063	1331	7.01	1364	7.59	1397	8.19	1430	8.82	1464	9.48	1496	10.15	1528	10.83	1558	11.52	1589	12.21	1620	12.91	1650	13.63	1682	14.37	1714	15.15
14000	2222	1429	8.66	1459	9.28	1490	9.92	1520	10.58	1552	11.28	1583	11.99	1612	12.70	1642	13.44	1670	14.18	1699	14.92	1727	15.67	1756	16.43	1785	17.23
15000	2381	1527	10.55	1556	11.23	1584	11.91	1612	12.60	1641	13.33	1670	14.07	1699	14.84	1727	15.61	1754	16.40	1781	17.19	1808	17.99	1835	18.79	1861	19.60
16000	2540	1624	12.70	1652	13.42	1679	14.15	1705	14.88	1732	15.64	1759	16.42	1786	17.22	1813	18.03	1840	18.86	1865	19.70	1891	20.54	1916	21.39	1940	22.24
17000	2698	1722	15.13	1749	15.90	1774	16.67	1799	17.44	1824	18.23	1849	19.04	1875	19.88	1900	20.73	1926	21.60	1951	22.48	1975	23.37	1999	24.26	2022	25.16
18000	2857	1820	17.85	1846	18.67	1870	19.49	1893	20.30	1917	21.13	1940	21.97	1964	22.84	1989	23.73	2013	24.63	2037	25.55	2060	26.48	2083	27.42	2106	28.37
19000	3016	1920	20.92	1943	21.76	1966	22.62	1988	23.48	2010	24.35	2033	25.23	2055	26.13	2078	27.04	2101	27.98	2124	28.94						
20000	3175	2018	24.30	2040	25.18	2062	26.09	2084	26.99	2105	27.90	2126	28.82														
21000	3333	2117	28.03	2138	28.99																						

Figure 9-33. Typical fan curve and multi-rating table (*Courtesy of the Trane Company*).

9.2.6 Fan Laws

Fan laws relate the performance variables for any dynamically similar series of fans, and apply to all fans. The performance variables are the fan size (impeller diameter), D; the speed of rotation, N; the gas (air) density; the volume flow rate, Q; the pressure, P_T or P_s; the power, W; and the fan efficiency. The fan laws are mathematical expressions that relate the above performance variables. Unless otherwise indicated, fan performance is based on dry air at standard conditions, i.e., 14.7 70°F and 0.075 lb/ft^3.

Special care must be exercised when applying the fan laws in the following situations:

1. Where any component of the system does not follow Equation (9.3).

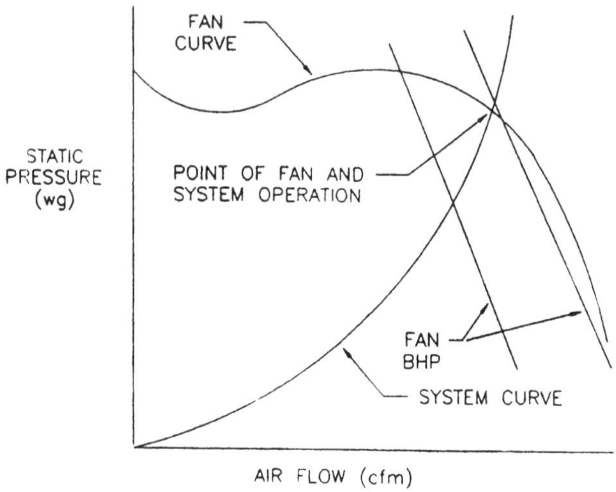

Figure 9-34. Fan and system curves.

2. Where the system has been physically altered or for any reason, operates on a different system curve.

For changes in the rotation speed, N, the flow rate varies with the speed of rotation, the pressure varies as the square of the speed of rotation and the power varies as the cube of the rotation speed. It is assumed that the fan size is constant, the air density is constant and the system itself is constant.

$$Q_2 = Q_1(N_2/N_1) \qquad \text{Eq. (9.7)}$$

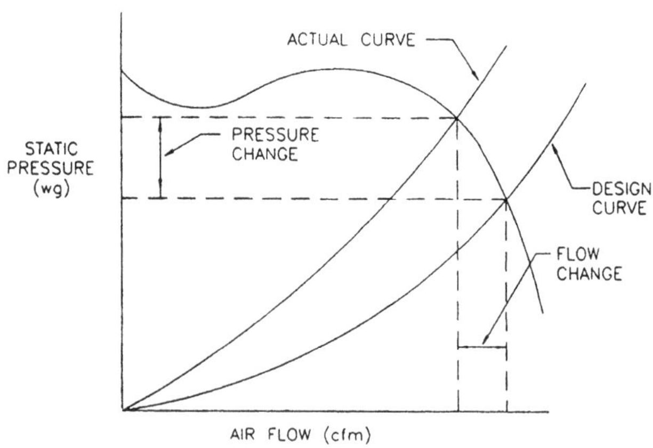

Figure 9-35. Variations from Design.

$$P_2 = P_1 (N_2/N_1)^2 \qquad \text{Eq. (9.8)}$$

$$W_2 = W_1 (N_2/N_1)^3 \qquad \text{Eq. (9.9)}$$

For changes in gas (air) density, the flow is not affected by a change in gas density. The pressure and power vary directly with the density. It is assumed that the fan size, airflow rate, fan speed and system are a constant.

$$Q_2 = Q_1 \qquad \text{Eq. (9.10)}$$

$$P_2 = P_1 (\rho_2/\rho_1) \qquad \text{Eq. (9.11)}$$

$$W_2 = W_1 (\rho_2/\rho_1) \qquad \text{Eq. (9.12)}$$

For changes in the fan size, D, the airflow and power vary as the square of the impeller diameter, the rotation speed varies inversely as the impeller diameter and the static pressure remains constant. This set of laws are used in the design of fans, but are rarely used in the design of HVAC systems. It is assumed that the rotation speed, the gas (air) density and the system are constant.

$$Q_2 = Q_1 (D_2/D_1)^2 \qquad \text{Eq. (9.13)}$$

$$W_2 = W_1 (D_2/D_1)^2 \qquad \text{Eq. (9.14)}$$

$$P_{s1} = P_{s2} \qquad \text{Eq. (9.15)}$$

$$N_2 = N_1 (D_2/D_1) \qquad \text{Eq. (9.16)}$$

Since the tolerances of fans are not proportional, slightly better performance is generally obtained when projecting from a given fan to a dynamically similar larger fan. The most frequent error in applying the fan laws is trying to predict the fan performance after changing some physical aspect of the system to which the fan is attached.

Example 9.2. A fan delivers 20,000 cfm of air at 1 inch wg static pressure at a rotation speed of 414 rpm and at a brake horsepower of 4.72. If the heating or cooling load were changed such that the system was circulating 22,000 cfm without any changes in the distribution system, what will be the resulting change in the fan rotation speed, brake horsepower and static pressure?

Rearranging Equation (9.7) to calculate rotation speed:

$$N_2 = N_1 (Q_2/Q_1) = 414 \text{ rpm}\left(\frac{22{,}000 \text{ cfm}}{20{,}000 \text{ cfm}}\right) = 455 \text{ rpm}$$

By Equation (9.7), $Q_2/Q_1 = N_2/N_1$, substituting into Equation (9.8),

$$P_2 = P_1 (N_2/N_1)^2 = \left(1.0'' \text{wg}\right)\left(\frac{22{,}000}{20{,}000}\right)^2 = 1.21'' \text{wg}$$

Substituting $Q_2/Q_1 = N_2/N_1$ into Equation (9.9),

$$W_2 = W_1 (N_2/N_1)^3 = W_1 (Q_2/Q_1)^3 = (4.72 \text{ Bhp})\left(\frac{22{,}000}{20{,}000}\right)^3 = 6.28 \text{ Bhp}$$

9.3 MAINTAINING AND OPERATING HVAC SYSTEMS FOR OPTIMUM ENERGY EFFICIENCY

HVAC and air distributions systems typically require routine maintenance to keep them operating at optimum efficiency. Significant energy savings can be achieved by proper routine maintenance and operation. Proper routine maintenance and operation should be the first step in any energy management program for HVAC and air distribution systems. The following are operations and maintenance procedures that will result in optimum energy efficiency for most HVAC and air distribution systems.

Replace Air Filters: Most AHUs include air filters to remove particulate matter from the circulated air stream. As the air flows through the filters, energy is lost in the form of a drop in pressure from the upstream side of the filters to the downstream side. The amount of pressure used as the air flows through the filters is a function of the air velocity through the filters and the resistance to airflow of the filters. The air velocity is directly related

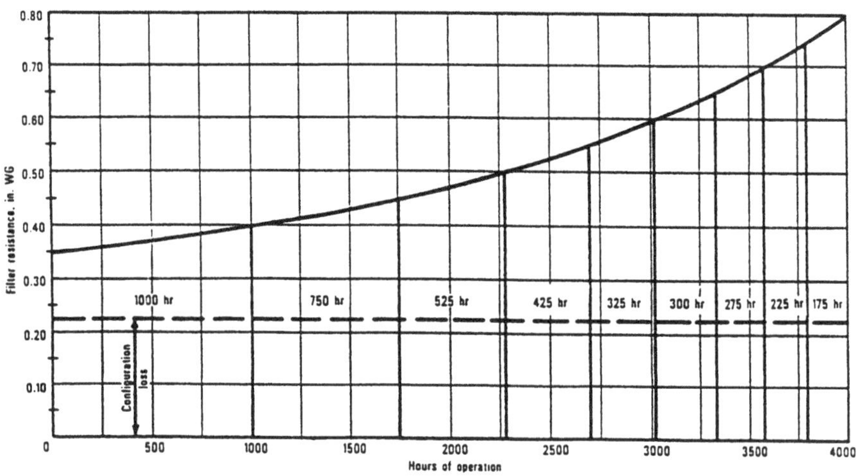

Figure 9-36. Typical filter resistance versus operating time.

to the airflow rate, which is usually set by the heating and cooling load, or is set by ventilation requirements.

For any given airflow rate, the resistance of a filter is a function of the type of filter, the filter media, the cleaning efficiency (arrestance) of the media, the surface area of the filtering media and the cleanliness of the media. The type of filter, the type of media, the arrestance and the surface area of the media are usually determined by the process requirements.

The cleanliness of the filter media for any given application is a function of the size and amount of suspended particles in the air stream. The cleanliness is also a function of the length of time the filters have been in service. When new filters are installed, they have an initial resistance to the airflow. The filter resistance increases as the hours of operation increase. A typical curve for a filter with an initial resistance of 0.35 inches wg for 2,500 cfm air flow rate is shown in Figure 9-36.

As may be observed from Figure 9-36, the filter resistance increases exponentially with time in service. The resistance increases by 50% of the inital resistance (0.35″ wg) to 0.525″ wg in about 2,400 hours. The resistance increases by an addition al 50 – 0.7″ wg in only 1,300 additional hours of operation. As a rule of thumb, the filters should be changed before the resistance reaches a value twice that of the inital resistance.

Equation (9.17), which is Equation (9.6) rearranged, may be used to calculate the horsepower consumed by a fan motor for a pressure loss, including pressure losses through filters. It assumes a constant velocity and does not include fan drive losses or motor efficiency losses.

$$HP = \frac{QP_S}{6{,}356\eta_F} \qquad \text{Eq. (9.17)}$$

where Q = air flow, ft^3/min
P_S = static pressure drop, inches wg
η_F = fan efficiency

Example 9.3. A fan with an efficiency of 60% will circulate 10,000 ft^3/min of air for 1,000 hours of operation. Presently, the filters are dirty and have a pressure drop of 0.7″ wg at 10,000 ft^3/min. If new filters with a pressure drop of 0.35″ wg are installed, how much energy would be saved? Assume constant pressure drops for both situations.

Using Equation (9.17) to calculate the energy consumption for the dirty filters,

$$HP = \frac{(10{,}000)(0.7)}{6{,}356(0.60)} = 1.83 \text{ hp}$$

For 1,000 hours of operation, the energy consumed would be

$$E = (1.83 \text{ hp})(1{,}000 \text{ hrs})(1.34 \text{ Kw/hp}) = 2{,}452.2 \text{ Kwh}$$

If new air filters with an intial pressure drop of 0.35″ wg were installed, the energy consumed would be

$$HP = \frac{(10{,}000)(0.35)}{6,356(0.60)} = 0.92 \text{ hp}$$

For 1,000 hours of operation, the energy consumed would be

$$E = (0.92)(1,000)(1.34) = 1,232.8 \text{ Kwh}$$

The energy savings would be

$$\Delta E = 2{,}452.2 - 1{,}232.8 = 1{,}219.4 \text{ Kwh}$$

Reset Discharge Air Temperature: Many heating and/or cooling systems supply conditioned air at a constant discharge temperature regardless of the heating or cooling load. Very few heating or cooling loads are constant. Most loads vary with time. When a heating or cooling load varies, an opportunity exists for saving energy by resetting the supply air temperature for some air systems. Energy may be saved when the load varies for multi-zone systems, dual duct systems and variable air volume systems.

Operating multi-zone, dual duct and variable air volume systems with a fixed discharge air temperature is an easy approach; however, this approach can result in significant energy loss through reheating or recooling of the air by either reheating the air at the load or by mixing hot and cold air streams at the load or conditioned space. By resetting the discharge air temperature according to the load or space requirements, the system energy consumption is minimized by matching the system heating and/or cooling output of the load.

Recalibrate Controls: All HVAC systems have controls to regulate temperatures, pressures, humidity levels, flow rates, etc. HVAC control systems may be electronic, electric, pneumatic or a combination of these systems. Overtime, all control systems become uncalibrated and no longer provide accurate control of the controlled variable. Pneumatic control systems are especially prone to becoming out of calibration.

Control systems should be recalibrated and adjusted on a regular basis. As part of a routine maintenance program for energy efficiency, the following checks and adjustments should be made to control systems.

- Calibrate thermostats and temperature sensors to ensure that the correct temperature is sensed and the thermostat cycles the HVAC system at the correct temperatures.
- Calibrate damper actuators and linkages to ensure proper damper position, especially the fully closed position.
- Adjust actuators on control valves to ensure that valves open and close at the required control settings.
- Check pressure sensors to ensure that they accurately measure the required pressure.

- Calibrate flow sensors to be sure they accurately measure the flow rate.
- Adjust all clocks and timers for proper time of day and time intervals.
- Check all compressed air lines and pneumatic components for air leaks.

Adjust Fan Drives: Earlier in this chapter, the efficiency of various types of fans was discussed. The efficiency of the fan itself is a function of the fan design. The overall energy efficiency of the fan and motor combination is a function of the fan efficiency and the efficiency of the motor and fan drive.

For fans that are not direct driven, the fan drive typically consists of a fan sheave, a motor sheave and one, or more, drive belts. For optimum fan efficiency, the fan drive should be periodically checked and adjusted. Drive belts should be adjusted for proper tension to reduce slippage and wear. Worn drive belts should be adjusted to produce the required fan rotational speed.

Install Heat Shields: Many processes operate at high temperatures and emit tremendous amounts of heat. The heat given off is typically transferred to the surrounding air in the space by convection heat transfer and to sur rounding objects by radiant heat transfer.

In addition to providing general ventilation, it is common practice to provide local ventilation for high heat producing processes. Local ventilation removes heat transferred to the air by convection, however, it does little to remove heat that is radiated to the surroundings. Because air temperature has no significant effect on radiant heat transfer, ventilation has very little effect on radiant heat transfer to the surroundings.

Radiant heat transfer from hot surfaces in a space may be reduced by insulating or placing radiation shields around the heat source. The most effective radiation shields are infrared reflectors, which should reflect the radiant heat back to the source, where it can be removed by local exhaust. The shield should completely surround the heat source and, to be effective, the reflective shield should be a bright color.

Balance Air Distribution Systems: When air distribution systems are designed and installed, they are designed such that they provide the required airflow to satisfy a heating or cooling load, If the air system serves multiple outlets, each outlet is designed and installed to provide a prescribed airflow rate.

After air distribution systems are in service for somc timc, they become out of adjustment and no longer provide the required airflow at each outlet. This is particularly the case for air systems that been modified since the original installation of the system. In order to provide the proper airflow rates at the air outlets, air distribution systems most be adjusted and rebalanced.

The balancing of an air distribution system includes the testing of the airflow rate at various locations in the distribution system, as well as adjustments to the system to obtain the desired airflow rates. Balancing of an air system typically includes adjusting the fan speed to provide the required airflow, and adjusting dampers in the duct branches of the system to obtain the desired airflow rates at the various outlets. After the testing and balancing of an air system has been completed, airflow at the branches, the outlets and the fan should be documented for future reference.

9.4 ENERGY CONSERVATION MEASURES FOR AIR DISTRIBUTION AND HVAC SYSTEMS

The energy conservation measures for air distribution and HVAC systems discussed in the previous section were operation and maintenance procedures that would improve the energy efficiency of existing systems. Most of those measures were little, or no, cost measures that could be performed on an ongoing basis to improve efficiency.

In addition to the low cost OM procedures, there are also modifications that could be made to air distribution and HVAC systems, which would improve energy efficiency. These modifications, however, require a more significant cost investment and are permanent. The modifications listed below are improvements that should be considered under any energy manage ment program for existing, or proposed, air distribution and HVAC systems.

9.4.1 Convert Constant Volume Air Systems to Variable Air Volume (VAV)

Many air distribution systems have constant volume despite the fact that the load varies. Constant volume systems operate continuously at peak airflow, which results in peak energy consumption by the fan motor.

If the load varies with time, an opportunity exists for significant fan energy savings by converting a constant air volume system to a variable air volume, system, which varies the airflow rate based upon the actual heating or cooling load. Existing fans and air distribution systems can be converted from constant volume to variable air volume with some modifications to the fan and air distribution system.

To convert a main supply fan from constant volume to variable volume, some type of fan capacity control is required. There are four control methods that were discussed earlier in this chapter; were discharge dampers, a fan bypass ar rangement, fan inlet vanes, and variable speed drives. Discharge dampers and fan bypass arrangements are not recommended as they produce little energy savings. The relative power consumed by various VAV control methods is shown in Figure 9-37. Variable speed drives are discussed in "Chapter 10: Electrical and Lighting Systems".

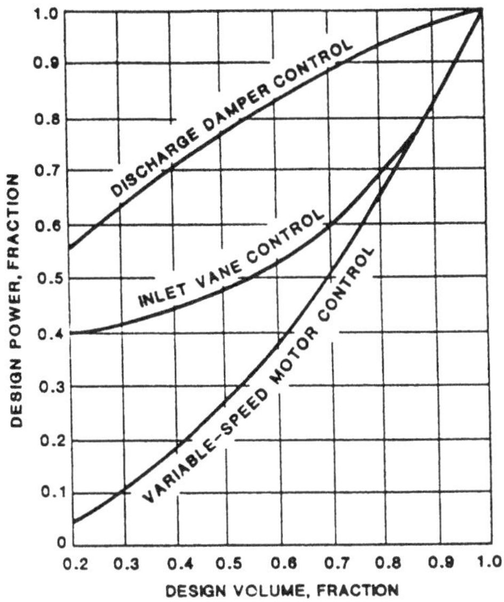

Figure 9-37. Fan power vs. volume characteristics.

Inlet vanes are an effective air capacity control method for fans. A centrifugal fan with inlet vanes is shown in Figure 9-20. Other types of fans, such as axial and vane axial fans, are also available with inlet vanes. Although the fan is still rotating at full rotational speed, energy is saved since the fan is not fully loaded. The inlet vanes also impart a pre-rotation swirl to the air as it enters the fan. The reduction in airflow and the pre-rotation swirl both reduce the load on the fan motor.

The airflow at the load is typically controlled by some type of terminal unit with an automatically controlled and actuated air damper. The damper opens and closes to control the airflow rate in response to a thermostat at the load. A variable air volume system schematic with multiple VAV air terminals is shown in Figure 9-19.

To control the airflow rate of the fan, it is advisable to install a static pressure sensor in the main supply air duct. As the terminal units open or close to vary the airflow rate at the load, the pressure in the duct varies accordingly. As a result, the airflow of the fan is readily controlled by measuring the static pressure in the duct. The pressure sensor is typically located about two-third of the distance from the fan to the most remote terminal.

Example 9.4. A fan supplies 10,000 ft^3/min of air to cool a process at a static pressure of 2 inches wg and operates for 3,000 hours per year. The fan and drive have an efficiency of 60%. It is determined that the fan could supply air at 10,000 ft^3/min for 1,000 hours, at 7,500

ft^3/min for 1,000 hours and 5,000 ft^3/min for 1,000 while still satisfying the cooling load of the process. How much energy would be saved annually if the air system was converted to that of variable air volume. Neglect motor efficiency losses.

Equation (9.17) may be used to calculate the motor horsepower for 10,000 ft^3/min and 2″ wg static pressure.

$$HP = \frac{(10,000)(2)}{6,356(0.60)} = 5.24 \text{ hp}$$

If the fan operates at a constant volume for 3,000 hours per year, the annual electrical energy consumption would be

$$E = (5.24)(3,000 \text{ hrs})(1.34 \text{ Kw/hp}) = 21,065 \text{ Kwh/yr}$$

If the fan and air distribution system was converted to that with variable air volume and operated as indicated, Equations (9.7), (9.8) and (9.9) of the fan laws could be combined and used to determine the fan operating characteristics at the various flow rates.

The horsepower at 7,500 ft^3/min would be

$$HP_2 = HP_1 (Q_2/Q_1)^3 = 5.24\left(\frac{7,500}{10,000}\right)^3 = 2.21 \text{ hp}$$

The horsepower at 5,000 ft^3/min would be

$$HP = 5.24\left(\frac{5,000}{10,000}\right)^3 = 0.67 \text{ hp}$$

The energy consumed at 10,000 ft^3/min for 1,000 hours would be

$$E = (5.24)(1,000)(1.34) = 7,022 \text{ Kwh}$$

The energy consumed at 7,500 ft^3/min for 1,000 hours would be

$$E = (2.21)(1,000)(1.34) = 2,961 \text{ Kwh}$$

The energy consumed at 5,000 ft^3/min for 1,000 hours would be

$$E = (0.67)(1,000)(1.34) = 898 \text{ Kwh}$$

The annual energy consumed by the variable air volume system would be

$$E = 7,022 + 2,961 + 898 = 10,881 \text{ Kwh/yr}$$

The total annual energy savings for converting to variable air volume would be

$$\Delta E = 21,065 - 10,881 = 10,184 \text{ Kwh/yr}$$

Converting from a constant volume air system to a variable air volume system would result in an annual fan motor energy savings of almost 52%.

9.4.2 Provide Local Exhaust for Processes

When a process is hot or discharges pollutants into the air, such as fumes, dust or mist, it is necessary to remove the heat or pollutants from the indoor air. Frequently, the heat and/or contaminants are removed by general building exhaust. The heat or contaminants are discharged into the air, where they are removed by the general exhaust system.

Allowing the heat or contaminants to be discharged into the indoor air results in higher exhaust airflow rates than necessary. As a result, energy is usually wasted in heating or cooling the incoming make-up air. Providing local exhaust for processes can reduce the amount of make-up air that must be heated or cooled. Equation (9.18) may be used to calculate the amount of heat required to raise the temperature of an air stream.

$$q = 1.08Q(T_2 - T_1) \qquad \text{Eq. (9.18)}$$

where Q = the air flow rate, ft^3/min
T_2 = the leaving air temperature, °F
T_1 = the entering air temperature, °F

The amount of exhaust air and the corresponding amount of make-up air can be significantly reduced by providing local exhaust at the source of

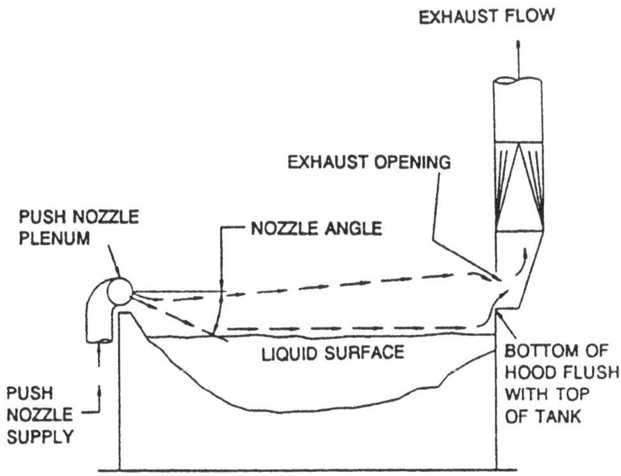

Figure 9-38. Local process ventilation.

the heat or pollutant. Local exhaust enables close capture of heat and pollutants before they enter the air in the conditioned space.

Local exhaust systems frequently have a hood that is located over the process or surrounds the process as much as possible. Occasionally, make-up air is supplied directly at the process to help contain the heat or pollutants, as shown in Figure 9-38. Local exhaust systems are able to contain heat and contaminants since the systems use the capture velocity of the air rather than volumetric airflow. The heat and pollutants are contained by the moving air stream as it enters the hood or exhaust collector. Since the area through which the air flows is restricted, the velocity of the air increases as it passes over the process entraining the heat or contaminants. Additional information about local exhaust systems may be found in ASHRAE (1995) and ACAIH (1998).

Example 9.5. A manufacturing plant has a process that gives of large amounts of noxious chemicals. The plant has a general exhaust system, which provides 5,000 ft^3/min of exhaust to control the contaminants. The exhaust air is replenished by an air handling unit with a steam heating coil that supplies air at 70°F. The plant is located where the average daytime winter temperature is 45°F. The process and the exhaust/make-up air systems operate 8 hours per day (160 hours per month).

If a local exhaust/make-up air system that requires 1,000 ft^3/min of exhaust and make-up air were installed for the process how much heating energy would be saved? Assume heat is required from November through March (five months).

Equation (9.18) may be used to calculate the amount of heat required to heat 5,000 ft^3/min of make-up air from 45°F to 70°F.

$$q = 1.08(5{,}000)(70 - 45) = 135{,}000 \text{ Btu/hr}$$

The energy to heat the air for the heating season would be

$$E = (135{,}000 \text{ Btu/hr})(160 \text{ hr/mo})(5 \text{ mo}) = 108 \times 10^6 \text{ Btu/yr}$$

If the exhaust air and make-up air requirements were reduced to 1,000 ft^3/min, the required heat would be

$$q = 1.08(1{,}000)(70 - 45) = 27{,}000 \text{ Btu/hr}$$

The energy to heat the reduced airflow for the heating season would be

$$E = (27{,}000)(160)(5) = 21.6 \times 10^6 \text{ Btu/yr}$$

The annual energy savings would be

$$\Delta E = 108 \times 10^6 - 21.6 \times 10^6 = 86.4 \times 10^6 \text{ Btu/yr}$$

In the above example, an average winter temperature was used to simplify the calculation process. For more accurate calculations, typical weather data should be used. Engineering Weather Data, Air Force Manual 88-29 contains typical year bin temperature data. The tabulated bin temperatures show the number of hours that a given temperature occurs for a typical weather year for various locations. A spread sheet may be used to perform the calculations for various temperature, to get the energy consumption for a typical year of weather.

9.4.3 Provide Make-up Air

Frequently, air for general exhaust and local exhaust is exhausted without any provisions for making up the exhausted air. This results in a negative pressure inside the building. In tightly sealed buildings, the negative pressure can result in negative pressures in the building ranging from 0.1 to 0.25 inches wg. Negative pressure conditions in a building can result in operational problems, as well as increased energy consumption.

Negative pressure conditions can reduce the airflow rate of exhaust systems, reducing the performance of exhaust fans. Inadequate performance of exhaust systems can result in a buildup of indoor air pollutants and also cause problems with combustion systems inside the building.

Negative pressures inside a building can also increase energy consumption beyond what would also be consumed if adequate make-up air units that heat the make-up air to the indoor space temperature were used. Negative pressures cause outdoor air to be drawn into (infiltrate) the building. During cold weather, infiltration can cause drafty conditions near the exterior walls, which causes discomfort to nearby occupants. Some areas near exterior walls may also be cooled by the infiltrating outside air below the indoor air dew point temperature, which can result in condensation on cold surfaces. The typical solution to these problems is to provide additional space heating near exterior walls. Heat from the additional units is usually drawn into the interior of the building due to the velocity of the infiltrating air. This often results in overheating in the building interior while the perimeter areas remain uncomfortable. Frequently, additional exhaust is added to reduce interior overheating, which causes a greater negative pressure and increased infiltration.

Make-up air units should be provided to supply tempered make-up air at a rate at least equal to the exhaust rate. Preferably, a slightly greater make-up airflow rate should be provided; it would result in a slightly positive pressure in the building and a significant reduction in infiltration of outdoor air into the building.

9.4.4 Air Side Economizer Cycles

Many buildings and processes require cooling year round, regardless of the outside air conditions. Frequently, mechanical cooling is provided in the form of direct expansion or chilled water cooling throughout the year. There can be significant savings in energy required for cooling by providing cooling with an air side economizer cycle when conditions are favorable.

An economizer simply uses outside air for cooling whenever outdoor air conditions are favorable. Air side economizers may be either a dry bulb type or enthalpy type. Dry bulb, or sensible, economizers provide cooling whenever the outdoor air dry bulb temperature is low enough. Enthalpy economizers provide cooling when the enthalpy (total energy) of the outside air is favorable. The system will not go into economizer if the total heat (sensible and latent) is greater than the total heat in the conditioned space. Unfavorable conditions may exist when the wet bulb temperature of the outside air is high even though the dry bulb temperature is suitable.

Some economizer systems may be set up such that the economizer operation is assisted by mechanical cooling. This situation may occur for enthalpy economizers, where the outside dry bulb temperature is suitable, but the wet bulb temperature is too high. The mechanical cooling would just provide the required latent cooling.

Air side economizers require the proper automatic controls for proper operation. Usually, this includes an outdoor air sensor, a mixed air sensor, a discharge air sensor, a sensor in the conditioned space, and automatically actuated outside and return air dampers. When an air system goes into economizer mode, some method to relieve the increased outdoor airflow rate from the conditioned space is required.

9.4.5 Evaporative Cooling

Many industrial and manufacturing buildings are mechanically cooled. Cooling may be provided for occupant comfort or it may be required by a particular process.

Although cooling may be either required or desirable, cooling large, non-office type buildings mechanically can be costly due to the energy consumed by the cooling equipment. An alternative to mechanical cooling of industrial and manufacturing type buildings is evaporative type cooling, which can be energy-efficient and cost-effective if current technology and equipment are used. In many applications, all of the cooling for a facility can be provided by evaporative cooling, or evaporative cooling can be supplemented by mechanical cooling. Evaporative cooling is quite energy-efficient and cost-effective where large amounts of outdoor air are brought into a facility and then exhausted outdoors.

In many cooling applications, cooling is achieved by passing air that is cooler than the skin over the body. Evaporative coolers are well suited to this application. In evaporative coolers, the air is cooled by evaporation of water. When water is evaporated into air, it results in direct evaporative cooling and a reduction in air temperature.

The leaving dry bulb temperature of the air and the amount of cooling are a function of the entering wet bulb temperature of the air. If the system brings in 100% outdoor air, or a large percentage of outdoor air, the performance of the system is quite dependent upon climatic conditions which are dependent upon geographic locations. Evaporative cooling performs best in geographic locations with high dry bulb temperatures and relatively low wet bulb temperatures. In locations with high coincident wet bulb temperatures, evaporative cooling is less effective.

Evaporative cooling equipment can be classified as either direct or indirect. Direct evaporative coolers cool air by direct contact with water, either by an extended wetted surface or with a series of water sprays. Indirect evaporative coolers cool air by transferring heat with a heat exchanger to either a secondary stream of air, which has been evaporatively cooled, or to water that has been evaporatively cooled by a heat rejection device such as a cooling tower. In all cases, indirect evaporative cooling rejects heat to a secondary air stream or water stream through a heat exchanger.

Direct evaporative coolers evaporated water directly into the conditioned air steam. The heat and mass transfer between the air and water lowers the air dry bulb temperature and increases the humidity of the air at a constant wet bulb temperature. When water and non-saturated air/water mixtures come in free contact, heat and water transfer occurs. The amount of heat and water transfer is a function of the difference between their respective temperatures and respective vapor pressures. In direct evaporative cooling, the air is cooled due to the conversion of sensible heat to latent heat. The conversion of sensible heat continues until the air has reached saturation and water temperatures and vapor pressures equalize. This is called *adiabatic saturation* since no external heat transfer occurs.

There are several types of direct evaporative coolers available: random media coolers, rigid media air coolers, slinger packaged air coolers and packaged rotary air coolers. Random media coolers contain pads made of wood or plastic fiber/foam. A water circulating pump circulates water from a sump to a distribution system, which distributes water to flow down through the pads as air passes through the pads. Rigid media coolers have sheets of rigid corrugated material as a wetted surface over which the air is passed. A "slinger" air cooler consists of a water slinger in an evaporative cooling section, and a fan section. Outdoor air is drawn through the cooling section which has a water spray, an evaporative filter pad and a pad that entrained moisture. Rotary coolers wet and wash an evaporative pad by rotating the pad through a water bath. A typical packaged rotary evaporative cooler is shown in Figure 9-39.

Indirect evaporative coolers include indirect package air coolers and cooling tower/coil systems. In indirect evaporative coolers, outdoor air or

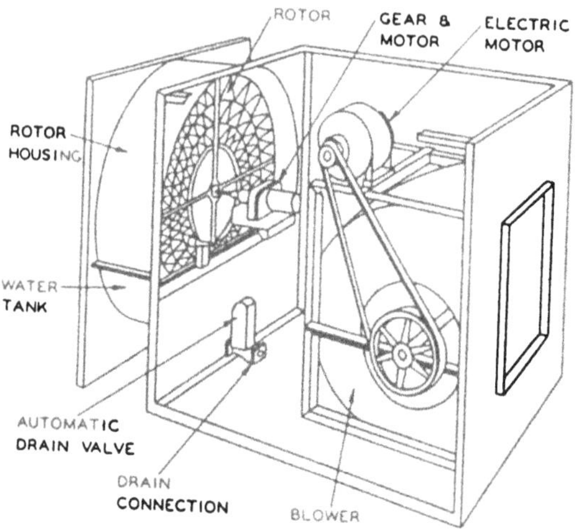

Figure 9-39. Typical package rotary evaporative cooler.

exhaust air from the conditioned space passes through one side of a heat exchanger. This air, which is called secondary air, is evaporatively cooled. The surfaces of the secondary side of the heat exchanger are cooled by the secondary air. On the other side of the heat exchanger, the primary air (air supplied to the conditioned space) is cooled by transferring heat through the heat exchanger to the secondary air. Only sensible heat transfer takes place between the two air streams.

Cooling tower/coil systems employ a water to air heat exchanger (coil) to transfer sensible heat from the primary air to water. The water is cooled by circulating it through a cooling tower. The water is evaporatively cooled in the cooling tower. Cooling towers are discussed in Chapter 8.

Since the performance of evaporative cooling systems is dependent upon local weather conditions, among other variables, the analysis of evaporative cooling systems is somewhat involved. For additional information on evaporative cooling systems, refer to ASHRAE (1996) and Watt and Brown (1997).

9.4.6 Convert from Steam Heating to Gas Heating

Many older industrial and manufacturing buildings are heated with steam. These systems typically consist of a steam boiler, a steam supply piping system, terminal heating units and a condensate return system. Terminal heating equipment typically consist of steam unit heaters and air handling units with steam heating coils.

Heating such buildings with steam can be very energy-inefficient. First, steam systems are notorious for leaking steam. Most of the leakage occurs in air vents and steam traps (see Chapter 5). Producing steam is also quite energy-inefficient. Many older boilers have efficiencies of

60%, of less. Just operating the boiler can result in energy losses of over 40%.

Large buildings and facilities can be more efficiently heated by converting from steam heaters to gas fired heaters and furnaces. Many gas heaters and furnaces are available with combustion efficiencies over 90%. The energy lost in the combustion process for a high efficiency gas furnace is a fraction of the losses for a steam boiler.

Converting to gas also eliminates the distribution system losses inherent in steam systems. Since heat is produced at the point of use, there are no distribution system losses for the gas heating systems.

9.4.7 Use Radiant Heaters

Many large buildings and facilities are heated with steam unit heaters, hot water unit heaters, and gas fired unit heaters and furnaces. Large industrial and manufacturing buildings frequently have high roofs, large interior volumes and high rates of outside air infiltration. Heating these buildings in this fashion is typically inefficient and energy wasteful due to the large volumes of the building, as well as high infiltration rates and temperature stratification inside the buildings. A more efficient way to heat such buildings is with infrared radiant heating.

Infrared heaters may be gas fired or electric and used for spot heating or to heat an entire building. Infrared heaters do not use the inside air to provide heat to the building and occupants. Infrared energy warms floors and objects within the building, which in turn releases heat to the air by convection. Human comfort is provided by controlling the mean radiant temperature of the space and the air dry bulb temperature. Since infrared heating consists of both radiant and convective heat transfer, air movement and stratification in the heated space is minimized. As a result, radiant heating can reduce heating energy consumption by 30% or more, with equal comfort. Energy savings also result from the fact that comfort can be achieved with radiant heat at lower indoor air temperatures.

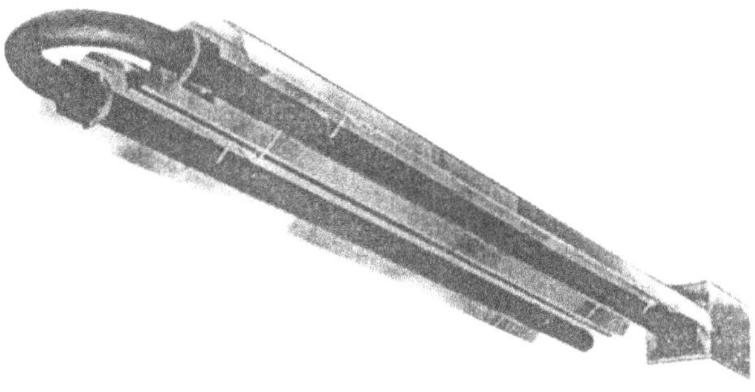

Figure 9-40. Gas fired low intensity radiant heater.

Figure 9-41. Gas fired high intensity radiant heater.

Radiant heaters are classified by the level of radiation intensity. Low intensity radiant heaters have source temperatures between 300°F and 1,200°F. A typical low intensity gas fired low intensity radiant heater is shown in Figure 9-40. Medium intensity heaters have source temperatures ranging from 1,200°F to 1,800°F. High intensity heaters have source temperatures ranging from 1,800°F to 5,000°F. A typical gas fired high intensity radiant heater is shown in Figure 9-41.

9.4.8 Heat Recovery

Many manufacturing and industrial buildings have high rates of exhaust. Frequently, processes require the removal of excessive heat or pollutants from the indoor air with local or general exhaust. The various types of exhaust systems were discussed earlier in this chapter.

When air is exhausted from a conditioned space, it carries with it the energy required to condition the air for the space. This energy may be in the form of either sensible heat or latent heat. Exhaust air usually includes both sensible heat and latent heat energy. Conditioning the make-up air needed as result of exhaust is probably the most significant annual cost of a ventilation system.

Whenever possible, the air that might otherwise be exhausted should be returned to the conditioned space. Air that contains airborne pollutants, such as vapors or particulate matter, may be cleaned and returned to the conditioned space. Heat may also be removed from the air stream by cooling coils or evaporative cooling.

When is not possible to clean the air and return it to the conditioned space, some sort of heat recovery system should be considered. Heat recovery systems transfer heat between the exhaust air and outside make-up air streams. Whenever large volumes of air are exhausted, heat conservation and recovery should be considered as it can result in substantial energy savings. The transfer of energy from exhaust air to replacement air may be economically feasible depending upon (1) the location of the exhaust and replacement air ducts, (2) the dry bulb and wet bulb temperatures of the two air streams and (3) the nature of the contaminants being exhausted.

For air-to-air energy recovery to be economically feasible, there are a number of considerations, including the following:

- The cost of energy. High energy costs favor high levels of energy recovery.
- The grade of waste energy. High grade (i.e., high temperature) waste energy is generally more economical to recover than low grade energy. Large temperature differences between the waste energy source and destination are most economical.
- Coincidence and duration of waste energy supply and demand. Energy recovery is most economical when the supply is coincident with the demand and both are relatively constant.
- Proximity of the supply and demand.
- Operating environment. High operating temperatures, or the presence of corrosives, condensables and particulates in either air stream result in high equipment and maintenance costs.
- Effects on heating and cooling equipment. Heat recovery equipment may reduce the size requirements for primary heating and cooling equipment.

Air-to-air energy recovery equipment may be one of two basic types; sensible and total heat recovery. Sensible heat recovery systems recover only sensible heat from the exhaust air stream. Total heat recovery systems, also called enthalpy recovery, transfer both sensible heat and latent heat between the two air streams.

The type of heat recovery equipment used is based upon the type of energy required (sensible or total), the proximity of the air steams, the amount of cross-contamination that can be tolerated, and the cost of equipment and systems. The most common types of energy recovery systems and equipment include fixed plate heat exchangers, run around loops, heat pipes, twin tower recovery loops and thermosiphion heat exchangers.

Fixed plate heat exchangers consist of alternate layers of plates, separated and sealed (referred to as heat exchanger core). The exhaust and incoming air streams flow between alternating spaces between the plates. The plates, which are made of a material that readily conducts heat, transfer sensible heat between the two air streams. A typical fixed plate heat exchanger is shown in Figure 9-42.

Fixed plate heat exchangers are typically made of aluminum because of its resistance to corrosion, ease of fabrication, heat transfer characteristics and cost-effectiveness. Heat exchangers made of steel alloys are available for tempe ratures exceeding 400°F.

Heat exchanger modules range in airflow capacities from 25 to 10,000 ft^3/min. One advantage of fixed plate heat exchangers is that it is a static device built such that there is little or no leakage between the air streams.

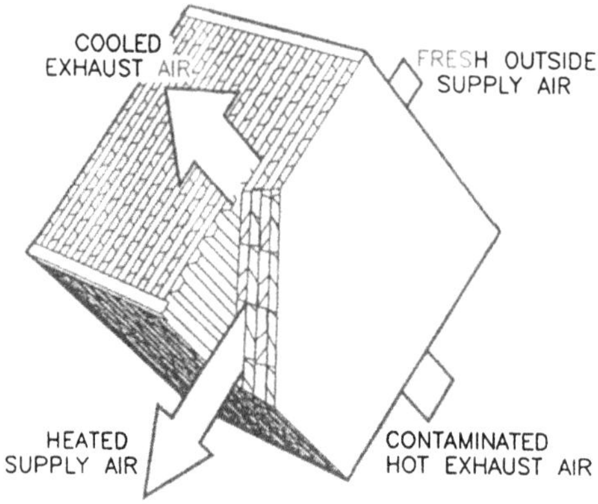

Figure 9-42. Fixed plate heat exchanger.

One disadvantage is that it will recover only sensible heat. Fixed plate heat exchangers can recover over 80% of the available waste exhaust sensible heat.

Rotary air-to-air heat exchangers can recover latent, as well as sensible heat, depending upon the heat transfer media used. A rotary air-to-air heat exchanger (heat wheel) has a revolving cylinder filled with an air permeable heat exchange media with a large internal surface area. Adjacent make-up and exhaust air streams each flow through one half of the heat exchanger.

The direction of the two air streams is opposite (counterflow). The heat transfer media type may recover only sensible heat or it may transfer total heat (sensible heat plus latent heat). A typical rotary air-to-air energy exchanger is shown in Figure 9-43.

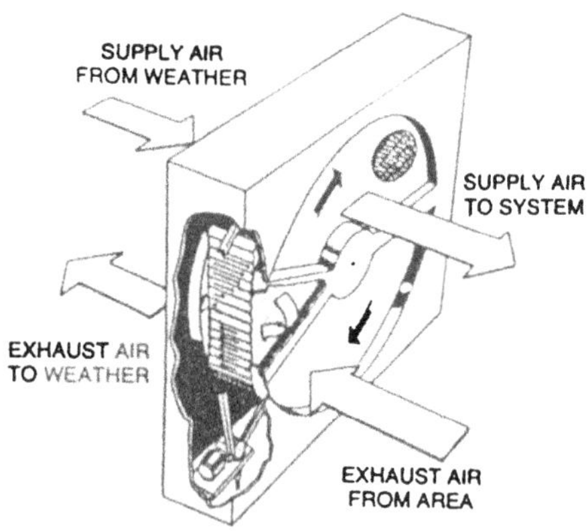

Figure 9-43. Rotary air-to-air energy exchanger.

Energy is transferred from one air stream to the other as the media is exposed to the air. Sensible heat is transferred as the medium picks up and stores heat from the hot air stream and releases it to the cold air stream as the wheel rotates. If the media is suitable for latent heat transfer, latent heat is transferred as the media (1) condenses moisture from the air steam with the higher humidity ratio with a simultaneous release of heat and (2) releases the moisture through evaporation into the air stream with the lower humidity ratio.

Air contaminants, dew point temperatures, exhaust air temperature and make-up air properties determine the type of media that is most suitable. Media for sensible heat transfer is typically aluminum, copper, stainless steel and monel. Media for total energy recovery may include any number of materials treated with a desiccant, such as lithium chloride or alumina. The rate of energy recovery is a function of the rotational speed of the wheel.

Run around energy recovery loops may be used to transfer heat when two air steams are not adjacent to each other or when there can be no cross contamination between the air streams. A typical run around energy recovery loop has an extended surface, as well as finned tube water coils in the make-up and exhaust air streams. The coils are connected with a closed loop hydronic system, which includes counterflow piping, a circulating pump and a circulated heat transfer liquid. A typical run around energy recovery loop is shown in Figure 9-44.

A principal advantage of the run around energy recovery loop is that the two air streams need not be adjacent, or even in close proximity, to each other. The distance between the two air steams is limited only by the economics of the piping and pumping systems. Run around energy recovery systems have no possibility of cross-contamination between the air streams. The main disadvantage of these systems is that they can transfer only sensible heat between the air streams.

Heat pipes are tubes that have been fabricated with an integral capillary wick, evacuated, filled with a refrigerant (heat transfer fluid) and

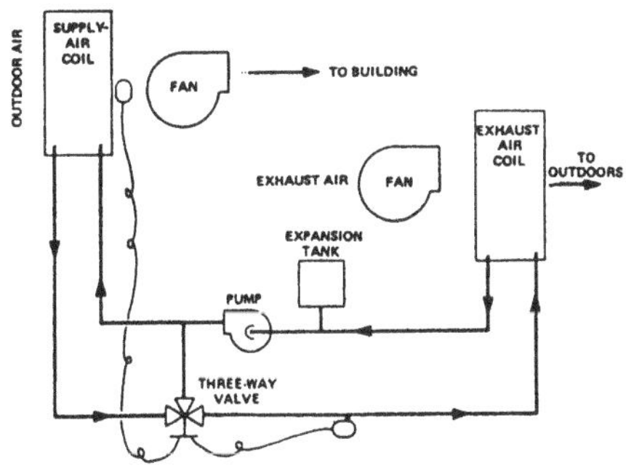

Figure 9-44. Runaround energy recovery loop.

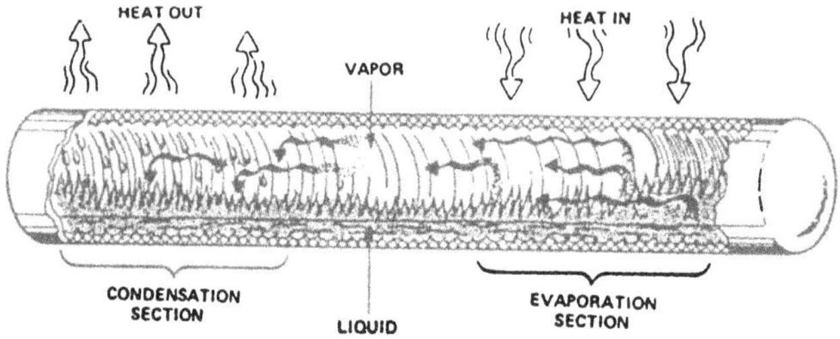

Figure 9-45. Heat pipe.

permanently sealed. A heat pipe is shown in Figure 9-45. Thermal energy applied to either end of the pipe causes the refrigerant liquid at that end to evaporate. The refrigerant vapor rises due to the difference in density between the vapor and the liquid to the outer end of the pipe, which causes it to condense into a liquid where it flows by gravity back to the evaporator end, thereby completing the cycle. The heat pipe thus operates in a close loop evaporation/condensation cycle, which is continuous as long as there is a temperature difference and the cold end is higher than the warm end. Multiple heat pipes are combined to form a heat pipe heat exchanger, as shown in Figure 9-46.

Twin tower energy recovery units are air-to-liquid, liquid-to-air enthalpy recovery systems. A twin tower energy recovery system is shown in Figure 9-47. In twin tower energy recovery systems, a sorbent liquid is circulated continuously between the exhaust and make-up air streams, alternately contacting both air steams directly in contactor towers. The circulated liquid transfers total heat, latent heat and sensible heat.

Since the circulated fluid comes in contact with each air stream and since it is a sorbent, twin tower systems are able to transfer both sensible

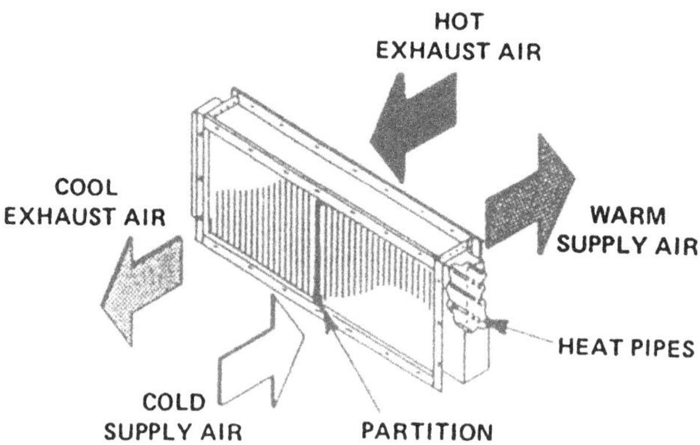

Figure 9-46. Heat pipe heat exchanger.

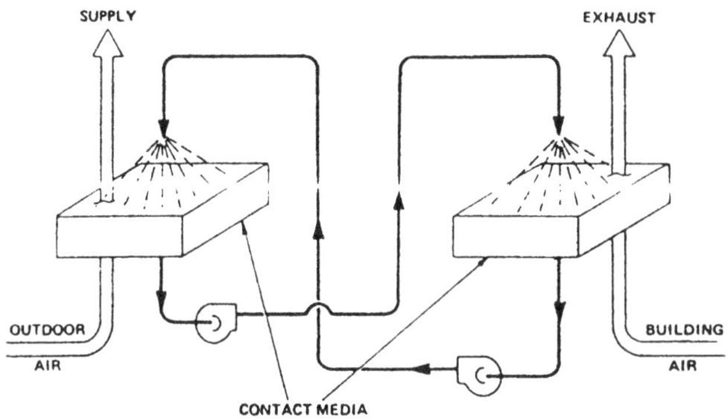

Figure 9-47. Twin tower energy recovery system.

and latent heat. Another advantage of these systems is that the two air steams do not have to be adjacent to each other, or even in close proximity. One disadvantage is that they are not suitable for high temperature applications. There is also the possibility of cross-contamination if the air steams contain large amounts of gas eous contaminants.

Thermosiphon heat exchangers have no moving parts. They are similar to heat pipes in that they employ a fluid that changes phase as part of the heat transfer process. Two phase thermosiphon heat exchangers are sealed systems that consist of an evaporator, a condenser, interconnecting piping

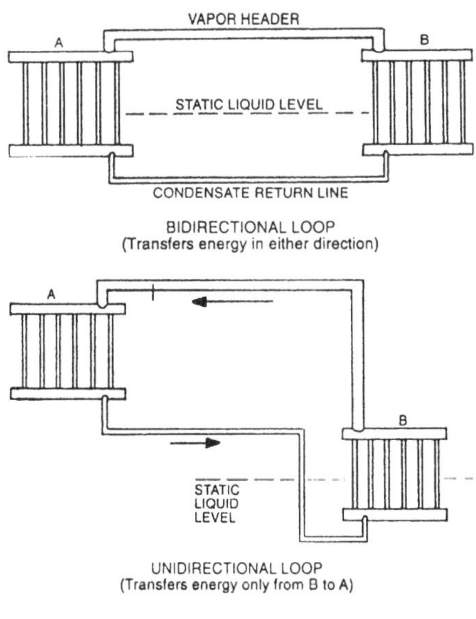

Figure 9-48. Thermosiphon loop heat exchanger.

and a heat transfer fluid. The heat transfer fluid changes phase as it is heated or cooled. The change of phase, which results from a temperature difference, and the force of gravity cause the heat transfer fluid to circulate between the evaporator (hot end) and the condenser (cold end). Figure 9-48 shows thermosiphon loop heat exchangers.

Although thermosiphon heat exchangers are similar to heat pipes, thermosiphon heat exchangers are different in two ways: (1) they have no wick and, therefore, rely only on differences in the fluid density caused by temperature differences, (2) thermosiphon tubes are dependent upon nucleate boiling to change from the liquid phase to the vapor phase.

9.5 ENERGY EFFICIENCY FOR NEW AIR DISTRIBUTION AND HVAC SYSTEMS

Air distribution and HVAC systems can be one of the largest consumers of energy at many facilities. Since air distribution and HVAC systems and equipment are so widely used and may consume significant amounts of energy, regardless of the type of facility, it is important that energy efficiency be a primary consideration when selecting new air distribution and HVAC equipment for new, or retrofit, installations. In some cases, local or national energy codes may dictate efficiencies of such equipment. Standards are also available, which provide guidelines for air distribution and HVAC equipment and systems (ASHRAE 1993).

Earlier in this chapter, energy conservation for air distribution and HVAC systems was discussed. Several of the energy efficiency measures discussed as retrofit opportunities for equipment, such as high efficiency motors, are applicable to new and retrofit installations. In addition to the previously discussed energy conservation measures for air distribution and HVAC systems, the following measures should be considered for new or replacement equipment and accessories.

9.5.1 Avoid Oversizing Equipment

Frequently, when air distibution and HVAC systems are sized and selected, they are sized by rules of thumb or short-cut load estimation methods. Since such methods are by nature quite inaccuarate, the result is that the loads are over-estimated to account for the inherent inaccuracies. In addition to the oversizing resulting from the simplified methods, designers frequently add a safety factor to the load which results in still greater oversizing of the equipment.

Most air distribution and HVAC equipment operates most efficiently at, or near, maximum capacity. The inefficiencies inherent in operating such equipment tends to be a smaller percentage of the total energy consumed by the equipment. Therefore, the equipment should be sized to match the actual load as closely as possible to maximize energy efficiency of equipment and systems.

Whenever a new or replacement system is being sized, the loads on the system should be reduced to the greatest extent possible. Frequently, energy conservation efforts can or have reduced the loads on a system. If the loads on the system can or have been reduced, the size and selection of air distribution and HVAC equipment should be adjusted accordingly.

9.5.2 Air Side Economizers

Whenever a new or replacement cooling system is to be installed, a dry bulb or enthalpy economizer should be included with the HVAC system. Air side economizers allow HVAC systems to provide "free" cooling with outdoor air when conditions are suitable. The most cost-effective time to provide air side economizer cooling is when a new air system is being installed, or when an existing system is being modified. Additional information on air side economizers may be found in Section 9.4.3 of this chapter.

9.5.3 Variable Air Volume

Unless a heating or cooling load, particularly a cooling load, is constant, the opportunity for varying the airflow in response to the load exists. If the load is variable, a VAV system may provide for the load. Significant fan energy savings can be achieved with VAV inlet guide vanes or with variable speed motor drives. The most cost-effective time to install VAV systems is when a new system is being installed or an existing system is being modified. Additional information about VAV systems can be found in Section 9.4.1 of this chapter.

9.5.4 High Efficiency Air Conditioning Equipment

When selecting new or replacement air conditioning equipment, such as air cooled condensing units, high efficiency units should always be selected. The efficiency of air conditioning equipment is usually expressed as the coefficient of performance (*COP*) or the energy efficiency ratio (*EER*). Equation (9.19) defines the *COP* for air conditioning equipment and Equation (9.20) defines the *EER*.

$$COP = \frac{q}{E_{\mathrm{Btu}}} \qquad \text{Eq. (9.19)}$$

$$EER = \frac{q}{E_W} \qquad \text{Eq. (9.20)}$$

where q = the cooling capacity, Btu/hr
E_{Btu} = energy input, Btu/hr
E_W = energy input, watts

The efficiency of air conditioning equipment may also be expressed as SEER, which is the EER adjusted for typical seasonal energy usage. Air conditioning equipment should be selected with an EER of 9.0 or higher.

9.5.5 High Efficiency Furnaces

Whenever new or replacement gas or oil fired furnaces are to be installed, they should always be selected with the highest efficiency. Steady state efficiency, utilization efficiency and annual fuel utilization efficiency (AFUE) were defined in Section 9.1.3 of this chapter. When selecting new or replacement furnaces, they should have an AFUE of 78% or more.

Chapter 10

Electrical and Lighting Systems

Almost all buildings and facilities consume electrical power. For many buildings and facilities, the cost of electrical energy is one of the largest operating expenses. Electrical energy is consumed in a number of ways, including electrical lighting and the operation of motors for all types of applications and processes. This chapter examines how electrical energy is consumed and how the consumption of electricity may be reduced. Major sources of losses are identified, as well as how losses may be reduced while still meeting the functional requirements of the electrical system.

10.1 BASIC PRINCIPLES OF ELECTRICAL AND LIGHTING SYSTEMS

10.1.1 Definitions and Terms Used in Electrical and Lighting Systems

See the Glossary for definitions of the terms used in electrical and lighting systems.

10.2.2 Basic Electricity and Electrical Systems

Electricity is electrons in motion. Electrical potential, or electromotive force (emf), causes free or loosely bound electrons to move along or through a medium. The forces that move electrons are magnetic forces.

The most basic elements of an electrical circuit include a method for producing an electrical potential, a load and interconnecting wiring to complete the circuit. A basic electrical circuit is shown in Figure 10-1. In order for electricity to flow, there must be a complete circuitous path, or circuit, for the electricity to flow through. The electrical power must return to the source from which it came. If there are any openings or breaks in the circuit, there will be no flow of electricity.

The flow of electrons through a circuit is referred to as electrical current, or just current. The current is capable of producing a number of effects, including heat and magnetism. Current is defined as the time rate of flow of an electrical charge and is typically measured in amperes (amps). An ampere is a rate of flow of electrical charge of one coulomb per second. As the current traverses the circuit, energy is expanded in the form of work. Current is measured by an instrument called an ammeter.

The source of electrical potential is typically provided by a battery or an electrical generator. It provides the electromotive force (emf) necessary for

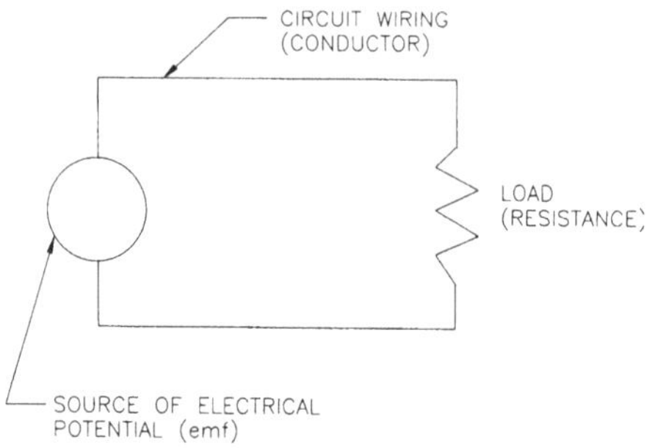

Figure 10-1. Basic electrical circuit.

the electrons to flow. An emf is also called a charge. According to the Law of Conservation of Energy, the energy that a charge receives from an emf source must be equal to that which is expended as the current flows through the complete circuit. Therefore, the emf of the source must equal the sum of the losses around the circuit. Electrical potential (emf) is measured in volts. Voltage is the electrical pressure that causes current flow in a circuit. One volt is when one coulomb of charge does one joule of work in moving from one point of another.

As the current flows through a circuit, it encounters resistance to the flow. As the current flows against the resistance, energy is consumed (a voltage reduction occurs). This reduction in voltage (voltage drop) is proportional to the rate of flow and the resistance. The relationship between voltage drop, current flow and resistance is defined by Ohm's Law, which states that the current in a metallic circuit is given by the emf in the circuit divided by the resistance of the circuit when the temperature is held constant. Ohm's Law is expressed mathematically by Equation (10.1).

$$I = E/R; E = IR \qquad \text{Eq. (10.1)}$$

where I = the current, amperes (amps)
E = the electromotive force (voltage) volts
R = the resistance, ohms (Ω)

An ohm (Ω) is a resistance across which there is a voltage drop when the current is one amp. The reciprocal of resistance is called conductance, hence wires are frequently called conductors since they are low in resistance.

The flow of electrical current through any resistance, no matter how small, produces heat. The heat results from the conversion of electrical energy to thermal energy. The amount of heat generated is exactly equal to the energy supplied electrically. The power dissipated in heat is given by Equation (10.2).

$$P = RI^2 = EI \qquad \text{Eq. (10.2)}$$

where P = the power in watts

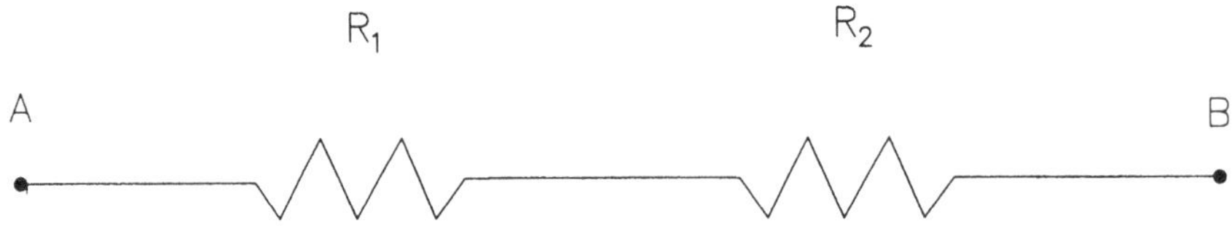

Figure 10-2. Resistances in series.

The simple circuit shown in Figure 10-1 contains only one resistance. This simple case rarely occurs in actual practice. Usually there are many resistances in a circuit. In order to simplify the analyses of circuits, multiple resistances are frequently combined into one single equivalent resistance. Typically, there are two situations for multiple resistances in electrical circuits: resistances connected in series and resistances connected in parallel. Each situation can be converted into a single equivalent resistance.

Figure 10-2 shows two resistances connected in series. The current flowing through one resistor must also be equal to the flow through the other. Therefore, the total equivalent resistance for a circuit with resistances connected in series is equal to the sum of the individual resistances and is given by Equation (10.3).

$$R_T = R_1 + R_2 + R_3 + \cdots\cdots\cdots + R_N \qquad \text{Eq. (10.3)}$$

where R_T = total equivalent resistance,
$R_1, R_2, \cdots\cdots R_N$ = individual resistances,

Since the current flow through each resistance is the same, the voltage drop across each resistance is a direct function of the resistance.

Figure 10-3 shows a circuit with the resistances connected in parallel. In circuits with parallel resistances, the voltage drops across the resistances

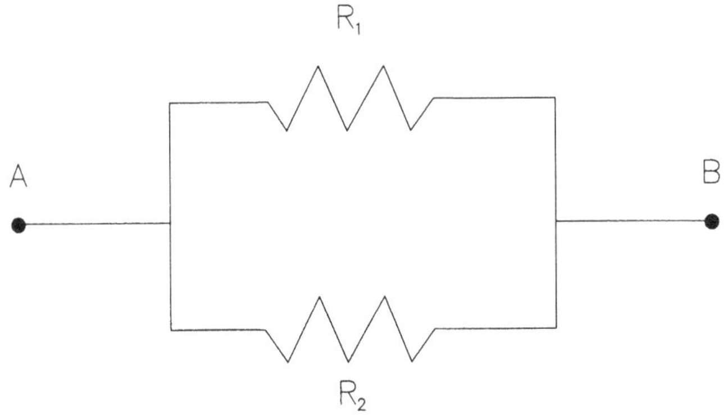

Figure 10-3. Resistances in parallel.

will always be equal and the current flow through each resistance will be a function of the resistance. The total equivalent resistance for resistances connected in parallel will always be less than the lowest resistance in the parallel combination. The reciprocal of the total equivalent resistance is equal to the sum of the reciprocals of the individual resistances and is given by Equation (10.4).

$$1/R_T = 1/R_1 + 1/R_2 + 1/R_3 + \cdots\cdots 1/R_N \qquad \text{Eq. (10.4)}$$

Up to this point, the flow of current in a circuit has been assumed to be in only one direction. However, this is not the case for most electrical systems. Almost all electrical power in facilities and buildings is alternating current.

As the name implies, direct current (DC) flows in only one direction—from a point of positive potential to a point of negative potential. A very common example of a direct current power source is a battery. The battery has a positive terminal and a negative terminal. The current flow is in one direction through a circuit connected to the battery, from positive to negative.

Whenever a conductor, such as a wire, is passed through a magnetic field, a current flow is induced in the conductor by the magnetism. The direction of current flow in the conductor is a function of the location of the field's magnetic poles (north and south) and the direction that the conductor is moving. If the conductor is passed through the magnetic field in one direction and then reversed and passed through the field in the opposite direction, the induced current will first flow in one direction and then in the opposite direction. The current flow direction alternates, hence the term alternating current (AC). The electrical current supplied by electrical utility companies in the United States to all industrial and commercial facilities is alternating current.

Electrical power from utility companies is produced by large generators. The generators have magnets that produce a magnetic field inside the generator. The magnetic field is produced by stators that are stationary magnets surrounding the inside of the generator. A rotor is located inside the generator and comprises conductors mounted in a cylindrical shape. As the rotor rotates in the magnetic field, each conductor is first passed through the magnetic field in one direction and then the other. The rotation on the magnetic field causes a current flow in one direction and then the other, hence alternating current. The magnitude of the AC current produced by a generator varies with time and results in a sine wave when current flow is plotted against time, as shown in Figure 10-4. The is called a waveform.

The voltage (emf) for alternating current follows the same pattern as the current; however, the time at which the maximum and minimum voltage peaks occur may not be exactly when the current flow peaks. Common voltages for AC circuits include 120 V, 208 V, 240 V and 480 V. Voltages up to 600 V are considered low voltages. The frequency of an AC current or

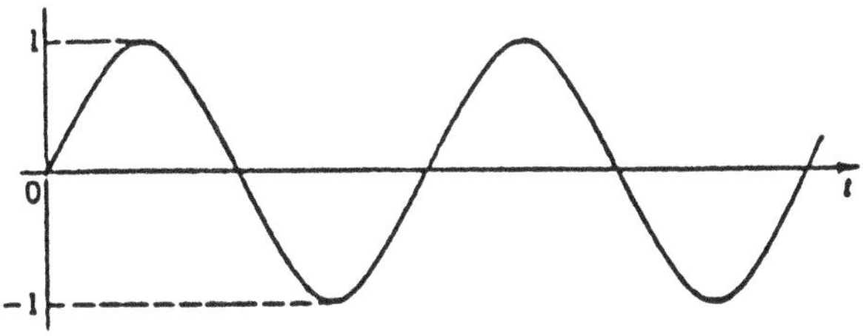

Figure 10-4. Waveform for alternating current.

voltage is the number of cycles that occur in one second. AC power in the Untied States is 60 cycles/second (60 hertz).

Since AC power follows a waveform as the current and voltage increase and decrease, all calculations for AC power is based on sine waves. There are four values of the AC waveform that are especially important:

1. Instantaneous value: The voltage and current in AC power is constantly changing from a maximum value to a minimum value. The value of the voltage or current at any given instant is known as the instantaneous value of that variable.
2. Maximum value: The voltage and current each reach a maximum positive and negative value instantaneously. These peak values are known as the maximum value.
3. Average value: The equivalent DC value that is equal to the area enclosed by the sine wave. For either the positive or negative halves of the sine wave, the average value is equal to 0.636 times the maximum value.
4. Effective value: The root mean squared (rms) value of an AC voltage or current is the value that would produce the same heating effect in a resistance as an equivalent direct current. It takes into account that the AC current is varying with time. The rms value of an AC sine wave is equal to 0.707 times the maximum value.

Generators typically have rotors with multiple conductors that are rotated through the magnetic fields. Each set of conductors produces AC power and a waveform. The waveforms of each set of conductors occur at different times. The time difference between the waveforms is measured in degrees of rotation of the rotor and is referred to as the phase angle (ϕ). The number of waveforms produced is referred to as the number of phases. AC power is typically single phase or three phase.

Power is defined as the time rate of doing work or expending energy. Whenever a force results in motion, work is performed. In electrical systems, the force (emf) is the voltage and the motion is the flow of electrons (current). The rate at which work is performed results in power. Power may be expressed in various types of units, including ft-lb/min and horsepower. In electrical systems, power is expressed as watts (W) or kilowatts (kW).

The calculation of power for DC circuits is straightforward. It is simply the voltage times the amperage and is expressed by Equation (10.2). In AC circuits, the calculation of power is not as straightforward since the resistance in AC circuits is complicated by the alternating current. When there is a flow of electricity through a conductor, a magnetic field is created around the conductor. The direction of the magnetic field depends upon the direction of current flow. When the flow of electrical current stops, the magnetic field collapses. The collapse of the magnetic field, however, is not instantaneous and, as it collapses, it induces a voltage in the conductor. This voltage is in the opposite direction of the current flow and opposes the voltage in the conductor. This resistance is called inductance. Since alternating current is rapidly changing direction (60 cps), the magnetic field and the induced current are constantly changing. As the current flow changes, however, the induced current flow lags behind the main current flow by a brief period of time. This results in an additional resistance to the main current flow.

Capacitance is similar to resistance in that it resists the flow of current. However, capacitance offers resistance as the result of changes in voltage across it. Capacitance in an electrical circuit allows electrons to be stored within the circuit. A capacitor has two conducting surfaces separated by a dielectric material (a material with almost infinite resistance). As the voltage increases, the capacitor stores current. As the voltage decreases, the capacitor discharges current in the direction of the original current flow.

Impedance is the total of all resistance to current flow offered by an AC circuit resistance, capacitance and inductance. Ohm's Law for an AC circuit is expressed by Equation (10.4).

$$I = \frac{E}{Z} \qquad \text{Eq. (10.4)}$$

where Z = circuit impedance, Ω

Since capacitance and inductance in AC circuits add impedance to the circuit, the power calculated by Equation (10.2) does not indicate the actual wattage supplied by the source (generator). In circuits where capacitance predominates, the current will lead the voltage. In circuits where inductance predominates, the current will lag behind the voltage. The measurement of difference in phase between the current and voltage is called power factor: It is the cosine of the angle between the current and voltage and is expressed as a decimal. Equation (10.5) expresses the power for AC circuits.

$$P = EI\sqrt{\phi}\,(PF) \qquad \text{Eq. (10.5)}$$

where ϕ = phase of the power supplied (i.e., single phase or three phase)
PF = power factor for the circuit

10.1.3 Electric Billing

An electric meter (a watt-hour meter) is essentially an electric motor that runs in exact proportion to the power being used. The watt-hour meter records the energy used over a billing period, usually a month. A chart-type meter indicates the highest rate at which power (demand) is used over the billing period.

Most electric bills from electric utilities include a number of costs. These costs may include an energy cost, a fuel adjustment charge, a customer related charge, a demand charge and a low power factor penalty.

Energy costs are typically billed on a sliding block approach for the energy charge part of a bill. As an example, the first 1,000 kWh is billed at one rate per kWh, another rate for the second 1,000 kWh, and so on. In most cases, the cost per kWh decreases as the total amount consumed goes up.

Frequently, utilities impose an additional charge for energy, called a fuel adjustment. The fuel adjustment reflects the increased cost of fuel and it can be a substantial charge.

Some electric utilities add what are called customer related costs. These costs are special charges that reflect part of the distribution investment, part of the operating and maintenance costs, costs for accounting and collection, and so on. If these costs are not a separate charge, they may be included with the energy cost.

Demand charges are used to pass on the cost of the utility's fixed investments in the production, transmission and distribution equipment required to meet the maximum rate of consumption by the customer. The charge is based on the rate at which electricity is consumed. The consumer's actual demand is computed as the average amount of energy consumed in a predetermined demand measurement interval, usually 15 minutes. Regardless of the interval, the highest demand recorded during a billing period becomes the actual demand for the month.

Lower power factor penalty is sometimes applied for low power factor within the consumer's electrical distribution system. Power factor is discussed later in this chapter.

Some of the charges listed above are not directly related to the amount of electricity consumed during the billing period. As a result, predicting the dollar cost savings for reductions in electrical consumption are typically not easily predicted. A reduction in electrical consumption is not necessarily reflected in an equivalent reduction in an electric bill. Since the charges listed above vary from one electric utility to another, predict ing actual cost savings for energy improvements is beyond the scope of this book.

10.2 ELECTRIC MOTORS AND MOTOR CONTROLS

One of the most common uses of electrical power is motors. Electric motors are devices that convert electrical energy into kinetic energy, usually in the form of a rotating shaft. Rotating shafts from motors are typically used to drive a fan, pump, comp ressor, etc.

It has been estimated that over 7% of the annual electrical power consumption in the United States is motor driven equipment. Electrical motors are almost exclusively used to drive equipment such as fans and other equipment. The majority of the motors operate on AC power.

Motors can be divided into two classifications: hermetic and non-hermetic. Hermetic motors consist of a stator and a rotor without shaft end seals or bearings. These types are used in sealed refrigeration compressor units. Non-hermetic are not in sealed units and are the most common type of motor. An example of a non-hermetic motors would be a fan motor. Since non-hermetic motors are the most common, they will be the only motor type discussed here.

Motors are also generally classified according to their enclosures. The type of enclosure is based on the environment in which the motor will operate. The National Electrical Equipment Manufacturers Association (NEMA) has classified motor enclosures by type. The most common are open drip proof (suitable for applications where solid or liquid drops may fall on the motor at any angle not greater the 15°), splash proof (applications where solid or liquid particles may come at the motor in a straight line), totally enclosed (prevents free flow of air between inside and outside of the enclosure, but it's not airtight) and explosion proof (motor is completely enclosed suitable for applications where explosive gases and dusts are present).

Motors are also classified by their horsepower. Motors that are less than 1 hp are referred to as fractional horsepower motors. Motors that are 1 hp or greater are referred to as integral motors.

Under normal operating conditions, the temperature rise results from conversion of electrical energy to mechanical work and the friction of the rotating parts. Motors designed for continuous service can carry the design load for reasonably long periods of time without overheating. Occasionally, a motor may be operated at a higher than rated horsepower load. This is normally acceptable provided the service factor limit is not exceeded. Standard open type motors have a service factor (SF) that allows a continuous overload above the rated nameplate horsepower without causing an excessive temperature rise due to overload. Table 10.1 lists service factors for fractional and integral horsepower motors. These service factors are applicable only when the voltage and frequency are at

TABLE 10.1 Motor Service Factor

Motor Size	Service Factor
1/20,1/12,1/8	1.40
1/6, 1/4,1/3	1.35
1/2,3/4,1	1.25
11/2 and up	1.15

the rated value. If an undervoltage is present, the listed service factors should be reduced by the square of the reduced voltage ratio.

Earlier in this chapter, inductance and power factor for AC circuits were discussed. Some motors have permanent magnets to produce the magnetic field in which the rotor rotates. Many AC motors, however, have coils to produce the magnetic field. Since the coils produce the magnetic field as the result of AC power, the AC power can result in significant inductance for the motor. The inductance of a motor can have a significant effect on the power factor for an AC circuit, particularly when large motors are involved. Typically, the larger the motor, the larger the impedance and the lower the power factor for the circuit. Many large motors have capacitors included with them or installed in the circuit to improve the power factor.

10.2.1 Single Phase AC Motors

Single phase AC power is typically available in 120 V, 208 V and 240 V systems. Some 120 V systems are available as 120 V, 2 conductor systems; however, most 120 V systems in commercial and industrial facilities are 3 conductor.

Most 208, 240 and 480 V AC circuits are 3 conductor. A typical 240 v, 3 conductor, single phase circuit is shown in Figure 10-5. A 3 conductor single phase circuit has two groups of conductors to produce 120 volts between each outside line and the center neutral line, which results in 240 V between the two outside lines. This arrangement can supply either 120 V or 240 V motors.

Single phase motors may be used for motors up to 5 hp, however, three phase AC power is preferred for motors larger than 3/4 hp. Single phase

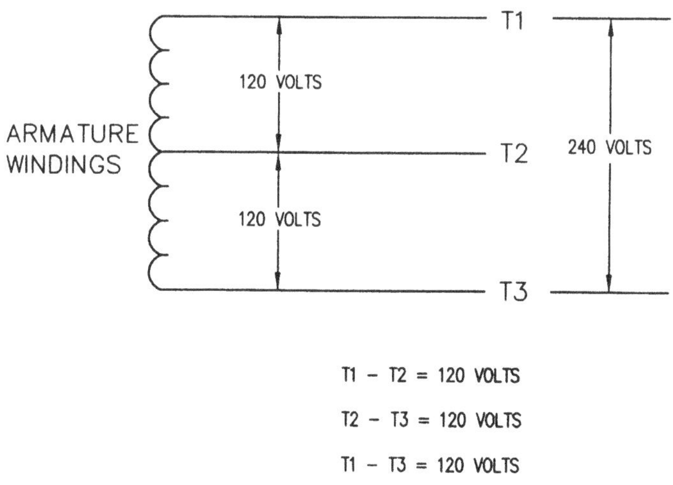

Figure 10-5. 120/240 Volt circuit.

TABLE 10.2 Single Phase Motors

Motor Type	Voltage	Horsepower
Shaded Pole	115,230	1/100 to 1/4
Split Phase	115,230	1/20 to 1/3
Capacitor Start	115,115/230	1/20 to 1.5
Split Capacitor	115,230	1/20 to 5

induction motors have only one winding in the stator to produce a magnetic field. The AC produces a magnetic field with alternating polarity; however, it does not move or revolve—it remains in a fixed position. Therefore, it is necessary to provide a method of producing torque to accelerate the motor to the full load speed. The most popular single phase motor types are the shaded pole, split phase, capacitor start and split capacitor. Table 10.2 lists the various types of single phase motors, voltages and sizes.

In shaded pole motors, the trailing edge of each pole is wound with a shading coil, which produces slip. The current in the main winding of the stator produces the slip. The slip gives the pole the torque necessary for rotation and helps establish the initial low starting torque, which turns the rotor and the load.

Shaded pole motors are typically high in slip, low in starting torque and low in running torque. They are inexpensive, can operate at variable speeds and have inherent overload protection. The efficiency of a shaded pole motor is quite low. Shaded pole motors are air cooled. The low starting torque of shaded pole motors limit their application to small direct driven equipment such as propeller fans and fan coil units. They are always fractional horsepower motors.

Split phase motors have two windings: a main stator winding and an auxiliary start winding. The start winding is mounted such that it is 30° electrically from the main winding. This produces the slip necessary for rotation. When the motor reaches a speed of approximately 70% of its full load rotational speed, a centrifugal switch is mounted on the rotor, and, in series with the start winding, removes the start winding from the circuit. The motor then operates as a standard induction motor.

The split phase motor has a low starting torque and requires a large starting current. The motors have inherent overload protection and no slip with constant speed. This type of motor operates with satisfactory noise levels and is relatively inexpensive. The high starting current may result in a flicker in lights served by the same AC system.

Split phase motors are classified as general purpose motors and are frequently used for small loads. The starting torque allows the motor to be used for direct or belt driven equipment, however, they are limited to loads of 1/3 hp or less.

The capacitor start motor is also similar to the split phase motor in that it has two windings: a start winding and a main stator winding. To improve the performance of the split phase motor, a starting capacitor is added in series with the starter winding and the centrifugal switch. When the motor reaches approximately 70–75% of the full load rotational speed, the capacitor and starter winding are removed from the circuit by a voltage relay. The motor then operates as a regular induction motor. This arrangement results in higher starting torque and quicker acceleration to operating speed. The increased starting torque is due to the low impedance capacitor.

Capacitor start motors are used on direct or belt driven equipment with loads up to 1/2 hp. They are frequently used for applications such as small direct driven or belt driven equipment.

The split capacitor motor is similar to the capacitor start motor in that it has a capacitor connected in series with a start winding. The capacitor is also in the circuit with the main winding. The capacitor and the starter winding remain in the circuit permanently, along with the main winding. The capacitor causes a phase lag in the starter winding, which produces the torque necessary for rotation.

These motors have a low starting torque, high slip, variable speed and inherent overload protection. Since the starting torque is low, the use of these motors is limited to direct connected loads with low starting torque. They are not suitable for belt driven applications. They need no overload protection, are quieter in operation than shaded pole motors and are quite efficient.

10.2.2 Polyphase AC Motors

The majority of motors that are 3/4 hp and larger applications are polyphase, or, specifically, three phase motors. Three phase motors are preferred for integral motors since they are self-balancing on three phase power. Three phase AC systems have three current carrying conductors. The phase of these conductors is 120° apart. During one complete cycle, each of the three conductors carries an AC current, which peaks at a different time.

The windings of an AC generator can either be connected together to form a delta configuration or a wye configuration. A three wire delta

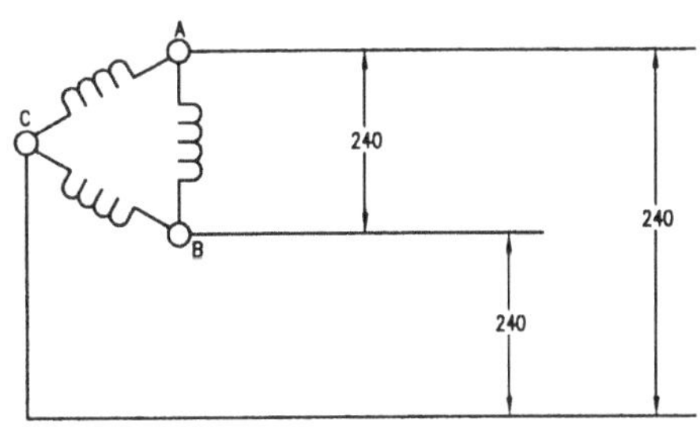

Figure 10-6. Three wire delta arrangement.

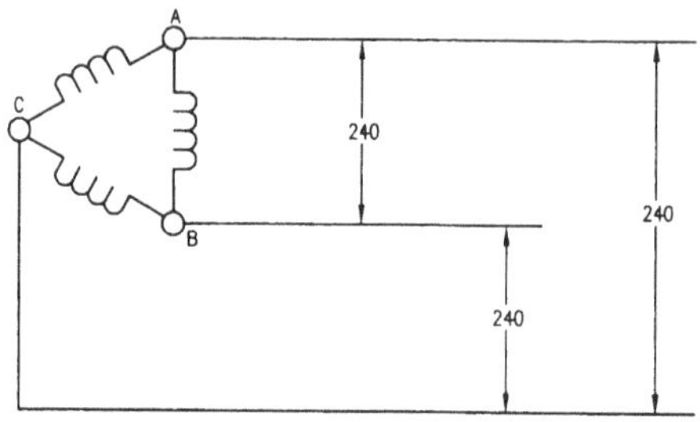

Figure 10-7. Four wire delta arrangement.

arrangement is shown in Figure 10-6. Three wire delta circuits produce three phase, 240 volt AC power.

Three phase AC systems can also have four conductors, three to carry the three phases of AC current and a fourth wire that is a neutral conductor. The neutral conductor is grounded and is connected to the center tap of an AC generator. There is a 120 volt potential between the neutral conductor and any one of the other three conductors. Three phase, 4 conductor AC systems can either be a delta configuration, as shown in Figure 10-7, or a wye configuration, as shown in Figure 10-8. Four wire delta circuits can produce 120 or 240 volt 3 phase power, and 4 wire wye circuits can produce 120 or 208 volt, 3 phase power. Four wire systems can also be used to produce 480 volt and higher AC power.

The most commonly used three phase motors in HVAC systems are synchronous, wound-rotor induction and squirrel cage induction.

Synchronous motors are designed to operate at synchronous speeds only; they are just constant speed motors. Their speed is unaffected by changes in the supply voltage or by the load on the motor. Since they operate at synchronous speeds, an external DC exciter is required for the motor starting and running torque. A DC generator is used to provide the required DC power. Synchronous motors have a limited number of applications. They are used only for large loads. When the motor comes up to operating speed, it runs with a slip of only 2–3%. Large horsepower synchronous motors operating at low speeds are

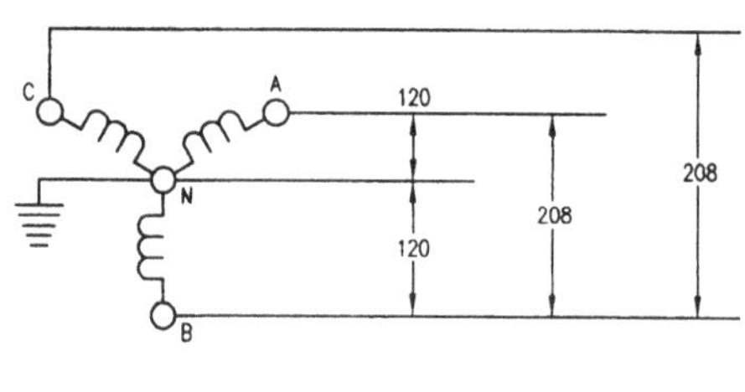

Figure 10-8. Four wire wye arrangement.

relatively low in cost. Some of the characteristics of synchronous motors include:

- low starting torque
- constant speed within the motors load limits
- nearly constant load requirements
- a possible leading power factor
- high efficiency

Wound rotor induction motors provide higher torque during starting and a range of adjustable speed operations. The speed of a wound rotor motor is controlled by varying the current supplied to the rotor. The current is varied with external resistors. They may be used where speed control is required. Typical applications include large fans and centrifugal compressors. Characteristics of wound rotor motors include:

- lagging power factor
- speed control by varying external resistance
- relatively high maintenance
- relatively high initial cost
- high starting torque and low starting current

The most commonly used polyphase motor is the squirrel cage motor. It is the most economical in terms of cost per horsepower.

The two major components are the stator (the stationary magnets and coils) and the rotor. The stator is a laminated iron core within and around the inside of the motor. The rotor is a laminated iron core with bar windings in various shaped slots around the periphery of the core. The rotor looks like a squirrel cage, hence the name squirrel cage motor.

The stator windings of a squirrel cage motor are arranged such that the stator will alternate between positive and negative as the current flow alternates. This creates a pulsing magnetic flux. The pulsations of the magnetic flux produce a rotating magnetic field. As the flux from the stator rotates, it cuts the bars or coils of the rotor, which induces a voltage within the rotor circuits. A current is generated within the rotor bars, which causes a second magnetic field. The interaction of the two magnetic fields produces the torque that accelerates the rotor.

The speed of a squirrel cage motor is constant and is determined by the number of poles that are active. On a two speed motor, two wind motor, the two windings determine the number of active poles. The windings are externally switched, which results in speed combinations such as 1,800/1,200, 1,800/900, 1,800/600 rpm. These motors are available in a number of speeds, however, they are not

TABLE 10.3 NEMA Squirrel Cage Induction Motor Classes

Class	Starting Torque	Starting Current	Slip	Full Load Efficiency	Typical Uses
A	Normal	High	Low	Beat	Uniform loads such as fans
B	Normal	Medium	Low	Good	Same as for Class A
C	High	Low	Medium	Fair	Uniform high torque loads
D	High	Low	High	Low	Intermittent high torque loads

variable speed. Among the characteristics of squirrel cage induction motors are:

- lagging power factor
- multi-speed motors are available
- high efficiency over a wide load range
- rugged and relatively inexpensive
- wide range of starting torque

Squirrel cage induction motors are available in four classes, which have been standardized by NEMA as shown in Table 10.3.

10.2.3 Motor Starters

Motor starters are devices that bring a motor from the off position up to the normal operational rotation at the design load. They are also used to stop the motor. In order to start a motor, it is frequently more involved than closing a switch and applying power to the motor. Part of a starter is a contactor that automatically connects or disconnects a motor from a power source. However, starters provide more functions than starting and stopping the motor.

Starters may also provide overload protection for the motor. Operating a motor in an overloaded condition results in overheating in the motor. A starter can include an overload relay that senses when the motor is operating in an overload condition and disconnects the motor from the power source when an overload condition occurs. The overload relay contains overload heaters, which are made of bi-metallic alloy elements. The temperature of the elements are proportional to the amount of current flow. If the current flow is too high, the elements flex and open the relay contacts.

Starters may also provide protection from damage resulting from single phasing in three phase circuits. Single phasing occurs when one of the three ungrounded conductors serving a three phase motor losses its voltage. In this situation, the motor will draw at least 1.7 times the design current flow through the two remaining ungrounded conductors. The current flow can be much higher if the motor stalls. When a single phasing situation occurs, the motor may be destroyed unless the two remaining ungrounded conductors are immediately disconnected. The overload heaters also serve to disconnect the motor when it is in a single phase condition.

Motor starters may also protect a motor from an undervoltage condition. If the voltage of the power supplied to a motor drops below about 65% of the design voltage, a coil disconnects the motor from the supply voltage.

Starters are frequently used to provide automatic control for motors. A magnetic starter provides the ability to control the starting and stopping of the motor automatically using an electrical interlock device or some other control device.

Motor starters can be either manual or magnetic type. Manual starters are normally used only for small fractional horsepower motors. Manual starters may be toggle, selector or push button switches. They are always across the line type, which means that whenever the starting switch closes, the motor receives whatever line voltage is available. The motor draws whatever current it requires to start and run the motor.

Magnetic starters are utilized for integral motors. They use an auxiliary circuit within the starter to energize a magnetic coil, which closes the starting switch. Control of the magnetic coil may be part of the starter itself or it may be remote. Infrequently started motors may have manual starters, however, motors that are started and stopped frequently, or automatically, require magnetic starters. A magnetic starter consists of two basic components: a contactor and an overload relay. The contactor consists of stationary contacts and an electromagnetic coil with an armature that controls the position of a moveable contact. When a voltage is applied to the coil, the moveable contacts seat against the stationary contacts. When the voltage is removed from the coil, a spring, or gravity, causes the contacts to move away from the stationary contacts. The overload relay holds either overload heaters or bi-metallic sensors. The overload relay senses the overheat that results from the overcurrent condition and breaks the starter's control circuit.

There are two basic classes of starters: across the line (full voltage) and reduced voltage. Across the line starters connect the motor directly to the full voltage, which results in 100% of the normal inrush current at locked rotor conditions, as well as full starting torque. Across the line starters are the lowest cost and simplest method for starting a 3 phase motor. They are especially well suited to small-size motors where the current draw is low, single phase motors up to 5 hp and three phase motors up to 7 hp. On larger motors, they draw full locked rotor current, which can cause a momentary voltage drop in the system. Across the line magnetic starters may be used with automatic control systems for automatically starting and stopping motors.

It is often desirable to start motors with a starter that draws less than full starting current as the across the line starters do. One type of reduced current starter is the part winding starter (incremental starter). Part winding starters are used to reduce line disturbances by connecting only part of the motor winding to the line voltage and connecting the second motor winding after a time interval of 1 to 3 seconds. A part wind motor has two separate three phase windings, each wound for the full voltage of the system. The application of full voltage to one of the windings results in about 60–75% of the full voltage full winding current flow and torque. When the second set of contacts close and the second winding is energized, the current flow and torque reach the normal operating levels.

Another starter used for reduced current starting is the star-delta starter. A delta wound motor has six leads and is started with a star delta starter to reduce the inrush current. These types of starters are used for polyphase squirrel induction or synchronous motors. The starter is initially connected to the motor in a star delta configuration, as shown in Figure 10.8. In this configuration, the motor draws about 33% of the locked motor current and provides about 33% of the full torque. After a preset interval, a timer energizes the second set of contacts and the motor windings are then in a delta configuration, as shown in Figure 10-7. After switching to the delta configuration, the full line voltage is applied to the motor. This increases the motor current draw and torque, bringing the motor up to full speed. The current flow then drops off when the motor reaches full speed conditions. These starters are frequently used on large motor applications. The normal inrush and starting torque are determined by the characteristics of the connection to the motor.

Auto-transformer starters also provide reduced current inrush starting. The auto-transformer typically has three voltage taps: 50%, 65% or 80% of the full line voltage. This results in an inrush current of 25 – 64% of the locked rotor current and full load starting torque. When the motor is started, a starting contactor connects the motor to the line voltage through the design voltage tap. At the same time, a timer is energized. After a short interval, the auto-transformer is removed from the circuit and the motor is connected to the full line voltage. Auto-transformer starters result in low current draw from the circuit. The starter uses transformers, timers and extra contacts to control the inrush current. These starters tend to generate heat within the starter itself. They should not be used in applications that require frequent starts.

Primary resistance starters reduce the voltage to the motor windings with a resistance wired in series with the motor stator windings. The inrush current varies directly with the voltage applied. The starting torque also varies as the square of the voltage at full load torque. A number of steps can be provided to reduce the inrush current accordingly. When a motor is started, it receives a reduced voltage. After a short interval, the run contacts are closed, bypassing the start contacts and resistors. The motor is then operating on full line voltage. Primary resistance starters are applicable to situations where full locked rotor inrush is undesirable. Applying the voltage to the motor in steps results in steps of locked rotor current. This allows the system to readjust before the next stage of load is applied.

10.3 ELECTRICAL DISTRIBUTION SYSTEMS

Electrical distribution systems consist of a number of components, including conductors, switches, overcurrent protection devices, switchgear, relays, capacitors and transformers, as well as electrical loads. From an energy management standpoint, conductors, transformers and capacitors have the most significant impact on energy consumption in an electrical distribution system.

Switching devices, which may include switchgear and relays, are used to switch electrical components and circuits on and off. They may also be used to disconnect an electrical component from the source of power. Switching devices can include isolating switches, knife switches, general use switches, motor circuit switches and safety switches.

Overcurrent protection consists of fuses and circuit breakers. These devices are designed to interrupt the flow of current in the event of a short circuit. Molded case circuit breakers are designed to protect conductors and distribution systems against short circuits and protect devices against overloads. They are available with voltage ratings up to 600 volts and current ratings up to 4,000 amps. Fuses may also be used to protect circuits against short circuits and overloads. NEMA defines a fuse as "a device that protects a circuit by fusing open its current-responsive element when a overcurrent or short circuit passes through it." Fuses are available with interrupting ratings up to 200,000 amps. Fuses may be plug type or cartridge type for low voltage applications.

10.3.1 Conductors

Conductors are the wires through which electrical current flows. The two most common types of conductors are aluminum and copper wire. Since the surface of aluminum oxidizes readily, terminal connectors that have been designed specifically for use with aluminum wire must be used.

The voltage rating and ampacity of a conductor is very much dependent upon the type of insulation and covering used, as well as the size of the wire and material. The voltage loss, or drop, for a given amperage is primarily a function of the wire size. Table 10.4 shows the voltage loss per million feet at various amperages based upon wire size. The voltage drop is always measured as the "line" voltage (from line to line) rather than from line to neutral. The amount of the voltage drop is a function of:

- thc rcsistancc (*R*) of the wire (material and size).
- the reactance of the circuit (*X*), which is determined by the size of the wires, the spacing of the wires, and whether the wires are in a magnetic or non-magnetic conduit.
- the magnitude of the current (*I*) in the wires.
- the power factor (*PF*) and the reactive power (*RF*).

The reactive power is given by Equation (10.6).

$$RF = (1 - PF)^{0.5} \qquad \text{Eq. (10.6)}$$

where PF = power factor expressed as a decimal

The line voltage drop for a single phase circuit is given by Equation (10.7) and for a three phase circuit by Equation (10.8).

$$E_{D1} = (I)(R)(PF) + (I)(X)(RF) \qquad \text{Eq. (10.7)}$$

$$E_{D3} = (1.732)(I)(R)(PF) + (1.732)(I)(X)(RF) \qquad \text{Eq. (10.8)}$$

where I = line current, amps
R = resistance of the circuit (both wires),
X = reactance of the circuit,

Conductors typically loose about 3 – 5% of the rated voltage at the maximum load.

Example 10.1. A No. 6 copper wire is 180 feet long and carries a current of 40 amps with a lagging power factor of 80%. The electrical power is 208 volt, three phase. What is the percent voltage loss in the wire?

From Table 10.4, the loss per million feet of wire per amp for a No. 6 wire with 3 phase power and 80% power factor is 735 volts/amp-10^6 feet. The voltage loss for 180 feet of wire carrying 40 amps is

$$E = (180\text{ ft.})(40\text{ amps})(735\text{ volts/ amp - }10^6\text{ ft.}) = \frac{(180)(40)(735)}{1 \times 10^6}$$

$$= 5.3\text{ volts}$$

The percent voltage loss is

$$\%\,\text{loss} = \frac{5.3}{208} \times 100\% = 2.55\%$$

In addition to wires and cables, conductors may also include busways and wireways. A busway is a standardized factory assembled unit containing current conducting buss bars, of either copper or aluminum, and insulators. Wireways are lay-in ducts and troug hs in which wires are merely laid.

10.3.2 Transformers

Transformers are devices that are used to change the AC voltage. They are typically used to reduce the voltage, or step the voltage down; however, they can also be used to increase the AC voltage. Transformers cannot be used on DC circuits. A transformer is typically used to reduce the voltage supplied by a utility company before it enters a facility or building, generally from 4,160 V to 480 V. Additional transformers may be located inside a facility to reduce the voltage from 480 V to 240, 208 or 120 systems. Some transformers have multiple taps (connections) for multiple reduced voltages.

Transformers consist of two (or more) separate coils separated by a dielectric material. There is a primary coil, which has the AC power supplied to it, and a secondary coil(s). The primary and secondary coils are not connected electrically since they are separated by a dielectric material. However, they are located such that the magnetic fields around the coils do "communicate" magnetically. The space between the primary and secondary coils may be either a liquid or a solid material. Transformers with liquid material are referred to as liquid-filled; transformers with a solid material are referred to as dry-type transformers.

Earlier in this chapter, the effect of alternating magnetic fields and induced current was discussed. Transformers operate on the principle of alternating magnetic fields. As current flow alternates in the primary coil, it induces an alternating current in the secondary coil. The secondary voltage is a function of the windings in the primary and secondary coils.

Transformers are a available for single phase or three phase AC systems. Transformer power capacity is rated in kVA (kilovolt-amperes).

As transformers operate, they generate heat due to the impedance produced by the magnetic fields in the coils. This impedance causes heat to be generated within the transformer, which is rejected to the surrounding air. Whenever a transformer is located in side an air conditioned space, the heat rejection rate must be considered in energy consumption of the HVAC system.

In summary, transformers are selected based on the primary voltage, secondary voltage, phase, kva rating, type of insulation, physical size, sound and heat generation.

10.3.3 Capacitors

A capacitor is an electrical device that stores electrons. It allows electrical energy to be stored for later use. A capacitor comprises two electrically conductive materials, which are usually in the form of two plates. The plates are separated by a dielectric (insulating) material. The magnitude of the electron storage is a function of the area of the two electron storage plates, the thickness of the dielectric material separating the plates and the applied voltage.

TABLE 10.4 Voltage Losses for Copper Conductors

		Ampere Rating			Volt Loss										
					Three Phase — 60 Cycle. lagging power factor					Single Phase — 60 Cycle. lagging power factor					
	Wire Size	Type T. TW (60°C Wire)	Type RH. THWN. RHW. THW (75°C Wire)	Type RHH. THHN. XHHW (90°C Wire)	Direct Current	100%	90%	80%	70%	60%	100%	90%	80%	70%	60%
Copper Conductors in Iron Conduit	14	15	15	15	6,100	5,280	4,800	4,300	3,780	3,260	6,100	5,551	4,964	4,370	3,772
	12	20	20	20	3,828	3,320	3,030	2,720	2,400	2,080	3,828	3,502	3,138	2,773	2,404
	10	30	30	30	2,404	2,080	1,921	1,773	1,540	1,340	2,404	2,221	2,003	1,779	1,547
	8	40	45	50	1,520	1,316	1,234	1,120	1,000	880	1,520	1,426	1,295	1,159	1,017
	6	55	65	70	970	840	802	735	665	590	970	926	850	769	682
	4	70	85	90	614	531	530	487	445	400	614	613	562	514	462
	3	80	100	105	484	420	425	398	368	334	484	491	460	425	385
	2	95	115	120	382	331	339	322	300	274	382	392	372	346	317
	1	110	130	140	306	265	280	270	254	236	306	323	312	294	273
	0	125	150	155	241	208	229	224	214	202	241	265	259	247	233
	00	145	175	185	192	166	190	188	181	173	192	219	217	209	199
	000	165	200	210	152	132	157	158	155	150	152	181	183	179	173
	0000	195	230	235	121	105	131	135	134	132	121	151	156	155	152
	250M	215	255	270	102	89	118	123	125	123	103	136	142	144	142
	300M	240	285	300	85	74	104	111	112	113	86	120	128	130	131
	350M	260	310	325	73	63	94	101	105	106	73	108	117	121	122
	400M	280	335	360	64	55	87	95	98	100	64	100	110	113	116
	500M	320	380	405	51	45	76	85	90	92	52	88	98	104	106
	600M	355	420	455	43	38	69	79	85	87	44	80	91	98	101
	700M	385	460	490	36	33	64	74	80	84	38	74	86	92	971
	750M	400	475	500	34	31	62	72	79	82	36	72	83	91	95

	800M	410	490	515	32	29	61	71	76	81	33	70	82	88	93
	900M	435	520	555	28	26	57	68	74	78	30	66	78	85	90
	1000M	455	545	585	26	23	55	66	72	76	27	63	76	83	88
Copper	14	15	15	15	6,100	5,280	4,790	4,280	3,760	3,240	6,100	5,530	4,936	4,336	3,734
Conductors	12	20	20	20	3,828	3,320	3,020	2,700	2,380	2,055	3,828	3,483	3,112	2,742	2,369
in Non-	10	30	30	30	2,404	2,080	1,910	1,713	1,513	1,311	2,404	2,202	1,978	1,748	1,512
Magnetic	8	40	45	50	1,520	1,316	1,220	1,100	976	851	1,520	1,406	1,268	1,128	982
Conduit Lead covered	6	55	65	70	970	840	787	715	641	562	970	908	825	740	648
cables or	4	70	85	90	614	531	517	466	422	374	614	596	538	486	431
installation in	3	80	100	105	484	420	410	379	344	308	484	474	438	397	355
fiber or other	2	95	115	120	382	331	326	303	278	250	382	376	350	321	288
non-magnetic	1	110	130	140	306	265	266	251	232	211	306	307	289	267	243
conduit, etc.	0	125	150	155	41	208	216	206	192	176	241	249	237	221	203
	00	145	175	185	192	166	176	170	160	148	192	203	196	184	171
	000	165	200	210	152	132	145	141	134	126	152	167	163	155	145
	0000	195	230	235	121	105	119	118	114	108	121	137	136	131	125
	250M	215	255	270	102	89	105	106	104	100	103	121	122	120	115
	300M	240	285	300	85	74	92	95	93	91	86	106	109	107	105
	350M	260	310	325	73	63	82	85	84	83	73	94	98	97	96
	400M	280	335	360	64	55	75	78	79	78	64	86	90	91	90
	500M	320	380	405	51	45	64	69	71	70	52	74	80	82	81
	600M	355	420	455	43	38	57	63	66	66	44	66	73	76	76
	700M	385	460	490	36	33	53	58	61	63	38	61	67	70	73
	750M	400	475	500	34	31	51	56	60	61	36	59	65	69	70
	800M	410	490	515	32	29	49	55	58	60	33	57	64	67	69
	900M	435	520	555	28	26	46	52	55	57	30	53	60	64	66
	1000M	455	545	585	26	23	43	50	54	56	27	50	58	62	64

When DC power is applied to a capacitor, one plate will become positive and the other negative. Electrons will flow from the negative plate to the positive plate until the plate is fully charged.

When the plate is fully charged, no current will flow. After the capacitor has discharged, both plates will be neutral.

When AC power is applied to a capacitor, the voltage and current are continuously changing direction. As the electrons flow in one direction, one plate of the capacitor is charged. When the current and voltage reverse, the charge on the capacitor becomes greater than the source voltage, and the voltage discharges. The electrical effect is for the electrons to prevent a charge in the voltages. This is a similar effect as with magnetism and current, except that capacitors tend to affect a change in voltage. This effect causes the voltage to lag the current, and the opposite effect is produced on the current. The opposition of a capacitor is called the capacitive reactance.

10.4 LIGHTING AND LIGHTING SYSTEMS

Almost all facilities and occupied buildings have some sort of electrical lighting system. In some buildings, lighting represents 30 – 50% of the operating energy costs.

A lighting system includes that portion of a branch circuit wiring system that supplies the lamps or ballasts, together with the associated controls, such as switches and/or dimmers. The system also includes the light sources, luminaries, shielding and optical control media, the entire space to be lighted and the nature of the illumination required.

10.4.1 Types of Lighting and Light Sources

There are several types of light fixtures and sources available. The type selected depends upon a number of variables, including the area to be lighted, the mounting height of the light fixtures and the task being performed in the area being lighted.

Lighting equipment is categorized by the light source within the luminaire, rather than by the luminaire itself. This is due to the fact that the electricity consumption for a given lamp wattage will be different for the different types of sources. There are three types of light sources in common use: incandescent, fluorescent and high intensity discharge (HID) lamps.

Incandescent lamps consist of material filaments that are heated within a glass enclosure until they glow. Light is produced by passing an electric current through a very thin wire filament, causing it to heat up to a point of incandescence, thereby producing light. Figure 10-9 shows typical incandescent lamps. They are generally less expensive, easier to dim with rheostats and warmer in color than other types of lamps.

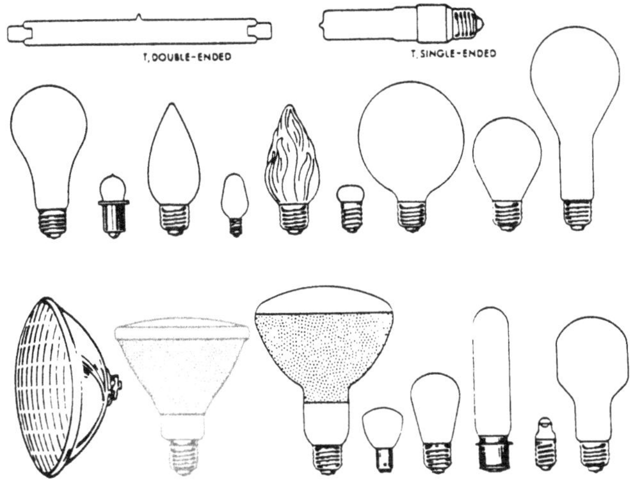

Figure 10-9. Typical incandescent lamps.

Incandescent lamps have a low efficacy (lumen output per watt) rating. Initial efficacy of typical incandescent lamps (25 – 1,000 watt) ranges from 10 to 23 lumens per watt. Only about 12% of the wattage used goes toward the production of light; the remainder of the energy goes to heat. They also have a comparatively short life—they are typically designed for approximately 1,000 hour of life.

A newer generation of incandescent lamps are the tungsten halogen and quartz lamps. These lamps are essentially high wattage incandescent lamps. They have extremely good lumen maintenance over a life of 2,000 hours or more. They emit a highly concentrated and easily controlled beam of light. Because of their characteristics, tungsten halogen lamps are used extensively in floodlighting, photographic, photocopy and special application lighting.

Fluorescent lamps are tubular, low-intensity, electric discharge lamps. A typical fluorescent tube is shown in Figure 10-10. They produce light by generating an electric arc, which passes through mercury vapor sealed within the tube. Light is produced by the fluorescence of phosphors activated by ultraviolet energy from a low pressure mercury arc. The lamp requires a ballast to limit the current and, in many instances, to transform the supply voltage. Lamp performance is influenced by the character of the ballast, luminaire, line voltage ambient temperature, burning hours per start and air movement. They are efficient producers of light with an efficacy of 50 – 80 lumens per watt, and produce little heat compared to incandescent lamps. They typically have a life of 9,000 to 20,000 hours.

Fluorescent lamps are the most common type of general lighting in buildings. They are available in many different lengths and wattages within the same length. The most common lengths are 4 foot and 8 foot tubes. The luminaries may or may not have plastic diffusers. Newer luminares have

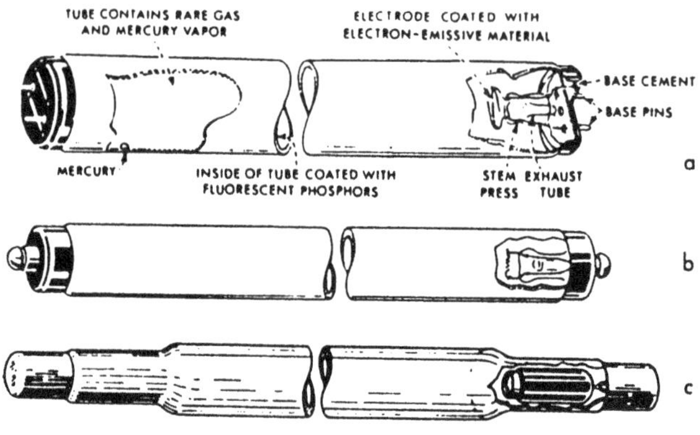

Figure 10-10. Typical fluorescent lamp.

egg crate reflectors. Newer fluorescent lamps include miniature fluorescent lamps called PL lamps, which have a small twin or quad tube arrangement on a plug in base with a ballast.

High intensity discharge (HID) lighting is a general category of lamps that includes mercury vapor, metal halide, low pressure sodium and high pressure sodium lamps. These lamps have a long life expectancy and consume comparatively little energy to produce a great deal of light from a relatively small source. They combine the form of an incandescent lamp with the efficiency of a fluorescent. The basic difference between HID lamps and fluorescent lamps is that HID lamps operate at a much higher arc pressure. Spectral characteristics differ from those of fluorescent lamps due to the higher pressure arc that emits a large portion of visible light. HID lamps require current limiting devices, which consume 10 – 20% additional power. HID lamps are commonly used for outdoor lighting, lighting in warehouses and large industrial spaces.

Mercury lamps produce light when an arc is struck in a quartz tube containing vaporized mercury. A typical mercury lamp is shown in Figure 10-11. They are available in 40 – 1,000 watt sizes and produce twice as much light as comparable incandescent lamps. Mercury lamps have about the same efficacy as fluorescent lamps, however, they are low in efficacy compared to other HID sources. Typical efficacy ranges from 30 to 63 lumens per watt, not including ballast loss. Since they have a life of 16,000 – 24,000 hours, they are often used when burning hours are long and service life is difficult. Clear mercury lamps cast a blue-green light.

Metal halide lamps are similar to mercury lamps, except that metal halide is added to provide more light and better color. Metal halide lamps have a higher efficacy than mercury lamps, 80 – 100 lumens per watt, not including ballast loss. They are available in 400 – 1,500 watt sizes and have a life of 1,500 – 15,000 hours. Because of their compact size, their light can be optically controlled. They are suitable for both indoor and outdoor applications.

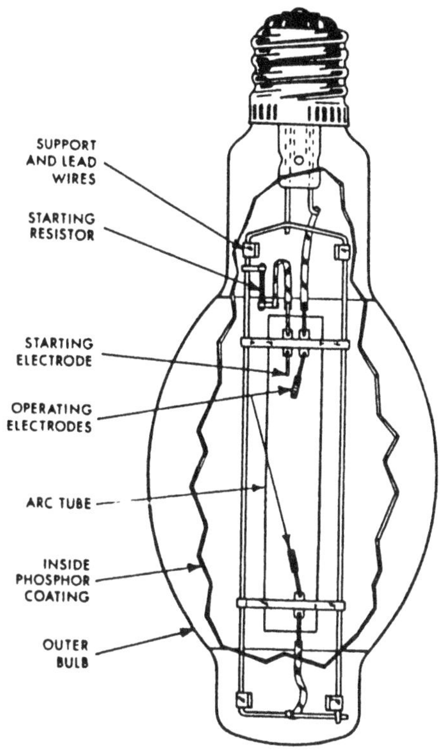

Figure 10-11. Typical mercury lamp.

High pressure sodium (HPS) lamps are a very efficient source of white light. Light is produced by passing electricity through sodium vapor. A typical HPS lamp is shown in Figure 10-12. They are available in 35 – 1,000 watt sizes and have a life of 10,000 – 20,000 hours. The efficacy of HPS lamps are from 80 to 140 lumens per watt, not including ballast power loss. HPS lamps produce a slightly yellowish/pink cast similar to that of a warm-white fluorescent lamp.

Low pressure sodium lamps (LPS) are similar to high pressure sodium lamps. They are available in 35 – 180 watt sizes and have a life of up to 18,000 hours. Typical efficacies for LPS lamps are from 137 to 183 lumens per watt, not including ballast power loss. Applications for LPS lamps are limited due to the monochromatic yellow color of the light they emit.

10.4.2 Lighting System Controls

Lighting controls are an important part of a lighting system. Control techniques may range from the simple on/off switch to elaborate computer controlled master switching systems for operating the lighting systems of

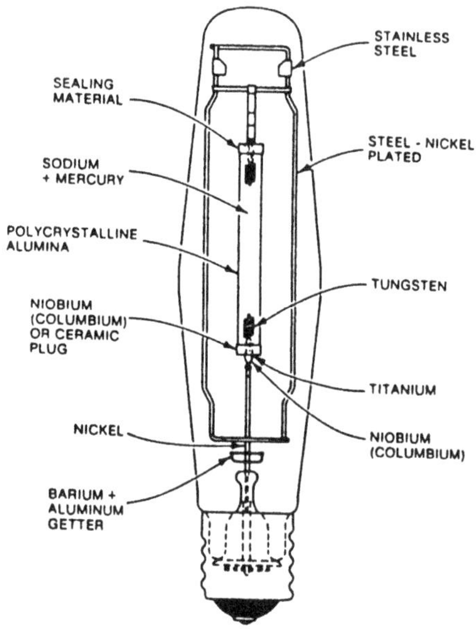

Figure 10-12. Typical high pressure sodium lamp.

large facilities. All lighting controls, whether a simple switch or a sophisticated computer based system, can normally be classified in two basic categories: (1) on/off controls and (2) level controls. Lighting controls can also be grouped into two general categories: centralized controls and local controls. Centralized controls are used in large facilities or areas where it is desirable to control large areas of the facility on the same schedule. An example of centralized control is a master parking lot light control. Localized controls affect only specific areas. Lighting control systems can also include a centralized control system, which is used to activate local controls. The local controls would allow area specific overrides of the central control.

On/off controls typically consist of wall switches, power lighting contactors, multi-pole lighting contactors, low voltage relays, timing controls and sensors. Wall switches are typically AC snap switches and can allow selective use of lighting, as well as different levels of lighting in a space. Contactors and relays can be used with central control systems to provide localized control of smaller lighting systems. Timing controls include a broad variety of time switches, which turn lighting systems on and off according to the time of day. Sensors are used to control lighting systems based upon a particular condition. They may be classified as either photoelectric sensors or presence sensors. Photoelectric sensors control lighting systems according to the ambient lighting level in the space or outdoors. Presence sensors are complete control systems that sense when a space is occupied and automatically turn on the lights for a preset time period. Presence sensors may be either passive infrared sensors of ultrasonic sensors. Infrared sensors detect and respond to changes in radiated heat caused by the presence and movement of a human body. An ultrasonic sensor creates an ultrasonic field in a space being monitored. When a person enters the space, the field is disturbed, the sensor detects the disturbance and activates the lighting system.

Lighting level controls primarily consist of dimmers. Dimmers were previously available only for incandescent lamps; however, now electronic dimmers are available for incandescent, fluorescent and HID lighting. The most common type of dimmer is the silicon controlled rectifier (SRC). It operates on a principle of switching the current on a proportional distance through each AC half cycle. An SRC is essentially a very fast switch.

10.5 MAINTAINING ELECTRICAL AND LIGHTING SYSTEMS FOR OPTIMUM ENERGY EFFICIENCY

Electrical and lighting systems typically require routine maintenance to keep them operating at optimum efficiency. Significant energy savings can be achieved by proper routine maintenance and operation. Proper routine maintenance and operation should be the first step in any energy management program for electrical and lighting systems. The following are operations and maintenance procedures that will result in optimum energy efficiency for most electrical and lighting systems.

Turn Electrical Systems and Equipment Off: One of the easiest ways to reduce energy waste in electrical equipment is to turn equipment and systems off when they are not needed or not in use. Table 10.5 shows typical losses for electrical equipment.

Local switches for lighting systems and electrical equipment should be labeled to encourage users to turn lights and equipment off when not needed. When practical, get persons working after hours to work in close proximity to lessen all energy requirements.

Reduce Illumination Levels: The designs of older lighting systems frequently included lighting levels that were lit two, or even three, times more than required. Lighting levels of existing systems should be checked and compared to those listed in ASHRAE (1993) and IES (1981).

In general work areas, non-uniform lighting systems can be applied when the average workstation size is less than one worker per 50 – 70 square feet. If workstations are larger, a uniform lighting system is generally more practical.

Lighting levels can be reduced by installing lower wattage lamps or by removing selected fixtures. When removing fluorescent or HID lamps, it is also necessary to disconnect the ballast to prevent the consumption of energy by the unused ballast.

Workstation Modification: A lighting pattern that throws light equally on all areas regardless of the task may waste up to 50% of the energy used for lighting. Most workstations can be relocated to take maximum advantage of an existing lighting system.

Try to group work activities requiring similar illumination levels together in the same area as much as possible. Move work surfaces and desks to positions that will use the installed luminaires to their greatest advantage. Relocate work stations requiring the highest illumination levels to areas near windows.

Periodically Clean Luminaires: The efficiency of a luminaire is very dependent upon the reflectance of interior surfaces. The reflectance of these surfaces is a function of the amount of dirt on the surfaces. All surfaces eventually collect dust, and, as a result, loose their lighting effectiveness over a period of time. Typical losses of lighting effectiveness are on the order of 10% – 25% over a two year period for a typical 8 – 12 hour per day usage.

Luminaire efficiency can be maintained by periodically properly cleaning the reflecting surfaces and shielding media of the luminaire. Luminarie manufacturers publish expected luminaire dirt depreciation factors to estimate the losses in efficiency due to the accumulation of dirt. A periodic luminaire cleaning program should be coordinated and scheduled along with a luminaire relamping program. Replace lens shielding that has yellowed or become hazy with a clear acrylic lens with good non-yellowing properties. Clean ceilings, walls and floors frequently to maintain reflective qualities of these surfaces.

Institute a Relamping Program: Earlier in this chapter, the normal life expectancy of the various types of lamps was discussed. Lamp efficiency deteriorates over the life of a lamp. The degradation of the light output of lamps is referred to as lamp lumen deterioration. The light output of lighting system should be checked regularly with a calibrated light meter by maintenance personnel.

The effects of lamp lumen depreciation may be reduced by a planned group relamping program. When the light output of a group of lamps has fallen to approximately 70% of the original light output, all the fixtures should be replaced at the same time. Knowing the rated life of the existing type of lamps enables prediction of when to replace lamps in the lighting system before the maximum failure occurs.

Typical maintenance programs replace lamps as needed as they burn out. This method is inefficient and costly since the lighting output of lamps may be reduced by as much as 50% before they are replaced.

A planned group relamping program saves energy and labor costs. A planned program replaces all the lamps in a given area at periodic intervals. From a labor standpoint, it is less costly to replace all the lamps in a given area at once rather than one at a time as they burn out. The optimum time for a relamping program occurs between 60 and 80% of the life of the lamps.

TABLE 10.5 Losses in Electrical Equipment

Component	Energy Loss at Full Load (%)
Outdoor circuit breakers (15–230 kV)	0.002–0.015
Generators	0.09–3.5
Medium voltage switch gear	0.005–0.02
Current limiting reactors (600 V–15 kV)	0.09–0.30
Transformers	0.55–1.90
Load break switches	0.003–0.025
Medium voltage starters	0.02–0.15
Busway (480 V and below)	0.05–0.50
Low voltage switch gear	0.13–0.34
Motor control centers	0.01–0.40
Cable	1.00–4.00
Motors	
1–10 hp	14.0–35.0
10–200 hp	6.0–12.0
200–1500 hp	4.0–7.0
1,500 hp and larger	2.3–4.5
Rectifiers (large)	3.0–9.0
Static variable speed drives	6.0–15.0
Capacitors (watts loss/var)	0.5–2.0

Power Transformer Management: Proper management of transformer operations and related factors can result in substantial energy savings. The following guidelines should be considered:

- Shade exterior transformer banks from solar radiation to reduce the buildup of heat and resultant losses.
- Ventilate transformer banks to keep them as cool as possible.
- De-energize transformers whenever possible, including unloaded transformers, transformers supplying unoccupied or unused areas, refrigeration chiller transformers during the heating season, heating equipment transformers during the cooling season, and heating and cooling system transformers during periods of time when neither heating nor cooling are required, such as spring and fall.
- De-energize transformers during weekend and holidays.

Outdoor Switchgear: Since heat can increase energy losses in switchgear, consider the following guidelines for outdoor switchgear:

- Provide shading from solar radiation.
- Remove or prune any shrubbery or obstructions that tend to obstruct natural ventilation through switchgear.

10.6 ENERGY CONSERVATION MEASURES FOR ELECTRICAL AND LIGHTING SYSTEMS

The energy conservation measures for electrical and lighting systems discussed in the previous section were operation and maintenance procedures that would improve the energy efficiency of existing systems. Most of those measures were little, or no, cost measures that could be performed on an ongoing basis to improve efficiency.

In addition to the low cost O&M procedures, there are also modifications that could be made to electrical and lighting systems to improve energy efficiency. These modifications, however, require a more significant cost investment and are permanent. The modifications listed below are improvements that should be considered under any energy management program for existing, or proposed, electrical and lighting systems.

10.6.1 Lamp and Luminaire Modifications

Consider replacing or modifying existing lamps and luminaries with those that have a higher efficiency and produce the same level of illumination at a lower rate of power consumption. Whenever possible, relocate luminaries to provide light on tasks as directly as possible. Table 10.6 lists the characteristics and average lumens per watt for various types of lamps.

The following is a quick method to determine the advisability of replacing one lighting system with another. It is only an approximation to determine whether a more detailed analysis is warranted.

1. Determine the lighting level required for the task and check the maintained lumen rating of the lamp presently in use. Consider whether the present level meets the needs, or if an increase or decrease in lighting level is desirable.
2. Determine if the color rendition is a concern for the lighted space. If not, lamp efficacy and life should be the primary considerations.
3. Select a light source that has the highest efficacy with color rendition, life (maintenance, schedule and cost) and application suitability. Refer to Table 10.6.
4. Make a rough estimate of the number of fixtures required based upon the lumen rating of the proposed system.
5. Select the appropriate luminaire for the source and application
6. Estimate the number of fixtures required, their cost, and the installation cost and sum together to get a total installed cost.
7. Determine the amount of power consumption reduction for each fixture, then multiply it by the number of fixtures.
8. Estimate the number of hours of usage for the system per year.

9. Multiply the number of hours of operation by the reduction in system power consumption to get annual energy savings. Multiply the annual energy savings by the cost per kWh to get the annual cost savings.
10. Divide the total installed cost by the annual cost savings to get the estimated payback in years.

A redesign of a lighting system may require the services of a qualified consultant.

10.6.2 Lighting System Controls

Initial cost considerations and lack of knowledge about the final arrangement of lighted spaces often result in the use of central panel boards as the only means of controlling large sections of lighting. This type of control for lighting frequently results in unnecessary energy consumption. Frequently, some areas are lit, even though they are not occupied, or in use, due to large sections of lighting being controlled by a single switch. The efficiency of lighting can be improved by modifying the lighting controls. Such modifications might include the use of photo-cells, occupancy sensors and dimmers.

When natural day lighting is available, consider the use of photo-cell switching to turn off banks of lights in areas where the natural light is sufficient for the task. Use photo-cell and/or time clocks of outdoor lighting whenever feasible.

New switching devices, which sense the presence of people, may be used to effectively turn lights off when spaces are unoccupied and lighting is not

TABLE 10.6 Approximate Range of Lamp/Ballast Characteristics

Type of Lamp	Wattages	Average Lumens Per Watt Including Ballast Losses: Initial	Mean	% Lumen Maintenance	Average Rated Life (Hours)	Warm-up/ Restrike (Minutes)
Low Pressure Sodium	18–180	62–143	100	10–18,000	7.00/0–1	
High Pressure Sodium	70–1000	79–130	71–118	90–91	20–24,000	3–4/.5–1
Metal Halide	175–1500	69–105	54–84	77–80	10–15,000	2–5/10–20
Mercury Vapor	50–1000	30–60	23–46	75–89	16–24,000	5–7/3–6
Fluorescent	40–215	63–100	42–88	66–88	12–20,000	Immediate
Incandescent	60–1500	15–24	13–23	90–95	750–2500	Immediate

Note: Approximate range of lamp/ballast combination characteristics are based on data applicable to most commonly used lamps within each category. Data are provided for general comparative purposes *only*. Actual performance of a lamp with regard to any of the factors indicated will vary due to the uniqueness of a specific installation. Efficiency of an installation depends on far more than the lamp or lamp/ballast efficiency.

Source: The National Lighting Bureau, Washington, D.C.

needed. Occupancy sensors were discussed in Section 10.4.2 of this chapter.

Lighting systems can be retrofitted with dimming controls that sense lighting levels and dim interior lights when sufficient natural light is available. Since illumination decreases with distance from the windows, electric lighting and controls should be "layered" parallel to the window pane so that rows of lights can be dimmed as needed. Solid state (digital) dimmers are available for fluorescent and HID lamps, as well as for incandescent ones. These solid state dimmers, unlike the old style resistance type dimmers, reduce power consumption as the lights are dimmed. Fluorescent fixtures cannot be totally dimmed and they do require a dimming ballast on most dimming systems. Solid state dimmers use a very fast switch (SCR) to dim lamps. Instead of using a voltage to change a capacitor and turn on the switch, the solid state dimmer counts a number of steps through each half-cycle, then turns on the switch.

10.6.3 Install High Efficiency Ballasts

A ballast is an electrical component used in lighting fixtures that performs the following functions:

- It applies voltage to the lamp's electrodes to preheat a rapid-start lamp before ignition.
- It starts the arc inside the lamp.
- It regulates the current and voltage in the lamp to maintain normal operation.

Most ballasts for fluorescent light fixtures are of the electromagnetic (or just magnetic) type. These ballasts operate on alternating current with a 60 Hz cycle. As a result, electrons flow on and off 120 times per second. When the flow is on, the arc inside the lamp is energized, causing the phosphor coating on the inside wall of the tube to glow.

The performance of fluorescent lamps can be improved and the energy consumption reduced with electronic ballasts. Electronic ballasts offer improved performance that includes reduced flicker, improved voltage and thermal regulation, and the ability to dim fluorescent lamps over a wide range without affecting lamp life.

Electronic ballasts increase the frequency of the on-off cycle of the arc to the 20,000 – 60,000 Hz range. As a result of the higher frequency, the phosphors remain energized throughout much more of the cycle and the lamp can effectively produce the same amount of light while consuming less power. Electronic ballasts also allow full range dimming of fluorescent fixtures. It is difficult to effectively dim fluorescent fixtures with magnetic ballasts.

Electronic ballasts have efficiencies of over 15% above that of the magnetic type. They also provide an additional energy savings of 13% by providing constant nominal lamp rated wattage throughout the lamp life. Electron ic ballasts have power factors near 99%.

10.6.4 Replace Oversized Motors

Original estimates for the loads on motors are usually conservative, resulting in motors that are oversized for the actual equipment load. Energy conservation measures for the equipment driven by the motor may also reduce the load on a motor, resulting in an oversized motor. The efficiency of a motor is a function of several variables, including the size of the motor with respect to the actual load. Oversizing a motor may result in inefficiency. The efficiency of a motor falls off rapidly at loads lighter than the rated full load.

Polyphase motors usually reach peak efficiency at loads slightly less than full loads. Motor efficiencies drop off sharply below half load, as shown in Figure 10-13. Lightly loaded motors also degrade the power factor in an electrical distribution system.

10.6.5 Install High Efficiency Motors

Motors under 125 hp have traditionally been purchased or specified on the basis of first cost, size and weight with little regard for efficiency. The

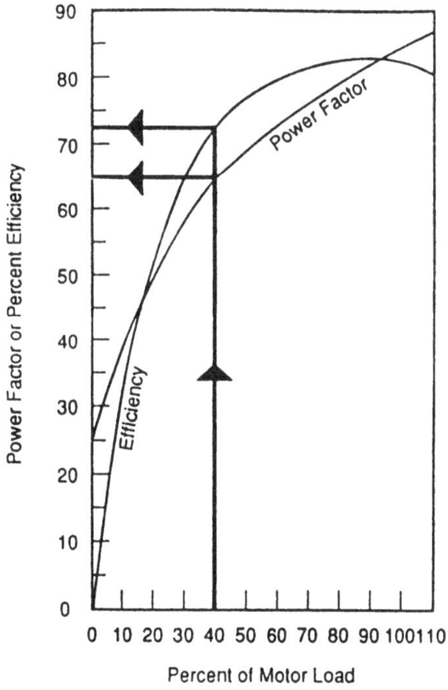

Figure 10-13. Percent motor efficiency versus motor load.

energy efficiency of a typical motor, under 50 hp, is about 91%. Smaller motors are inherently less efficient than large motors. Factors affecting motor efficiency include:

- size of the motor with respect to the actual load
- type of motor
- motor design speed
- type of bearings

The only way to improve motor efficiency is to reduce motor losses. Motor losses can be classified into five major areas:

- Core losses, which result from the energy required to magnetize the core material (hysteresis) and include losses due to eddy currents that flow in the core.
- Windage and friction losses, which are caused by air resistance and bearing friction.
- Stator losses, which result from current flow through the stator windings. These losses are frequently referred to as I^2R losses.
- Rotor losses, which are I^2R losses in the rotor winding. Rotor losses are variable losses and typically represent 20 – 25% of total motor losses.
- Stay load losses, which result from the leakage of flux induced by the load currents.

Consideration should be given to replacing existing motors with high efficiency motors, especially if they are near the end of their useful life. The efficiency of high efficiency motors is typically 3 – 4% higher than standard motors. Although the increase in efficiency for a high efficiency motor is only 3 – 4%, high efficiency motors still offer substantial energy savings. The average service life of an electric motor is about 18 years. Over the life of a motor, a 3 – 4% increase in efficiency can result in significant savings.

10.6.6 Variable Speed Drives

In most motor installations, motors are sized to provide the maximum foreseeable power output required. The power required to drive a load, such as a fan or a pump, is proportional to the cube of the rotational speed. If the rotational speed of a load is constant, the power input remains constant at the maximum power.

The most common type of load on a motor has a constant torque. Since the power is proportional to the torque times the rotational speed, a constant torque load makes the power needed directly proportional to the speed. A centrifugal pump or fan typically has a variable torque characteristic. The load torque is proportional to the square of the rotational speed and the power is proportional to the cube of the speed.

Since the power consumed by a motor is proportional to the cube of the rotational speed, significant savings could be achieved if the speed of the motor was matched to the output requirements of the device driven by the motor. For example, if a motor is slowed to 90% of the design speed, an energy savings of 27% would result. The only way to change the shaft speed of an AC induction motor is to change the number of poles on the motor, change the amount of motor slip or to vary the frequency of the AC power supplied to the motor.

Variable speed drives are devices that match the motor speed to that of the application. An electric variable speed drive is an electronic device that rectifies the incoming AC voltage and current into DC, then inverts the DC back in to AC at a different voltage and frequency. Electric variable speed drives are also called variable frequency drives. Centrifugal fans and pumps are usually very good applications for variable speed drives. Energy savings from variable air volume systems was discussed in Chapter 9, and for variable flow pumping systems in Chapter 6.

There are three types of electronic AC variable speed drives:

- Variable Voltage Inverter (VVI): It utilizes a silicon controlled rectifier (SCR) front end to convert incoming AC power to a varying DC bus voltage.
- Current Source Inverter (CSI): It utilizes SCRs in the rectifier and inverter sections. The SCRs produce a variable voltage for a DC buss. The DC bus consists of only an inductor.
- Pulse Width Modulation (PWM) Drives: These drives use a full wave diode bridge rectifier to convert the incoming power to DC. PWM drives are the most common type of variable frequency drive since they have a higher efficiency, higher power factor, larger speed ranges and lower motor heating.

10.6.7 Improve Power Factor

The electrical power that must be supplied to any induction load, such as a motor, transformer fluorescent lamp, etc., is comprised of real and reactive power. Reactive power establishes magnetic fields surrounding conductors and provides magnetizing currents for transformers and rotating machines.

Inductive loads, such as motors, transformers, lighting ballasts, which include reactive power, frequently cause the current to lag the voltage. The lagging current is the magnetizing current required for operating electromagnetic devices. Current that is not in phase with the voltage does not produce real work. The lagging current does, however, result in additional real burden on the electrical system. The lagging current is measured in kilovolt amperes (kVAR). Power factor is a ratio of the current drawn that produces real work to the total current drawn. The power factor can range from 1.0 to zero. A power factor of less than 1.0 results when electrical loads draw reactive current that produces real work.

Facilities that have lightly loaded electric motors, which draw excessive amounts of reactive power typically have low power factors. Low power factors result in increased energy losses in the electrical distribution system. Improving power factor reduces total current flow by reducing the flow of reactive current. Power factor improvements also increase system voltages by reducing the total current draw. Additional costs for low power factor may result if the power demand is recorded (in kVA) and charged on the electrical bill, the electric rate has a kVAR or power factor penalty clause and if there are problems with voltage regulation of chronic low voltage.

Referring to Figure 10-14, the power factor is a measure of the active power to the apparent power and is given by Equations (10.9) and (10.10).

$$PF = P/Q = \cos\theta \qquad \text{Eq. (10.9)}$$

$$S = (P^2 + Q^2)^{1/2} \qquad \text{Eq. (10.10)}$$

where P = active power, kW
Q = reactive power, kVAR
S = apparent power, kVA

The energy used in magnetic fields established by reactive current is known as reactive power and is measured in kilovars (kVAR). The apparent power is the square root of the sum of the square of the active power and the square of the reactive power as given in Equation (10.10).

Generally, electric utility billings for industrial facilities include demand and power factor penalty charges. Improving power factor reduces energy and demand charges. It also eliminates power factor penalties, lowering overall electric bills. A facility can often determine its power factor from its electrical bill. Power factor may also be measured directly using instruments that read power factor, or calculated using Equations (10.9) and (10.10).

Power factor may be improved by de-energizing motors and transformers that are unloaded. In addition, there are four types of devices that may be used to correct power factor:

- capacitors
- tuned harmonic filters
- static volt-ampere-reactive compensators (VAR)
- synchronous machines

Capacitors are frequently used as corrections for poor power factor. They are the simplest and easiest solution to low power factor. Placing capacitors in an electrical circuit causes current to lead the voltage, the

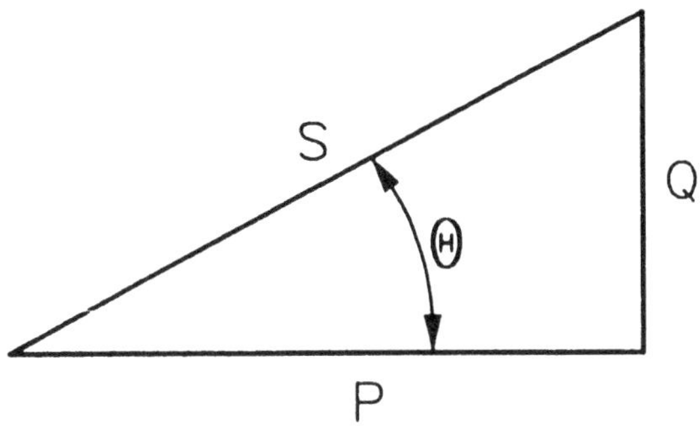

Figure 10-14. Power factor triangle.

opposite of reactance. In most electrical systems, there are few capacitive components. Adding capacitors to an electrical system is an effective way to increase system power factor. Capacitors are considered kVAR generators because they supply reactive current. However, switched capacitor banks can cause problems, such as switching surges as capacitors being switched in and out of the system.

Tuned harmonic filters, which are inductors and capacitors, may also be used to correct low power factor where harmonics are present in the system. The filters can be tuned to specific harmonic frequencies, reducing the effect of harmonic frequencies on the electrical system. Tuned filters are usually based upon the system impedance and sized on present harmonic distortion levels. Automatically switched harmonic filter systems are also available. The complexity of these systems make them unsuitable for some applications.

Synchronous machines include generators and motors with condensers. Generally speaking, if the machine is driven by a mechanical source, it acts as a generator. Synchronous motors can supply or consume reactive power. There are three locations where power factor correction is typically applied to electrical distribution systems. If the main objective is to reduce power factor penalties and the facility owns the primary distribution, power factor correction may be applied at the connection to the utility at the point of entrance to the facility. When inductive loads are widely distributed throughout a facility, applying power factor correction at a point on a bus serving many inductive loads may be more effective. When there are many inductive loads widely scattered throughout a facility, applying the power factor correction at the inductive load downstream of the motor starter may be more effective. This method, however, can be much more costly than the other methods.

Generally speaking, a power factor high enough to avoid large utility penalties, but less than 1.0, is desirable. If the power factor was greater than 1.0, overvoltage conditions could result. Therefore, excessive power factor correction measures should not be taken.

Example 10.2. A 230 volt electrical distribution system has a lagging power factor of 0.70. If the power factor were to be increased to a lagging power factor of 0.90, what size capacitor would be needed?

The current drawn by the load is

$$I = \frac{p}{E(PF)} = \frac{10,000}{(230)(0.7)} = 62.1 \text{ amps}$$

The apparent power factor is

$$Q = (S^2 - P^2)^{1/2} = 14,283 \text{ kVA}$$

Rearranging Equation (10.10) to solve for the reactive power when the $PF = 0.7$,

$$Q_{0.7} = (S^2 - P^2)^{1/2} = [(14283)^2 - (10000)^2]^{1/2} = 10{,}200 \text{ VAR}$$
$$= 10.2 \text{ kVAR}$$

A capacitor draws no average power, therefore, the capacitor load must also be 10 kW. The net reactive power of the combination of the active power and new power factor can be obtained by referring to Figure 10.14 and trigonometry to solve for the new reactive power.

$$Q_{0.9} = P \tan[\cos^{-1}(PF)] = 10 \tan(\cos^{-1} 0.9) = 4.84 \text{ kVAR}$$

The difference in reactive power that must be made up by the capacitor is

$$Q_C = Q_{0.9} - Q_{0.7} = 4.84 - 10.2 = -5.36 \text{ kVAR}$$

Therefore, a capacitor that adds 5.36 kVA is required for a power factor of 0.9.

10.6.8 Reduce Peak Electrical Demand

Most electrical utilities bill not only for the amount of electricity consumed, but also for the rate at which power is consumed. Utility customers usually pay for the rate at which they use power by paying a demand charge in addition to an energy charge on their electrical bill.

The power demand is the load integrated over specified time intervals of 15, 30 or 60 minutes, as determined by the utility. The peak demand is the maximum demand in kW measured during a billing period. A high demand for just one demand interval can raise the demand charge for the entire billing period. Controlling peak demand is an important and effective energy management strategy. Demand is typically controlled or reduced by one, or more, methods. Methods for reducing demand include load-shedding, cogeneration and cool thermal storage systems.

Load-shedding is a demand limiting energy management program, which shuts off non-essential electrical loads when the demand reaches a predetermined limit. As part of a load-shedding program, all electrical loads should be prioritized so they can be shut off in decreasing order of importance. Peak loads that cannot be predicted should be monitored. Equipment operation can be programmed to automatically de-energize selected loads at peak demand and reactivate them when the reduced demand permits. Available automatic load limiting devices range in complexity from a simple thermal sensor to a computer based energy management system. Energy management systems are discussed in Chapter 16.

Cogeneration systems generate electricity on-site to reduce demand and provide waste heat that can be recovered for other process applications. Cogeneration systems can handle selective or peak loads, or they can provide all the site power requirements. Cogeneration systems are discussed in Chapter 13.

Cool thermal storage systems generate chilled water or ice during off-peak hours for later use during peak demand hours. Cool thermal storage systems level out or eliminate peak cooling loads by operating the cooling equipment during off-peak periods. Thermal storage systems are discussed in Chapter 15.

10.7 ENERGY EFFICIENCY FOR NEW ELECTRICAL AND LIGHTING SYSTEMS

Electrical and lighting systems are typically one of the largest consumers of energy at many facilities. Since electrical and lighting systems and equipment are so widely used and may consume significant amounts of energy, regardless of the type of facility, it is important that energy efficiency be a primary consideration when selecting new electrical and lighting equipment for new, or retrofit, installations. In some cases, local or national energy codes may dictate efficiencies of such equipment. Standards are also available, which provide guidelines for electrical and lighting equipment and systems ASHRAE (1993).

Earlier in this chapter, energy conservation for electrical and lighting systems was discussed. Several of the energy efficiency measures discussed as retrofit opportunities for equipment, such as high efficiency motors, are applicable to new and retrofit installations. In addition to the previously discussed energy conservation measures for electrical and lighting, the following measures should be considered for new or replacement equipment and accessories.

10.7.1 High Efficient Motors

When selecting and specifying new equipment that is driven by an electric motor, high efficiency motors should always be specified. Although there is a cost premium for high efficiency motors, the cost premium is usually recovered in a short period of time.

When sizing new motors, oversizing should be avoided. Oversized motors operate less efficiently and reduce electrical system power factors.

10.7.2 Lighting Systems

Lighting types for new lighting systems should be selected based upon the highest efficacy, that will provide the necessary quality of light and acceptable color-rendering. New lighting systems should be designed and installed with selective switching to allow unused areas to not be lit. Whenever possible, lighting systems should provide individual task lighting for tasks that that require a higher level of light than required for general lighting levels.

New fluorescent lighting fixtures should have electronic ballasts with two levels of illumination. New systems should have high efficiency fluorescent lamps. New 32 watt fluorescent tubes should be used instead of the standard 40 watt tubes. The new 32 watt tubes use 15% less electrical power while producing about 8% less light. Since the human eye responds to light exponentially, rather than linearly, the difference in light output is often unnoticeable.

10.7.3 Variable Speed Drives

For motor driven equipment that has variable loads, such as centrifugal fans and centrifugal pumps, variable speed drives should be considered. The power to drive a rotating machine is proportional to the cube of the rotational speed. Significant energy savings can be accomplished by varying the rotational speed in response to varying load conditions.

10.7.4 Computer Based Energy Management Systems

Computer based energy management systems (EMS) are microprocessor based devices, which can be programmed to provide both demand control and energy management. Microprocessor based EMS can have any, or all, of the following capabilities:

- scheduling
- duty cycling
- optimum start/stop
- monitoring
- temperature setback
- lighting controls

Chapter 11

Compressed Air Systems

The U.S. Department of Energy (DOE) has determined that air compressors and compressed air systems are one of the largest consumers of energy in industrial and manufacturing applications. Compressed air is used in practically all phases of manufacturing. It is used in stamping processes, air wrenches, aerating equipment, conveying equipment, pneumatic controls; it is also used as part of chemical and manufacturing processes. Most industrial facilities utilize compressed air as a utility, or form of energy, like natural gas and electricity.

Compressed air is an expensive form of energy compared to the cost of other forms of utility service. A typical compressed air system operates about 8,000 hours per year and produces about 4 cfm of air per horsepower input at a discharge pressure of 100 psig. When the amortized cost of an installed compressed air system is added to the operating cost, the use of compressed air is four to five times that of electrical energy.

Energy saving improvements to compressors and compressed air systems can result in savings of 20–50% of the energy consumed by compressed air systems. Energy saving opportunities for compressed air systems typically occur in at least three general areas. First is the compressor itself, and compressor operation and control. Another area is the compressed air distribution system, and the third is how the compressed air is used and misused.

This chapter discusses compressed air systems that typically produce compressed air in the range of 80–150 psig. Devices that produce air at pressures up to 12 psig are considered blowers or fans and are discussed in Chapter 9.

11.1 GENERAL COMPRESSOR OPERATION AND BASIC THEORY

11.1.1 Definitions and Terms Used for Compressors and Compressed Air Systems

When working with air compressors and compressed air systems, there are a number of terms that are unique to the systems. See the Glossary for useful definition of terms that are commonly used when discussing compressed air systems.

11.1.2 Basic Laws Relating to Compressed Air Systems

The First Law of Thermodynamics is often called the law of conservation of energy. It states that during any cycle, a system undergoes the cyclic integral of the heat, which is equal to the cyclic integral of the work. In other words, energy cannot be created or destroyed during a process, such as the compression and delivery of a gas, although it may change from one form of energy to another. The First Law of Thermodynamics is discussed in Chapter 2.

The Second Law of Thermodynamics implies that it is impossible to construct a machine that operates on a thermodynamic cycle that has a thermal efficiency of 100%. It may be stated in several ways:

1. Heat cannot, itself, pass from a cooler body to a warmer body.
2. Heat can be made to go from a body at a cooler temperature to a body at a warmer temperature only if external work is done.
3. The available energy of an isolated system decreases in all real processes.
4. Heat or energy, of itself, will only flow from a higher state to a lower state.

Boyle's Law states that, at a constant temperature, the volume of an ideal gas varies inversely as the absolute pressure and may be stated as follows

$$\frac{V_2}{V_1} = \frac{P_2}{P_1} \qquad \text{Eq. (11.1)}$$

$$P_2\, V_2 = P_1\, V_1 \qquad \text{Eq. (11.2)}$$

Charles' Law states that, at a constant pressure, the volume of an ideal gas varies directly as the absolute temperature and may be stated as follows:

$$\frac{V_2}{V_1} = \frac{T_2}{T_1} \qquad \text{Eq. (11.3)}$$

$$\frac{V_2}{T_2} = \frac{V_1}{T_1} \qquad \text{Eq. (11.4)}$$

Amonton's Law states that, at a constant volume, the absolute pressure of an ideal gas will vary directly with the absolute temperature.

$$\frac{P_2}{P_1} = \frac{T_2}{T_1} \qquad \text{Eq. (11.5)}$$

$$\frac{P_2}{T_2} = \frac{P_1}{T_1} \qquad \text{Eq. (11.6)}$$

Dalton's Law states that the total pressure of a mixture of ideal gases is equal to the sum of the partial pressures.

$$P = P_a + P_b + P_c + \ldots\ldots\ldots \qquad \text{Eq. (11.7)}$$

Amagat's Law, which is similar to Dalton's Law, states that the volume of a mixture of ideal gases is equal to the sum of the partial volumes that the constituent gases would occupy if each existed alone at the total pressure and temperature of the mixture.

$$V = V_a + V_b + V_c + \ldots\ldots\ldots \qquad \text{Eq. (11.8)}$$

When analyzing the operation of compressed air systems, it is frequently possible to consider air as a perfect (ideal) gas without significant error. A perfect gas is a gas that obeys the following law:

$$PV = mRT \qquad \text{Eq. (11.9)}$$

where P = the absolute pressure, lbf/ft^2
V = the volume, ft^3
m = the mass, lbm
R = a constant, lbf-ft/lbm- °R
T = the absolute temperature, °R

For air R = 53.34 ft-lbf/lbm- °R

Air behaves like a perfect gas when it is at, or near, standard pressure and temperature. However, as the pressure and temperature change, air does not behave as would be predicted by the perfect gas law. The perfect gas law assumes that there is no intermolecular reaction (attraction or repulsion) and that the relative distance between molecules is great. For real gases, such as air, repulsion or attraction forces between molecules causes the volume to be greater or lesser than would be predicted by the perfect gas law. To account for this deviation, it is necessary to introduce a correction factor known as the compressibility factor (Z).

$$Z = \frac{PV}{RT} \qquad \text{Eq. (11.10)}$$

Compressibility is experimentally derived from data about the actual behavior of a particular gas under PVT changes. Compressibility factors

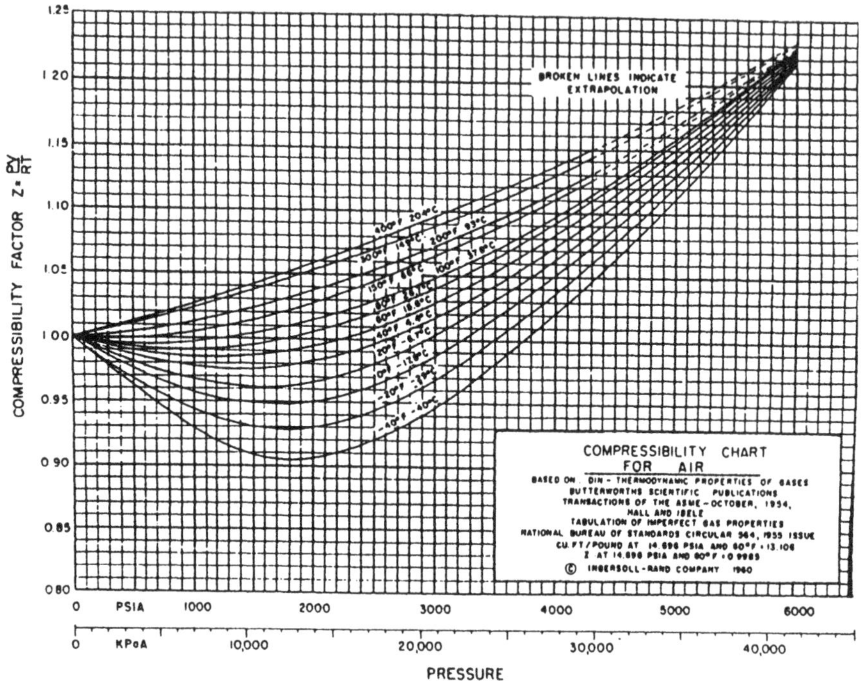

Figure 11-1. Compressibility chart for air.

for gases (such as air) are presented in charts. A compressibility chart for air is shown in Figure 11-1.

When Z is greater than 1.0, the repulsion forces between molecules results in a volume greater than would be predicted by the perfect gas law. When Z is less than 1.0, the attractive forces result in volumes that are less than those for a perfect gas.

11.1.3 Basic Types of Air Compressors

There are two general classifications for air compressor types: positive displacement and dynamic. The classifications have to do with the method by which air is compressed.

Positive displacement compressors are those in which successive volumes of air are confined within an enclosed space and elevated to a higher pressure. The flow of air from a positive displacement compressor is intermittent, which results in some pulsations in the discharge pressure. There are two methods used to compress air:

1. Trap consecutive quantities of air inside an enclosure, reduce the volume thus increasing the pressure, then push the compressed air out of the enclosure.

2. Trap consecutive quantities of air inside an enclosure, carry it without volume change to a discharge opening, compress the air by backflow from the discharge system, then push the compressed air out of the enclosure.

Reciprocating compressors use method 1. Sliding vane, liquid piston, straight lobe and helical lobe compressors use method 2.

Dynamic compressors are rotating continuous-flow machines in which a rapidly rotating element accelerates the air as it passes through the element, converting the velocity head to pressure, partially in the rotating element and partially in stationary diffusers or blades. Dynamic compressors compress the air by mechanical action of rapidly rotating impellers or bladed rotors that impart velocity and pressure to the flowing air. The velocity pressure is partially converted into static pressure by stationary diffusers or blades. Dynamic compressor types include centrifugal compressors, axial compressors and mixed flow compressors. Mixed flow compressors are dynamic machines that combine the characteristics of centrifugal and axial flow types.

11.1.4 Factors Affecting Compressor Operation

Compressibility and the deviation in behavior from an ideal gas was discussed earlier. The generally accepted method for correcting air volumes and compressor horsepower is to use compressibility factors (Z). Compressibility factors for air may be obtained from a compressibility chart, such as that shown in Figure 11-1. An increase in compressibility between the compressor inlet and discharge will increase the compressor power per unit of air flow. A decrease in Z during compression will decrease power per unit of airflow.

Whenever an air compressor is installed at a location that is more than 1,000 feet above sea level, the effect of altitude must be taken into account. A compressor operating to provide 100 psig air at an altitude would have a capacity reduction of about 6% at 100 psig and a reduction in power consumption of 4% compared to the same compressor at sea level. The same compressor operating at 10,000 feet would have a capacity reduction of about 30% at 100 psig and power consumption reduction of about 20%. When evaluating the performance of compressors at altitudes other than sea level, the compressor manufacturer should be consulted for specific performance data.

Most components in compressors have physical limitations on how much they can raise the pressure of air at a given flow rate. Limitations vary with the type of compressor; however, the most significant limitations include discharge temperature, pressure increase (differential), compression ratio and energy consumption. In order to raise the pressure beyond that allowed by the compressor limitations, it is frequently necessary to increase the pressure of the air in stages, or steps. Compressing air in stages is referred to as multi-staging. Multi-staging may be used to save power and to limit pressure differential. Typically, intercooling is provided between stages, which reduces the final air discharge temperature at the final pressure.

All compressors use work resulting from rotational motion to compress air. The performance of compressors is directly related to their rotational speed. The affinity laws may be used to predict the performance of dynamic compressors at various rotational speeds. The affinity laws for dynamic compressors are:

$$Q_2/Q_1 = n_2/n_1 \qquad \text{Eq. (11.11)}$$

$$P_2/P_1 = (n_2/n_1)^2 \qquad \text{Eq. (11.12)}$$

$$HP_1/HP_2 = (n_1/n_1)^3 \qquad \text{Eq. (11.13)}$$

where n = rotational speed, RPM
Q = air flow rate, cfm or ft^3/min
P = pressure, psig or lb/ft^2 gage
HP = power, horsepower

The performance of dynamic compressors is also affected by the surge limit of the compressor. Surge occurs at low airflow and the compressor operation becomes unstable. Below the surge point, or capacity, the discharge pressure exceeds that produced by the machine and the flow reverses direction momentarily. The flow reversal is momentary since the discharge pressure immediately drops, at which time the compressor resumes normal operation. Since the flow direction alternates, the compressor tends to surge. No compressor should be operated at a capacity close to the surge limit.

Atmospheric air contains small amounts of moisture in the form of vapor. The amount of moisture in air is frequently expressed as relative humidity. It may also be expressed as a humidity ratio, which is the ratio of the weight of the moisture to the weight of the air expressed in pounds of moisture per pound of dry air. Moisture in the air does not affect the polytropic head; however, it does affect the discharge pressure of the air. An increase in moisture content of the air will reduce the compression ratio and also the final pressure for a given polytropic head.

As air is compressed, its temperature increases. When multi-stage compressors are used, the opportunity exists for cooling of the air between the stages. Cooling of the air during compression reduces the power by lowering the average temperature within the compressor, which increases the average density of the air and decreases the polytropic head. Cooling of the air between stages is discussed later in this chapter.

Stonewalling is a phenomenon that can occur in dynamic compressors. Stonewalling occurs when the airflow rate is so high that the air velocity approaches the sonic velocity of air. When the air reaches the sonic velocity, a shock wave develops, which restricts flow. The design of most compressors is such that the air velocity will never approach sonic velocity.

11.2 AIR COMPRESSOR TYPES AND CHARACTERISTICS

11.2.1 Reciprocating Air Compressors

Reciprocating type compressors are positive type compressors and are the largest type of compressor used to compress air. They come in the widest range of sizes. Reciprocating compressors are similar in design to an internal combustion engine, in that the compressor has pistons that move back and forth inside a cylinder to compress the air. As a result, reciprocating compressors deliver a pulsating flow of air.

There are two basic types of reciprocating air compressors: single acting and double acting. In single acting compressors, the piston performs compression on only one side during the power (compression) stroke. Double acting compressors are designed such that the piston can compress the air on either side. As a result, a double acting compressor provides a compression stroke as the piston moves in either direction.

Reciprocating compressors may also have single or multiple stages of compression. Multi-stage reciprocating compressors usually have two stages of compression in series. Single stage compressors are usually used to produce compressed air in the 25 to 100 psig range. Two stage reciprocating compressors produce compressed air in the 100 to 250 psig range. Three stage compressors that can produce air pressures up to 2,500 psig are also available.

Earlier in this chapter, the need for cooling between stages of compression was discussed. Cooling the air increases the density of the air, which reduces the power required for compression. Heat in the cylinders of reciprocating compressors result from the work of compression, as well as the friction of the piston and rings on the cylinder wall and the rod packing on the rod. Large amounts of heat can be generated when moderate and high compression ratios are involved. Reducing the temperature with cooling produces several benefits, including:

1. Reducing the cylinder wall and cylinder head temperatures reduces losses in capacity and horsepower per unit volume. Cooling will result in a greater weight of air in the cylinder to be compressed.
2. Reducing cylinder wall and cylinder head temperatures will remove more heat from air during compression, which will lower its final temperature, reducing the power required.
3. Reduced temperatures of compressor components result in longer service life for the components.

4. Reduced cylinder wall temperatures promote better lubrication, which reduces friction, reduces work lost to friction and results in longer service life of lubricated parts.

Cooling methods for reciprocating compressors include air cooled, water cooled and still cooled methods. Air cooled compressors are those with moderate and high compression ratios in small capacity compressors. Compression temperatures can be as high as 400°F. To facilitate cooling by transferring heat to ambient air, compressor cylinders usually have fins to increase heat transfer. Air cooled units are usually not recommended for continuous or heavy duty service applications.

Compressors used for continuous duty or heavy duty service are usually water cooled. Heat rejection rates for water cooled compressors range from 125 to 700 Btu/Bhp-hr. Inlet water temperatures should be about 15° to 20°F higher than the air inlet temperature to prevent moisture from condensing out of the air. Water outlet temperatures should not exceed 120°F. Equation (11.14) may be used to estimate the required cooling water flow rate.

$$Q_w = \frac{(q_w/Bhp)(Bhp)}{(500)\ (T)} \qquad \text{Eq. (11.14)}$$

where Q_w = water flow rate, gpm or gal/min
Bhp = brake horsepower input, HP
q_w/Bhp = heat rejection rate per Bhp, Btu/Bhp-hr
T = water temperature rise, °F

Most compressor manufactures recommend a 15° to 20°F temperature rise for normal operating conditions.

Reciprocating air compressors may also be still-cooled. Still-cooled cylinders have jackets that are filled with a non-recirculating liquid. The liquid equalizes the temperature around the cylinder bore instead of actually cooling the cylinder. Still-cooling is not used when cooling is critical to compressor operation.

The controlled variable for reciprocating air compressors is usually the discharge air pressure. The most common method of capacity control for reciprocating compressors is start-and-stop control. This control method operates the compressor at full capacity until the set point pressure is reached, at which time the compressor is shut off. An air receiver is required to maintain a constant pressure in the air distribution system. When the compressor starts, it is usually *unloaded*. Unloading consists of venting the cylinder discharge air to atmosphere, holding the inlet valves open or adding large clearance volume to the cylinder.

The capacity of reciprocating compressors may also be controlled while the compressor is operating at constant speed. Capacity control at constant speed may be accomplished by blocking the inlet, intake throttling, external bypassing and inlet valve unloading. Blocking the inlet consists of closing a valve in the compressor inlet line when the capacity is to be reduced and opening the valve for full delivery when the airflow is required. Intake throttling involves closing a valve on the compressor inlet in steps between full open and full closed. As the valve is opened and closed in response to the demand for compressed air, the throttling valve changes the density of the air as it enters the compressor, resulting in changes in the airflow rate. External bypassing is a capacity control method in which compressed air is piped back to the compressor inlet. The amount of air bypassed is inversely related to the demand for air. Inlet valve unloading is also known as free -air unloading and is the most common constant speed capacity control method. This capacity control method simply holds the inlet valves open during both suction and discharge strokes. Air is drawn into the cylinder on the intake stroke and pushed back out through the inlet valve on the compression stoke without compressing the air. All of the constant speed control methods control the capacity of the compressor, with it operating at full speed, fully loaded part of the time, and fully or partially unloaded at other times.

Reciprocating air compressors may also have variable speed control. This method is suitable whenever the compressor driver is capable of operating at different speeds in response to the demand for air.

11.2.2 Rotary Air Compressors

Rotary air compressors are also positive displacement machines. The types of rotary air compressors in common use include rotary vane and rotating screw. Rotary type compressors are classed as constant-volume variable pressure units.

A rotary sliding vane compressor consists of a cylindrical housing with heads and a rotor assembly with vanes that slide in and out as the rotor turns. The rotational axis of the rotor is not centered within the housing so that the vanes slide in and out as the rotor rotates. The vanes trap pockets of air between the rotor and the cylindrical housing. As the rotor turns and the vanes slide into it, the pocket volume decreases and the air is compressed.

Rotating sliding vane compressors do not have intake and discharge valves. Airflow into and out of the compressor is controlled by the locations of intake and discharge ports in the compressor housing. As air is compressed in a sliding vane compressor, the compression of the air continues until a discharge port is uncovered by the leading vane of each pocket. The location of the discharge port, and therefore the discharge pressure, is factory set by the compressor manufacturer.

Sliding vane compressors typically operate at a constant rotational speed. Capacity control is accomplished by blocking the inlet, intake throttling and by external bypass. These control methods are similar to the capacity control methods described for reciprocating compressors in the preceding section of this chapter. Variable speed control is also possible with certain types of compressor drives.

A rotating screw compressor consists of a pair of rotors that mesh together in a dual bore housing. A main, or driving rotor, meshes with the helical grooves of a secondary, or driven, rotor. The main rotor is driven by the compressor drive while the secondary rotor is driven by the main rotor without metal-to-metal contact. An oil film provided by oil flooding produces a seal between the rotors. In some compressors, the secondary rotor may be driven by synchronous gears.

In a screw compressor, there is no compression or reduction of air volume while the rotors are turning. The rotors just move the air from the inlet to the outlet. The air is compressed by the backflow of air into the housing from the discharge line at the time the discharge port is uncovered. The air is then displaced into the discharge system as the rotors continue to rotate. There are no inlet and discharge valves. Like rotating vane compressors, airflow into and out of the compressor is controlled by the locations of intake and discharge ports in the compressor housing.

Both sliding vane compressors and screw compressors are available in single and multi-stage units. Multi-stage compressors are usually available with intercoolers between the stages.

11.2.3 Dynamic Air Compressors

A dynamic compressor is a rotary continuous airflow unit in which the mechanical action of rotating vanes, or impellers, imparts velocity and pressure to the flowing air. Dynamic compressors are available in two common types: centrifugal compressors and axial compressors.

Centrifugal air compressors are machines in which air is compressed by the mechanical action of rotating impellers, which impart a velocity to the air by centrifugal force within a confined housing. The impeller is mounted on a shaft that is positioned within a housing. A diffuser surrounds the impeller and acts to convert the centrifugal kinetic energy to the high velocity air to potential energy (pressure) at a higher pressure level. A volute further reduces the air velocity and converts more kinetic energy to pressure.

Centrifugal air compressors are large airflow capacity machines. They typically provide airflow rates between 1,000 and 30,000 cfm at pressures in the 100–125 psig range. Centrifugal compressors operate at high rotational speeds and produce a smooth, non-pulsating flow of compressed air.

Centrifugal compressors are available in single or multiple stages. Cooling may be provided between stages. The cooling medium used for centrifugal compressors is usually air or water. Air cooling is usually limited to smaller capacity compressors. Water cooling may be accomplished with external heat exchangers. Water cooling is more effective because of ease of control and a greater specific heat (heat absorbing capacity per unit volume). Water cooling is usually used for continuous duty compressors. Water cooling may also be accomplished by injection cooling, which involves spraying controlled quantities of water into the air as it passes between stages of compression. Water injection provides both sensible cooling and latent cooling of the air.

Centrifugal compressors must be controlled so the output of the compressor approximates the demand for compressed air. During periods of low demand for compressed air, care must be taken to ensure that the airflow rate does not drop below the surge limit of the compressor. Centrifugal compressors can operate down to flow rates of 40% of peak flow. Surge control is typically provided by having a minimum flow regulator to throttle the flow of air. Throttling is usually accomplished with an inlet throttling valve or inlet guide vanes.

General capacity control is provided by three methods: variable speed, blow-off of excess output and throttling the intake. These controls may be used individually for compressor control, or they may be used in combination.

Variable speed control is the simplest way to control the compressor output. Variable speed motor drives are available to control both airflow rate and discharge air pressure. By modulating the rotational speed of the compressor, the output can be matched to the air system demand while maintaining system pressure at a virtually constant level. When using variable speed control, care must be taken to avoid operation at any critical speeds of the compressor.

Blow-off control simply involves discharging excess capacity to the surroundings or recirculating the air through the compressor. Recirculating involves throttling excess air back to the compressor intake at reduced pressure and putting it through the compressor again. Recirculated air must be cooled to prevent excessive intake and discharge air temperatures. There is no power consumption savings by discharging or recirculating excess capacity since the compressor is operating at full capacity even when there is a reduced demand for air.

Intake throttling will reduce the discharge pressure for a given volume or a reduced volume for a given discharge pressure. Intake throttling control may be used with the compressor operating at a constant speed or at variable speeds. Intake throttling is accomplished with inlet guide vanes. The inlet guide vanes restrict the flow of air into a centrifugal compressor and also provide a pre-rotation swirl to the air in the same direction as the impeller rotation. By providing a swirling action to the air as it enters the compressor, the performance characteristics of the first impeller is altered, which lowers the surge limit. Compressor power consumption is reduced by using inlet guide vanes for control, however, the savings are not as great as variable speed control.

Axial compressors are also dynamic compressors. Unlike centrifugal compressors in which the air is compressed by centrifugal force as it flows radially outward from the impeller, the flow of air in an axial compressor is parallel to the axis of rotation. An axial compressor consists of multiple sets of rotating blades (rotor blades), which are separated by stationary blades (stator blades). As the rotor blades rotate, air is forced through the compressor in an axial direction. The stator blades then convert the kinetic energy (velocity) of the air to potential energy (pressure). The stator blades also deliver the air to the next set (stage) of blades at the proper angle for the next stage of compression.

Axial compressors can produce very high rates of airflow at very high pressures. The head-capacity curve of an axial compressor is much steeper than that of a centrifugal compressor, and the operating range between normal flow and surge is much narrower. The efficiency of an axial compressor is usually better than an equivalent centrifugal compressor, depending upon the size of the compressor and operating conditions. Operating speeds for axial compressors is considerably higher than for centrifugal machines. Axial compressors produce air that is free of pulsations.

11.3 AIR COMPRESSOR DRIVES

Air compressor drives are the devices that provide power to operate the machine and to compress the air. A drive includes the connecting medium between the power producing device and the compressor, such as gears, couplings, sheaves and belts. In addition to driving the compressor at the required rotational speed, the drive must also start the compressor from rest, accelerate it to the required rotational speed and keep the unit operating under varying load conditions. Sources of power to drive air compressors include electric motors, steam driven engines and turbines, and fuel powered engines.

11.3.1 Electric Motors

Electric motors are the most commonly used device used to drive air compressors. There are a number of different types of electric motors for driving air compressors, including squirrel-cage induction, wound rotor induction motors and synchronous motors. Motors can be in an open drip-proof (ODD) housing or in a totally enclosed fan cooled (TEFC) housing.

Squirrel-cage induction motors are the most widely used motor for compressor drives, especially under 200 HP. They are the most economical in terms of cost per horsepower. Above 1,000 or 1,250 HP electric motors become less cost-effective.

The two major components are the stator (the stationary magnets and coils) and the rotor. The stator is a laminated iron core within and around the inside of the motor. The rotor is a laminated iron core with bar windings in various shaped slots around the periphery of the core. The rotor looks like a squirrel cage, hence the name squirrel-cage motor.

The stator windings of a squirrel-cage motor are arranged such that the stator will alternate between positive and negative as the current flow alternates. This creates a pulsing magnetic flux. The pulsations of the magnetic flux produce a rotating magnetic field. As the flux from the stator rotates, it cuts the bars or coils of the rotor, which induces a voltage within the rotor circuits. A current is generated within the rotor bars, which causes a second magnetic field. The interaction of the two magnetic fields produces the torque that accelerates the rotor.

The speed of a squirrel cage motor is constant and is determined by the number of poles that are active. On a two speed motor, two wind motor, the two windings determine the number of active poles. The windings are externally switched, which results in speed combinations such as 1,800/1,200, 1,800/900, 1,800/600 rpm. These motors are available in a number of speeds, but they are not variable speeds. Characteristics of squirrel-cage induction motors are:

- lagging power factor
- multi-speed motors
- high efficiency over a wide load range
- rugged and relatively inexpensive
- wide range of starting torque

Wound rotor induction motors provide higher torque during starting and a range of adjustable speed operations. The speed of a wound rotor motor is controlled by varying the current supplied to the rotor. The current is varied with external resistors. They may be used where speed control is required. Applications include large fans and centrifugal compressors. Characteristics of wound rotor motors include:

- lagging power factor
- speed control by varying external resistance
- relatively high maintenance
- relatively high initial cost
- high starting torque and low starting current

Synchronous motors are designed to operate at synchronous speeds only; they are constant speed motors only. Their speed is unaffected by changes in the supply voltage or by the load on the motor. Since they operate at synchronous speeds, an external DC exciter is required for the motor starting and running torque. A DC generator is used to provide the required DC power. Synchronous motors have a limited number of applications, and are used only for large loads. When the motor comes up to operating speed, it runs with a slip of just 2–3%. Large horsepower synchronous motors operating at low speeds are relatively low in cost. Some of the characteristics of synchronous motors include:

- low starting torque
- constant speed within the motors load limits
- nearly constant load requirements
- a possible leading power factor
- high efficiency

Electric motors for air compressors should not be selected for less than the full-load continuous compressor operation. The fact that the compressor may be unloaded part of the time should not be considered.

Electric motors, "motor accessories and electrical systems are discussed in greater detail in Chapter 10: Electrical and Lighting Systems".

11.3.2 Steam Driven Engines and Turbines

When steam generated for other uses is available at a suitable pressure and quantity, it may be used to drive air compressors. For steam to be suitable for driving air compressors, it must be available in sufficient quantities, available when compressed air is needed, and available in the range of 250 psig and 500°F. Steam drivers for air compressors may be steam engines or steam turbines.

Steam engines are typically used for reciprocating air compressors. Steam engines are best suited for applications in which the steam engine satisfies a plant heat balance need and/or the flexibility of variable speed control is a decided process advantage. Steam engine driven compressors are of the continuous heavy duty type. The engine is usually built integral with the compressor, but some are coupled. There are three basic arrangements for steam engine compressor drivers:

1. Horizontal straight line design with a single steam cylinder. These engines are available up to about 150 hp.
2. Horizontal four-corner design with two steam cylinders, with the two steam cylinders at one end and two compression cylinders at the other end. These engines are available in a wide range of sizes between 150 and 1,500 hp.
3. An *L* arrangement with vertical and horizontal compression cylinders. These engines are available from 900 to 4,000 hp depending upon the number of cylinders.

Steam turbines may also be used to drive large capacity air compressors. They are suitable as drivers for both positive displacement, as well as dynamic compressors. The rotational speed of stream turbines can match the speeds of centrifugal and axial compressors, which makes them suitable for direct drive applications.

There are two general classifications for stream turbines used as compressor drives: impulse type and reaction design. In the impulse type, the steam is fully expanded in the nozzles or stationary blades. In the reaction design, there is a pressure drop throughout both fixed and rotating vanes.

Control of steam turbines is provided by one, or more, control valves on the steam entering the turbine. The valves modulate the flow of steam to control the rotational speed of the turbine.

Steam systems and energy management for steam systems are discussed in "Chapter 5: Steam Systems".

11.3.3 Fuel Powered Engines

Fuel powered engines and turbines may also be used to drive large capacity air compressors, such as centrifugal and axial air compressors. These engines and turbines may use gasoline, fuel oil, kerosene and other fuels.

Turbine engines are suitable for high speed, large capacity air compressors. They are most suitable for driving certain helical-lobe compressors and axial flow compressors. Turbine engines are essentially constant speed drivers. The efficiency of turbines falls off rapidly when the speed is reduced. Single shaft turbines have an efficiency of around 10%. Most turbine engines use gasoline or kerosene for fuel. Gas turbines are lightweight drivers, capable of rapid start-up and loading with no standby losses. Gas turbine engines require little, or no, water cooling.

Fuel powered engines may also be used as drivers for large volume air compressors. Most engines are gasoline powered, 4 stroke or 2 stroke, heavy duty internal combustion engines. At least one manufacturer builds a gas-fired naturally aspirating rotary engine. Gas powered engines may also be supercharged. Supercharging increases the density of the air as it enters the engine, which improves the thermal efficiency and produces a greater output per unit space and per unit weight. Gas powered engines are variable speed drivers that offer considerable flexibility in compressor capacity control. The thermal efficiency of naturally aspirated engines ranges from about 28–32%. Supercharged engines can have thermal efficiencies in the range of 34–40%. Water cooling is usually required for gas engine drivers. These machines reject heat at the rate of approximately 2,000 Bhp for naturally aspirated engines and around 1,200 Bhp for supercharged engines.

Fuel powered engines and turbines can be suitable drivers for air compressors under some circumstances, including:

1. Availability of relatively inexpensive fuel.
2. High efficiency of prime mover resulting from high load factors and low efficiencies of other types of drivers.
3. Electric power source that is very expensive, limited or undependable.

These types of situations may exist at facilities such as refineries and petrochemical plants.

Although there are a number of energy saving opportunities for fuel fired engines and turbines, a discussion of such opportunities is beyond the scope of this publication. The reader is referred to specialty publications dealing with engines and turbines.

11.4 ACCESSORIES FOR AIR COMPRESSORS AND COMPRESSED AIR SYSTEMS

The purpose of an air compressor is to take air at atmospheric pressure, or some reduced pressure, and increase the pressure of the air to a higher pressure. The compressor must also provide the higher pressure air in the needed quantity based upon the demand. Air compressors, for the most part, change only the pressure of the air and perhaps increase the leaving temperature of the air as a result of the compression process. After the air is compressed, it is usually necessary to condition the air before it is suitable for the intended use. This section discusses some of compressed air system accessories used to condition compressed air.

11.4.1 Air Dryers

Earlier it was indicated that atmospheric air contains small amounts of moisture in the form of vapor. Air entering a compressor carries with it a certain amount of moisture depending upon ambient air conditions. As long as the moisture remains in the form of vapor, it causes little problems in the compressed air system. Since the air discharged from a compressor has been heated by the compression process, it is held in suspension in vapor form. However, as the air is distributed throughout the air distribution system, it is naturally cooled to ambient temperatures. As the air is cooled in the air distribution system, water vapor will condense out of the compressed air unless steps are taken to remove the moisture. Problems associated with excessive moisture condensation in a compressed air system include:

- washing away of required lubrication
- increased wear and maintenance
- sluggish and inconsistent operation of automatic air valves and cylinders
- malfunctioning and increased maintenance of control instruments
- spoilage and damage to products and processes
- rusting of parts exposed to compressed air
- freezing of air lines exposed to outdoor temperatures
- loss of efficiency of air tools

When air is discharged from an air compressor, an aftercooler may be used to cool the air and remove moisture to some extent. Aftercoolers, however, may not remove sufficient moisture from compressed air to avoid many of the problems listed above.

Frequently, it is necessary to remove moisture from compressed air to a greater extent than can be accomplished by an aftercooler. In these applications, it is necessary to remove moisture with a device called an air dryer, or just dryer. The amount of moisture that can be removed as a vapor and remain mixed with air is a function of the temperature of the mixture and the volume. Pressure is not a factor.

There are three basic methods by which moisture is removed from compressed air: chemical drying, adsorption and refrigeration. In all methods, aftercooling of the air and adequate condensate removal must be provided. The primary function of a dryer is to remove moisture; however, it may also remove entrained lubricating oil from the air to some extent.

Chemical type dryers use chemical materials that combine with or absorb moisture from the air when the material is brought into close contact with the air. Chemical dryers are available in three different types. One type uses deliquescent materials in the form of pellets or beads. Deliquescent chemicals can obtain a dew point depression of 20–35°F below the air temperature entering the compressor. The chemical dissolves as it absorbs the moisture, therefore it must be replaced periodically.

Another type of chemical dryer uses a salt impregnated matrix. When the warm water vapor in the compressed air is exposed to the salt, the salt forms hydrated crystals. The hydrated salt crystals can be returned to their original form by exposure to hot air. Depending upon the type of salt used for drying, these types of chemical air dryers can achieve dew point depressions of 30–50°F.

A third type of chemical air dryer uses an ethylene glycol liquid to absorb the moisture from the compressed air. The diluted glycol may have the absorbed moisture removed (regenerated) by heating it to drive off the moisture. These types of chemical air dryers can achieve dew point depressions of up to 40°F.

Adsorption is another method by which compressed air may be dried. It is a process by which the moist air is passed through an extremely porous material called a desiccant. The porous material holds the moisture in its pores, thereby removing it from the air. The desiccant can be regenerated by passing dry hot air through the material. The hot air used to regenerate the adsorption material is vented to the atmosphere after it has removed the moisture from the drying material. Adsorption dryers can achieve dew points as low as -100°F.

Refrigerated air dryers are the most common type of air compressed air dryer. Refrigerated air dryers operate on the principle of cooling the air to a temperature below its dew point, which causes moisture to condense out of the air. These dryers usually employ a direct expansion refrigeration system to cool the compressed air. Refrigerated air dryers can cool the air to as low as 35°F. Lower temperatures would subject the condensed moisture to freezing temperatures. After the air has been cooled, it is frequently necessary to reheat it. Reheating the air has several advantages, which include:

- The air volume is increased, therefore less free air is required.
- The chance of air line condensation is reduced.
- Sweating of the cold air pipe leaving the dryer is eliminated.

Refrigerated air dryers also remove some lubricating oil from the compressed air.

11.4.2 Air Filters and Separators

Atmospheric air is drawn into air compressors for compression. Even the cleanest atmospheric air includes significant amounts of dirt and other

TABLE 11.1 Air Filter Performance

Filter Type	Filter Action Efficiency (%)	Particle Size (Microns)
Dry	100	10
	99	5
	98	3
Viscous Impingement	95	20
	85	10
Oil Bath	98	10
	90	3

impurities. Unless the impurities contained in atmospheric air are removed, significant problems in the compressed air system may result. Impurities in the air may cause unnecessary wear on the compressor and other components. Filters are usually installed on the intake of compressors to remove impurities before they enter the compressed air system. There are three basic types of air filters used in compressed air systems: dry, viscous impingement and oil bath.

Dry filters contain a dry filtering media made of either felted cloth or a special lightly oil coated paper. Dry type filters simply strain contaminants from the air as it passes through the filter media. The densely spaced media blocks the passage of the particulate matter.

Viscous impingement filters have filter media in cells or frames. The filter media may be woven fibrous material or packed wire. The filter media is coated with a liquid, such as oil, which causes the particulate matter to adhere. These types of filters are not suitable for extremely dusty areas or for non-lubricated air compressors.

Oil bath filters remove particulate mater from the air by a scrubbing action. These filters draw unfiltered air through an oil reservoir where the particulate matter adheres to the oil. The air then passes through a screen mesh, or other media, which removes the oil and dust particles. The dust is then washed down and collected in a sump where it is manually or automatically removed as sludge.

Table 11.1 provides a summary of the performance of the various types of filters.

Filters are used to remove solid particulate matter from air. Separators are used to remove liquids from air. Separators are typically used to remove liquid water and oil that have become entrained in the compressed air. Although separators cannot remove as much moisture as an air dryer, they can remove significant amounts of liquid. They can be installed any place in the compressed air system where it is desirable to remove entrained liquids.

The most common types of separators are those which separate liquids from the air by rapidly changing the direction of the air as it passes through the separator. The liquids are separated by centrifugal force as the air changes direction. Separator types include impingement, centrifugal and cyclone. All separators should be equipped with traps or drains to remove the liquids from the separator.

11.4.3 Air Receivers

Air receivers are large volume storage tanks that store compressed air to supplement compressor output when temporary demand for air is greater than the compressor output capacity. A receiver acts as a temporary reservoir to prevent excessive temporary pressure drops during sudden short-term, high-demand periods. Receivers are typically located immediately downstream of the aftercooler.

Receivers should always be provided for applications using reciprocating air compressors. Receivers are also recommended for other types of positive displacement compressors. The receiver tends to dampen pulsations from positive displacement compressors, especially reciprocating compressors.

A receiver volume is typically selected to provide at least 12 seconds of air demand for small compressors and at least 8 seconds of demand for large compressors.

11.4.4 Aftercoolers

Aftercoolers condense and remove moisture from the compressed air by cooling it after it is discharged from the compressor. Aftercoolers perform the same function as an air dryer (remove moisture from compressed air), however, they are used to pre-dry the air before it enters an air dryer. They can economically remove large amounts of moisture and lower the dew point temperature of the compressed air to near that of the cooling medium. Air leaving the aftercooler is near saturation conditions at the aftercooler discharge temperature. The cooling medium used for aftercoolers is usually water; however, it may be ambient air.

Aftercoolers that use ambient air as the cooling medium are occasionally called dry coolers. Dry coolers have a much lower heat transfer rate than aftercoolers using water as the cooling medium; therefore, they are suitable only for use with smaller capacity air compressors.

Water cooled aftercoolers are the most common type of aftercooler for compressed air systems. There are two basic designs for water cooled aftercoolers: shell-and-tube design and water tube design. Shell-and-tube aftercoolers cool the compressed air as it flows through tubes surrounded by the cooling water. In water tube aftercoolers, the cooling water flows through the tubes, which are surrounded by the compressed air. Both designs are effective heat exchangers; however, the water tube design may be preferable if the water is dirty or from an open cooling tower due to ease of tube cleaning.

11.5 MAINTAINING AND OPERATING COMPRESSED AIR SYSTEMS FOR OPTIMUM ENERGY EFFICIENCY

For air compressors and compressed air systems to operate at peak efficiency and minimum energy consumption, it is necessary that they receive regular maintenance and service. Proper routine maintenance should be the first step in any energy management program for compressed air systems. There are few utility systems for which maintenance can have such a significant effect on energy consumption as for compressed air systems. Annual maintenance costs for compressed air systems can be expected to be in the range of 5% of the initial capital cost of the system. These maintenance costs are also typically less than 20% of the annual cost to operate the compressed air system. The following are maintenance procedures that will result in optimum energy efficiency for most compressed air systems.

Test for Standard Performance: When the compressed air system is in good repair and operating properly, tests of the system should be performed to establish a basis for standard performance. Ideally, these tests should be performed when the system is new and in peak operating condition. These tests should be repeated periodically to determine the system performance at that time. Any change in performance that will affect the energy consumption of the system will become evident when the tests are performed. System tests may include the following:

- Compressor tests: Operate each compressor at full load while measuring the airflow rate, discharge air pressure, discharge air temperature and power input to the compressor drive.
- Leakage tests: With all compressed air loads off (typically at night or on weekends), these test measure the flow of compressed air over a period of several hours. With all loads off, this will give a good estimate of the amount of leakage from the system.
- Measure pressure differential across filters: High pressure loss through filters indicate dirty clogged air filters.
- Measure pressure at various locations in the distribution system: When new loads are added to the system, air consumption may increase significantly. Low pressure at remote locations in the system may indicate undersized piping or restrictions in the distribution system, such as partially closed valves.

Maintain Accurate Records: At a minimum, the result of periodic performance tests should be recorded. Other records that should be maintained include dates of compressor oil changes, dates of filter replacement and/or servicing, and any additions and/or modifications to the compressed air system.

Perform Routine Checks and Inspections: A program should be established to check the performance of an air compressor, accessories and the system on a regular basis. Some of the checklist items include:

- Valves: Remove and inspect at the end of the first three months of service. Afterward, valves should be inspected and cleaned every six months.
- Filters: Inspect air intake filters for cleanliness and pressure drop at least once a week
- Receiver and intercooler: In humid weather, receivers and intercoolers should be drained daily and, in less humid conditions, weekly.
- Oil level: The compressor oil level should be checked and oil added as required at least once a week.
- Belts and drives: Check belts and drives for tightness at least every two to three months.
- Air leakage: Check for air leaks at least monthly.
- Shaft seals: Check for leakage at least once a week.
- Compressor vanes: Check vanes on dynamic compressors for wear at least every 8,000 operating hours, or sooner if recommended by compressor manufacturer.
- Bearings: Inspect whenever major repairs requiring compressor shutdown and disassembly are made on compressor.
- Cooling water: Check water flow and temperatures weekly.
- Drive motor: Check motor current draw at least once a week.
- Controls: Check compressed air pressure controls for calibration at least once a month.

Review Usage of Compressed Air: Compressed air is an expensive form of energy when compared to other energy related utilities, such as electricity. For industrial tasks, electricity or hydraulic power can be more efficient and cost-effective than compressed air. Total energy consumption for a particular end use may drop by as much as 80% when converted to another utility.

Review Operating Schedule: Frequently, compressed air systems are operated 24 hours a day, even though there are no operations and no demand for compressed air. This is a very inefficient and costly practice. In multiple compressor operations, avoid operating more compressors than necessary for the actual demand. Automatic control systems can sequence compressors to match the actual load.

Reduce Compressed Air Pressure: In many compressed air systems, the supply pressure of the system is set at an arbitrarily high pressure and then reduced at the point of use. Although it is necessary to have a high enough pressure for the intended use and to offset pressure losses in the distribution system, the discharge pressure is frequently set higher than necessary. Energy can be saved by reducing the pressure to the minimum acceptable level. A 10% reduction in pressure will result in a reduction in energy consumption of about 4% of the energy input to the compressor driver. Figure 11-2 shows the approximate savings in reducing pressure for single stage compressores, and Figure 11-3 shows the approximate savings for two stage compressors.

Reduce Compressed Air Leakage: Air system leaks are unintended losses that can consume between 30 to 50% of compressor capacity. Leaks can occur anywhere between the compressor discharge and the point of use. Most compressed air systems have leaks that are at least 20% of compressor capacity. Sources of leaks include:

- threaded pipe joints
- valve stems
- filters
- connectors
- check valves
- flanged connections
- traps and drains
- air horses
- actuators on pneumatic controls
- relief valves

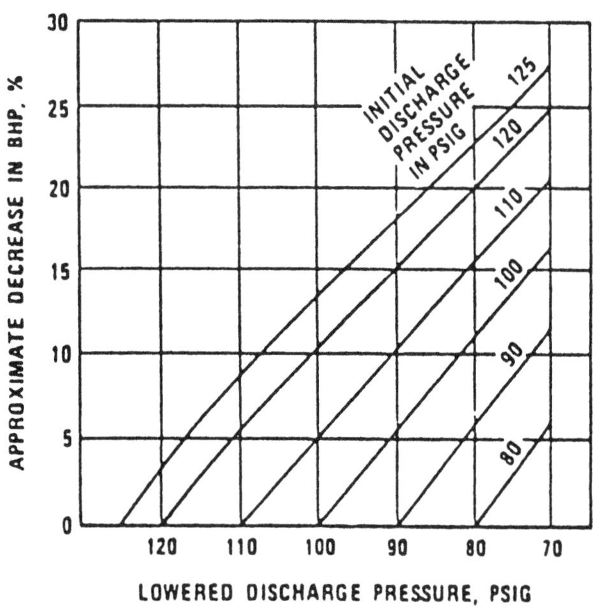

Figure 11-2. Reduced pressure savings for single stage reciprocating and rotary screw compressors.

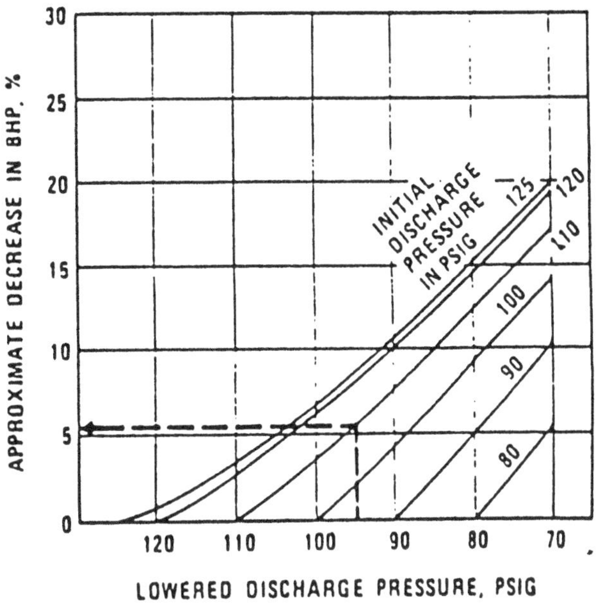

Figure 11-3. Reduced pressure savings for double stage reciprocating and rotary screw compressors.

Table 11.2 lists conservative estimates for various size holes in a system operating at 100 psig and the associated costs, assuming it costs \$0.40 to provide 1,000 ft^3 of air and assuming 4,000 operating hours per year.

The time to check for system leaks is during off-hours, such as at night or on weekends. During off-hours there is little, or no, demand for compressed air. If there is no demand, all compressed air produced is for leaks. During off-hours, it is also easier to hear the hissing sound caused by larger leaks. One way to check for an air leak is to simply brush soapy water onto a suspected area and to look for any air bubbles caused by a leak. This procedure, while simple and easy, can be very time-consuming and labor-intensive. Another method to check for air leaks is to use an acoustic

TABLE 11.2 Annual Cost of Air Leaks

Equivalent Hole Diameter (in)	Leakage Rate (scfm)	Air Loss ($ft^3 x 10^3 yr$)	Cost per year (\$)
1/64	0.25	60	24
1/32	0.99	238	95
1/16	3.69	950	380
1/8	15.86	3,806	1,522
1/4	63.44	15,226	6,090
3/8	142.74	34,258	13,703

detector. These detectors detect the high frequency hissing sound that may be emitted by compressed air leaks.

Provide Adequate Ventilation: Adequate ventilation for air compressors is very important. On a motor driven air cooled compressor, the heat radiated to the surrounding air can be equal to 65% of the electrical power input. On a water cooled compressor, the radiated heat can be equal to about 15 – 25% of the energy input.

Maintain Air Intake Filters: Even the cleanest atmospheric air contains large amounts of suspended contaminants. Air filters have one of the highest pressure drops of any component in the compressed air system, even when clean. As filters are used, they become clogged and the pressure drop increases. An added 1 psi drop in a 100 psig compressed air system would result in an approximately 10% increase in energy consumption for the system. Air filters require regular service and must be cleaned on a regular schedule. Dry filter media can usually be cleaned by carefully blowing debris off the filter surface or by washing with a mild solution of detergent and water.

Compressor Drives: Compressors with V-belt drives should have the belts replaced or adjusted periodically to ensure that power is properly transmitted from the drive motor to the compressor. If the belts are too tight, the compressor and motor bearings will be overloaded and cause the bearings to wear out prematurely. Worn bearings will increase friction and waste power. Loose belts will slip, waste power, overheat and quickly wear out. Belts should be checked for proper tension at least every 500 hours of operation.

Lubrication: Air compressors require lubricating oil for both friction reduction and to provide a seal between moving parts. As oil is used in a compressor, it becomes contaminated or corrosive, which increases friction and power consumption. A lack of proper lubrication will eventually result in serious damage to a compressor. To maintain optimum performance of an air compressor, lubricating oil must be changed on a regular basis and the oil level must be maintained between oil changes.

Traps and Drains: Traps are devices that permit liquids to pass but do not allow gases, such as compressed air, to pass. Occasionally, a trap may stick or fail in the open position. If they stick of fail in the open position, traps will pass significant amounts of compressed air. If they stick or fail in the closed position, liquids will not pass and will build up in the system, which can reduce system performance. Clogged drains should be inspected and repaired on a regular basis to ensure proper operation. Air receivers and aftercoolers should be blown-down and drained on a regular basis.

Example 11.1 A two stage reciprocating air compressor with a 100 HP electric motor provides 110 psig compressed air. The compressor operates for 6,240 hours per year. If the pressure of the system was lowered from 110 psig to 95 psig, how much energy would be saved?

Referring to Figure 11-3, the intial pressure (110 psig) is located at the bottom of Figure 11-3. A line is drawn vertically until it reaches the reduced pressure (95 psig). A second line is drawn horizontally to the left axis where a approximate reduction in brake horsepower of 5.4% is read. The savings in brake horsepower is then

$$\text{Power Savings} = (5.4\%)(100\ HP) = 5.4\ HP$$

The annual savings in electirical power then is

$$\begin{aligned}\text{Annual Power Savings} &= (5.4\ HP)(0.746\ kW/HP)(6240\ hrs/yr) \\ &= 25,137\ kWh/yr\end{aligned}$$

11.6 ENERGY CONSERVATION MEASURES FOR COMPRESSED AIR SYSTEMS

The energy conservation measures for compressed air systems discussed in the previous section were operation and maintenance procedures that would improve the energy efficiency of existing compressed air systems. Most of those measures were little, or no, cost measures that could be performed on an ongoing basis to improve efficiency.

In addition to the low cost O&M procedures, there are also modifications that could be made to compressed air systems to improve energy efficiency. These modifications, however, require a more significant cost investment and are permanent improvements. The modifications listed below are improvements that should be considered under any energy management program for existing, or proposed, compressed air systems.

11.6.1 Reduce Usage of Compressed Air

Compressed air is used widely in industrial and manufacturing facilities. However, for many applications, electric or hydraulic systems can perform the same task at a lower energy cost and a lower capital cost. In existing facilities, this will usually require some capital cost to modify, or extend, other utility systems to replace the use of compressed air.

11.6.2 Automatic Controls

For many compressed air systems, the output of the compressor is controlled by pneumatic pressure switches, which measure the discharge air pressure. For single compressor applications, this method has been reasonably effective. In multiple compressor applications, however, pressure switches have some serious limitations. For pressure switches to function, it may be necessary to have a total system pressure differential (control range) of 15–20 psi between the start and stop pressure. There is also a practical limitation on the number of compressors that may be sequenced using pressure switch control.

Microprocessor based electronic controls can perform the necessary logic to control multiple compressors. Compressor sequence control can include the integration of time, demand and compressor operating characteristics into a complex program that will provide the most efficient operation. Control systems can range from simple sequencers to complex operating and control programs. These control systems can also control other system components, such as air dryers.

Microprocessor controls can also be programmed such that high capacity, high efficiency compressors are used to provide the base compressed air load, while other, less efficient units are brought on line only when demand requires greater capacity.

The air pressure drop through intake air filters can also be monitored with microprocessor control systems. The control system can alert the appropriate personnel when filters need servicing or replacement.

It is also possible to save energy through proper timing of the purge cycles of desiccant air dryers. Microprocessor control systems can control the timing to optimize energy consumption by basing the purge cycles on actual air consumption, rather than arbitrary time periods.

11.6.3 Replace Existing Drive Motors with High Efficiency Motors

In installations with existing air compressors, older compressors were probably purchased with standard efficiency electric motors. Since the average operating time of most compressors is around 4,000 hours per year, compressor drive motors consume significant amounts of electrical energy.

Since compressor drive motors consume large amounts of electrical energy, significant energy savings can be achieved by replacing standard efficiency motors with high efficiency motors. Table 11.3 lists typical efficiencies of standard motors and high efficiency motors for various sizes.

Replacing large horsepower motors can be a large capital expenditure. A life cycle cost analysis should be performed before replacing a motor. Whenever a motor burns out and needs replacement, a high efficiency replacement motor should be installed. The cost premium is

TABLE 11.3 Typical Efficiencies for Electric Motors

Motor HP	Standard Efficiency	High Efficiency
5	82.5%	89.5%
10	84.0%	91.0%
50	89.5%	94.1%
100	91.0%	95.1%
200	91.7%	96.2%

usually quite small to upgrade from a standard motor to a high efficiency motor.

11.6.4 Install Heat Recovery

Air compressors naturally generate large amounts of heat as they compress air. Heat is a natural by-product of the compression process. Adiabatic compression of air to 100 psig results in air temperatures between 350 and 500°F. Multi-stage compressors usually have intercoolers between stages as indicated earlier in this chapter. When compressed air is cooled to ambient temperatures, 60–90% of the energy of compression may be removed and used for other purposes. Since approximately 80% of the energy used to compress air is available as heat, the compressed air is almost like a by-product. Typical uses for recovered heat include (reheat of compressed air after a refrigerated air dryer) regeneration heat for air dryers, space heating, domestic water heating, boiler make-up water heating, boiler combustion air heating and process heating. Refer to Chapter 14 for additional information on heat recovery.

11.6.5 Relocate Compressor Air Intake

Air compressors operate most efficiently when they have unrestricted access to intake air and when the air is cool. Frequently, the air taken in by an air compressor is from the equipment room, which houses the air compressor, as well as other heat generating equipment. If the room is completely enclosed by walls, the airflow to the compressor may be restricted. Taking air from the enclosure also results in warm air being taken into the compressor.

Air intakes for compressors should be located such that air is taken directly from outdoors. This eliminates any unnecessary restrictions to free air intake. The air taken is also cooler. Cooler air is more dense than warmer air, resulting in a higher mass flow rate and less energy consumed in providing air at a given pressure and flow rate. Table 11.4 gives the energy savings potential for providing lower intake air temperatures. The feasibility of extending indoor air intakes to the outside should be investigated whenever possible.

Example 11.2 An air compressor with a 150 HP electric motor draws air from an equipment room where the average temperature is 80°F. The compressor operates and supplies 1,000 cfm of compressed free air at 70°F for 6,240 hours per year. If the compressor air intake was relocated so air is taken from outdoors where the average air temperature is 50°F, how much energy would be saved?

From Table 11.4, the volume of air at 80°F, which must be taken in to provide 1,000 cfm of free air at 70°F is 1,020 cfm. The volume of air at 50°F

TABLE 11.4 Energy Savings Potential in Providing Lower Temperature Intake Air

Temperature of Air Intake, °F	Intake Volume Required to Deliver 1,000 cu ft free air at 70°F	Percent HP Saving or Increase Relative to 70° Intake
30	925	7.5 Saving
40	943	5.7 Saving
50	962	3.8 Saving
60	981	1.9 Saving
70	1,000	0
80	1,020	1.9 Increase
90	1,040	3.8 Increase
100	1,060	5.7 Increase
110	1,080	7.6 Increase
120	1,100	9.5 Increase

is 962 cfm. Since the compressor would be compressing a smaller volume of air at 50°F, the percent savings in horsepower would be:

$$\Delta\%\ HP = 0.057 = 5.7\%\ \text{savings in horsepower}$$

The compressor operates for 6,240 hours per year and consumes energy at a rate that is 5.7% less. The annual savings in electrical power to drive the compressor would be:

$$\begin{aligned}\text{Annual power savings} &= (0.057)(150\ HP)(0.746\ kW/HP)(6,240\ hr/yr) \\ &= 39,800\ kWh/yr.\end{aligned}$$

11.6.6 Modify Air Piping Distribution System

When compressed air systems are designed and installed, they are usually sized and routed for the present demand plus, perhaps, some future growth in demand. After several years of service, the demand has usually been increased and the location of air loads has changed. As a result, the distribution systems frequently become inadequate for the compressed air requirements.

Air distribution systems that are more than several years old should be analyzed to determine actual airflow requirements and modified or improved as required. Main air headers should be significantly oversized, not just to reduce pressure drop, but also to provide a better system air storage capacity for more economic control of compressor operation. Whenever possible, a looped distribution system should be installed. This allows the system to provide flow to a concentrated load from two directions Figure 11-4 shows the various arrangements and ranks them in terms of energy efficiency.

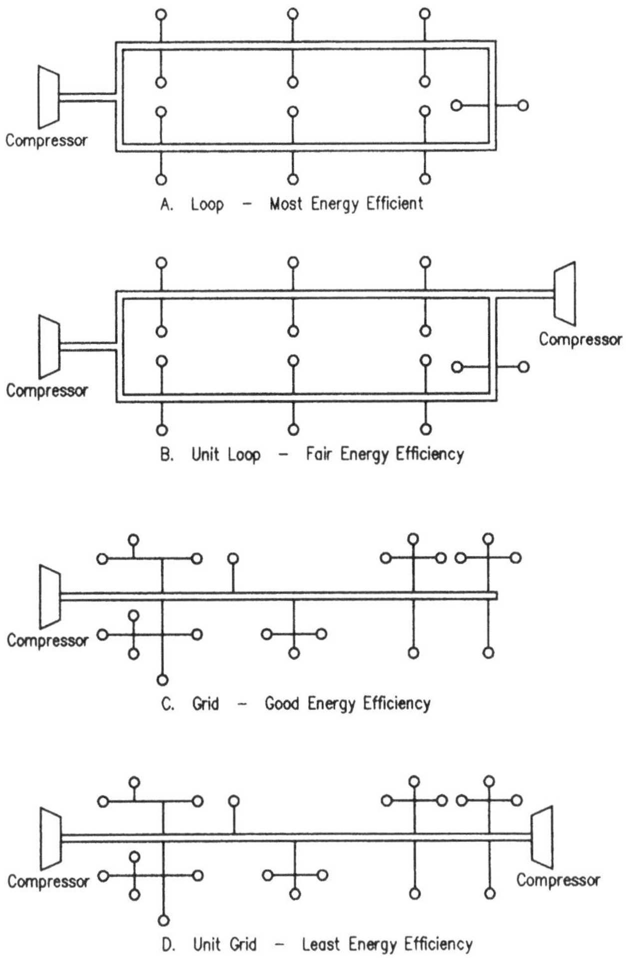

Figure 11-4. Piping distribution system efficiency.

11.6.7 Install Additional Air Receivers

Occasionally, compressed air systems have compressed air loads that require large amounts of air for a brief period of time, and are located some distance from the air compressor and main receiver tank. A common mistake is to add additional air compressors to the system to handle the brief peak demand. This scenario wastes energy for two reasons. First, the additional compressor must be operating to meet the peak load which consumes additional energy when it is operating. Second, the sudden high demand for air may result in air passing through the air dryer too quickly to be properly dried.

A more efficient, and least costly, way to handle a short-term peak demand for air is to install additional air receivers in the air system near loads that require large amounts of air for brief periods of time. Additional receiver tanks are best suited for peak demand time periods of 10–20 seconds in duration. Equation (11.15) may be used to determine the required volume of air receivers.

$$V = \frac{tCP_0}{(P_1 - P_2)} \qquad \text{Eq. (11.15)}$$

where V = receiver volume, ft^3
t = operating time of air load, minutes
P_0 = atmospheric air pressure, psia
P_1 = initial receiver pressure, psig
P_2 = final receiver pressure, psig

This method of sizing receivers is a conservative one since it assumes no airflow into the receiver while the air is being discharged.

11.6.8 Install Variable Speed Drives

The output capacity of many air compressors is controlled by one of two methods: start-and-stop control or unloading of the compressor while it is operating at a constant speed. Both of these methods tend to be costly from an energy cost standpoint.

When air compressor motors with large electric motors are started, they require a relatively large inrush of current to start the motor and accelerate the motor/compressor up to operating speed. Most commercial buildings and industrial facilities pay for the amount of electricity consumed, as well as a demand charge on their electrical bill. Frequently, starting large motors can result in very significant electrical demand charges. Starting and stopping compressors also results in pressure variations in the compressed air system.

While unloading the compressor results in less load on the driver and less energy to operate the driver, the driver is still operating at a constant speed and consuming energy. On large compressors, the power required to drive the machine unloaded and over come the internal friction losses can be significant.

The power required to drive a compressor is a function of the torque and the speed at which it is driven. During reduced load periods, the capacity of the compressor can be reduced by reducing its speed. This results in a direct reduction in power consumption. Adjustable speed drives for compressors should be considered for part load control when:

- The air demand for the system is highly variable.
- The need for constant air pressure in the system is important.
- The compressor will operate for a large number of hours annually.

- Electrical energy costs are high.
- Electrical demand costs are high.

The compressor manufacturer should always be consulted before installing a variable speed. drive. See Chapter 10: Electrical and Lighting Systems for additional information on adjustable speed drives.

11.6.9 Install a Reheater

Compressed air leaving a refrigerated air dryer is usually near 40°F, or lower, and saturated. At saturation conditions, the relative humidity is near 100%. Water vapor may continue to condense in the air system after the dryer and may also cause moisture to condense on the outside of the cold piping just after the dryer.

Increasing the temperature of the compressed air leaving a refrigerated air dryer significantly reduces the relative humidity of the compressed air. Increasing the air temperature also increases the volume of the air.

Energy savings will result if the heat from the compression process is used to reheat the compressed air. The heat can be recovered from the compressor intercoolers and used to reheat the compressed air. Some compressor manufacturers also offer heat recovery from the compressor oil system.

11.7 ENERGY EFFICIENCY FOR NEW COMPRESSED AIR SYSTEMS

Air compressors and compressed air systems are one of the largest consumers of energy at many facilities. Since air compressors and compressed air system equipment consume significant amounts of energy, regardless of the type of facility, it is important that energy efficiency be a primary consideration when selecting new compressed air equipment for new, or retrofit, installations.

Earlier in this chapter, energy conservation for compressed air systems was discussed. Several of the energy efficiency measures discussed as retrofit opportunities for compressed air equipment, such as high efficiency motors, are applicable to new and retrofit air compressor installations. In addition to the previously discussed energy conservation measures for compressed air systems, the following measures should be considered for new and replacement air compressors and accessories.

11.7.1 Develop an Air Demand Profile

In general, constant speed compressors operate most efficiently when operated at, or near, full capacity (full load). An air compressor, or combination of compressors, should be selected such that they operate at

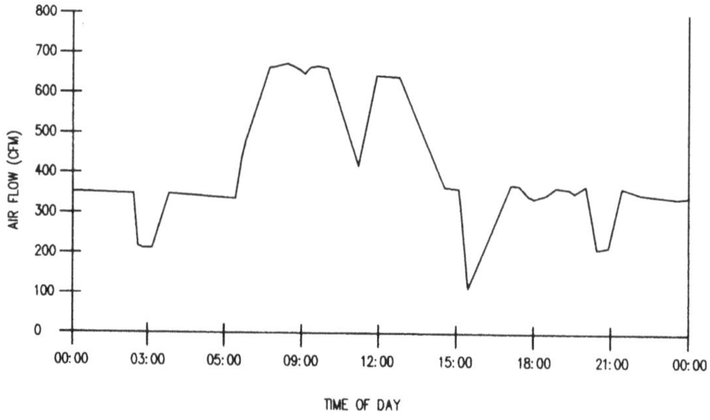

Figure 11-5. Typical air demand profile.

peak efficiency, which is near full load. Most compressed air loads are not constant, but are highly variable. Simply adding all connected pneumatic loads together to get a peak air load will result in an oversized compressor, which will operate at less than peak efficiency. In order to determine the actual maximum load on a proposed new, or an existing, system an air demand profile for the system should be developed.

A demand profile is a graphic representation of the demand for compressed air of a system over a time interval, usually a day. A *load factor* represents the percent time a particular pnumatic load is operating and the percent of maximum airflow under normal operating conditions. A load factor should be determined for each pneumatic load for each time increment, such as an hour. The total of all load factors should be summed for each time increment for the time interval. The total sum of demand factors should be plotted on a graph for each time increment for the time interval. For example, the total sum of demand factors for each hour for a day should be plotted graphically, as shown in Figure 11-5.

Air demand profiles may also be determined indirectly for existing systems by measuring and recording the power input to the compressor for each time increment. The compressor performance data may then be used to determine the total system demand factor for each time increment for the time interval.

The air demand profile may then be used to determine the actual peak demand for the compressed air system. Based upon the actual system peak demand, the optimum compressor may then be selected.

11.7.2 Selection of Compressor Type

The power required to drive an air compressor is a function of the compressor torque and the speed at which the compressor is driven, or rotated. The compression ratio (the ratio of the discharge pressure to the

TABLE 11.5 Suggested Compressor Types 100 psig Operation

		Reciprocating			
Size (HP)	Airflow (cfm)	Single Acting	Double Acting	Rotary Positive Displacement	Dynamic
5–30	20–120	1 & 2 stage	1 stage	Rotary screw	NNM
30–150	100–750	1 & 2 stage	2 stage	Rotary screw or Lobe type	NNM
150–300	600–1,600	1 & 2 stage	2 stage	Rotary screw	Centrifugal
300+	1,300+	NNM	2 stage	Rotary screw	Centrifugal

NNM = Not normally manufactured.

inlet pressure) and the internal friction to drive the compressor determine the compressor torque. The rotational speed of the compressor is a function of the required compressed airflow rate. Therefore, the discharge pressure and the flow rate of the compressor determine the brake horsepower input to the compressor.

Compressors may be compared using the Volume/Power Effectiveness Ratio (VPER). The VPER is the amount of air compressed per brake horsepower for a compressor and is the power required by the compressor less the motor efficiency loss and drive power transmission losses. A similar measure of compressor performance is the Compression Specific

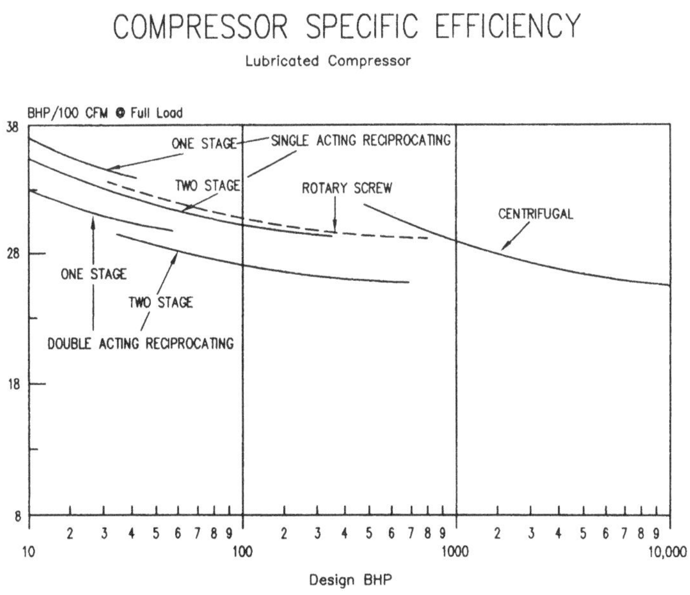

Figure 11-6. Relative full load power required of typical lubricated compressors 100 psig at sea level.

Efficiency (CSE), which is the reciprocal of the VPER multiplied by 100 to get a percentage. Both the VPER and CSE are useful for comparing the performance of compressors at full load operation. New compressors should have a VPER of at least 4. Existing compressors with a VPER of less than 4 are good candidates for replacement.

When selecting a compressor for a particular application, the compressor must have adequate capacity to provide air for the required discharge pressure and at the required airflow rate. A system demand profile, as previously discussed, should be used to determine the peak load and discharge pressure. Consideration should be given to selecting a compressor that is most efficient at the peak demand requirements. Table 11.5 may be used to make a general selection of compressor type, or combination of compressor types, based upon the horsepower or air flow at 100 psig.

A measure of compressor efficiency, which may be used to compare compressors, is the specific efficiency. The specific efficiency of a compressor is the brake horsepower input to the compressor per 100 cfm of air at the inlet conditions when the pressure is raised to 100 psig. Figures 11-6 and 11-7 show typical specific efficiencies for various types of compressors. Actual performance data from the compressor manufacturer should be consulted when comparing compressors for a specific application. Care should be taken to obtain the actual output motor brake horsepower and not just the nameplate data. Air compressor manufacturers frequently publish performance curves for their equipment. An air

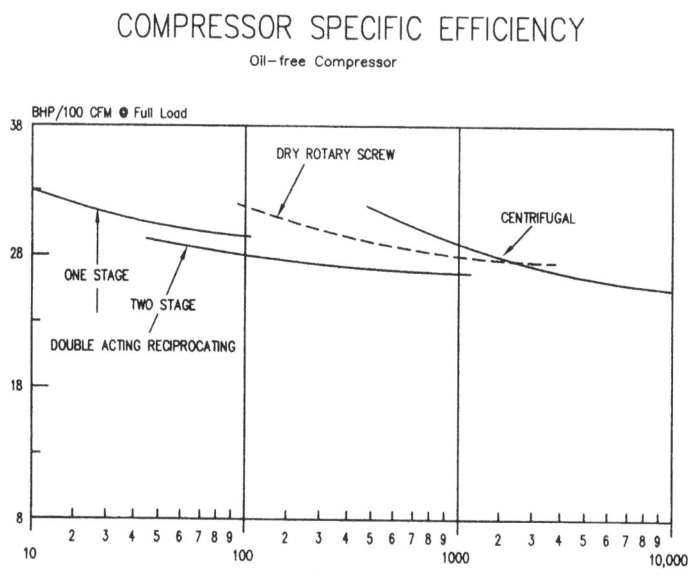

Figure 11-7. Relative full load power of typical oil-free compressors 100 psig at sea level.

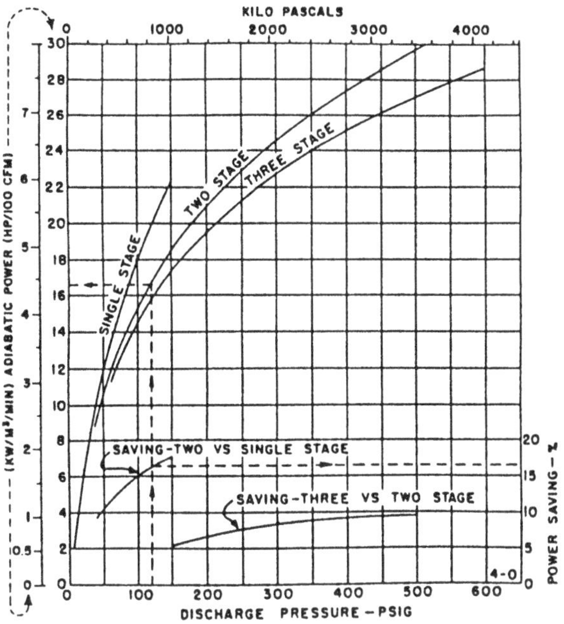

Figure 11-8. Theoretical energy savings possible through multi-staging reciprocating air compressors courtesy of Ingersoll-Rand, Co.

compressor performance curve relates the brake horsepower input for a compressor to the air delivery rate of the compressor for a given discharge pressure.

Whenever an air compressor is to increase the air pressure by 50 psi, or more, multi-staging should be considered for a new compressor. Significant energy savings can be achieved by providing multiple stages, with intercooling between stages for pressure increases of 50 psi or more. Figure 11-8 shows the theoretical savings for compressors with multiple stages.

Most compressed air systems operate at less than peak flow/pressure majority of the time. The compressor performance curve may be used to determine the compressor power consumption at any given flow rate. Even though a demand profile may be developed for a compressed air system, it is likely that the compressor(s) will be operating at part load part, or all, of the time. The air system demand profile, which lists air demand at various times, along with the compressor performance curves, may be used to estimate the compressor energy consumption for a desired time period, such as a day. Compressors should be compared on the basis of the full operating profile of the application and not just at full load conditions. Part load performance has a significant impact on total energy usage.

11.7.3 Air Compressor Controls

Although an air compressor, or combination of air compressors, serving air distribution system must be capable of providing the peak compressed air demand, air compressors are usually required to operate at part load conditions for a significant portion of their operating life. To provide the actual air demand of a system at any given time, the compressor(s) must be controlled to provide the reduced system demand. Most controls for air compressors provide compressor control by monitoring the system air pressure. As a result, the system pressure differential (control range) from the desired system set point pressure. There are various methods by which compressors are controlled. Control methods for the various types of compressors were discussed earlier in this chapter.

Since most compressors operate most efficiently when they are operating near peak capacity, a compressor control system should be set up to bring on multiple compressors such that the online compressors are operating near peak capacity. The compressor control system should be set up to have high capacity, high efficiency compressors carry the system base load under normal operating conditions. Smaller, less efficient compressors should be brought online as necessary to handle short-term peak demands.

The control system should also bring compressors with large electric motors online gradually, rather than all at once. The effect of large electrical demands was discussed earlier. Whenever possible, compressors should be started in gradual steps until the compressor has reached full capacity.

The control of compressors may be achieved by manually controlling the compressors with transfer switches. Manually controlling compressors usually does not result in optimum energy efficiency. To optimize compressor control, a solid state master controller should be used. The master controller, in turn, sends signals to the individual controller on each compressor. Microprocessor based master controllers select the combination of compressors that will result in the most efficient operation, and also stage the starting of compressors to keep electrical demand to a minimum.

11.7.4 Air Distribution Systems

The primary function of a compressed air distribution system is to provide a required amount of compressed air at a given location at a given pressure. As the air passes through the piping of the distribution system, interaction between the air motion and the pipe wall results in friction, which results in air pressure drop. The pressure drop is a function of the velocity of the air in an exponential relationship. The power required to drive a compressor is proportional to the compression ratio of the air. The required compression ratio for a compressor is directly related to the discharge pressure. For any given pressure required by an air load, the required compressor discharge pressure must be the pressure required by the load plus the pressure losses through the distribution system.

Section 11.6.6 discussed modifications to existing distribution systems to make the system more energy efficient. Figure 11-4 shows various piping arrangements and their relative energy efficiency. New air distribution systems should be designed and installed in accordance with Figure 11-4. Compressed air systems are usually designed and sized in terms of free air or standard cubic feet per minute (scfm). Equation (11.16) may be used to convert between actual airflow in cfm and airflow at standard conditions in scfm.

$$scfm = cfm\left(\frac{P}{14.7}\right)\left(\frac{520}{T+460}\right) \qquad \text{Eq. (11.16)}$$

where P = absolute pressure, psia
T = air temperatue, °F

Since many air loads are intermittent and of short duration, receiver tanks may be strategically located throughout the system to handle localized short-term demand. Receiver tanks also allow some compressors to operate more efficiently during periods of low air demand. Receiver tanks were discussed in Section 11.6.7. When sizing main compressed air headers, it is advisable to significantly oversize the main headers. The main reason for oversizing headers is not only to reduce pressure drop, but also to provide compressed air storage for large short-term air loads. Oversizing the main headers may be a more cost-effective way to provide air storage than providing additional receiver tanks.

Quick connect fittings for hoses should be avoided. These fittings can cause significant air leaks. The pressure loss through quick connect fittings may also be as high as 5–10 psi.

Valves should be judiciously placed in the air distribution system. They should be placed at each piece of air consuming equipment, as well as major sections of piping, to isolate equipment and sections of piping for maintenance purposes. More valves than necessary, however, should not be installed. Valves represent a significant pressure loss in the system in addition to a major source of air leakage if not properly maintained. Since air piping is usually installed high, close to the ceiling or in an inaccessible location, maintenance on valves, such as repacking, is rarely performed. The type of valve installed must also be considered. The pressure loss through a globe valve can be more than 30 times that of the same size gate valve or butterfly valve. When designing compressed air distribution systems, the following guidelines should be considered:

- The discharge piping from the compressor should be at least as large as the compressor discharge connection and it should run directly to the aftercooler.
- The main riser, main header and some selected branch headers should be oversized to provide air storage and for future growth of air demand.

- The entire distribution system and conditioning equipment should be sized and selected for a combined total pressure drop of no more than 10% of the compressor discharge pressure. The air velocity in the piping should be limited to a maximum of 4,000 feet per minute. Crane Company (1974) may be used to determine the pressure drop of compressed air in piping.
- Looped air distribution systems are preferable to gridded distribution systems.
- Fittings and valves that offer the least resistance to flow should be used in distribution systems.
- Welded pipe and fittings should be used as welded fittings are less likely to leak than threaded or flanged fittings.
- Subheaders and branches should be as close as possible to the intended point of use.
- Outlets and branch lines should be taken from the top of the main to minimize water carryover.
- Piping to air loads should follow the most direct route, with the fewest offsets as is reasonably possible.

11.7.5 Air Compressor Drives

Whenever a new compressor is to have an electric motor as the driver, a high efficiency motor should be selected. Table 11.3 listed the efficiencies for both standard and high efficiency motors for various sized motors. Although usually there is a cost premium for high efficiency motors, the additional cost will be recovered in energy savings in a few years.

Just as in the case of the compressors themselves, it is important to consider the efficiency of motors at part load. The manufacturer of the motor supplied with the compressor can usually provide performance curves for their motors, which shows the efficiency of the motor at various load levels. The demand profile for the compressed air system should be used to determine the load profile for the compressor motor. Using the motor performance curve along with the system demand profile, the annual energy cost for the compressor drive motor can be estimated for both standard and high efficiency motors. The difference in the annual operating cost of the motors may be used to perform an economic analysis of the motor choices.

11.7.6 Compressed Air System Aftercoolers

In compressed air systems where air dryers are preceded by aftercoolers, it is usually cost-effective to oversize the aftercooler. Since the aftercooler uses air or water to cool the air and remove moisture, the cost of removing the moisture is usually less than removing the moisture with an air dryer. Oversizing the aftercooler removes more moisture from the compressed air and, therefore, less must be removed by the air dryer.

If water is used as the cooling medium in an aftercooler, the flow of cooling water through the aftercooler should be controlled. Water flow may be controlled by a modulating control valve, which either bypasses water around the aftercooler or modulates the flow of water through the cooler in response to the cooling load. The best location for a control valve is on the water discharge line from the aftercooler.

11.7.7 Heat Recovery

Earlier in this chapter, heat recovery for existing compressors and compressed air systems was discussed. Heat recovery should also be considered for new compressed air systems.

New air compressors can be purchased with a heat recovery option that is less expensive than a separate heat exchanger. Rotary, helical screw and flooded oil compressors over 50 horsepower are most suitable for heat recovery. Since compressors must be cooled when they are operating, the heat recovery option is a relatively easy modification to the standard compressor design. If a suitable use for the recovered heat is available, the cost of the heat recovery option will be quickly paid back in energy cost savings.

11.7.8 Air Dryers

The need for compressed air dryers and various types of dryers were discussed earlier in this chapter. The use of air dryers can significantly increase the amount of energy used and energy costs for a compressed air system. When choosing an air dryer, it should be selected to provide drying only to the extent that drying in actually necessary. Air should not be dried to a lower dew point than absolutely necessary. Usually, the extent of drying is determined by the lowest ambient temperature to which the compressed air system will be exposed. If the compressed air distribution system runs outdoors above grade, the air should be dried to a dew point lower than the lowest expected outdoor temperature. However, if the system is completely within a heated building, the air may need to be dried to a dew point of only about 50°F.

Occasionally, a particular process will require air that has a very low moisture content or dew point. When very dry air is needed only at a few points in the system, local dryers should be installed at the point of application. This will allow the main part of the air system to be dried to meet just ambient air temperature conditions.

When selecting dryers that require heat to regenerate them, consideration should be given to dryers that can use heat recovered from the air compressor or heat recovered from other sources. Heat recovery systems are discussed in Chapter 14.

Chapter 12

Refrigeration Systems

Refrigeration is a mechanical thermodynamic process by which heat is removed at a rate such that the temperature of a material, or space, is maintained lower than the temperature of the surroundings. Refrigeration is used extensively in the food and beverage industry, chemical industry, in the production of ice and in the production of cryogenic materials. It is also used in almost all comfort air conditioning systems. The operation of refrigeration systems in many industries and facilities represents a significant part of the energy budget. This chapter deals with energy management for mechanical refrigeration systems.

12.1 BASIC PRINCIPLES OF REFRIGERATION SYSTEMS

Refrigeration is defined as any process of heat removal. It is that branch of science that deals with the process of reducing and maintaining the temperature of a space or material below the temperature of the surroundings. Typically, refrigeration deals with maintaining temperatures down to -200°F. Systems that produce temperatures below -200°F are referred to as cryogenic systems.

12.1.1 Definitions and Terms Used in Refrigeration Systems

For definitions of important terms used in refrigeration systems, see the Glossary.

12.1.2 Basic Refrigeration Processes

Refrigeration is a thermodynamic cycle, which removes heat from a material or space to be maintained at a low temperature and discharges the heat in a region of higher temperature. Since heat normally flows from a high temperature region to a low temperature region, heat cannot flow in the opposite direction without the addition of mechanical energy in the form of work. The refrigeration process provides the mechanical energy that causes heat to flow from a low state to a high state.

The rate at which heat must be removed from the refrigerated space or material in order to produce or maintain the desired temperature conditions is called the refrigeration load or cooling load. The rate of heat removal is typically expressed in Btu per hour or in tons of refrigeration.

A ton of refrigeration is equal to 12,000 Btu/hr and is actually an ice-melting equivalent. A refrigeration system having a capacity of one ton has the cooling capacity equivalent to the melting of one ton of ice in a 24-hour period.

When analyzing refrigeration cycles, pressure-enthalpy and pressure-entropy diagrams are frequently used. A schematic diagram for a basic vapor compression cycle is shown in Figure 12-1. A typical pressure-enthalpy diagram for a vapor compression cycle is shown in Figure 12-2. The refrigerant leaves the evaporator at point 1 as a low pressure, low temperature saturated vapor, where it is compressed reversibly and adiabatically. At point 2, it leaves the compressor as a high temperature, high pressure, superheated vapor and enters the condenser, where it is first desuperheated and then condensed at a constant pressure. At point 3, the refrigerant leaves the condenser as a high pressure, medium temperature saturated liquid and enters an expansion valve, where it expands irreversibly and adiabatically (constant enthalpy). At point 4, the refrigerant leaves the expansion valve as a low pressure liquid and enters the evaporator, where it is evaporated reversibly at a constant pressure to the saturated vapor state at point 1.

The principal components in a vapor compression direct expansion refrigeration system are an evaporator, whose function is to provide a heat surface through which heat passes from the refrigerated space or material into the vaporizing refrigerant; a vapor compressor, whose function is to remove the vapor from the evaporator and to raise the temperature and pressure of the refrigerant vapor to a point such that the vapor can be condensed with a normally available media (such as air or water); a

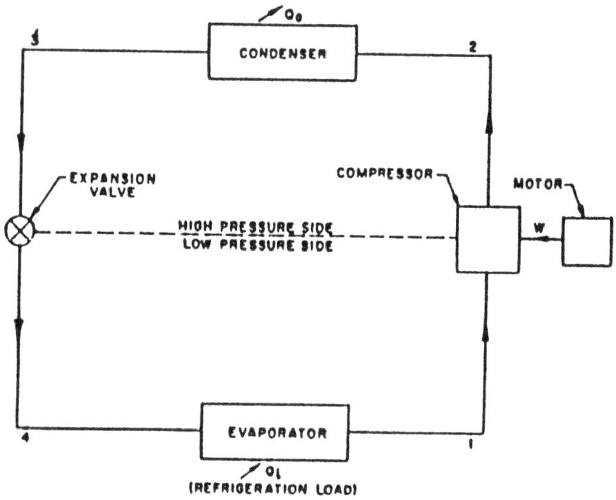

Figure 12-1. Schematic diagram for basic vapor compression refrigeration cycle.

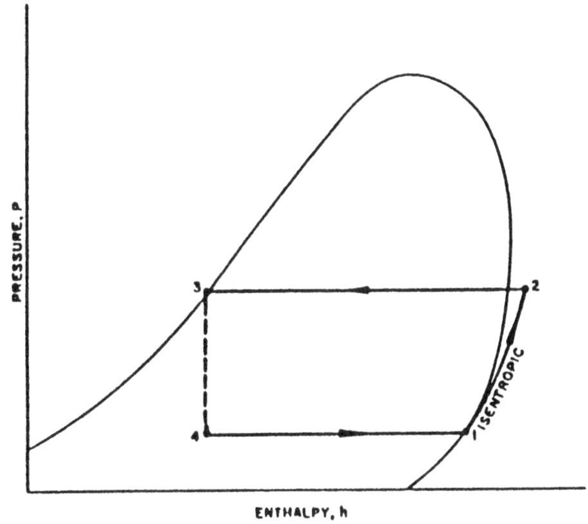

Figure 12-2. Pressure-enthalpy diagram for basic vapor compression cycle.

condenser, whose purpose is to provide heat transfer between the hot refrigerant vapor and the condensing medium; and a refrigerant flow control (expansion or float valve), whose function is to meter the proper amount of refrigerant to the evaporator and to reduce the pressure of the liquid refrigerant entering the evaporator so that the liquid will vaporize in the evaporator at the desired low temperature.

The refrigeration schematic presented in Figure 12-1 shows an expansion valve for a refrigeration flow control valve. In direct expansion systems with thermostatic expansion valves, the valve controls the flow of refrigerant entering the evaporator in response to the superheat of the refrigerant gas leaving the evaporator. It keeps the entire evaporator active, without permitting unevaporated refrigerant liquid to be returned through the suction line to the compressor.

Refrigerant flow control may also be provided by a float valve. Refrigerant float control valves are used in systems with flooded evaporators. Flooded evaporators are evaporators that are completely filled with liquid refrigerant that evaporates as it absorbs heat and cools the space or material to be cooled. The float valve is a float operated valve that admits liquid refrigerant to the evaporator in response to the cooling load so that a constant liquid level is maintained in the evaporator. Flooded evaporators are frequently used in liquid chillers. Chillers are discussed in Chapter 7.

The efficiency or performance of refrigeration systems is defined by the coefficient of performance (COP). The COP is the ratio of the refrigeration system output divided by the energy (work) input to the system. The output is the amount of heat absorbed by the system. The input is the amount of energy required to produce the output. Referring to Figures 12-1 and 12-2, the COP of a refrigeration system is given by Equation (12.1).

$$\text{COP} = \frac{{}_4Q_1}{{}_1W_2} = \frac{h_1 - h_4}{h_2 - h_1} \qquad \text{Eq. (12.1)}$$

where ${}_4Q_1$ = the heat absorbed from the refrigerated space or material, Btu/hr

${}_1W_2$ = the energy equivalent of work supplied to the compressor, Btu/hr

h_1 = enthalpy of the refrigerant entering the compressor, Btu/lb

h_2 = enthalpy of the refrigerant entering the condenser, Btu/lb

h_4 = enthalpy of the refrigerant entering the evaporator, Btu/lb

Note that as the refrigerant flows through the expansion valve, it expands irreversibly and adibatically; therefore, the enthalpy of the refrigerant remains constant and $h_3 = h_4$.

12.1.3 Refrigerants

A refrigerant is a fluid used for heat transfer in a refrigerating system. The refrigerant absorbs and transfers it at a higher pressure with a change of state. The refrigerant absorbs heat when changing from a liquid to a vapor (heat of vaporization). Likewise, it gives up heat when condensing from a vapor to a liquid.

A refrigerant must satisfy many requirements, some of which do not directly relate to its ability to transfer heat. There is no substance that is an ideal refrigerant, although theoretically there are few limitations on the number of substances that can be used in refrigeration systems. A refrigerant must satisfy many requirements. The properties of a substance that determine its suitability as a refrigerant can be divided into two general categories: physical and thermodynamic. Physical properties include compatibility with other materials, safety, toxicity, flammability, leakage tendency, chemical stability under conditions of use, cost and availability. Thermodynamic properties include working pressures and temperatures, specific volume, enthalpy, latent heat of vaporization, specific heat, thermal conductivity and viscosity. From a performance standpoint, working temperatures and pressures and latent heat of vaporization are the most significant.

A desirable refrigerant should have an evaporation pressure as high as possible and, simultaneously, a condensing pressure as low as possible. High evaporating pressures imply high vapor densities, and, as a result, a greater system capacity for a given compressor. However, the efficiency at higher pressures is lower, especially as the condenser pressure approaches the critical pressure. The saturated evaporating temperature establishes the pressure criteria for the condenser and the compressor discharge

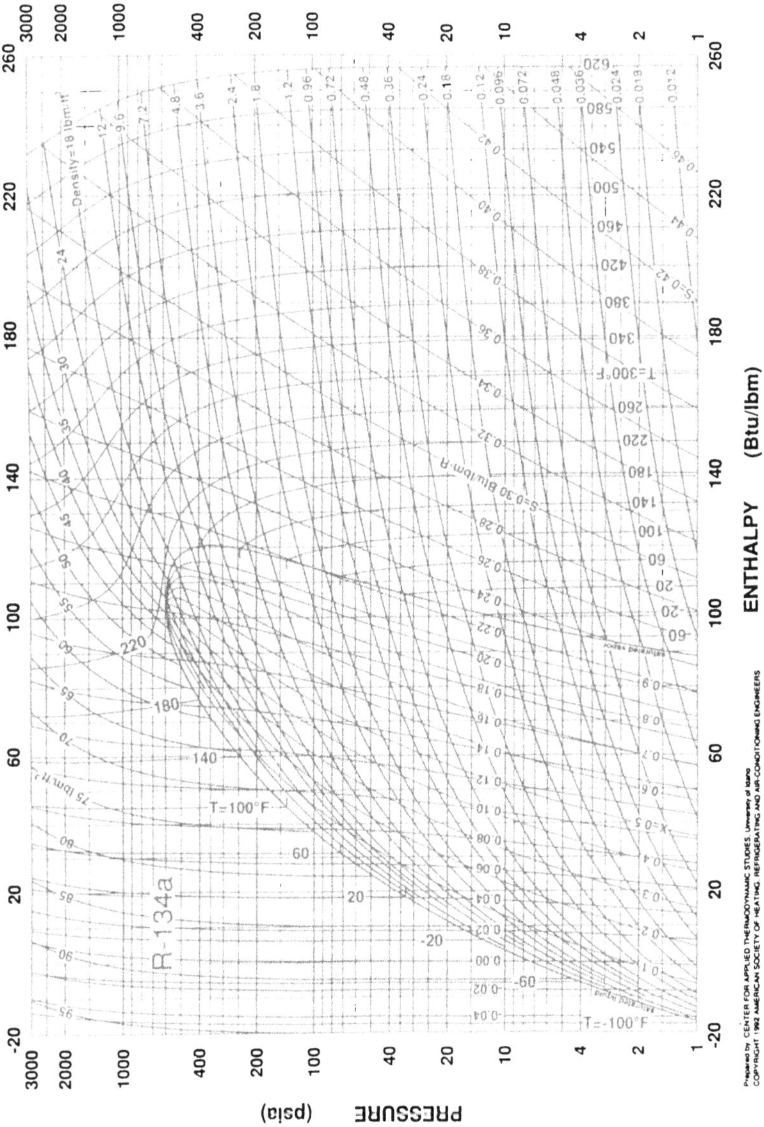

Figure 12-3. Pressure enthalpy diagram for refrigerant 134a.

pressures. Collectively, the difference between the condensing and suction temperatures affects the size and drive power of the compressor.

The amount of refrigeration effect (cooling) provider by a refrigerant is directly related to its latent heat of vaporization. The latent heat of vaporization (h_{fg}) is the difference between the enthalpies of the saturated vapor and liquid conditions of a refrigerant at any given temperature. A typical pressure enthalpy diagram for Refrigerant 134a is shown in Figure 12-3.

12.2 REFRIGERATION SYSTEM COMPONENTS

Figure 12-1 presents a schematic diagram of a refrigeration system, which includes the main components in the system. The main components include a compressor, a condenser, an evaporator and refrigeration system controls.

12.2.1 Compressors

The refrigerant compressor is a mass mover that transfers refrigerant vapor from a low pressure area at evaporating conditions to a high pressure area at condensing conditions. The temperature difference between the compressor discharge (condensing) and compressor suction (evaporating) is referred to as the thermal lift of the compressor. The thermal lift directly affects the compressor power requirements. As thermal lift is increased, the compressor power requirements are increased.

There are two basic classifications for refrigeration system compressors: open and hermetic. An open compressor has an external drive and may be direct driven through a coupling, or belt driven. An open compressor unit is shown in Figure 12-4. Open compressors can use several drives, including electric motors, internal combustion engines or turbines. A hermetic compressor has the drive and compressor built into an integral

Figure 12-4. Open compressor unit (*Courtesy of Carrier Corporation*).

Figure 12-5. Hermetic compressor unit (*Courtesy of Carrier Corporation*).

housing, as shown in Figure 12-5. The drive is almost always an electric motor. The motor and compressor have a common shaft and bearings. The drive motor is generally cooled by the refrigerant suction gas passing through the motor windings. Motors may also be water cooled.

Another basic classification of refrigerant compressors is positive displacement or dynamic. Positive displacement compressors increase the pressure of the refrigerant vapor by reducing the volume of a compression chamber through work applied to the compressor drive shaft. Positive displacement compressors include reciprocating, rotary (rolling piston, rotary vane, single screw and twin screw) and scroll. Dynamic compressors increase the pressure of the refrigerant vapor by a continuous transfer of angular momentum from a rotating impeller to the rotating member to the vapor followed by the conversion of this momentum into a static pressure. Dynamic compressors include centrifugal compressors.

The most common types of compressors used in refrigeration systems are reciprocating compressors. Reciprocating compressors have pistons that travel back and forth in cylinders compressing the refrigerant vapor. A reciprocating compressor is shown in Figure 12-6. Valve rings are mounted in a plate at the top of the cylinder. The plate has a suction (inlet) valve and a discharge valve. The refrigerant vapor enters a suction manifold and passes through the suction valve into the cylinder a s the piston is on the down (intake) stroke. When the piston is on the up (compression) stroke, the discharge valve opens and the refrigerant vapor is compressed as it enters a discharge manifold.

Since most refrigeration loads vary with time, it is necessary to provide some sort of capacity control for a compressor. There are five methods by which reciprocating compressor are typically controlled: on/off control, suction valve unloading, cylinder head bypass, speed control and hot gas bypass.

The simplest method of capacity control for a reciprocating compressor is on/off control. The on/off control is usually controlled by a thermostat

Figure 12-6. Reciprocating compressor.

located in the refrigerated space. As the temperature in the space rises above the set point temperature, the compressor is cycled on. When the temperature drops, the compressor is cycled off. On/off control is best suited for smaller capacity refrigeration systems.

Suction valve unloading control unseats the compressor suction valves of certain cylinders so they can remain open. As a result, compression cannot take place. This method is an efficient way to control capacity since the passage of the refrigerant vapor in and out of the unloaded cylinder through the suction valves results in small losses.

Cylinder head bypass control is used on multi-cylinder reciprocating compressors. One or more banks of cylinders may be inactivated by bypassing refrigerant vapor from the cylinder discharge to the intake port. A check valve is installed to separate inactive cylinders from active cylinders. The refrigerant vapor passes through the inactive cylinders, but is not compressed. As a result, the only loss is through the valves, cylinders and connections.

Speed control is an effective method for controlling the capacity of reciprocating compressors. Compressor capacity is closely related to the rotational speed of the compressor. Speed control is typically obtained with a multi-speed compressor motor, which provides two or three speeds. Speed control can also be obtained with a variable speed drive for the motor.

Hot gas bypass controls the compressor by artificially loading it. This method bypasses heat in the form of hot refrigerant gas from the compressor discharge to the compressor suction. The discharge is connected to the suction side with a constant pressure valve that admits hot gas to the compressor suction as the evaporator pressure drops. As a result, the suction pressure is maintained at a constant level.

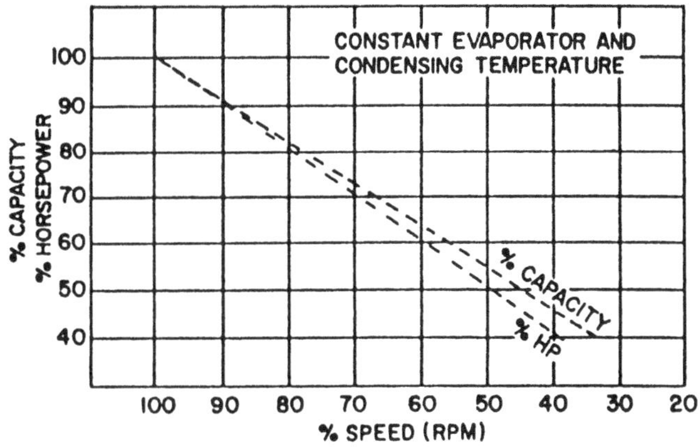

Figure 12-7. Reciprocating compressor performance versus percent speed.

There are three factors that affect the performance of a reciprocating compressor: rotational speed, suction pressure discharge and the type of refrigerant used in the system.

The performance of a reciprocating compressor is almost directly proportional to the rotational speed of the compressor. Referring to Figure 12-7, it can be observed that the compressor capacity and horsepower increase as the rotational speed of the compressor increases.

The compressor suction pressure also affects the performance of a reciprocating compressor. Since the refrigerant pressure is a function of the temperature, the evaporator temperature governs the capacity of the compressor. Referring to Figure 12-8, it may be observed that, as the compressor suction temperature increases at a constant discharge pressure, the capacity increases.

The horsepower also increases with the suction temperature. This results from the fact that the compressor handles an increased weight of refrigerant.

Finally, the discharge pressure of a reciprocating compressor affects its performance. Since the discharge pressure is directly related to the saturated condensing temperature, it is actually the condensing temperature that affects the compressor performance. Referring to Figure 12-9, it can be observed that the capacity of a reciprocating compressor is inversely related to the condensing temperature. As the condensing temperature increases, the compressor capacity decreases. The compressor horsepower, however, increases with condensing temperature.

Rotary compressors employ a circular, or rotary, action to compress the refrigerant gas. They are positive displacement compressors and are usually direct driven. The performance of rotary compressors is characterized by high volumetric efficiency due to a small clearance volume and correspondingly low re-expansion loss. The most common types of rotary compressors include rolling piston compressors, rotating vane compressors, screw compressors and scroll compressors.

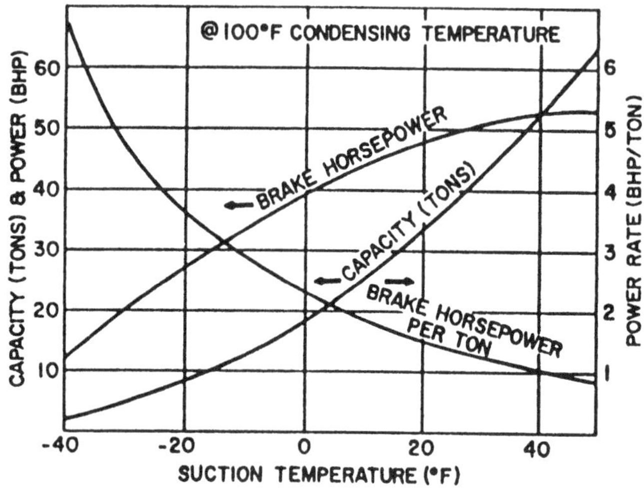

Figure 12-8. Reciprocating compressor performance versus suction temperature.

Rotary vane compressors are typically used in smaller refrigeration systems. These compressors are usually in the 3–50 HP range and may be single or two stage. Single stage rotary vane compressors can produce saturated suction temperatures in the range of -40 to +45°F. Two stage compressors can be used for temperature applications as low as -60°F.

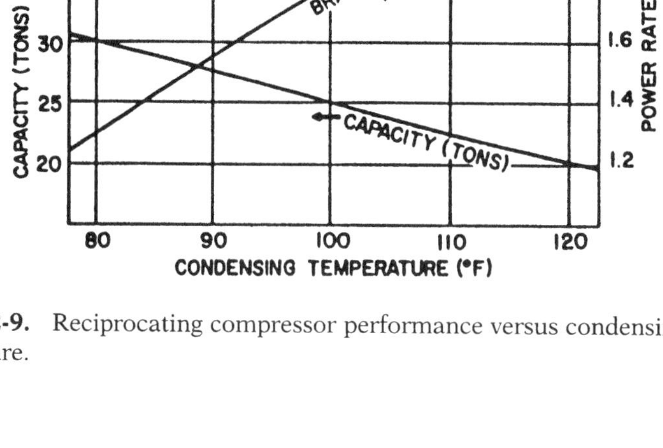

Figure 12-9. Reciprocating compressor performance versus condensing temperature.

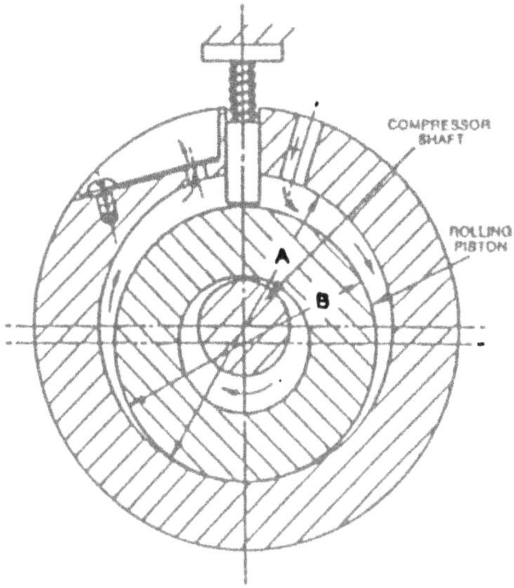

Figure 12-10. Fixed vane rolling piston compressor.

Rolling piston, or fixed vane, compressors are also used in smaller refrigeration systems with compressors up to 3 HP. A rolling piston compressor uses a roller mounted on an eccentric shaft. A single vane or blade mounted in the housing reciprocates as the eccentric roller rotates, compressing the refrigerant vapor. A rolling piston and vane is shown in Figure 12-10.

Screw compressors are part of a broad class of rotary positive displacement compressors. They are available with a single screw or twin screw configuration. Both configurations use fluid injection to cool and seal the compressor. Screw compressors can operate at pressure ratios above 20 to 1 for a single stage. Screw compressors have capacities ranging from 20 – 1,300 tons of refrigeration.

A single screw compressor consists of a single cylindrical main rotor that works with a pair of gate rotors. Figure 12-11 shows a typical single screw compressor. The main rotor has six helical grooves with cylindrical peripheral and globoid (hourglass shaped) root profile. The two identical gate rotors each have teeth that match the main rotor located on opposite sides of the main rotor. The casing enclosing the main rotor has two slots, which allow the teeth of the gate rotors to pass through them. Two diametrically opposed ports discharge the compressed refrigerant vapor into a common discharge manifold. Capacity control can be achieved with both rotational speed modulation and suction throttling. Variable compressor displacement, which is the most common capacity control method, usually is accomplished with two movable slide valves in the compressor casing. At part load, each slide valve produces a slot that delays the point at which compression begins.

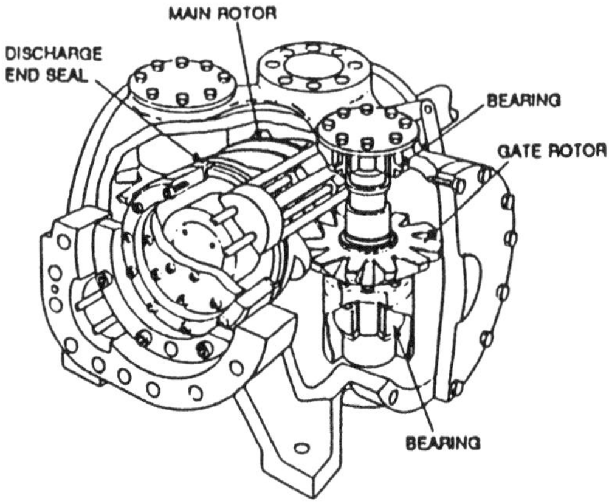

Figure 12-11. Single screw compressor.

A twin screw compressor consists of two mating helically grooved rotors. One rotor is a male rotor with lobes while the other is a mating female rotor with flutes or gullies. Male and female rotors are shown in Figure 12-12. The two rotors are mounted in a stationary housing with inlet and outlet ports. The flow of the refrigerant vapor is primarily in an axial direction. The capacity of twin screw compressors is controlled as single control and capacity slide valves. Slide valves are the most common method of control.

Scroll compressors are also positive displacement compressors. They have two interfitting, spiral shaped scroll members which rotate with an orbital motion. The scroll members are typically a geometrically matched pair, assembled 180° apart out of phase. Each scroll member is open on one end of the vane and bounded by a base plate on the other end. The two scrolls form pockets between their respective base plates and various lines of contact between their walls. Compression occurs as the result of sealing of the suction gas in pockets of given volume at the outer periphery of the scrolls and progressively reducing the size of those pockets as the scroll

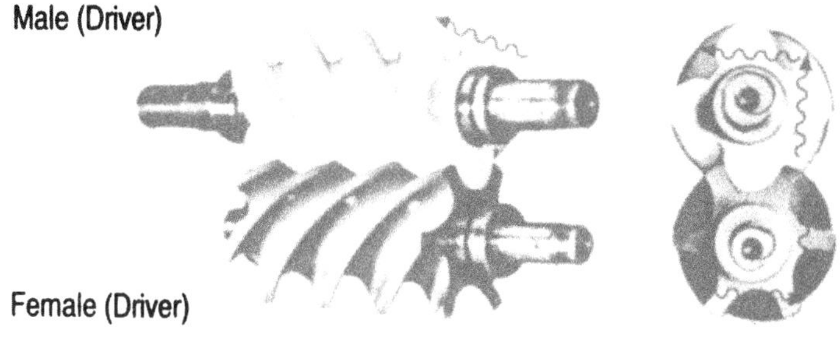

Figure 12-12. Twin screw compressor rotors.

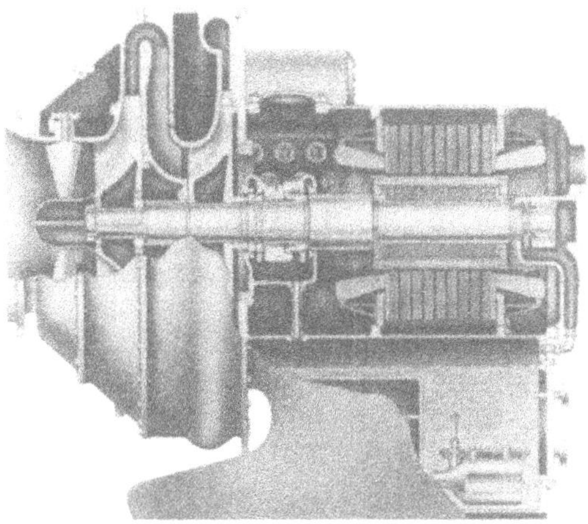

Figure 12-13. Cross-section of a centrifugal compressor.

relative motion moves them inward toward the discharge port. Capacities of scroll compressors range from 1 to about 15 tons.

Centrifugal compressors are large capacity dynamic compressing machines. They are frequently used as part of a chiller to provide refrigeration in a chilled water system. Chillers are discussed in Chapter 7.

Centrifugal compressors use centrifugal force to compress the refrigerant vapor and provide a continuous flow of refrigerant gas at large rates of flow. A cross-section of a centrifugal compressor is shown in Figure 12-13. The compressor draws the refrigerant gas into the center, or eye, of a rapidly rotating impeller in an axial direction. It is then discharged radially at a high velocity where the dynamic pressure is converted to a static pressure in a radial diffuser and scroll outboard of the impeller.

Capacity controls for centrifugal compressors are required to match the compressor output to the refrigeration load. The most common methods of capacity control are:

- Variable pitch inlet guide vanes, which impart a swirl to the refrigerant vapor as it enters the rotating impeller. They are adjustable pre-rotation vanes that swirl the vapor in the direction of the impeller rotation. Inlet guide vanes are used when the driving speed is constant. They produce new performance curves without a change in rotational speed. Inlet guide vanes provide efficient capacity control and are the most common method of capacity control.
- Suction dampers are throttling devices that reduce the inlet pressure of the refrigerant vapor, which increases the volume flow for a given mass flow.
- Speed control changes the rotational speed of the compressor in response to changes in refrigerating load. Refrigeration capacity is

directly related to compressor speed, but the compressor's ability to produce pressure is a function of the square of the speed. Decreasing the pressure reduces the work input to the compressor, therefore, speed control is an energy-efficient method of capacity control.

- Hot gas bypass is a false loading technique. This method passes refrigerant gas from the compressor discharge to its suction through a pressure reducing valve. The bypassing of the hot refrigerant gas produces an artificial load on the compressor, keeping the flow of gas entering the compressor constant, even though the flow from the evaporator is decreasing as the refrigeration load decreases. There is no reduction in power input to the compressor as the bypass valve is opened. This is a relatively energy-inefficient method to control the capacity of a centrifugal compressor.

12.2.2 Condensers

After the refrigerant vapor is compressed, it is a hot, high pressure vapor. The vapor then goes to a condenser where heat is removed from the refrigerant to make it a saturated liquid. The condenser rejects heat from the refrigerant to a cooler medium, usually ambient air or water. The most common types of condensers are air cooled, water cooled and evaporative (air and water cooled).

An air cooled condenser removes heat from the refrigerant by passing ambient air over an extended surface coil through which the refrigerant vapor is circulated. A typical air cooled condenser is shown in Figure 12-14. An air cooled condenser consists of a coil, casing, fan(s), and motor(s). It condenses the refrigerant gas by means of sensible heat transfer between the refrigerant and ambient air. For a given coil surface area and airflow, the capacity of an air cooled condenser varies directly with the difference

Figure 12-14. Air cooled condenser (*Courtesy of York International*).

in temperature between the refrigerant condensing temperature and the entering air dry bulb temperature.

For a refrigeration system to function properly, the condensing pressure and temperature must be controlled. An increase in condensing temperature causes a reduction in capacity and requires additional power. Low condensing pressures hinder the flow of refrigerant through liquid feed devices, which starves the evaporator and causes a loss of capacity. There are two basic methods of controlling the capacity of air cooled condensers: air side control and refrigerant side control.

Air side controls vary the flow of air through the condenser. Airflow may be controlled by cycling of multiple fans, the modulation of a volume damper installed in the fan discharge, a face damper on the coil or by a variable speed fan. Air side condenser controls may also be a combination of the above methods.

Refrigerant side controls include modulation of the active coil surface area, or flooding of sections of the condenser. Modulation of the coil surface area is accomplished by flooding the coil with liquid refrigerant. Flooding the coil reduces the surface area available for heat transfer. Multiple valving arrangements provide the required amount of flooding to meet varying capacities. Sections of the coil may be blocked off during light loads with solenoid valves that block off one of the condenser sections as well as its pump down to suction.

Water cooled condensers are used for larger refrigeration systems and larger chillers. A water cooled condenser consists of heat transfer tubes mounted within a steel shell. The refrigerant is condensed by the flow of water through the condenser. The heat is then rejected to the atmosphere as the water is cooled in a cooling tower. Cooling towers are discussed in Chapter 8. The most common types of water cooled condensers are shell-and-tube, shell-and-coil, tube-in-tube and brazed plate.

Shell-and-tube condensers circulate the cooling water through tubes while the refrigerant is contained by the shell and condenses outside the tubes. A shell-and-tube condenser is shown in Figure 12-15. The tubes may have a single or multi-pass circuit. Shell-and-tube condensers are available in sizes from 1 to 10,000 tons of refrigeration.

Shell-and-coil water cooled condensers are similar to shell-and-tube except that the containment of the water and refrigerant is reversed. Shell-

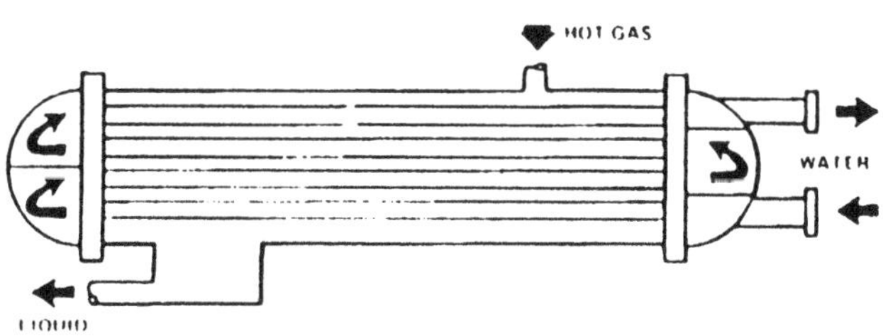

Figure 12-15. Shell-and-tube condenser.

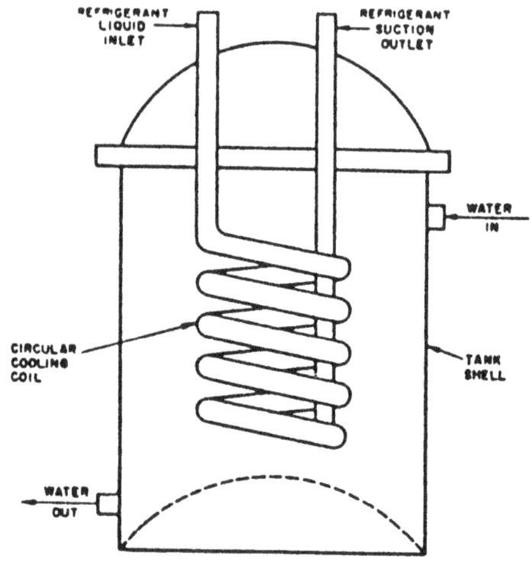

Figure 12-16. Shell-and-coil condenser.

and-coil condensers have the cooling water circulated through one or more coils. The refrigerant is contained by the shell and condenses outside the tubes. A shell-and-coil condenser is shown in Figure 12-16. The capacities of shell-and-coil condensers range from 0.5 to 15 tons.

Tube-in-tube condensers consist of one or more assemblies of tubes, one within the other. The refrigerant vapor is condensed in either the annular space between the tubes or in the inner tube. The other tube space contains

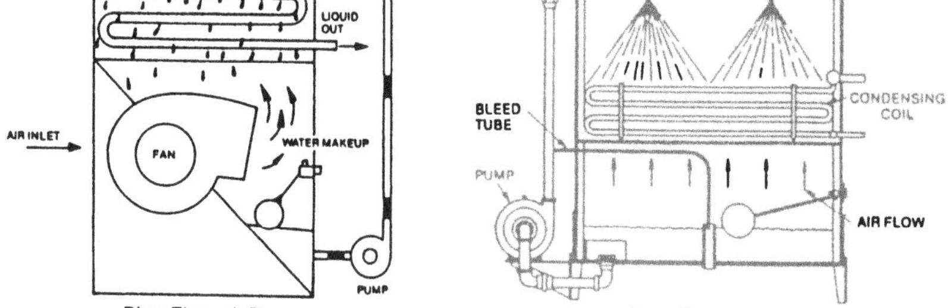

Blow-Through-Type

Draw-Through-Type

Figure 12-17. Evaporative condenser.

the circulated cooling water. Tube-in-tube condensers have refrigerating capacities ranging from 0.3 to 50 tons.

Brazed plate condensers are constructed of plates brazed together to form an assembly of separate channels. They are similar to plate and frame heat exchangers. The plates are usually stainless steel through which heat is transferred from the refrigerant to the cooling water. Capacities of brazed plate condensers range from 0.5 to 100 tons.

An evaporative condenser is essentially an air cooled condenser with water sprayed on the coil surface to increase the transfer of heat from the refrigerant inside the coil. Spraying water onto the coil reduces the coil surface temperature due to evaporation of the water in the circulated ambient air. An evaporative condenser is shown in Figure 12-17. In comparison with an air cooled condenser, an evaporative condenser requires less coil surface area and less airflow to reject the same amount of heat. Greater operating efficiencies can be achieved with an evaporative condenser operating at a lower condensing temperature.

12.2.3 Evaporators

The evaporator is a heat exchanger in which heat is transferred from the refrigerated space or material to the refrigerant. As the heat exchange occurs, the liquid refrigerant changes phase and evaporates from a liquid to a vapor.

Evaporators may be classified as either flooded or direct expansion. Flooded evaporators have the medium being cooled flowing within the tubes while the liquid refrigerant is outside the tubes. Direct expansion (DX) evaporators have the refrigerant flowing within the tubes while the medium being cooled is outside the tubes. DX evaporators are also called dry-type evaporators.

Flooded evaporators are typically shell-and-tube heat exchangers. They are frequently used in chillers for producing chilled water. The evaporator consists of a cylindrical shell and a number of straight tubes arranged in

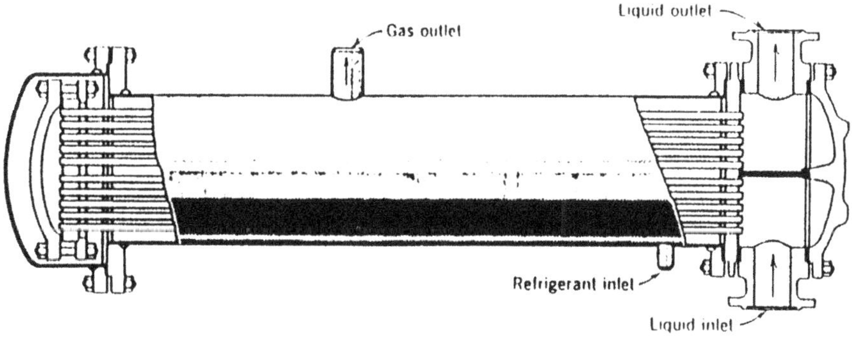

Figure 12-18. Flooded evaporator.

parallel and held in place at the ends by sheets. A typical flooded evaporator is shown in Figure 12-18. For chilled water applications, the water is cooled as it is circulated through the tubes while the liquid refrigerant is contained by the shell. As the water is cooled, the refrigerant evaporates and is removed at the top of the shell. The flow of liquid refrigerant into the evaporator is usually controlled by the level of liquid refrigerant in the shell in response to the rate at which it evaporates. Flooded evaporators are available with refrigerating capacities ranging from 10 to several thousand tons.

Direct expansion evaporators are also used for liquid chilling. They are used exclusively for the cooling of air in refrigerated spaces. There are three principal types of DX evaporators: bare-tube evaporators, plate-surface evaporators and finned coils. Bare-tube and plate-surface evaporators can be classified together as prime surface evaporators. The entire surface of these types of evaporators is essentially in contact with the vaporizing refrigerant inside. In finned tube evaporators, the refrigerant carrying tubes are the only prime surface and the fins are secondary heat transfer surfaces.

Bare-tube evaporators are typically large ceiling-hung bare pipe coils that employ natural convection of air in the refrigerated space. They are used in refrigerated storage rooms where the circulation of large quantities of low velocity air is desirable. They may also be used with centrifugal fans to provide high velocity cold air for blast cooling or freezing applications.

Plate-surface evaporators for larger refrigeration applications consist of formed tubing installed between two metal plates that are welded together

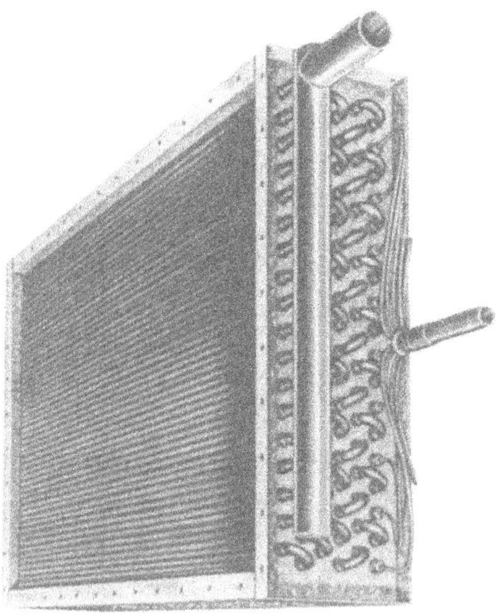

Figure 12-19. Finned coil evaporator.

at the edges. The space between the welded plates may be filled with a eutectic solution or be evacuated to improve thermal contact between the plates and the tubes carrying the refrigerant. Evacuating the space between the plates causes atmospheric pressure to be exerted outside the plates to press them together and hold them firmly against the tubing.

With bare-tube evaporators, much of the air circulated over the coil passes through the space between the tubes and is not cooled. Finned coils are bare-tube coils that have fins to increase the rate of heat transfer. The fins are attached to the tubes and serve as secondary heat transfer surfaces, increasing the outside heat transfer surface area of the evaporator. A finned coil evaporator is shown in Figure 12-19. Increasing the heat transfer surface area improves its heat transfer efficiency and its ability to cool air and other gases.

12.2.4 Refrigeration System Controls

The control of refrigerant flow in refrigeration systems is necessary to match the capacity of the system to the load as closely as possible. The primary control devices used to control refrigeration systems include pressure control switches, pressure transducers, temperature control switches, fluid flow sensing switches, differential control switches and float switches.

Pressure control switches operate one or more sets of electrical contacts and are typically used to open or close water or refrigerant flow control valves. They are also used as safety devices to verify the required pressure such as an oil pressure switch. Pressure transducers are often used as a substitute for pressure switches. The pressure transducer is an analog device that produces a continuous low voltage output signal proportional to the applied pressure.

Temperature switches are switches that open or close at a specified temperature. Temperature control switches may be indirect or direct switches. An indirect temperature switch is a pressure control switch in which the pressure response element is replaced with a temperature responsive element. A direct temperature control switch typically has bimetallic disks that activate electrical contacts when the temperature increases or decreases.

Other types of switches include flow sensing switches, differential control switches and float switches. Flow sensing switches are typically used to "prove" the flow of air or a liquid. Differential control switches are used to maintain given differences in pressure or temperature between two pipe lines or spaces. Float switches have a float that operates one or more sets of contacts, and are used to maintain liquid levels.

Control valves are used to start, stop, direct and modulate the flow of refrigerant to satisfy the system capacity requirements in accordance with load requirements. The most common types of control valves are thermostatic expansion valves, electric expansion valves and constant pressure expansion valves.

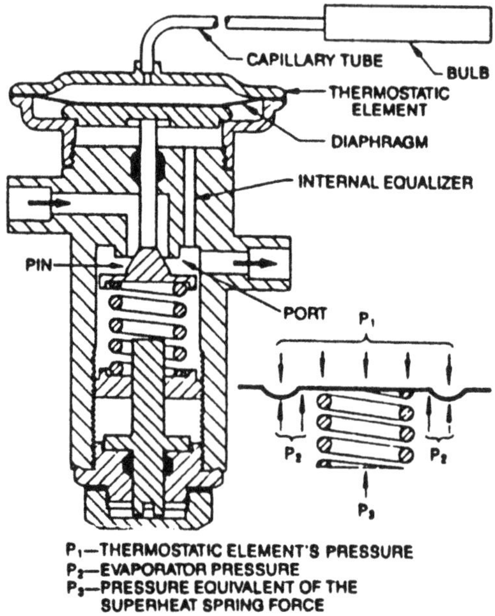

Figure 12-20. Typical thermostatic expansion valve (TXV).

Thermostatic expansion valves (TXV) are used in direct expansion (DX) refrigeration systems. The TXV controls the flow of liquid refrigerant entering the evaporator in response to the superheat of the vapor leaving the evaporator. The function of the TXV is to keep the evaporator active without permitting liquid refrigerant to return through the suction line to the compressor. The TXV controls the mass flow of the refrigerant entering the evaporator so it is equal to the rate at which it can be completely

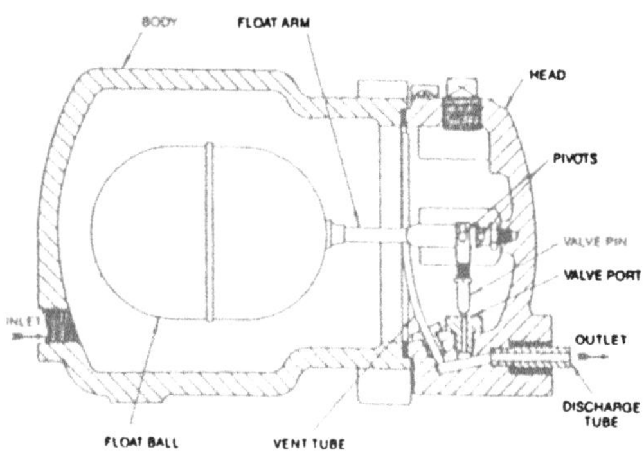

Figure 12-21. High side float valve.

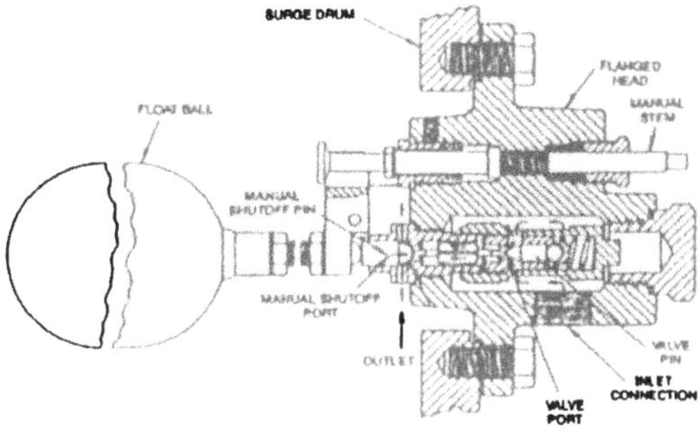

Figure 12-22. Low side float valve.

absorbed by the heat. A typical thermostatic expansion valve is shown in Figure 12-20.

Electric expansion valves are similar to thermostatic expansion valves. They are electrically conrtrolled devices that control the flow of refrigerant. They may be heat operated valves, magnetically modulated valves, pulse width modulated valves or step motor driven valves.

A constant pressure valve is operated by operator or valve outlet pressure. It regulates the mass flow of liquid refrigerant entering the evaporator and maintains its pressure at a constant value. Constant pressure valves are suitable only for constant load applications.

Float valves are used to control the flow of liquid refrigerant to the evaporator in flooded evaporators. Float valves may be high side float valves or low side float valves. A high side float valve is shown in Figure 12-21. A high side float valve controls the mass flow of refrigerant liquid entering the evaporator so it is equal to the rate at which the refrigerant vapor is pumped from the evaporator to the compressor. High side float valves are usually used on refrigeration systems with a single evaporator, compressor and condenser. A low side float valve performs the same function as a high side float valve, however, it is connected to the low pressure side of the system. A low side float valve is shown in Figure 12-22. When the evaporator liquid level drops, the float opens the valve, admitting more liquid refrigerant. Low side float valves can be used with multiple evaporators such that some evaporators may be controlled by other low side float valves and some thermostatic expansion valves.

12.3 MAINTAINING AND OPERATING REFRIGERATION SYSTEMS FOR OPTIMUM ENERGY EFFICIENCY

Refrigeration systems typically require routine maintenance to keep them operating at optimum efficiency. Significant energy savings can be achieved by proper routine maintenance and operation. Proper routine maintenance and operation should be the first step in any energy management program for refrigeration systems. The following are operations and maintenance procedures that will result in optimum energy efficiency for most refrigeration systems.

Perform Routine Maintenance Regularly: Routine maintenance on refrigeration systems should be performed on a regular basis. To ensure optimum energy efficiency, the following maintenance steps should be undertaken on a regularly scheduled basis.

- clean condenser and evaporator coils and tubes
- adjust expansion valve superheat on direct expansion systems
- adjust refrigerant float valves for proper operation on flooded refrigeration systems

Reduce Refrigeration Loads: The lower the refrigeration load, the lower the amount of energy consumed by a refrigeration system. Refrigerated spaces should be well insulated. The seals on doors to refrigerated spaces should be maintained to reduce infiltration of warm air and moisture into the refrigerated space. Doors to refrigerated spaces should be kept closed when not in use.

Ensure Adequate Air flow Over Evaporator: Air velocity and distribution across an evaporator coil significantly affects the performance of the evaporator coil. If air circulation is not adequate, heat is not carried away from the product to the evaporator at a rate sufficient to allow the evaporator to perform at peak efficiency. It is also important that the circulation of air be evenly distributed in all parts of the refrigerated space and over the evaporator coil.

All obstructions that might block or disturb the flow of air through the evaporator coil should be removed.

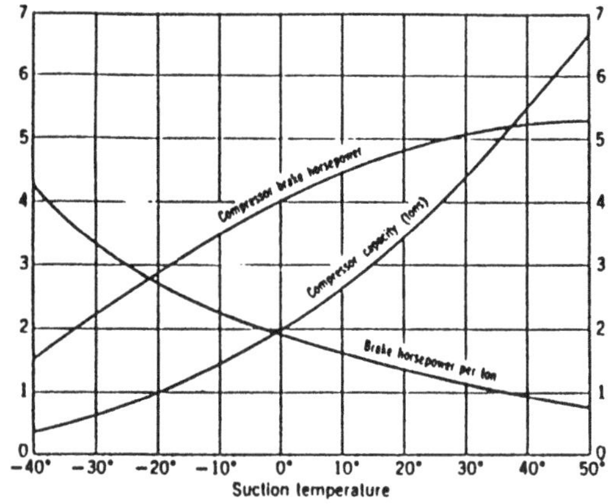

Figure 12-23. Compressor performance versus suction temperature.

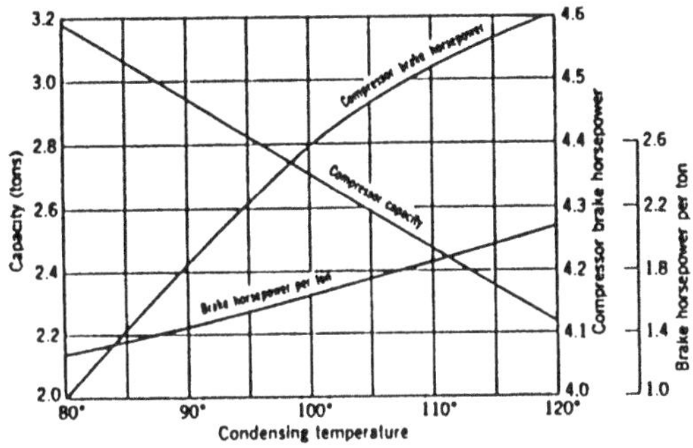

Figure 12-24. Compressor performance versus condensing temperature.

Reduce Thermal Lift: The efficiency of a vapor-compression refrigerating system varies considerably with both vaporizing and condensing temperatures. The vaporizing temperature has the greatest effect. The power required by a compressor depends upon the lift required to raise the pressure of the refrigerant from the pressure at the required suction temperature to that at the required condensing temperature.

The condensing temperature is a function of the temperature of the condensing medium. It is equal to the temperature of the condensing medium plus the temperature difference between the condensing refrigerant and the condensing medium. If the vaporizing temperature remains constant, the efficiency of the cycle decreases as the condensing temperature increases. The mass flow per unit mass to produce one ton of refrigerating capacity is less at a higher vaporizing temperature.

Less thermal lift results in less energy consumed per ton-hour of operation. As a rule of thumb, each 1°F lower head results in 1% more refrigerating capacity. Figure 12-23 shows the effect of suction temperature on compressor performance, and Figure 12-24 shows the effect of condensing temperature on compressor performance.

12.4 ENERGY CONSERVATION MEASURES FOR REFRIGERATION SYSTEMS

The energy conservation measures for refrigeration systems discussed in the previous section were operation and maintenance procedures that would improve the energy efficiency of existing refrigeration systems. Most of those measures were little, or no, cost measures that could be undertaken on an ongoing basis to improve efficiency.

In addition to the low cost OM procedures, there are also modifications that could be made to refrigeration systems to improve energy efficiency. These modifications, however, require a more significant cost investment and are permanent improvements. The modifications listed below are improvements that should be considered under any energy management program for existing, or proposed, refrigeration systems.

12.4.1 Floating Head Pressure Control

Most refrigeration systems have the compressor discharge (head) pressure set at a constant pressure high enough to ensure that the refrigerant pressure between the condenser and thermal expansion valve (TXV) is high enough to prevent flash gas formation. Constant head pressure control is simple, but it results in additional compressor energy consumption during most operating conditions.

Floating head pressure control allows the condenser temperature and pressure to float downward with the ambient temperature. The lower the condenser temperature, the more efficient the refrigeration cycle. Floating head pressure saves energy by reducing the compression ratio of the compressor when ambient temperatures are low.

Energy savings will result whenever ambient temperatures are low enough for condensing to take place below the constant pressure set point. Energy savings are related to the portion of annual compressor run time that occurs during low ambient conditions.

12.4.2 Liquid Subcooling

Liquid subcooling involves the cooling of the liquid refrigerant between the condenser and the TXV at a constant pressure. The refrigerant leaving the condenser is a saturated liquid. Cooling the liquid refrigerant below the saturation temperature while the pressure remains constant results in a subcooled liquid.

When the liquid refrigerant is subcooled before it reaches the refrigerant control valve, the refrigerating effect per unit mass is increased. The increase in refrigerating effect is equal to the change in the refrigerant enthalpy resulting from the heat removed during the subcooling. Because of the greater refrigerating effect per unit mass, the mass flow rate of the refrigerant per unit capacity is less for a subcooled cycle than for a saturated cycle. As a result, the refrigerating effect per unit mass is increased without increase in the energy input to the compressor. Subcooling will increase the cycle COP and reduce the power per unit capacity.

Subcooling of the liquid refrigerant may be accomplished several ways, including:

- Installing a liquid subcooler.
- Subcooling the refrigerant while it is stored in the liquid receiver tank.
- Subcooling the refrigerant while it passes through the liquid line giving off heat to the surroundings.

The increase in system capacity and efficiency resulting from liquid subcooling is frequently more than enough to justify the additional cost of a subcooler, especially for low temperature applications. For a constant condenser surface area, approximately a 1/2% increase in capacity will result for each 1°F of liquid subcooling.

Example 12.1. A refrigeration system using R134a as a refrigerant vaporizes the refrigerant in the evaporator at 20°F. The refrigerant leaves the condenser and enters the TXV as a saturated liquid at 100°F. If the liquid refrigerant was subcooled so it enters the TXV at 80°F, what would be the percentage increase in the refrigerating effect?

Figure 12-25 is a pressure-enthalpy diagram for R134a, which shows the state points for this example. Point C is the state point where the refrigerant leaves the evaporator as a saturated vapor at 20°F. The enthalpy for Point C is

$$hc = 104.6\,\text{Btu/lbm}$$

For the cycle with a saturated liquid refrigerant, point A_{sat} is the state point where the refrigerant leaves the condenser as a saturated liquid at a temperature of 100°F and a pressure of 138.8 psia. Point B_{sat} is the state point where the refrigerant leaves the expansion valve at 20 psia. Since the expansion of the refrigerant through the TXV is an adiabatic process, the enthalpy of the refrigerant remains constant as it passes through the valve. Therefore, the enthalpy at state points A_{sat} and B_{sat} is

$$h_{A-sat} = h_{B-sat} = 44.9\,\text{Btu/lb}$$

For the cycle with saturated liquid refrigerant, the refrigerating effect due to the enthalpy change in the evaporator as the refrigerant is vaporized would be

$$q_{sat} = h_C - h_{B-sat} = 104.6 - 44.9 = 59.7\,\text{Btu/lb}$$

If the liquid refrigerant was subcooled to 80°F at a pressure of 138.8 psia, point A_{sub} would be the state point of the refrigerant leaving the subcooler and point B_{sub} would be the state point of the subcooled refrigerant after leaving the TXV. The enthalpy of points A_{sub} and B_{sub} would be

$$h_{A-sub} = h_{B-sub} = 37.3\,\text{Btu/lb}$$

For the cycle with subcooled liquid refrigerant, the refrigerating effect due to the enthalpy change in the evaporator as the refrigerant is vaporized would be

$$q_{sub} = h_C - h_{B-sub} = 104.6 - 37.3 = 67.3\,\text{Btu/lb}$$

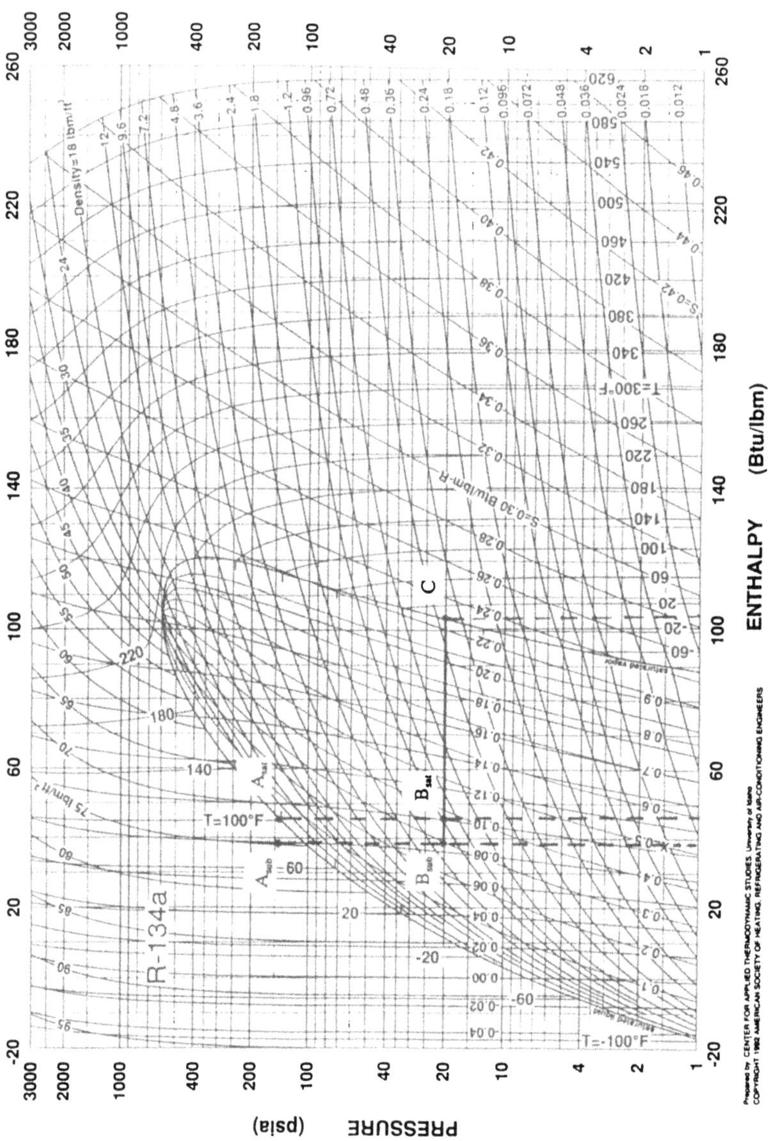

Figure 12-25. Pressure enthalpy diagram for Example 12.1.

The percent change in refrigerating effect due to subcooling the liquid refrigerant would be

$$\%\,\text{change} = \frac{67.3 - 59.7}{67.3} \times 100\% = 11.3\%$$

12.4.3 Liquid Refrigerant Pumping

Liquid refrigerant pumping (LRP) and liquid injection de-superheating are two patented methods of decreasing compressor energy consumption and increasing condenser capacity in an existing direct expansion (DX) reciprocating vapor compression refrigeration system. Conventional refrigeration systems maintain a constant refrigerant pressure as it is discharged from the compressor. This discharge (head) pressure is designed to allow condensation of the refrigerant at low ambient temperatures. As a result, energy is wasted during mild and cool ambient conditions because the condenser temperature and pressure are unnecessarily high. If the head pressure is allowed to float, the compressor discharge pressure can be lowered according to the ambient temperature. Unfortunately, the amount the compressor discharge pressure can be lowered is limited to the minimum pressure difference required across the thermal expansion valve (TXV) to maintain a liquid seal at the valve inlet and ensure stable valve operation without excessive vapor superheat.

The LRP system uses a small pump to pressurize the liquid refrigerant in the liquid line, as shown in Figure 12-26. A centrifugal pump is used to increase the pressure of the liquid refrigerant by 8 – 20 psi. The amount of pressure added is equal to the pressure loss between the condenser outlet and the TXV inlet. On systems with receivers, the pressure is added

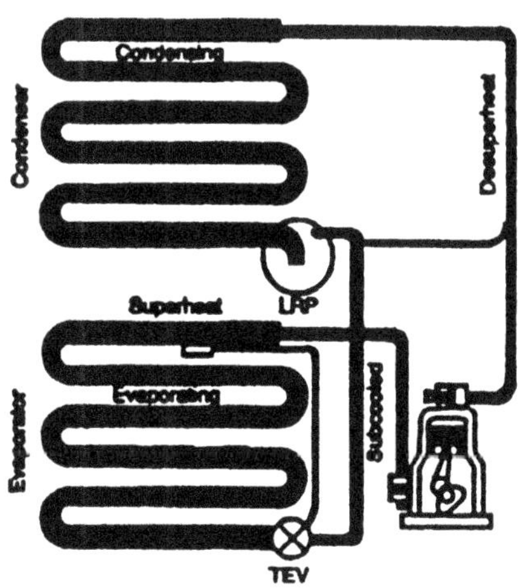

Figure 12-26. Refrigeration system with liquid refrigerant pumping.

between the receiver and TXV. By increasing the pressure of the liquid refrigerant, the associated saturated temperature is increased due to the pressure increase while the actual temperature remains constant. With LRP, the liquid refrigerant is below the saturation temperature for the increased pressure. The LRP has the effect of subcooling the liquid refrigerant without changing its temperature. It reduces the amount of vapor refrigerant passing through the TXV to the evaporator that produces no refrigerating effect. The LRP ensures that only liquid refrigerant reaches the TXV. By increasing the pressure of the liquid refrigerant, the condensing pressures can be allowed to fall to their lowest attainable pressures. This allows the head pressure to float.

Advantages of a liquid refrigerant pumping system include:

- Elimination of liquid line flashing by replacing liquid line pressure losses.
- Reduction in compressor energy consumption since pumping liquid refrigerant is more efficient than compressing refrigerant vapor.
- Increase in evaporator capacity along with net refrigeration effect.
- Lower compression ratios.
- Ensuring minimum pressure drop across the TXV while allowing the condensing pressure to remain low.

LRP systems can also increase refrigeration system energy efficiency by de-superheating the vapor refrigerant. Some of the liquid refrigerant from the LRP pump can be injected into the compressor discharge line or at the inlet to the condenser. Reducing the superheat of the vapor entering the condenser results in higher condenser efficiency, lower condensing temperature and greater compressor efficiency. De-superheating the refrigerant vapor entering the condenser can result in an energy savings of 6 – 12%.

The energy savings resulting from the installation of an LRP is very application-specific and depends on a number of factors, including:

- Ambient temperature. A low condensing temperature is required when decreasing the compressor head pressure, which is only possible during cool weather. The ambient temperature must be low enough not only to extract sufficient heat from the condenser coils, but also make up for increases in energy consumption by the LRP pump.
- System load profile. If the system refrigeration load profile is directly related to the ambient temperature, the total energy consumption by the system will be at a maximum when the maximum LRP energy savings are available. A system that is independent of ambient temperatures would result in greater energy savings.
- Operating hours. Greater hours of operation result in greater energy savings. If the above factors can be optimized during periods of peak electrical demand, further savings may result.

Little, if any, savings will result if the refrigeration system already has floating head control or systems that do not operate much during low ambient temperature conditions necessary to accrue floating head savings.

12.4.4 Replace Air Cooled Condenser With Evaporatively Cooled

An air cooled condenser removes heat from the refrigerant by circulating ambient air over an extended surface coil through which the refrigerant is circulated. All heat transfer between the coil and the ambient air is by sensible heat transfer. For a given surface area and air quantity, the capacity of an air cooled condenser, for the most part, is directly proportional to the difference between the refrigerant condensing temperature and the entering dry bulb temperature of the ambient air.

The primary advantage of air cooled condensers is their simplicity. They consist of a fan and motor, or fans and motors, a coil and an enclosure. Typically, they have a relatively large fan horsepower to move large volumes of air to achieve the necessary cooling of the refrigerant.

An evaporative condenser consists of a condensing coil, fan and motor, water distribution system, water sump, a water circulating pump and unit casing. A typical evaporative condenser is shown in Figure 12-17. Heat rejected by the refrigerant through the coil is dissipated by water sprayed over the coil as ambient air is circulated through the coil.

In an evaporative condenser, heat rejection occurs by two methods: sensible heat transfer and latent heat transfer. Latent heat is transferred as the circulated water is evaporated into the circulated air. There is also sensible heat transfer to the circulated air as in the air cooled condenser. The combined sensible and latent transfer of heat is a more effective means of heat dissipation. As a result, the total energy required to (pump horsepower and fan horsepower) reject heat is less than for an air cooled condenser under the same operating conditions. The evaporative condenser generally costs less and requires less space than an air cooled condenser.

12.5 ENERGY EFFICIENCY FOR NEW REFRIGERATION SYSTEMS

Refrigeration systems can be one of the largest consumers of energy at many facilities. It is important that energy efficiency be a primary consideration when selecting new refrigeration equipment for new, or retrofit, installations. In some cases, local or national energy codes may dictate efficiencies of such equipment.

Earlier in this chapter, energy conservation for existing refrigeration systems was discussed. Several of the energy efficiency measures discussed as retrofit opportunities for equipment, such as liquid refrigerant pumping, are applicable to new and retrof it installations. In addition to the previously discussed energy conservation measures for refrigeration systems, the following measures should be considered for new or replacement equipment and accessories.

12.5.1 Install Multiple Compressors

Frequently, refrigeration system compressors are oversized to ensure adequate refrigeration capacity. Even properly sized compressors typically operate at peak capacity for only short periods of time. As a result, many compressors usually operate at only part load. Oversized, or underloaded, compressors are less efficient, operate at a lower suction temperature and require more kW per ton of refrigeration.

Refrigeration compressors should be sized to match the actual load as closely as possible. Capacity control can readily be achieved by using multiple compressors with capacity control systems such that one compressor reaches its minimum capacity before another compressor starts to unload. The use of multiple compressors has the following advantages:

- Single speed motors may be selected and operated continuously at their best efficiency.
- Standby equipment is available, which allows partial load operation if one of the machines breaks down.
- Compressors may be started in sequence to limit the current inrush and demand if time delay devices are employed.

12.5.2 Liquid Subcooling Heat Exchanger

Earlier in this chapter the advantages of subcooling the liquid refrigerant between the condenser and the thermal expansion valve were discussed. The amount of subcooling can be increased by including a heat exchanger

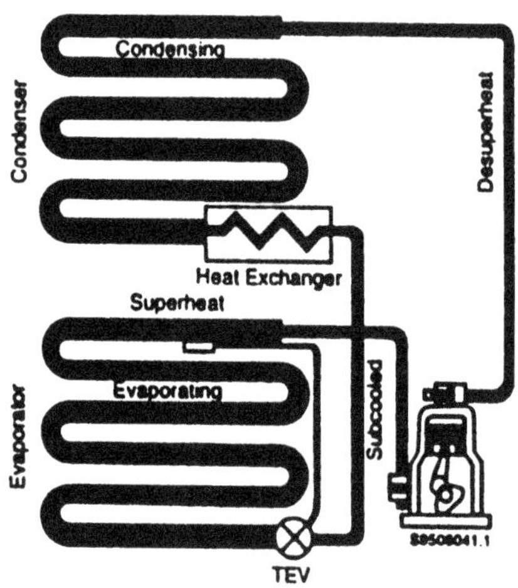

Figure 12-27. Schematic of refrigeration system with liquid cooling heat exchanger.

in the refrigerant line to subcool the refrigerant. Although these heat exchangers are not suitable as a retrofit in an existing system, they are available in some new systems. A schematic diagram of a refrigeration systems with a liquid cooling heat exchanger is presented in Figure 12-27.

Liquid subcooling heat exchangers use a heat sink to remove additional energy from the refrigerant. An external heat sink, such as ambient air, water from a small cooling tower or ground water may be used. There are three types of subcoolers available. These are (1) systems in which suction-line heat of the vapor-compression system acts as a heat sink; (2) systems that include a small (but more efficient) secondary vapor compression system for subcooling; and (3) systems with an external heat sink.

If the system uses an external heat sink to reject heat from the refrigerant, a smaller compressor can be used than is feasible with a suction-line subcooler. This increases the overall system performance and efficiency since the compressor and condenser can be downsized, which leads to higher overall efficiency, lower electrical demand and reduced energy consumption.

Refrigerant subcooling works best where year-round outdoor temperatures are relatively high and relatively constant. Most installations of subcooling heat exchangers are custom-designed to obtain optimum system performance.

Chapter 13

Cogeneration Systems

In a conventional power generation system, considerable energy is wasted in the form of heat rejection to the atmosphere through cooling towers, ponds, lakes or rivers. Significant energy savings may be achieved if the thermal energy that would otherwise be wasted during power generation is recovered and put to good use. Cogeneration systems are systems that make use of the waste heat that results from the generation of power.

Cogeneration may be broadly defined as the coincident generation of combined heat and power. These systems simultaneously produce electrical or mechanical (power) and useful thermal energy from a single energy steam, such as coal, natural gas, liquefied gas, biomass or solar. By capturing and applying heat from an effluent stream as power is produced, cogeneration systems operate at efficiencies greater than those achieved when heat and power are produced in separate or distinct processes.

13.1 BASIC PRINCIPLES OF COGENERATION

The main components of a cogeneration system may include a prime mover, electrical generation equipment and heat recovery equipment. Prime movers can be reciprocating engines, combustion turbines or steam turbines. Electrical generation equipment may include generators, power conditioning equipment and electrical distribution equipment. Energy recovery for cogeneration systems usually involves recovery of heat from the prime mover for other process heating, comfort heating or domestic hot water heating applications.

13.1.1 Definitions and Terms Used in Cogeneration Systems

For definitions of terms used in Cogeneration Systems, check the Glossary.

13.1.2 Regulatory Issues for Cogeneration

Frequently, a cogeneration facility generates more electrical power than it needs or uses. When excess power is generated, cogenerating facilities often sell the excess power to the local electrical utility. To encourage the use of cogeneration, a number of laws were developed in the United States to encourage facilities to use cogeneration by removing or limiting barriers to cogeneration.

In 1978, a series of laws referred to as the National Energy Acts were passed. One of the laws, the Public Utilities and Regulatory Policies Act (PURPA), was passed to encourage the development of cogeneration and renewable energy resources. PURPA removed several significant barriers, which may have limited the development of energy-efficient and economic cogeneration.

Under PURPA, the following benefits are mandated for a qualifying generator:

- Electric utilities must purchase excess power from cogenerators and small power producers (SPPs) at the price that the utility avoids by not producing it (avoided cost). A utility must allow congenerators to be paralleled (interconnected) with its grid if the cogenerator or SSP pays for it and complies with the utility's safety and protective requirements.
- The utility must provide supplemental power for facilities that do not cogenerate 100% of their power, standby power for emergency power during a cogenerator's outage, and maintenance power at nondiscriminatory rates during a planned outage for maintenance.
- Cogenerators and SSPs selling power to a utility are exempt from the "rate of return" (allowable return on investment) of utility regulations of the Federal Energy Regulatory Commission (FERC) and Securities and Exchange Commission (SEC).
- Utilities may wheel (transmit) the cogenerated power over their grid to a location remote from the cogeneration plant and, under certain conditions, can be ordered by the FERC to wheel it.

For a cogenerator to obtain the benefits listed above, it must become a qualifying facility (QF) by showing that it meets ownership and efficiency requirements that have been established by the FERC. The requirements are as follows:

- An ownership standard, which requires that no more than 50% of a facility be owned by electric utilities or electric utility holding companies, either singly or in partnership.
- An operating standard, which requires that a minimum of 50% of the useful output of a cogeneration system be in the form of thermal energy.
- An efficiency standard, which requires that the efficiency of a cogeneration system be at least 45% for topping cycles. If the amount of useful thermal energy produced by the cogeneration system exceeds 15%, then the efficiency threshold decreases to 42.5%. There

is no efficiency standard for a bottoming cycle or renewable resources. The FERC efficiency is defined as

$$\frac{\text{Power Output} + \frac{1}{2}\text{ Thermal Output}}{\text{Lower Heating Value of Fuel Input}}$$

- A cogenerator's retail power and steam rates may be regulated by state bodies under a broad authority and subject to local sales or gross receipt taxes.
- An SSP plant must be no greater than 80 MW unless powered by waste products or renewable energy.

In addition to the regulations listed above, cogenerators may also be subject to other laws, such as state laws and regulations and emission regulations.

13.1.3 Cogeneration Cycles

There are two broad classifications of cogeneration cycles. The classifications, which are based upon the primary use of the energy consumed by the cogeneration system, are: topping cycles and bottoming cycles.

A topping cycle is one in which the energy input to the cogeneration system is first used to produce useful power, and the rejected heat from the production of power is used for other purposes. An example of a topping cycle would be one that generates electricity from high pressure steam and uses the exhausted steam or other hot gas for process heat.

A bottoming cycle is one in which the energy input to the cogeneration system is first applied to another thermal process and the rejected heat is then used to produce power. The thermal energy is first used for a process application, then the exhaust energy is used to produce electricity at the bottom of the cycle.

A combined cycle is a combination of a topping cycle and a bottoming cycle. Power is produced in a topping cycle, typically a gas turbine generator. The heat exhausted from the turbine is used to produce steam, which is then expanded in a steam turbine to generate more electric or shaft power.

13.2 PRIME MOVERS

Cogeneration is the thermodynamically sequential production of two useful forms of energy from a single source. The useful forms of energy are thermal and power. Power may be electrical power or shaft power. In a cogeneration system, a prime mover is that system component that converts thermal or chemical energy into power. Commercially available prime movers include internal combustion reciprocating engines, combustion turbines and steam turbines.

13.2.1 Reciprocating Engines

Reciprocating engines include a variety of internally fired piston-driven engines. They are commonly used as prime movers in smaller cogeneration systems, under 15 MW and 27,000 brake horsepower (bhp). They are quite efficient and are available in a wide variety of sizes. They use most types of liquid and gaseous fuels, including methane gas, natural gas, fuel oil gasoline and diesel oil.

Major engine components include a starting system, ignition system, fuel-air mixing system, combustion chamber, and an exhaust gas collection and removal system. Figure 13-1 shows many of the components of a reciprocating engine. In addition to these components, some engines also include a turbocharger to increase the power output.

Reciprocating engines operate on one of two basic cycles: the Otto cycle or the diesel cycle. The Otto cycle is also known as the spark-ignition cycle. Spark-ignition engines use volatile liquids or gases as fuel and have compression ratios between 4:1 and 12:1. A typical spark-ignition engine internal combustion engine is shown in Figure 13-2. These engines apply an externally supplied spark to a compressed mixture of air and fuel in each engine cylinder. A reciprocating piston linked to a crank provides compression of the fuel-air mixture; the burning fuel expands and

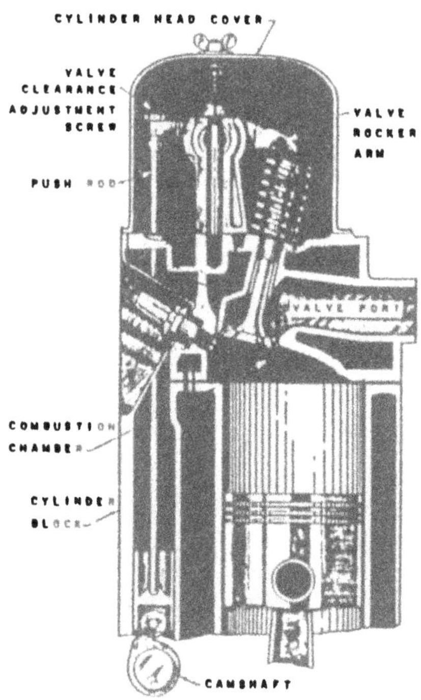

Figure 13-1. Components of a reciprocating engine.

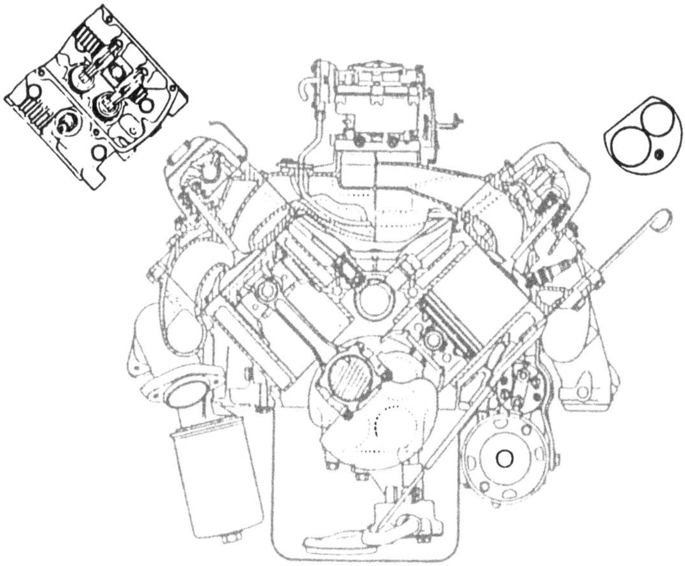

Figure 13-2. Typical spark-ignition internal combustion engine.

performs work, applying force to the piston and causing it to move. Spark-ignition engines are available in sizes up to 18,000 bhp.

Diesel engines are similar to spark-ignition engines; however, diesel engines do not utilize an external spark to ignite the fuel. These engines inject the fuel directly into the combustion chamber where the fuel-air mixture spontaneously ignites. Diesel engines are also called compression ignition engines and use liquid fuels of low volatility. The heat for spontaneous combustion occurs during the compression stroke, as the work of compression causes the temperature of the fuel-air mixture to increase. No ignition devices are used. Diesel engines have compression ratios ranging from 11.5:1 to 22:1 and compression pressures between 400 and 700 psi. Engine speed is controlled by varying the rate at which fuel is injected into the cylinders.

Reciprocating engines are generally 20 – 40% efficient in converting fuel to shaft energy. Other important performance characteristics of reciprocating engines include the engine's power rating, rate of fuel consumption and thermal output. The performance of an engine is dependent upon a number of design factors, including engine speed, method of aspiration, operating cycle, emissions controls and engine displacement. Environmental factors such as altitude and temperature also affect performance. The mechanical efficiency of an internal combustion engine is a function of several factors, including the compression ratio, which is defined as the ratio of the volume of uncompressed cylinder gases to the minimum cylinder compression stroke. The rate of fuel consumption is the greatest contributor to operating cost. It is influenced by the combustion cycle, engine speed, compression ratio and method of aspiration.

As part of the cogeneration process, heat is recovered from a reciprocating engine as it produces power. A reciprocating engine will reject heat from the following sources: radiation from the engine block and other hot surfaces, exhaust gases, lubricating oil, jacket water and, for turbocharged and aftercooled engines, the engine aftercooler. A measure of the recoverable heat from an engine is the thermal-to-electric ratio, which is the ratio of the useful thermal output to the electric power being generated. Heat recovery characteristics will vary as a function of engine speed, engine output, air-fuel ratio (rich versus lean), engine manufacturer and heat rejection strategy.

For most reciprocating engines, the recoverable thermal energy is heat that is rejected in the exhaust gases and jacket. Almost all of the jacket heat can be recovered by circulating a cooling fluid through the engine jacket. There are two methods for removing engine jacket heat: forced circulation cooling and ebullient cooling. A forced circulation cooling system removes heat from the engine by circulating a liquid coolant (such as water) through the jacket, which increases in temperature as it flows through the jacket. No phase change takes place as the coolant absorbs heat from the jacket. A pressurized high temperature forced circulation system can produce water temperatures as high as 260°F. Ebullient cooled engines reject heat to the coolant through a phase change in the coolant. The coolant enters the engine as a pressurized liquid at a temperature near its boiling point. Heat rejected from the engine is transferred to the coolant, where the coolant changes from the liquid phase to the vapor phase. The majority of the heat absorbed by the coolant results from the change of phase. After the vapor coolant leaves the engine, the heat is removed for use by thermal processes as it changes back to the liquid phase, where it is recirculated to the engine jacket. Ebulliently cooled engines can produce saturated steam at pressures up to 15 psig.

Heat can also be recovered from the engine exhaust gases and lubricating oil. Engine exhaust gases can reach temperatures as high as 1,200°F. Heat can be recovered from engine-lubricating oil at temperatures up to 160°F.

13.2.2 Combustion Turbines

Combustion turbines are also internal combustion engines; however, they operate on the Brayton cycle, consisting of adiabatic compression, constant pressure heating and adiabatic expansion. Turbine engines are available in sizes from 50 to 173,000 Bhp and can produce up to 1,000 MW of power. They can burn a wide range of liquid or gaseous fuels.

Combustion turbines consist of an air compressor, which increases the incoming air pressure; a combination fuel-air mixing and combustion section (combustor); and a expansion power turbine section that extracts energy from the combustion gases as they exit the turbine. Figure 13-3 shows a schematic diagram of the combustion turbine components. The major components include an air compressor, a combustor and a power

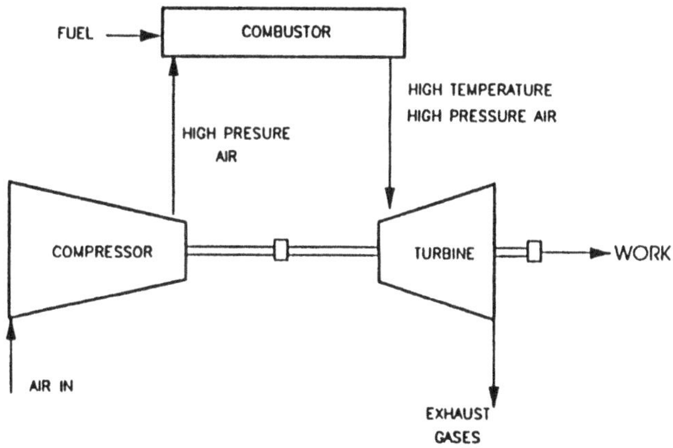

Figure 13-3. Schematic of combustion turbine components.

turbine. Incoming atmospheric air is compressed to about 175 to 600 psi in the compressor section. Fuel is then injected into the compressed air stream and ignited in the combustor. These high pressure hot gases are then expanded through a turbine, which provides power to drive the air compressor, as well as power to drive the load.

Combustion turbines offer several advantages over other internal combustion engine drives, including:

- small size, high power to weight ratio
- ability to burn a variety of fuels, though more limited than reciprocating engines
- ability to meet stringent pollution standards
- high reliability
- available in self-contained packages
- instant power, no warm-up required
- no cooling water required
- vibration-free operation
- easy maintenance
- low installation costs
- clean dry exhaust
- lubrication oil not contaminated by combustion oil

The mechanical efficiency of a combustion turbine is a function of the ratio of the turbine inlet and exhaust temperatures. The efficiency of a combustion turbine can be increased by increasing the turbine inlet temperature or by lowering the exhaust temperature. The mechanical efficiency of turbines also increases as the size of the turbine increases. Since the recoverable heat from a turbine will decrease relative to power output, the thermal-to-electric ratio will also vary as a function of turbine size.

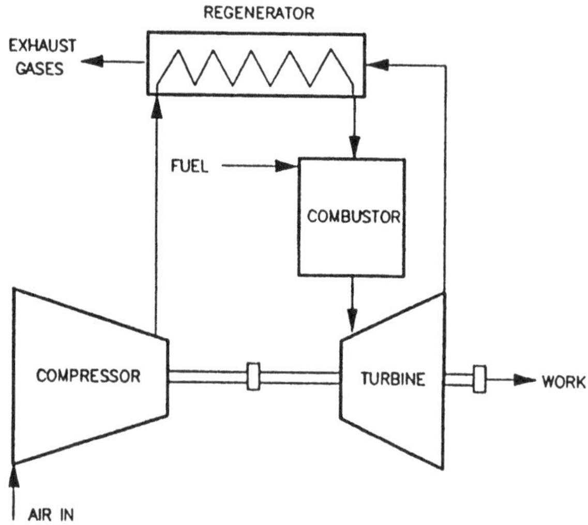

Figure 13-4. Gas turbine cycle with regenerator.

Increasing the efficiency of a turbine by lowering the temperature can be achieved by adding a regenerator to the cycle. The regenerator is a heat exchanger that utilizes waste heat from the exhaust gas of the turbine to preheat air discharged from the compressor. By preheating the compressed air, it is possible to reduce the amount of fuel required, which results in higher fuel economy and higher efficiency. A schematic diagram with a regenerator is presented in Figure 13-4. Recovering heat from the turbine exhaust to preheat combustion air reduces the amount of heat that is available for process heat recovery. The resulting gain in efficiency must also be balanced against the additional cost of the regenerator.

Gas turbine efficiency may also be improved by the addition of a reheat combustor to the cycle, as shown in Figure 13-5. The reheat

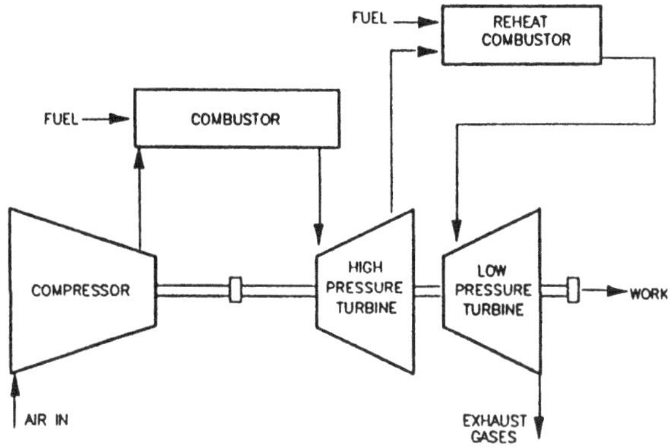

Figure 13-5. Gas turbine cycle with reheat combustor.

combustor provides a source of higher temperature heat in the cycle. In the reheat cycle, the combustion gases are partially expanded in the turbine and exhausted to a second combustion chamber where additional fuel is injected into the cycle and fired. As a result, the maximum temperature of the cycle is increased. The gases are then expanded in a second low pressure turbine, which produces additional power from the cycle. The reheat cycle increases the mechanical and thermal efficiencies of the cycle by increasing the average temperature at which heat is delivered to the cycle.

Gas turbine capacity and efficiency may also be increased by introducing water vapor into the compressor discharge air. This process is referred to as humid air turbine (HAT) cycle. Injecting either water or steam into the turbine combustor will increase the mass flow rate and the capacity of the turbine. The increased mass flow increases the turbine output and lowers combustion temperatures.

13.2.3 Steam Turbines

Steam turbines generate power by expanding high pressure steam through a set of blades in the turbine. A typical steam turbine based system consists of several major components, including a heat source, a steam powered turbine and a heat sink. A typical steam turbine system is shown in Figure 13-6. Steam turbine systems operate on the Rankine cycle.

The heat source in a steam turbine cycle is typically a steam boiler, which burns a liquid, gaseous or solid fuel and produces high pressure steam. Steam pressures can range from 150 to 3,000 psig and temperatures can range from about 370° to more than 1,000°F. Steam boilers are discussed in "Chapter 4: Steam and Hydronic Boilers." Steam distribution systems are discussed in "Chapter 5: Steam Systems."

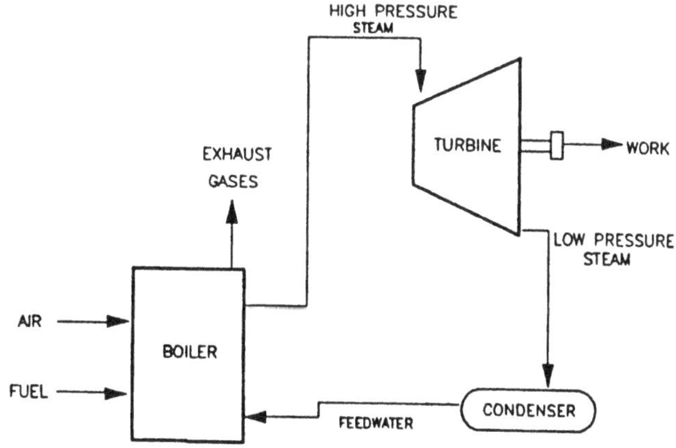

Figure 13-6. Steam turbine cycle.

A heat sink is a device for rejecting heat from the steam as it is exhausted from the turbine. The heat sink is commonly referred to as a condenser, since the steam condenses to liquid steam condensate in the heat sink. The heat sink can reject heat to a thermal process in the case of congeneration systems.

Steam turbines operate on the basic concept of expanding steam as it flows through sets of nozzles. As the steam flows through the nozzles, it accelerates and forms a high speed jet. The kinetic energy of the jet is then transferred to a series of rotating blades connected to a shaft to produce mechanical power or work.

There are two basic classifications of steam turbines that are based upon the condition of the stream as it exits the turbine: noncondensing and condensing. A noncondensing turbine, also referred to as a backpressure turbine, exhausts steam at atmospheric pressures or higher. Since much of the steams heat is latent heat of vaporization, is exhausted and then used for process heat recovery, the backpressure cogeneration system process efficiency or total energy efficiency can be quite high. Backpressure turbines can be a substitute for a steam pressure reducing valve. The turbine provides pressure reduction while producing power. Backpressure turbines, however, do not match the electrical load requirements well since the electrical output is controlled by the thermal load.

Condensing turbines exhaust steam after it has been expanded at pressures that are less than atmospheric pressure. As a result, the steam is directly and completely condensed within the turbine before being discharged. Condensing turbines are capable of producing more power than a backpressure turbine for a given set of steam conditions. Since the exhaust steam conditions are at lower temperatures, the difference between the inlet and outlet temperatures is greater, which results in higher turbine efficiencies. Condensing temperatures below 200°F are possible; however, temperatures that are low limit the usefulness of the heat recovered as part of the cogeneration process.

Steam turbines are also classified according to the path by which steam flows through a turbine. The classifications include straight flow, extraction turbine and induction turbine. In straight flow turbines, the steam path is continuous through the turbine from the inlet to the exhaust, and all of the steam is exhausted at a single point. Extraction turbines exhaust steam from the turbine at different conditions at several points as the steam flows through the turbine. An extraction turbine is shown in Figure 13-7. An advantage of this type of turbine is that it allows extraction of the quantity of steam required at each temperature or pressure needed by the thermal process. Depending on the cycle requirements, the extraction turbine's final exit conditions can either be backpressure or condensing. Induction turbines are designed to accept steam at multiple points in the turbine. An induction turbine is shown in Figure 13-8.

A steam turbine may also be a reheat turbine, which is a combination of an extraction turbine and an induction turbine. In a reheat turbine, steam

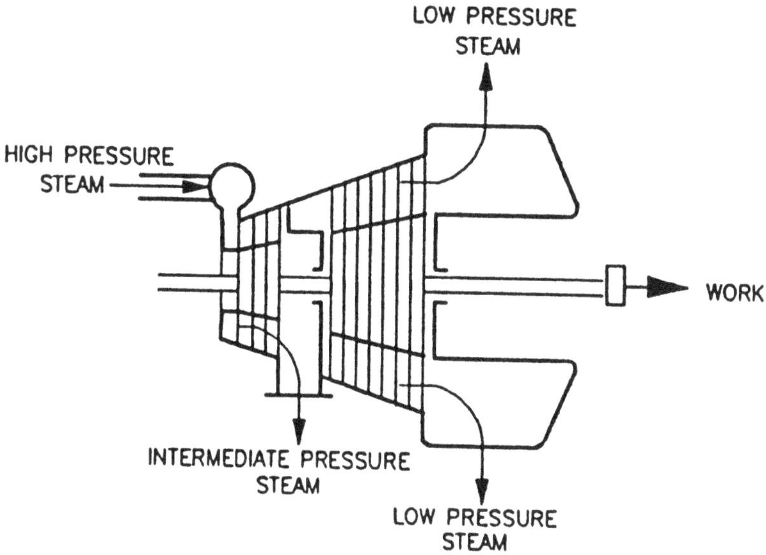

Figure 13-7. Extraction turbine.

is exhausted from the turbine at some intermediate point, heated, and then returned to the turbine for further expansion. A reheat turbine is shown in Figure 13-9.

The mechanical efficiency of steam turbines is in the range of 50–80%. Optimum turbine performance occurs at about 95% of the rated

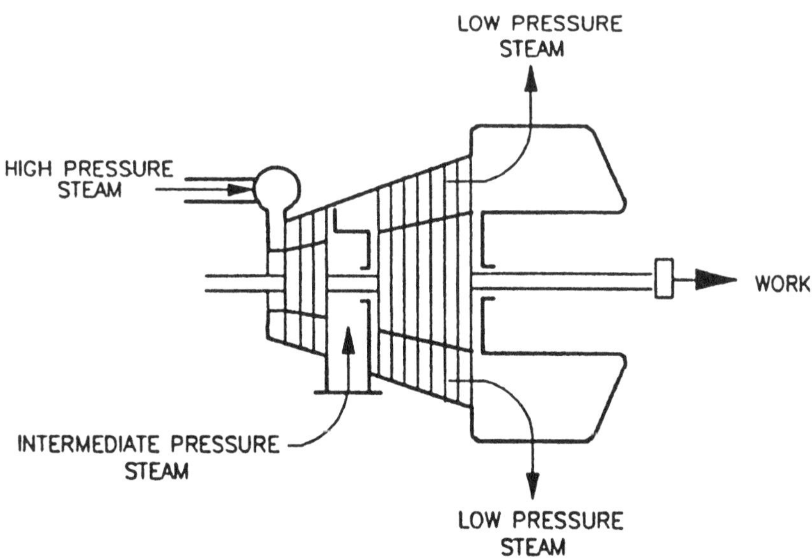

Figure 13-8. Induction turbine.

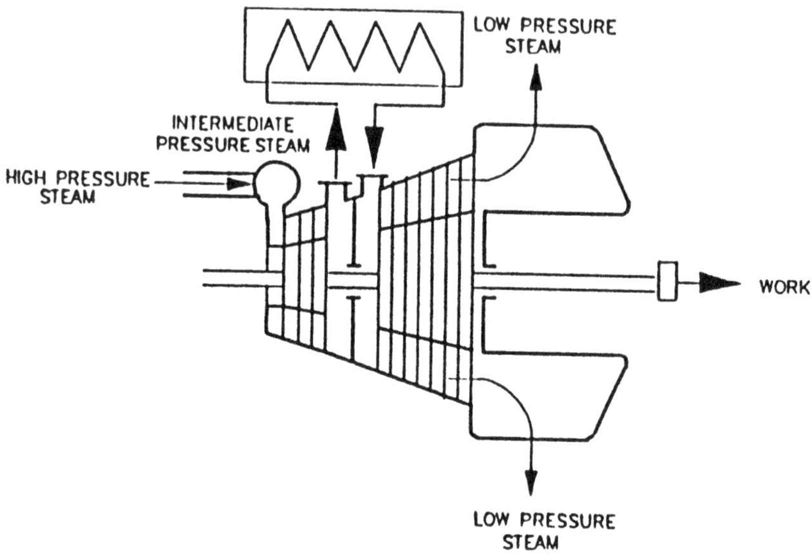

Figure 13-9. Reheat turbine.

load for the turbine. The part load characteristics are unique to each turbine. A steam turbine's maximum efficiency is limited to the Carnot efficiency, which is a function of the temperatures of the cycle's heat source and heat sink. The Carnot efficiency is reduced by mechanical inefficiencies, steam losses and imperfections in the steam flow path. Increasing the throttle pressure and/or temperature or decreasing the condensing pressure and/or temperature will increase the Carnot cycle efficiency of a turbine.

When a steam turbine is used in a topping cogeneration cycle, the turbine is either a backpressure or an extraction condensing type, which makes downstream, low pressure thermal energy available for use in thermal processes. Bottoming cogeneration cycles typically use condensing turbines, which yield more output power.

13.3 ELECTRICAL SYSTEMS FOR COGENERATION

Although cogeneration systems can be used to produce mechanical power, they are usually used to produce electrical power. Typically, the cogeneration system drives an electrical generator, which may provide all, or part, of the electrical needs of a facility. Frequently, cogeneration systems produce only part of the electrical needs of a facility. At other times, the cogeneration system may produce excess electrical power. As a result, a typical cogeneration system must accommodate the flow of power in two directions: both from and to an outside electrical utility. The most effective situation is usually one in which the cogenerator's generator is interconnected to an outside electric utility grid.

13.3.1 Utility Considerations

It is unusual that the electrical power generated by a cogeneration system exactly matches the needs of a facility. Typically, additional power must be purchased from an outside utility or excess power must be sold to an outside utility. As a result, cogeneration systems are usually interconnected to an outside utility grid. Interconnection with an outside electric utility grid provides a number of benefits to cogenerators, including the availability of supplemental and standby power, increased reliability and operating flexibility.

However, interconnection with an electric utility grid requires safety on the electric grid and the ability to meet operating problems of the electric grid and it's generating system. Utility interconnection concerns include:

- Power quality: A cogeneration system that is interconnected to a utility grid must not degrade the quality of power supplied by the utility as measured by the voltage, frequency, stability, power factor and harmonic content.
- Power safety: Interconnected generators must not have the potential to energize a utility circuit that is not being powered by the utility. Most utilities require that a cogenerator install an external disconnect switch that is accessible to utility personnel and that can be used to disconnect and lock out the cogeneration system.
- Grid operations: Interconnection to the utility grid must not jeopardize the utility's ability to manage grid operations. Utility operations of a grid can be very complex, and require the coordination of relays, switches and fault control.
- Dispatchability: During periods of emergency or grid safety, utilities require the right to request that a cogenerator be isolated from the utility grid.

13.3.2 Electric Power Generators

All of the prime movers previously discussed can produce power in the form of a rotating shaft. Most cogeneration systems convert the shaft power to electrical power.

There are several criteria used when selecting alternating current generators for cogeneration systems: (1) machine efficiency in converting mechanical input into electrical power at various loads; (2) electrical load requirements, including frequency, power factor, voltage and harmonic distortion; (3) phase balance capabilities; (4) equipment cost; and (5) motor-starting current requirements. These characteristics of a generator are a function of the type of generator. There are two basic types of generators used to produce electrical power in cogeneration systems: synchronous and induction generators.

Induction generators are similar in construction to induction motors. The generator draws excitation current from the utility's electrical system and produces power when driven above its synchronous speed. In a typical induction generator, full output occurs at 5% above synchronous speed. Induction generators do not have an internal source of reactive power. The generator will produce only electrical power when connected to some external reactive source. Typically, the reactive source is power from the utility grid. Induction generators have the following characteristics:

1. Induction generators operate at the same voltage and frequency as the power source supplying the excitation power. As a result, they are easy to synchronize with the utility's power grid.
2. Induction generators interconnected to the utility grid and using cogenerated power on-site to reduce utility purchases will worsen the power factor of the load served by the utility. An induction generator will increase the reactive load on the utility, as well as reduce the real power drawn from the utility.

The rated capacity of an induction generator is the most important consideration when selecting a generator; however, the generator's voltage, efficiency, harmonics, voltage and frequency stability, short-term overload capability, phase imbalance and fault current contribution are also important considerations. Typically, induction generators have efficiencies between 95% and 100%. Smaller induction generators, or generators operating at part load, may operate at efficiencies as low as 85% or less. Figure 13-10 shows the typical generator efficiency as a function of the load.

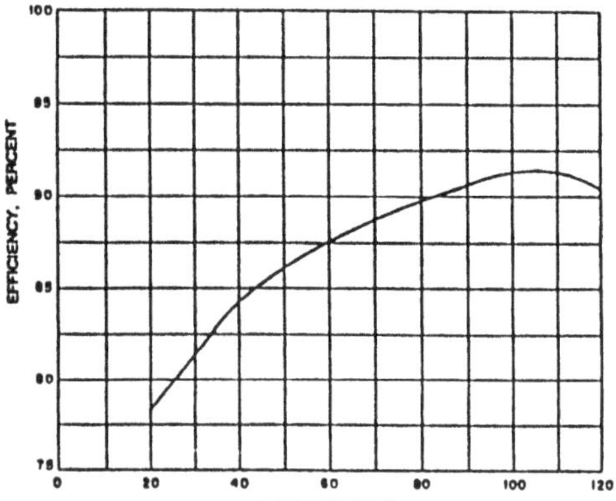

Figure 13-10. Typical generator efficiency.

Unlike induction generators, synchronous generators do not need an outside source for excitation. They provide their own source of reactive power and can operate independent of, or isolated from, any external source of power. They are the most common type of generator used for cogeneration systems and are always employed in larger applications. Since synchronous generators are operated in isolation, it is necessary to precisely control the generator and, therefore, the engine speed. When operated in parallel with the utility grid, the grid's frequency and voltage have a stabilizing effect on the generator.

13.3.3 Interconnecting Equipment

Occasionally, a facility will have two, or more, cogeneration systems connected to a common load, or grid. When two, or more, electrical power generators are connected to a common load or grid, they must be interconnected or paralleled. Interconnection, or paralleling, is the synchronous operation of two, or more, electrical power sources serving a common load. An interconnected generator will be in phase and in voltage synchronization with the utility grid.

An automatic synchronization system includes an automated monitoring system that constantly monitors frequency, voltage and phase angle. It provides a signal to a circuit breaker if conditions are acceptable. There are two types of automatic synchronizers: phase-lock and anticipatory. A phase-lock synchronizer monitors the bus, as well as the generator that is to be paralleled to the bus. It initiates closure when the generator is matched to the bus. An anticipatory synchronizer causes closure at a minimum phase difference.

13.3.4 Wheeling

Wheeling involves at least one electric utility interfacing with a cogenerator. It is the transfer by direct transmission, or displacement, of electric power from one utility or generator to another over the facilities of an intermediate utility. Electric utilities commonly "wheel" electrical power from one utility to another and exchange power supplies during local peak loads and emergencies.

There are two methods of wheeling applicable to cogenerators: (1) transmission of excess power to sister plants in the same state (intrastate); and (2) transmitting power from their states, which have lower buyback rates, to other states that have much higher buyback rates.

If electric utilities accept wheeling for cogenerators, they can become common carriers. As a common carrier, a utility would set the tariff for transmitting electricity from cogenerators without discrimination as long as certain conditions are met. The utility cannot prevent any cogenerator from wheeling power.

13.4 HEAT RECOVERY FOR COGENERATION SYSTEMS

Cogeneration is the concurrent production of power (either mechanical or electrical) and process heat in a single integrated utility system. In order for a power generation system to be classified as a cogeneration system, it must recover heat from the power generation process. The source of heat in cogeneration systems is the prime mover.

13.4.1 Reciprocating Engine Heat Recovery

Reciprocating engines include spark ignition (Otto) cycle engines and diesel engines. Both cycles burn volatile liquid or gaseous fuels, compress and burn air/fuel mixtures as part of the combustion process to produce power. The products of combustion can reach temperatures as high as 1,200°F. As a result, a significant amount of heat is available for heat recovery from a reciprocating engine. The approximate distribution of energy from the combustion of fuel is

- Shaft power 33%
- Heat convection and radiation 7%
- Heat rejected in water jacket 30%
- Heat rejected in exhaust 30%

These amounts will vary with engine design and speed; however, it can be observed that over 60% of the fuel energy is converted to heat in a reciprocating engine. Figure 13-11 shows an energy balance for a naturally aspirated reciprocating engine.

In reciprocating engines, heat may be reclaimed from the jacket cooling system, lubrication system, engine exhaust and aftercoolers. These engines require extensive cooling to remove excess heat not conducted to the power train during combustion and heat resulting from friction. The primary sources for heat recovery in reciprocating engines are exhaust gases and jacket cooling water.

Reciprocating engine exhaust gases typically have temperatures ranging from 700°F to 1,000°F. Exhaust gas temperatures will vary as a function of engine type and loading. Exhaust gas heat represents nearly one third of the energy input of the fuel. Heat can be recovered with a suitable heat exchanger to produce either hot water or low-pressure steam. While a significant amount of heat is available for recovery from engine exhaust gases, it is necessary that stack temperatures be maintained in the range of 300°F, or more, to prevent the condensation of moisture in the exhaust gases. As a result, it is not possible to recover as much as 50% of the heat available in exhaust gases.

Jacket cooling water represents another significant source of heat for recovery. Approximately, one third of the energy input to a reciprocating engine is converted to heat rejected from the engine block to the

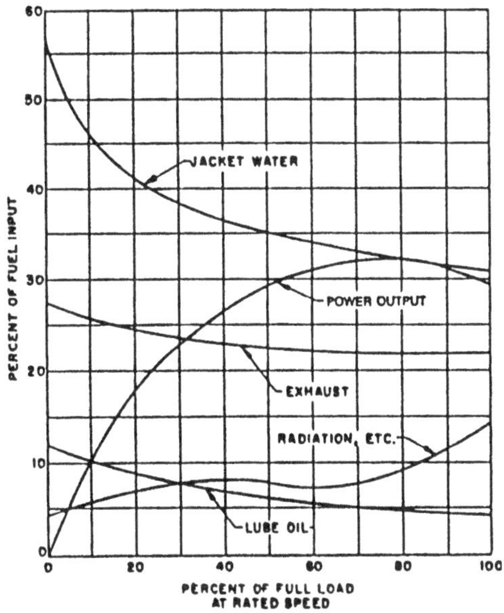

Figure 13-11. Energy balance for naturally aspirated reciprocating engine.

circulated cooling water. Engine jacket cooling water can reach temperatures as high as 260°F. Reciprocating engine jacket cooling systems can also produce low-pressure steam at pressure up to 15 psig. Most reciprocating engines produce liquid coolant temperatures in the range of 245°F to 250°F.

13.4.2 Combustion Turbine Heat Recovery

Gas turbine engines typically have a fuel to electrical efficiency between 12% and 35%. The rest of the fuel energy is lost as heat in the exhaust gases and through radiation heat losses, or to internal coolants in larger engines.

Although the exhaust stack temperatures for combustion turbines must be maintained above 300°F to prevent condensation, significant amounts of heat are available for recovery from the engine exhaust. As a result, overall thermal efficiencies for combustion turbines can be over 60%. Exhaust stack temperatures for combustion turbines is typically in the range of 800°F–1,000°F.

The fuel efficiency of combustion turbines may be improved with the use of duct burners, which are supplemental burners in the turbine exhaust. Combustion turbines operate with large amounts of excess air. As a result, turbine exhaust gases contain large quantities of oxygen. Exhaust gases with an oxygen content of 16% or more can support additional combustion with the addition of more fuel. Duct burners are installed directly in the exhaust gas stream and inject additional fuel in to the exhaust gases for combustion. Duct burners increase the temperature of the exhaust gas stream to temperatures as high as 1,600°F–1,800°F. The higher temperature exhaust gases result in improved fuel efficiency, and the duct burner efficiency can exceed 90%.

Heat can be recovered from the exhaust of a combustion turbine with a heat recovery steam generator (HRSG). An HRSG is an unfired boiler that uses turbine exhaust gases to produce steam. HRSGs can be equipped with economizers, superheaters, reheaters and duct burners to increase the production of steam.

13.4.3 Steam Turbine Heat Recovery

In a typical Rankine cycle used to generate power, steam is generated in a heat source (boiler), expanded in a turbine to generate power and heat is rejected to a heat sink (condenser). A typical steam turbine cycle is shown in Figure 13-6.

Since heat must be rejected to a heat sink as part of the cycle, there are significant amounts of heat available for recovery from the condenser. The efficiency of a Rankine steam cycle is a function of the average temperature at which heat is supplied from the heat source and the average temperature at which heat is rejected. Reducing the average temperature at which heat is rejected in the condenser will increase the efficiency of a Rankine cycle. Recovering heat from the condenser in a steam turbine cogeneration cycle will improve the overall efficiency of the cogeneration cycle.

High pressure steam turbines operate at inlet steam temperatures of 800°F and higher. If the turbine discharge pressure is at, or near, atmospheric pressure, the steam must be cooled below its saturation temperature (212°F) in order for it to condense from a vapor to a liquid. Large differences between the maximum cycle temperature and minimum cycle temperature will result in higher cycle efficiencies and larger amounts of heat that is available for recovery.

13.5 APPLICATIONS FOR RECOVERED HEAT

By definition, cogeneration systems recover heat as the result of the production of mechanical or electrical power. Significant amounts of heat are available for recovery from cogeneration systems. The recovered heat may have numerous applications, including space heating, domestic water heating, process heating, process drying and cooling.

In order for a cogeneration system to be economically viable, the energy produced by it must be utilized to the maximum extent. Heat recovered from the system must be used and not wasted. Applications for the use of recovered heat include thermal distribution and storage, heat activated

TABLE 13.1 Temperatures Normally Required for Various Heating Applications

Application	Temperature °F
Absorption refrigeration machines	190 - 245
Space heating	120 - 250
Water heating (domestic)	120 - 200
Process heating	150 - 250
Evaporation (water)	190 - 250
Residual fuel heating	212 - 330
Auxiliary power producers, with steam turbines or binary expanders	190 - 350

cooling and desiccant cooling. Table 13.1 shows the temperatures normally required for various heating applications.

13.5.1 Thermal Distribution Systems

Thermal distribution systems are used to transport thermal energy from a point where thermal energy is generated or stored to a point where thermal energy is used. Thermal distribution systems are typically hydronic systems and use hot or chilled water as the circulated medium to transport the thermal energy.

Chilled water systems are hydronic systems in which the water is at a temperature lower than the ambient temperature. Chilled water systems are discussed in "Chapter 7: Chillers and Chilled Water Systems."

Hot water distribution systems are classified according to the maximum water temperature in the system. Low temperature water systems have operating temperatures below 250°F. Medium temperature hydronic systems have water temperatures between 250°F and 350°F, and operating pressures ranging from 125 to 150 psig. High temperature systems have operating temperatures over 350°F, up to a maximum of 450°F, and have pressures up to 300 psig.

13.5.2 Thermal Energy Storage

Thermal energy storage (TES) systems store thermal energy during peak production periods for later use. These systems can decouple power generated from the production of dispatchable power while fully using the thermal energy available from the prime mover.

The medium most commonly used for thermal energy storage in cogeneration systems is water. Low temperature thermal energy systems store thermal energy at temperatures below ambient temperatures (chilled water), which may be generated from heat activated chillers. High temperature thermal storage can be used to store high temperature energy from sources such as a turbine exhaust stream. Thermal energy storage is discussed in greater detail in Chapter 15.

13.5.3 Heat Activated Chillers

Waste heat from the cogeneration process may be recovered and used to produce chilled water. Some absorption chillers have been specifically designed for use in cogeneration systems and heat recovery applications. These chillers operate on jacket cooling water or directly on exhaust gas. Typically, hot water (less than 200°F) or low pressure steam (less than 15 psig) can be used in absorption chillers to produce chilled water. These chillers have cooling capacities between 50 and 200 tons of refrigeration.

Absorption chillers operate on a process that uses certain chemical solutions, which have an attraction for the refrigerant, to create a pressure difference and compress the refrigerant. Absorption refrigeration equipment operates on heat cycles during which a secondary fluid (an absorber) is used to absorb a primary fluid (a gaseous refrigerant), which is vaporized in a evaporator. In an absorption chiller, heat is used to vaporize the gaseous refrigerant. These chillers have few moving parts and require little energy input in the form of work. The process does require significant amounts of energy input in the form of heat.

Absorption chillers suitable for use with cogeneration systems are available as single and two stage systems. Single stage chillers can utilize hot water up to 300°F and low pressure steam up to 15 psig as the energy source. More efficient two stage absorption chillers require steam pressures at approximately 115 psig. Absorption chillers are discussed in more detail in "Chapter 7: Chillers and Chilled Water Systems."

13.5.4 Desiccant Cooling

Heat recovered from a cogeneration system may also be used to provide desiccant cooling. Desiccants operate by removing water vapor from air (latent cooling). As the desiccant removes moisture from the air, the air temperature increases slightly. Desiccant cooling provides only latent cooling; there is no sensible cooling as part of the process.

Desiccant coolers use a chemical that absorbs moisture from air as the air is passed through the chemical. After operating for a period of time, the chemical desiccant becomes saturated with water vapor and must be recharged. Desiccants may be recharged by heating the desiccant to drive the moisture out of the chemical. Heat recovered from a cogeneration system my be used to recharge the desiccant material after it has become saturated with moisture.

13.6 COGENERATION SYSTEM FEASIBILITY

When considering a cogeneration system, there are a number of important factors that should be analyzed before a project is actually undertaken. Determining whether a cogeneration system is a feasible project requires an evaluation of a large number of factors. Before deciding to undertake a fairly costly detailed engineering feasibility study, a cursory rule of thumb analysis may be performed.

A quick rule of thumb preliminary method for determining whether a more detailed feasibility study of cogeneration should be performed is: If the cost of electricity expressed in \$/kWh is more than 0.013 times the cost of fuel expressed in \$/10^6 Btu, a more detailed study should be considered. If it is 0.026, or more, there is an excellent chance that a cogeneration system would have a simple economic payback period of three years, or less.

If a preliminary analysis indicates that a cogeneration system may be economically feasible, a more detailed engineering analysis should be conducted. The final decision regarding the development of an on-site system is dependent upon an engineering study that is detailed enough to provide reasonably accurate estimates of the project's capital budget, including all construction costs and resulting operating costs.

A more detailed engineering analysis should include a preliminary sizing of the cogeneration system, a conceptual design, a load curve analysis and reasonably accurate cost estimates. Additional information about feasibility analyses for cogeneration systems may be found in ASHRAE (1996) and Orlando (1996).

13.6.1 Conceptual Design of Cogeneration Systems

When performing a conceptual design and preliminary sizing of a cogeneration system, there are a number of important considerations, including:

- peak electrical demand
- minimum (or base-load) electrical demand
- peak thermal demand
- minimum (or base-load) thermal demand
- maximum economic return with available funds
- marginal cost/benefit ratio

Plant operating considerations include:

- rated output
- thermally dispatched (tracks facility thermal load)
- electrically dispatched (tracks facility electrical load)
- maximum economic return
- utility dispatched
- peak shaving

An important item that must be included in the design of a cogeneration system is the need for redundancy in the plant. Redundancy can take the form of additional units, additional standby capacity, or both.

When conducting a preliminary design of a cogeneration system, factors that must be considered include:

- The compatibility of the existing process or mechanical systems with the recoverable heat that is available from the cogeneration system.
- The availability of space for siting of the cogeneration system, and the ability to interconnect with the facility's thermal and electrical systems.
- The adequacy of the existing systems, including (1) powerhouse ancillary systems such as water treatment and condensate treatment; (2) major components such as boilers, emergency generators and chillers; and (3) existing mechanical and electrical distribution systems.
- Maintenance requirements of the cogeneration systems.
- Structural capacity of the building, including ability to support the weight of the new cogeneration equipment, and vibration isolation requirements.
- Acoustics, including the sound generated by the equipment, and any requirements for acoustical attenuation.
- Selection of the prime mover, including the required steam or water temperature, fuel or energy availability, efficiency of the prime mover, equipment availability, maintenance requirements and start-up requirements.

13.6.2 Load Duration Curve

When conducting an engineering feasibility study, it is necessary to develop load distribution curves for the electrical and thermal loads. Even at the early planning stages, it is important that reasonably accurate estimates be made of the following:

- fuel consumption (if cogeneration system is a topping cycle)
- amount of supplemental electricity that must be purchased
- amount of supplemental boiler fuel (if any) that must be purchased
- amount of excess power available for sale
- electrical capacity required from the utility for supplemental and standby power
- electrical capacity represented by any excess power if the utility offers capacity credits

Obtaining estimates of the performance values of multiple time-varying loads can be difficult. These values may also be complicated by the utility rate structure that may be based on time-of-day or time-of-year purchase and sale of power.

A load duration curve should be developed for the thermal demand that will be served by the proposed cogeneration system. Multiple thermal

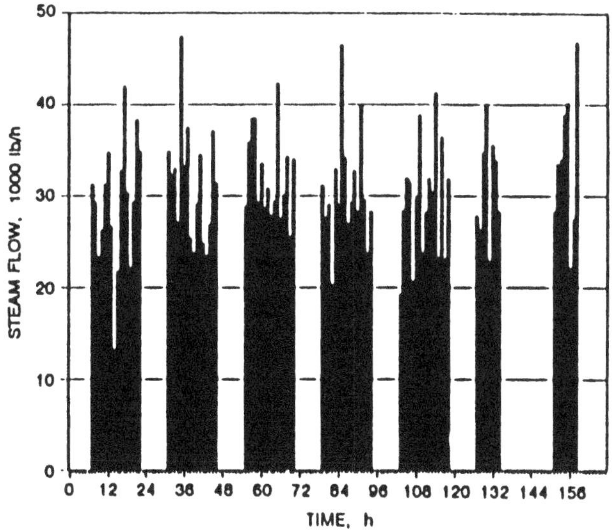

Figure 13-12. Hypothetical steam load profile.

profiles may be developed for multiple thermal loads. A profile should be generated for each thermal load to develop a total thermal load profile for the system. Load profiles should be developed for various steam demands at the appropriate steam pressure and temperature for each application. Load profiles should also be developed for process and domestic hot water. Data must include daily fluctuations due to normal variations in process needs, as well as seasonal weather effects and operating schedules. A hypothetical steam load profile is shown in Figure 13-12.

A power profile, either mechanical or electrical, must also be developed in a manner similar to that for thermal load profiles. These demand curves should include hour-by-hour power demands for the facility for "typical" days (or weeks) for each season or month of the year.

Chapter 14

Heat Recovery and Waste Heat Management

Waste heat may be defined as the energy associated with waste streams of air, exhaust gas, and/or fluids that leave the boundaries of a plant or building and enter the environment. The rejection of waste energy is often referred to as thermal pollution.

The main reason for attempting to recover waste heat is economic, like most energy conservation efforts. Waste heat that is successfully recovered and used directly replaces purchased energy, reduces energy consumption and thereby reduces energy costs. A second economic benefit that may result is the reduced capital costs for energy conversion equipment with less capacity. The economics of waste heat depends upon five factors: (1) there must be an appropriate use for the waste heat; (2) there must be an adequate quantity of waste heat; (3) the heat must be of adequate quality for the intended purpose; (4) heat must be transferred from the waste heat stream to the material or work piece where it is to be used; and (5) the waste heat must be used profitably.

If has been estimated that up to 50% of the energy consumed in the United States was discharged as waste heat to the environment. Unfortunately, only a fraction of this energy can be practically recovered. Typically, only exhaust streams with temperatures greater than 200°F can have waste heat economically recovered. However, energy exhaust streams that are above 200°F are estimated to contain an energy equivalent of up to 15% of the total industrial energy consumption. It has also been estimated that typical process industries could save 20% of their fuel requirements through heat recovery and waste heat management.

14.1 BASIC CONCEPTS OF HEAT RECOVERY AND WASTE HEAT MANAGEMENT

Heat recovery and the management of waste heat involves recovering energy, which would otherwise be discharged to the environment for reuse. The most basic classification of energy recovery is by application. Heat recovery applications include process-to-process, process-to-comfort and comfort-to-comfort. In process-to-process applications, heat is recovered from an exhaust stream and transferred to an incoming process stream. Process-to process recovery devices recover only sensible heat and do not transfer latent heat as moisture. In process-to-comfort applications, waste heat is captured from a process exhaust and used to provide comfort heating, such as building heating. Comfort-to-comfort applications involve recovery of heat from building exhaust systems and transferring of the heat to pre-condition building make-up air. These applications can involve the recovery of both sensible heat and latent heat.

Heat recovery systems may also be classified according to the method by which heat is transferred from the waste heat source to the point of application. The two methods of heat recovery are heat exchangers and heat pumps. Heat exchangers are passive or active devices that directly or indirectly transfer heat from a hot exhaust stream to a cooler incoming stream.

No mechanical work is required by the heat exchange system other than a pump or fan used to circulate a heat exchange medium, such as water or air. Heat pumps are heat exchange systems that transfer heat from a low temperature waste heat source to a higher temperature application. Heat pump systems typically have direct expansion refrigeration systems, and relatively significant amounts of mechanical work must be input to the heat pump system in order to accomplish the transfer of heat.

14.1.1 Definitions and Terms Used in Heat Recovery Systems

For definitions and terms used in heat recovery systems, check the Glossary.

14.1.2 Waste Heat

When considering waste heat recovery, a number of issues regarding the nature of the heat source should be included in the analysis process. These issues include the temporal availability of the waste heat, the quantity of the waste heat and the quality, or temperature, of the waste heat source.

The temporal availability of the heat has to do with whether of not the waste heat is available when needed. If the time period at which the waste heat is available does not match the time period for which the heat could be used, either heat must be stored , or it cannot be used and must be dumped. Storing heat for later use can add significant costs to a proposed waste heat recovery system. Thermal storage is discussed in "Chapter 15: Thermal Energy Storage Systems."

The quantity of waste heat available must also be adequate to justify the cost of heat recovery equipment and systems. Small amounts of waste heat may not justify the cost of heat recovery equipment. The quantity of available waste heat must also be considered with the quantity of heat required for other uses. A large mismatch between the quantity of waste heat available and the quantity of heat needed could make waste heat recovery economically unattractive.

TABLE 14.1 High Temperature Waste Heat Sources

Type of Device	Temperature, °F
Nickel refining furnace	2,500 – 3,000
Aluminum refining furnace	1,200 – 1,400
Zinc refining furnace	1,400 – 2,000
Copper refining furnace	1,400 – 1,500
Steel heating furnace	1,700 – 1,900
Copper reverberator furnace	1,650 – 2,000
Open hearth furnace	1,200 – 1,300
Cement kiln (dry process)	1,150 – 1,350
Glass melting furnace	1,800 – 2,800
Hydrogen plants	1,200 – 1,800
Solid waste incinerators	1,200 – 1,800
Fume incinerators	1,200 – 2,600

Finally, the quality of the available waste heat must be high enough to justify the cost of heat recovery systems and equipment. The quality of waste heat refers to the temperature of waste heat. Sources of waste heat can be grouped into three temperature ranges: high, medium and low temperatures. The temperature groups, or classifications, are as follows:

- High temperature range: 1,100°F$\leq$ T $\leq$ 3,000°F
- Medium temperature range 400°F $\leq$ T$<$1,100°F
- Low temperature range 80°F $\leq$ T$<$400°°F

The recovery of heat from a high temperature source is typically more economically feasible than medium or low temperature sources. However, heat recovery from high temperature sources can cause special problems, since special materials and careful design may be required in the heat recovery equipment. Tables 14.1, 14.2 and 14.3 list sources of high, medium and low temperature heat, respectively.

TABLE 14.2 Medium Temperature Waste Heat Sources

Type of Device	Temperature, °F
Steam boiler exhaust	450 – 900
Gas turbine exhaust	700 – 1,000
Reciprocating engine exhaust	600 – 1,100
Reciprocating engine exhaust (turbocharged)	450 – 700
Heat treating furnace	800 – 1,200
Drying and baking ovens	450 – 1,100
Catalytic crackers	800 – 1,200
Annealing furnace cooling systems	800 – 1,200

TABLE 14.3 Low Temperature Waste Heat Sources

Type of Device	Temperature, °F
Process steam condensate	130 – 190
Cooling water from:	
Furnace doors	90 – 190
Bearings	90 – 190
Welding machines	90 – 190
Injection molding machines	90 – 190
Annealing furnace	150 – 450
Forming dies	80 – 190
Air compressors	80 – 120
Pumps	80 – 190
Internal combustion engines	150 – 250

Medium and high temperature waste heat can be recovered for applications such as pre-heating combustion air for boilers using air pre-heaters, furnaces using recuperators, ovens using recuporators and gas turbines using regenerators. Low temperature waste heat can be used to pre-heat boiler feedwater or boiler make-up water.

14.2 HEAT EXCHANGERS AND HEAT RECOVERY SYSTEMS

Heat exchangers are used to transfer heat from a high temperature source (waste heat) to a low temperature application or process. They are usually used to transfer heat from a high temperature fluid to a low temperature fluid. Heat exchangers are characterized as single or multi-pass gas to gas, liquid to gas, liquid to liquid, evaporator, condenser, parallel flow, counterflow or crossflow.

In a single pass arrangement, each fluid flows past and transfers heat to/from the other fluid only once before exiting the heat exchanger. In multi-pass heat exchangers, each fluid flows past and transfers heat to/from the other multiple times. In parallel flow heat exchangers, both fluids flow parallel to each other in the same direction. Counterflow heat exchangers are arranged such that the fluids flow in opposite directions as they pass through the heat exchanger. In crossflow heat exchangers, the heat transfer fluids flow perpendicular to each other.

The term evaporator applies to a heat exchanger in which heat is transferred to a evaporating (boiling) liquid. A condenser is a heat exchanger in which heat is removed from a condensing vapor.

The essential parameters that should be considered when choosing a heat exchanger for a particular application include:

- Temperatures of the heat transfer fluids
- Flow rates of the heat transfer fluids

- Chemical composition of the heat transfer fluids
- Minimum and maximum temperatures of the heat transfer fluids
- Amount of contaminants in the heat transfer fluids.

14.2.1 Performance of Heat Exchangers

The amount of heat transferred between two fluids passing through a heat exchanger is a function of the temperature difference between the two fluid streams and the overall heat transfer coefficient between the two fluid streams. As the two fluids flow through the heat exchanger, the temperatures though the heat exchanger, the temperatures of the fluid streams change as heat is transferred. As a result, the temperature difference between the streams is not a constant, but a variable.

The log mean temperature difference (LMTD) is used to account for the variable temperatures in the fluid streams as they flow through a heat exchanger. Referring to Figure 14-1, the log mean temperature difference is the temperature difference at one end of the heat exchanger, less the temperature difference at the other end of the heat exchanger, divided by the natural logarithm of the ratio of these two temperature differences. The log mean temperature difference is expressed as

$$T_{\text{LMTD}} = \frac{(T_{h2} - T_{c2})(T_{h1} - T_{c1})}{\ln[(T_{h2} - T_{c2})/(T_{h1} - T_{c1})]} \qquad \text{Eq. (14.1)}$$

where T_{h1} = entering temperature of the hotter fluid, °F
T_{h2} = leaving temperature of the hotter fluid, °F
T_{c1} = entering temperature of the cooler fluid, °F
T_{c2} = leaving temperature of the cooler fluid, °F

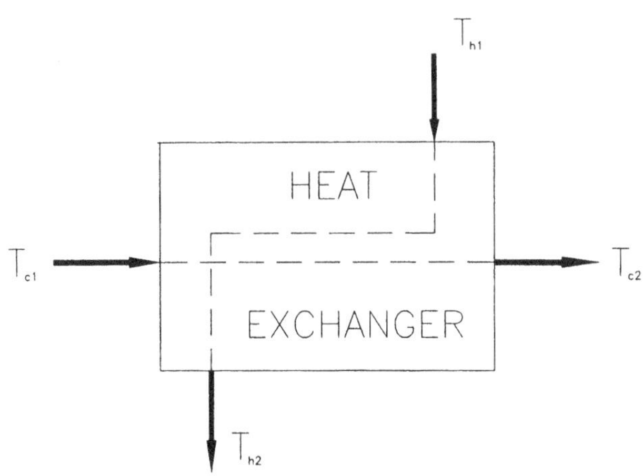

Figure 14-1. Heat exchanger schematic.

Equation (14.1) is based upon three important assumptions, which are:

- The specific heat of the two fluids remains constant.
- The overall heat transfer coefficient between the two fluids remains constant.
- The flow rates of the two fluid steams remain constant.

The heat transfer rate between the two fluids in a heat exchanger may be calculated using Equation (14.2).

$$q = U_m A \Delta T_{LMTD} \qquad \text{Eq. (14.2)}$$

where U_m = the overall heat transfer coefficient between the two fluids, Btu/hr.-ft.^{2}-°F
A = the heat transfer surface area, ft.^{2}

To estimate the performance of a heat exchanger, ΔT_{LMTD} must be determined. In order to determine ΔT_{LMTD}, it is necessary to know the entering and leaving temperatures of each fluid stream. Typically, these are not known and must be calculated by a trial and error iteration procedure. When the fluid inlet and outlet temperatures are to be evaluated for a given heat exchanger, the iterative procedure can get quite involved. An analysis of the performance of a heat exchanger may be more easily under taken utilizing a method based upon the effectiveness (ε) of the heat exchanger in transferring heat at a given rate. The effectiveness of a heat exchanger is defined as

$$\varepsilon = q_a / q_{max} \qquad \text{Eq. (14.3)}$$

where q_a = actual heat transfer rate, Btu/hr.
q_{max} = maximum possible heat transfer rate, Btu/hr.

The actual heat transfer may be determined by the change in temperature of either the hot fluid or the cold fluid. For a parallel flow heat exchanger the actual heat transfer would be

$$q_{act} = m_h c_h (T_{h1} - T_{h2}) = m_c c_c (T_{c2} - T_{c1}) \qquad \text{Eq. (14.4)}$$

where m_h = the mass flow of the hot fluid, lb/hr.
c_h = the specific heat of the hot fluid, Btu/lb-°F
m_c = the mass flow of the cold fluid, lb/hr.
c_c = the specific heat of the cold fluid, Btu/lb-°F

For a counterflow heat exchanger the actual heat transfer would be

$$q_{act} = m_h c_h (T_{h1} - T_{h2}) = m_c c_c (T_{c1} - T_{c2}) \qquad \text{Eq. (14.5)}$$

The maximum heat transfer would occur if one of the fluids were to undergo a temperature change equal to the maximum temperature difference in the heat exchanger, which is the temperature difference in the entering temperatures of the hot and cold fluids. The fluid that might undergo this maximum temperature difference governs the maximum heat transfer. It is the one that has the minimum product of the fluid mass flow rate times the fluid specific heat (mc). This is due to the fact that the heat given up by one fluid must be equal to the heat absorbed by the other fluid. The value of the mass flow rate of a fluid times its specific heat is often referred to as the capacity rate. Therefore, the maximum possible heat transfer is

$$q_{max} = (mc)_{min}(T_{h1} - T_{c1}) \qquad \text{Eq. (14.6)}$$

For a parallel flow heat exchanger where the hot fluid has the minimum capacity rate, the effectiveness is

$$\varepsilon_h = \frac{c_h(T_{h1} - T_{h2})}{c_h(T_{h1} - T_{c1})} = \frac{T_{h1} - T_{h2}}{T_{h1} - T_{c1}} \qquad \text{Eq. (14.7)}$$

For a parallel flow heat exchanger where the cold fluid has the minimum capacity rate, the effectiveness is

$$\varepsilon_C = \frac{c_c(T_{c2} - T_{c1})}{c_c(T_{h1} - T_{c2})} = \frac{(T_{c2} - T_{c1})}{T_{h2} - T_{c1}} \qquad \text{Eq. (14.8)}$$

For a counterflow heat exchanger where the hot fluid has the minimum capacity rate, the effectiveness is

$$\varepsilon_h = \frac{c_h(T_{c1} - T_{c2})}{c_h(T_{h1} - T_{c2})} = \frac{(T_{h1} - T_{h2})}{(T_{h1} - T_{c2})} \qquad \text{Eq. (14.9)}$$

For a counterflow heat exchanger where the cold fluid has the minimum capacity rate, the effectiveness is

$$\varepsilon_c = \frac{c_c(T_{c1} - T_{c2})}{c_c(T_{h1} - T_{c2})} = \frac{T_{c1} - T_{c2}}{T_{h1} - T_{c2}} \qquad \text{Eq. (14.10)}$$

Since the performance of a heat exchanger for any given hot and cold fluid inlet and outlet temperatures is a function of the heat transfer area, the overall heat transfer coefficient and the minimum capacity rate, the performance of a heat exchanger can be described by these three factors. The heat transfer performance of a heat exchanger may be defined by the number of heat transfer units (NTU), and the NTU is given by Equation (14.11).

$$\text{NTU} = \frac{AU}{c_{min}} \qquad \text{Eq. (14.11)}$$

where A = heat transfer surface area, ft.^{2}
U = overall heat transfer coefficient, Btu/hr.-ft.^{2}-°F
$c_{\min}$ = the minimum capacity rate, Btu/hr.-°F

In addition to the NTU, the ratio of the minimum specific heats (c_R) is used to quantify the performance of a heat exchanger. Equation (14.12) defines c_R.

$$c_R = c_{\min}/c_{\max} \qquad \text{Eq. (14.12)}$$

The effectiveness of a heat exchanger can be calculated if the NTU and c_R are known. This simplifies the performance analysis for a heat exchanger. The relationship between NTU, c_R and for a heat exchange is typically presented graphically, as shown in Figure 14-2. Graphs such as these may be used to determine the effectiveness, which, in turn, may be used to calculate the heat transfer and fluid temperatures for a heat exchanger.

Fouling, a process that occurs in a heat exchanger, affects the performance of the heat exchanger. Fouling refers to an accumulation of dust or condensates on heat transfer surfaces inside the heat exchanger. By increasing the resistance to fluid flow and generally decreasing heat transfer coefficients, fouling reduces heat exchanger performance. Fouling may be classified into five different types, as follows:

- Particulate fouling: The disposition of particles suspended in fluid streams such as sand, muds, corrosion debris and dust.

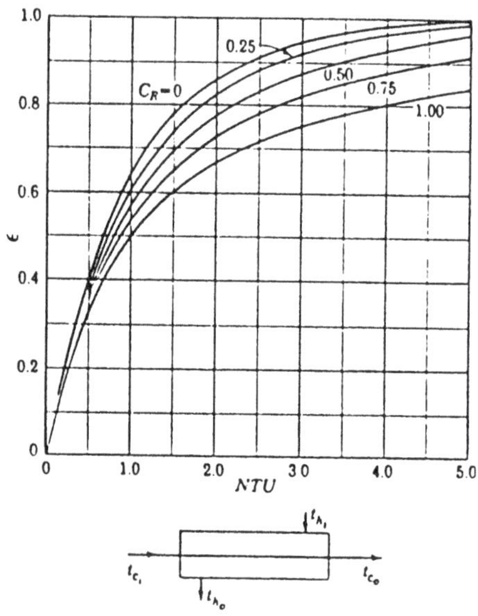

Figure 14-2. Effectiveness-NTU relationship for counterflow heat exchanger.

- Reaction fouling: The fouling caused by a chemical reaction at the heat transfer surface, namely cracking and polymerization of hydrocarbons.
- Solubility fouling: The precipitation and deposition of dissolved material caused by solubility changes with temperature such as wax from kerosene, calcium carbonate and calcium sulfate from cooling waters.
- Corrosion fouling: A special case where in situ deposits are formed with the assistance of the heat exchanger surface material.
- Biological fouling: The growth of microorganisms on heat transfer surfaces, such as slime formation in cooling towers.

14.2.2 Liquid to Liquid Heat Exchangers

As the name implies, liquid to liquid heat exchangers are used to transfer heat from one liquid stream to another. These heat exchangers may also be used to transfer heat from a vapor stream to a liquid stream.

Liquid to liquid heat exchangers are typically more compact than other types of heat exchangers since the fluid streams are more dense than for gases. Liquid to surface heat transfer coefficients also tend to be higher than for gases, which results in higher rates of heat transfer for a given temperature difference. There are two types of heat exchangers commonly used to transfer heat between liquid streams: shell-and-tube and plate type.

Shell-and-tube heat exchangers are the most common type of liquid to liquid heat exchangers. The shell-and-tube heat exchanger is very flexible when it comes to choice of materials, temperature and pressure limitations. Shell-and-tube heat exchangers usually have a U-tube design, as shown in Figure 14-3. The U-tube design consists of straight lengths of tubes bent into a U-shape, which gives it the name U-tube. Shell-and-tube heat exchangers may also have a straight tube design, as shown in Figure 14-4.

Plate heat exchangers are also commonly used for liquid to liquid heat transfer applications. Plate heat exchangers consist of multiple gasketed plates, through which heat is transferred. A plate heat exchanger is shown in Figure 14-5. The hot fluid is on one side of the plates while the cold fluid

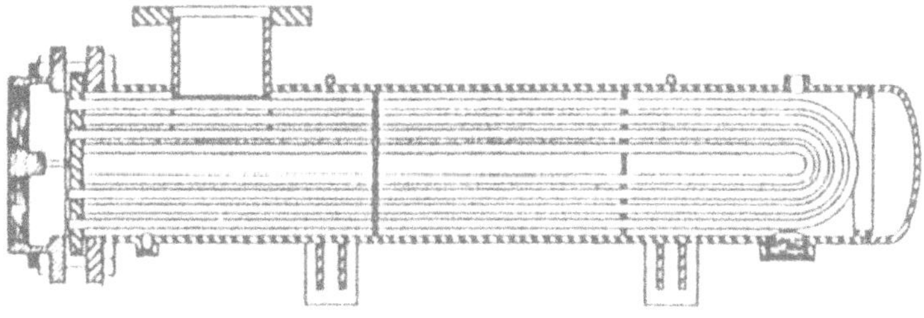

Figure 14-3. U-tube shell-and-tube heat exchanger.

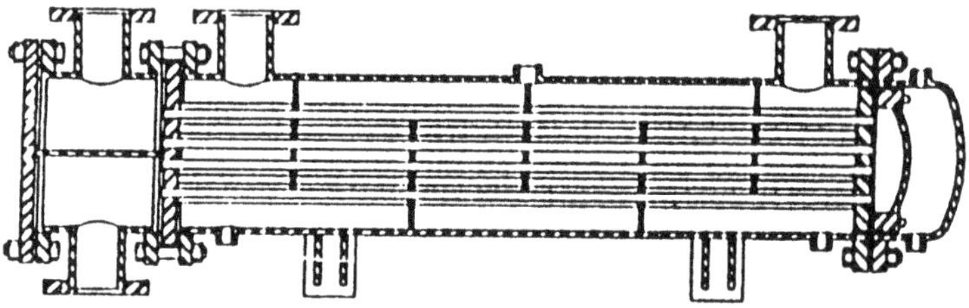

Figure 14-4. Floating head straight tube heat exchanger.

is on the other. Plate heat exchangers have the hot or cold fluid on either side of multiple plates. The plates are usually made of material that readily conducts heat, such as stainless steel. Since the plates provide relatively large surface areas through which heat is transferred, they can transfer large amounts of heat. These types of heat exchangers are also well suited for applications where the difference in temperature between the hot and cold fluids is relatively small.

Usually, the plates in plate heat exchangers have a corrugated herringbone or chevron pattern pressed into each plate. The pattern produces highly turbulent flow of the heat transfer fluid, which results in higher overall heat transfer coefficients and higher rates of heat transfer. The turbulent flow also reduces fouling on the surfaces of the plates.

14.2.3 Gas-to-Gas Heat Exchangers

Gas-to-gas heat exchangers are classified according to their principle of operation as (1) recuperative and (2) regenerative. In a recuperative design, heat is transferred directly from the hot gases on one side of a heat transfer

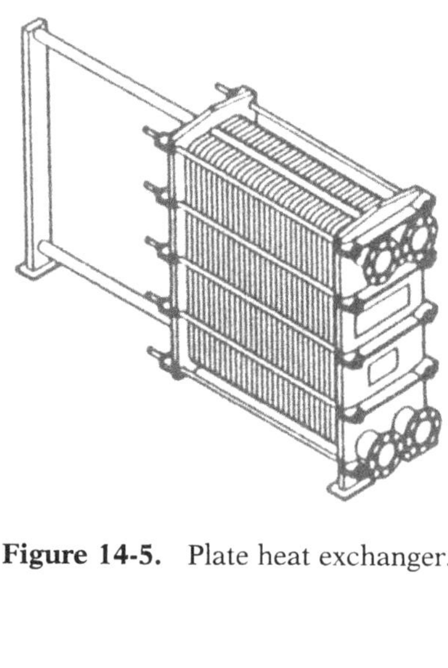

Figure 14-5. Plate heat exchanger.

surface to the gas on the other side. In a regenerative heat exchanger, heat is transferred from the hot gases to a cooler gas through some intermediate heat conduction medium.

Gas-to-gas heat exchangers are commonly used as recuperators to heat combustion air by exchange with flue gases. They are also commonly used to transfer heat from exhaust air to incoming makeup air.

Recuperators are closed heat exchangers and several different types are available, the simplest being a radiation recuperator. A typical radiation recuperator is shown in Figure 14-6. A radiation recuperator is designed for high temperature applications. The hot flue gases flow through an inner passage while the cold air to be heated flows through an external annulus. In a flue gas recuperator, the two gases usually flow in the same direction. The inner tube is often made of high temperature materials, such as stainless steel or steels of high nickel content.

Convective recuperators are used when radiation heat transfer is not effective. Frequently, convective type recuperators are shell-and-tube type heat exchangers that are generally more compact and have a higher effectiveness than radiation recuperators due to the larger heat transfer area, which is a result of the use of multiple tubes and multiple passes of gases. A typical convective type recuperator is shown in Figure 14-7. For maximum effectiveness of heat transfer, combinations of radiation type and convective type are used. The hot gases first pass through the radiation section, followed by the convective section.

The most common type of gas-to-gas heat exchanger is the air-to-air heat exchanger. Air-to-air heat exchangers may transfer both sensible and

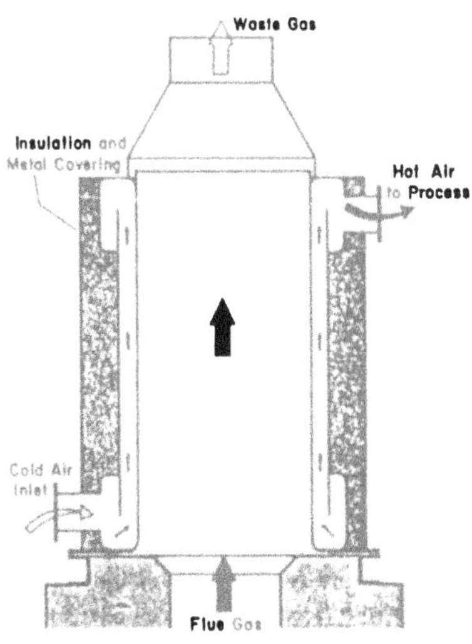

Figure 14-6. Radiation type recuperator.

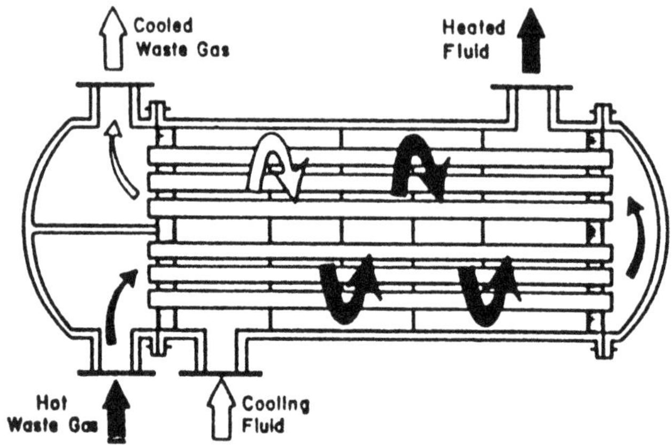

Figure 14-7. Convective type recuperator.

latent heat (moisture), or they may transfer only sensible heat. Air-to-air heat exchangers are frequently used to preheat outdoor make-up air.

Air-to-air heat exchangers, which transfer only sensible heat, are usually the fixed plate type, as shown in Figure 14-8. These heat exchangers have alternate layers of plates separated and sealed (referred to as the heat exchanger core), which form exhaust and make-up air stream passages. Heat is transferred directly from the warm air stream through the separating plates to the cool air stream.

Fixed plate heat exchangers are suitable for applications in which there can be no cross-contamination between air streams. Little or no leakage occurs between the exhaust and make-up air streams in plate type heat exchangers. Fixed plate heat exchangers do not transfer moisture between the two air streams; however, the energy associated

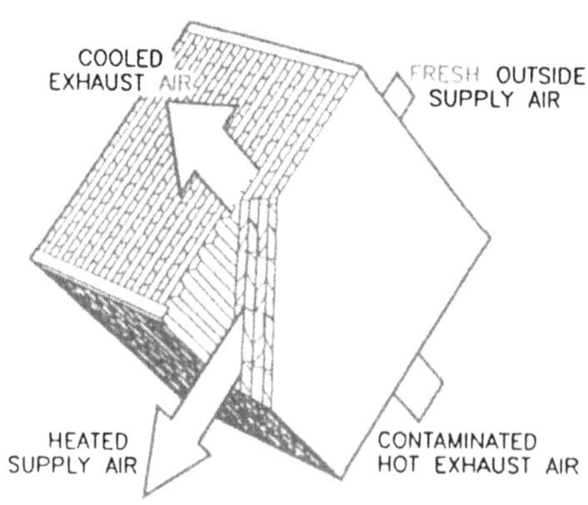

Figure 14-8. Fixed plate crossflow heat exchanger.

with the latent heat in the exhaust air steam may be transferred. Normally, both latent heat of condensation from moisture condensed as the temperature of the warm (exhaust) air, steam drops below its dew point] and sensible heat are conducted through the separating plates to the cool make-up air stream.

Plate type heat exchangers are also used as passive type air preheaters or passive gas-to-gas regenerators. They are used in applications where there cannot be any cross-contamination between the gas streams. A passive gas-to-gas regenerator is shown in Figure 14-9. Typically, these heat exchangers are used to remove heat from flue gases and transfer the heat to incoming combustion air. They come in packaged form for ease of installation.

Air-to-air or gas-to-gas heat exchange may also be accomplished with a heat wheel or rotary heat exchanger. A rotary heat exchanger is shown in Figure 14-10. A rotary heat exchanger (heat wheel) is a revolving cylinder filled with an air-permeable medium having a large internal surface area. Adjacent hot air and cold air streams each flow through one half of the heat exchanger in a counterflow arrangement. As the wheel, or disk, slowly rotates, heat is transferred to the cold air stream.

Rotary heat exchangers may be selected to recover only sensible heat or total heat (sensible heat plus latent heat). Sensible heat is transferred as the medium picks up and stores heat from the hot air stream and releases it to the cold one. Latent heat is transferred as the medium (1) condenses moisture from the air stream with the higher humidity ratio and a simultaneous release of heat; and (2) releases the moisture through evaporation (and heat pick-up) into the air stream with a lower humidity ratio. As a result, the moist air is dried while the drier air is humidified.

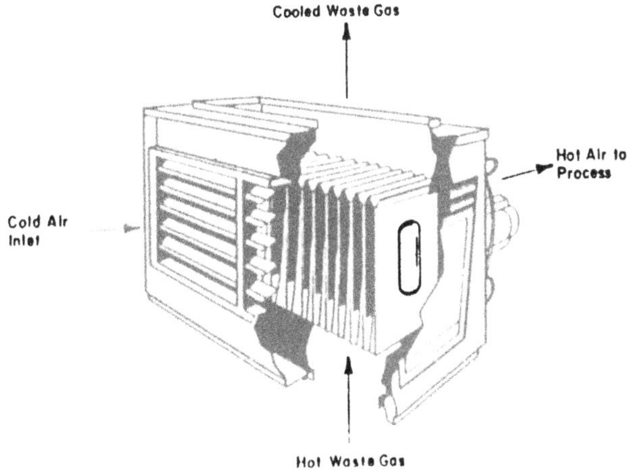

Figure 14-9. Passive gas-to-gas regenerator.

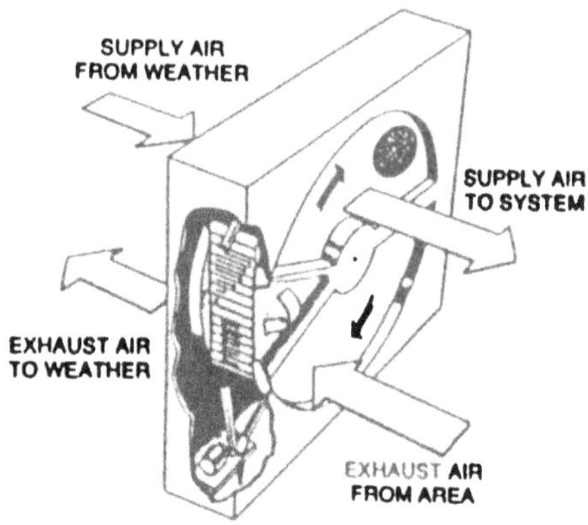

Figure 14-10. Rotary wheel heat exchanger.

Heat wheels are available in four types, which include:

- A metal frame packed with a core of knitted mesh stainless steel or aluminum wire.
- A laminar wheel, which is fabricated from corrugated metal and is composed of many parallel flow passages.
- A laminar wheel, which is constructed of a ceramic matrix of honeycomb configuration.
- A laminar wheel, which has flow passages coated with a hygroscopic material so that latent heat may be recovered. The hygroscopic material is often called a desiccant.

In rotary heat exchangers, cross-contamination between the two air streams can be a problem. Cross-contamination, or mixing, between the hot and cold air streams occurs in all rotational energy exchangers by two mechanisms: carryover and leakage. Carryover occurs as air is entrained within the volume of the rotation medium and is carried from one air stream to the other. Leakage occurs as a result of any static pressure differential across the two air streams from a region of higher static pressure to a lower static pressure region.

14.2.4 Gas-to-Liquid Heat Exchangers

A common gas-to-liquid heat exchanger is an economizer used on a boiler. An economizer is used to recover heat from the boiler flue gases and transfer it to boiler make-up water. An economizer is usually constructed as a bundle of finned tubes, installed in the boiler breaching, through which boiler feedwater flows as it is heated by the hot boiler flue gases. A boiler economizer is shown in Figure 14-11. The tubes are usually connected in series but can also be arranged in bundles to control the

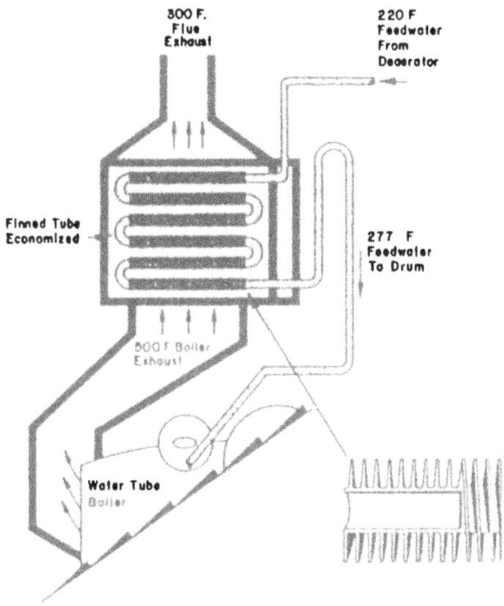

Figure 14-11. Boiler economizer.

liquid side pressure drop. The air side pressure drop is controlled by the spacing on the tubes and the number of tube rows.

In most recuperators, the gas is circulated over the tubes through which the liquid is flowing. Occasionally, the gas flow through the tubes with the liquid surrounding the tubes. This arrangement is referred to as a fire-tube heat exchanger.

The energy savings that could be achieved with a boiler economizer are discussed in "Chapter 4: Steam and Hydronic Boilers."

Another commonly used gas-to-liquid heat exchanger is a waste heat boiler. Waste heat boilers are unfired generators of steam or other fluid vapors. A waste heat boiler is shown in Figure 14-12. They are usually water tube boilers in which the hot exhaust gases from gas turbines, incinerators, furnaces, etc. are used to generate steam.

Water is vaporized in the tubes and collected in a steam drum from which it is drawn off for use as process or heating steam. Waste heat boilers may have either natural or forced circulation. Forced circulation boilers have a pump to circulate water into the boiler.

Waste heat boilers may be classified according to their application as follows:

- Total heat recovery: The cooled gas is vented directly to the atmosphere and the boiler is used to recover as much of the waste heat as possible.
- Partial heat recovery: The gas is cooled down to a temperature suitable for use in some process application.
- Process-control heat recovery: Excess heat is removed from the process stream to prevent further chemical reaction. The waste heat

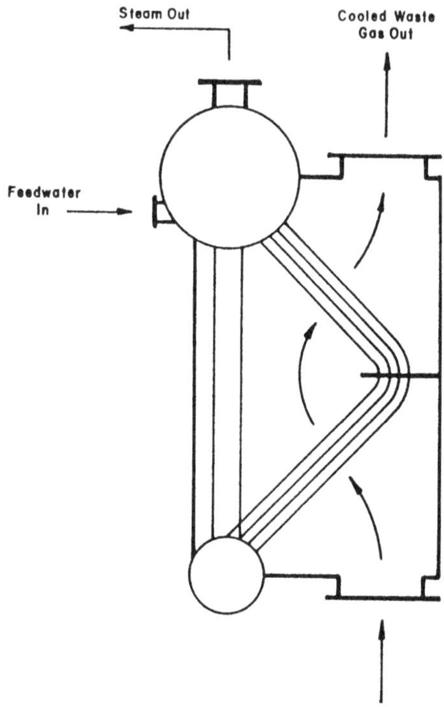

Figure 14-12. Waste heat boiler.

boiler is used to recover a portion of the heat available while providing stability for the preceding process.

14.2.5 Heat Recovery Systems

Heat exchangers may be used to directly transfer heat from one fluid stream to another. Heat exchangers can be used when the two fluid streams are in close proximity to each other. Occasionally, the two fluid streams are not close and the direct exchange of heat is not feasible. Heat recovery systems may be used for such applications. These systems typically consist of two (or more) heat exchangers, one for each fluid steam, and an intermediate heat transfer medium. Heat recovery systems include run-around heat recovery loops, heat pipes, themosiphons and heat recovery towers.

A typical run-around heat recovery loop has two extended surface fined tube water coils: one in the hot fluid stream, the other in the cold fluid stream. The coils are connected via counterflow piping through which an intermediate heat transfer fluid (usually water with anti-freeze) is circulated by a pump. A run-around heat recovery loop is shown in Figure 14-13. These heat recovery systems recover only sensible heat. The overall effectiveness of run-around heat recovery loops is typically between 45 and

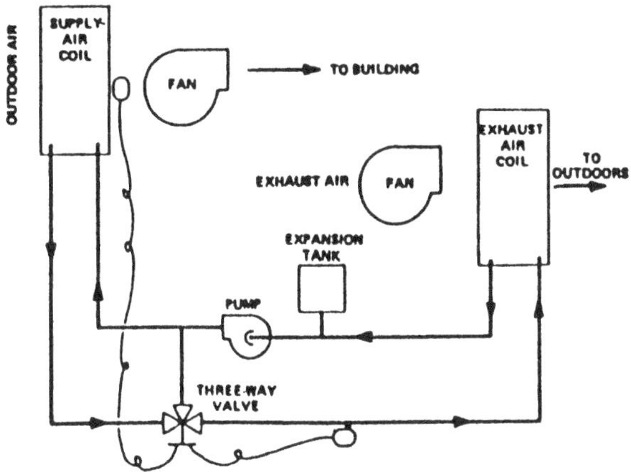

Figure 14-13. Run-around heat recovery loop.

65%; however, the effectiveness can be as high as 70% and the payback period can be less than three years.

Run-around heat recovery loops are more commonly used in building (HVAC applications than industrial type applications.) In HVAC applications, the warm and cold fluid streams would be air. The warm fluid stream is usually exhaust air and may contain significant amounts of moisture. Moisture must not freeze in the exhaust coil air passages when the exhaust air is cooled, due to the removal of heat. A dual purpose three way control valve is usually installed in the piping system to control the exhaust air temperature and keep it from freezing.

A heat pipe heat exchanger is also a heat recovery system. They are passive energy recovery devices comprised of a sealed pipe partially filled with water or some type of refrigerant type heat transfer fluid. A schematic of a heat pipe heat exchanger is shown in Figure 14-14.

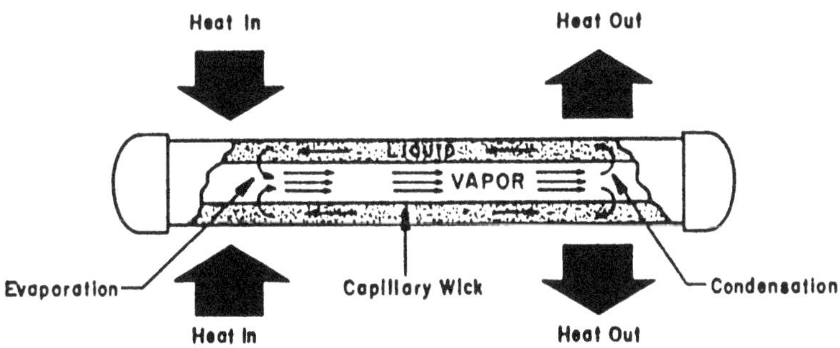

Figure 14-14. Schematic of a heat pipe heat exchanger.

As the hot gas or air flows through the evaporator side of the heat exchanger, heat is absorbed by the internal working fluid as it evaporates. As the fluid evaporates, a pressure differential causes the vapor to flow to the cooler condensing section at the other end of the heat pipe.

The cold gas or air passes the condenser side of the heat pipe at the other end of the pipe. At the condensing end of the pipe, the working fluid gives up latent heat to the cold air or gas as the fluid condenses.

Heat pipe tubes may be fabricated with an internal capillary wick structure, evacuated, filled with a suitable working fluid and permanently sealed. The wick material causes the working fluid to migrate by capillary action back to the evaporator section.

Multiple heat pipes are typically mounted together to form a bundle assembly, as shown in Figure 14-15. However, each individual pipe is a separate sealed element. The rate of heat transfer for a heat pipe heat recovery system may be controlled by adjusting the slope of the pipe bundle. Heat pipe assemblies usually have tilt controllers to adjust the slope of the heat pipe.

Thermosiphon heat exchangers are similar in operation to heat pipe heat exchangers. These heat exchangers also employ a heat transfer medium that undergoes a phase change as it transfers heat. They are sealed systems that consist of a condenser, interconnecting piping, and an intermediate working fluid that is present in both liquid and vapor phases. There are two basic types of thermosiphon heat exchangers in use—the sealed tube, as shown in Figure 14-16, and the coil type, as shown in Figure 14-17.

Sealed tube thermosiphon heat exchangers usually have the evaporator and the condenser at opposite ends of a bundle of straight, individual thermosiphon tubes. The heat transfer medium is evaporated by the heat at

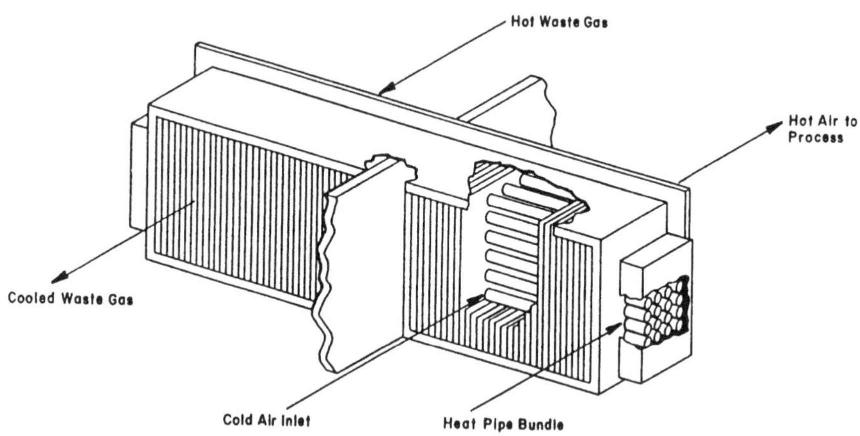

Figure 14-15. Heat pipe heat recovery bundle.

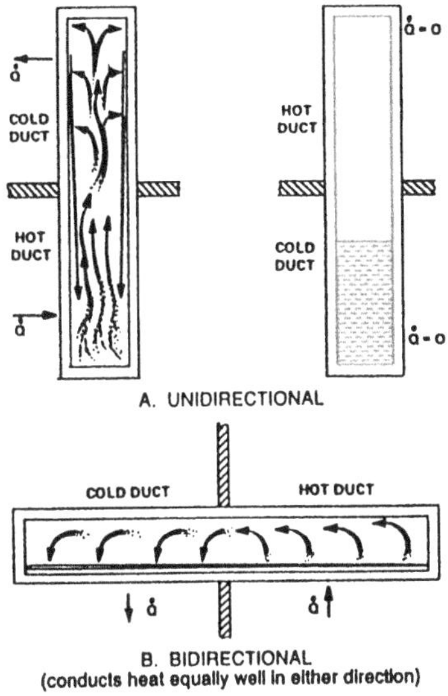

Figure 14-16. Sealed tube thermosiphon.

the hot end and rises due to the difference in fluid density to the condenser end. In the condenser end, the medium condenses as it transfers heat to the cold fluid. The condensed liquid then flows back to the evaporator end due to gravity.

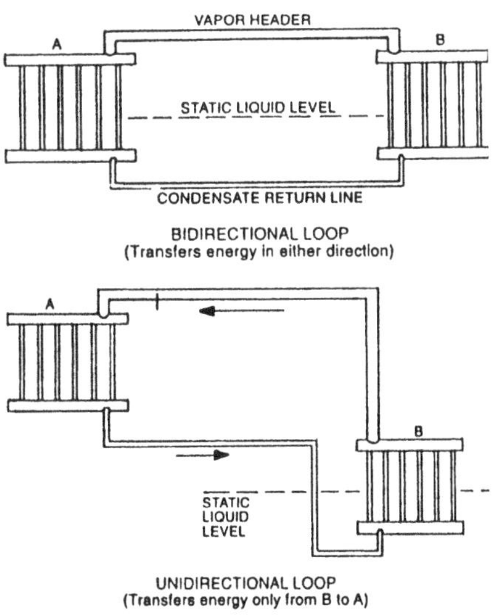

Figure 14-17. Thermosiphon coil loops.

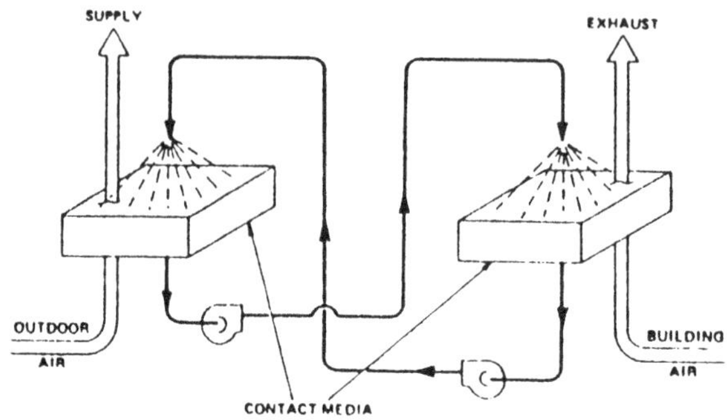

Figure 14-18. Twin tower enthalpy recovery loop.

The coil type thermosiphon heat exchangers have coils at opposite ends of interconnecting piping. One coil is the evaporator and is in the hot fluid stream. The other coil is the condenser and is located in the cold fluid stream. Heat is transferred by the same method as for the sealed thermosiphon.

Twin tower heat exchangers may be air-to-liquid or liquid-to-liquid heat transfer systems. These systems circulate a sorbent liquid continuously between the cold fluid stream and the hot fluid stream, alternately contacting both fluid streams directly in contactor towers. A twin enthalpy recovery loop is shown in Figure 14-18. Pumps circulate the sorbent liquid between supply and exhaust contactor towers.

Twin tower enthalpy recovery systems are primarily used for comfort air conditioning. They are not suitable for high temperature applications. They may operate with temperatures as high as 115°F and temperatures as low as -40°F. The sorbent in the solution acts as an effective antifreeze.

14.3 HEAT PUMPS

According to the Second Law of Thermodynamics, energy will naturally flow from a higher state to a lower state. In other words, heat energy will only flow from a higher temperature region to a lower temperature region. In order for heat to flow in the opposite direction, a system that operates on a thermodynamic cycle must be used, and energy in the form of work must be input into the system. A system or device that transfers heat from a lower temperature region to a higher temperature region is called a heat pump.

A heat pump is a device that operates on a thermodynamic cycle and absorbs energy at a low temperature and transfers the energy through the application of external work to a higher temperate. There are four basic types of heat pump cycles:

- Closed vapor compression cycle. This heat pump is shown in Figure 14-19 and it uses a conventional separate refrigeration cycle that may be single-stage, compound, multi-stage or cascade.

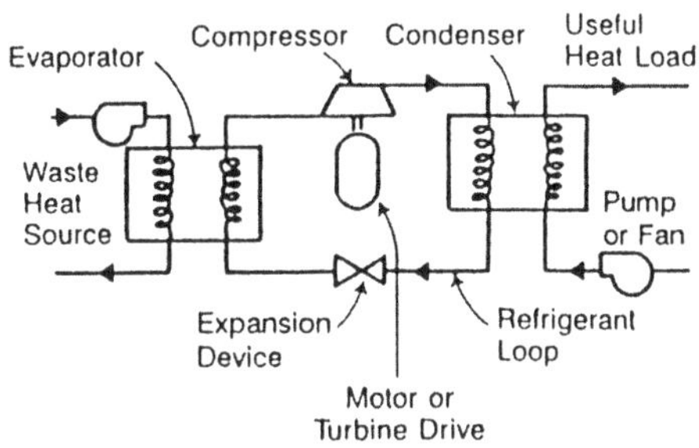

Figure 14-19. Closed vapor compression cycle.

- Mechanical vapor recompression (MVR) cycle with heat exchanger. In this cycle, process vapor is compressed to a temperature and pressure sufficient to reuse directly in a process, as shown in Figure 14-20. Typical applications for this cycle include evaporators (concentrators) and distillation columns.
- Open vapor recompression cycle. This heat pump cycle pumps heat, in the form of steam, to a higher pressure by compressing the lower pressure steam, as shown in Figure 14-21. A typical application of this process is in an industrial plant with a series of steam pressure levels and an excess of steam at a lower than desired pressure.
- Heat driven Rankine Cycle. This type of heat pump cycle is similar to a cogeneration cycle, and is shown in Figure 14-22. This cycle is most applicable where large quantities of heat are wasted and where energy costs are high. The Rankine Cycle is usually a closed system; however, the heat pump portion of the cycle may be either open or closed. Cogeneration systems are discussed in Chapter 13.

14.3.1 Types of Heat Pumps

Heat pumps may also be classified by (1) heat source and sink, (2) heating and cooling distribution fluid, (3) thermodynamic cycle, (4) size and configuration, and (5) limitation of the heat source and sink.

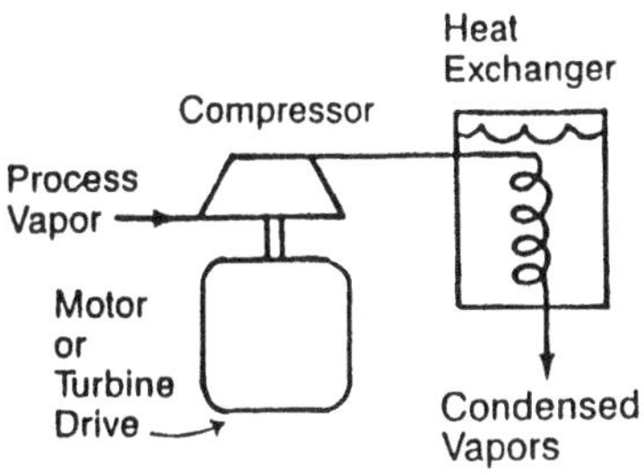

Figure 14-20. Mechanical vapor recompression cycle with heat exchanger.

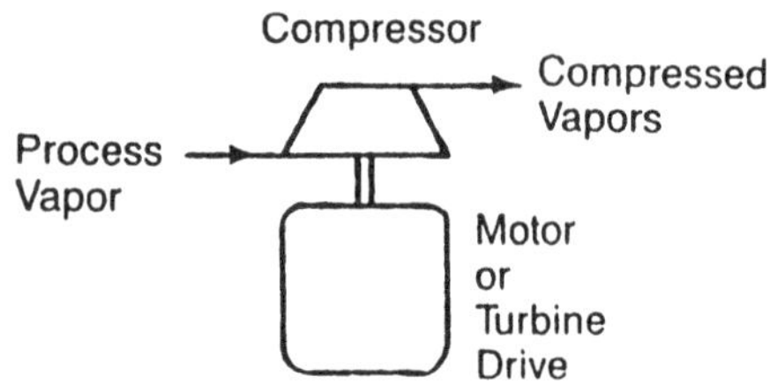

Figure 14-21. Open vapor recompression cycle.

Air-to-air heat pumps are the most common type. They are used to transfer heat from a source that is a gas (typically air) to a sink, which is a gas (also typically air). Air-to-air heat pumps are frequently available as packaged factory units, depending upon the capacity of the system and the temperature applications.

Water-to-water heat pumps use water, or some type of liquid, as the heat source and heat sink. Water is typically used to transfer heat. Water-to-water heat pumps are typically chillers that transfer heat from a chilled water loop to a condenser water loop.

Water-to-air heat pumps transfer heat from a liquid (usually water) to a gas (usually air). These heat pumps typically use water as the heat source or sink. They may transfer heat from water to air, or they may transfer heat from air to water.

14.3.2 Heat Pump Performance

Since a heat pump operates on a cycle that transfers heat from a low temperature region to a high temperature region, the addition of work to the cycle is required in order to accomplish the transfer of heat. The efficiency of an ideal heat pump is expressed as the coefficient of performance (COP) and is defined by Equation (14.13).

$$COP_{\text{ideal}} = \frac{Q_L}{Q_H - Q_L} = \frac{T_H}{T_H - T_L} \qquad \text{Eq. (14.13)}$$

where Q_L = heat transfer from the cold region
Q_H = heat transfer from the hot region

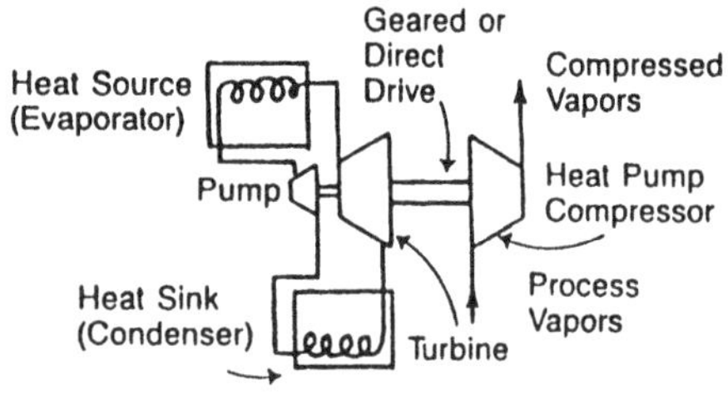

Figure 14-22. Heat driven rankine cycle.

T_L = absolute temperature of the cold region
T_H = absolute temperature of the hot region

Equation (14.13) describes the COP for an ideal heat pump. The COP of an actual heat pump is always less than that of an ideal heat pump. There are several factors that prevent an actual heat pump from achieving the COP of an ideal heat pump, including:

- The compressor efficiency is not 100%, but is actually in the 65 – 85% range.
- Losses occur from friction in lines, compressors and valving.
- Higher condenser temperatures and lower evaporator temperatures than theoretical are required to achieve practical heat flow rates from the heat source to the load.

For a heat to be economically attractive for a particular application, the heat pump should have a COP greater than 3. An additional constraint on the use of heat pumps in high temperature applications is that heat at temperatures above 230°F cannot be supplied directly to the heat pump due to the limits imposed by present compressor design and refrigerant technology.

14.4 APPLICATIONS OF HEAT RECOVERY SYSTEMS

In an ideal heat recovery system, all components work year-round to recover all heat before adding supplemental heat. Any excess heat is either stored or recovered and used, with no heat rejected. Such an idealized heat recovery system is referred to as a balanced heat recovery system.

14.4.1 Heat and Mass Balances

Rarely do the heat recovery sources match the heat loads so that the heat recovery system is a balanced system. Usually, there is some mismatch between the heat recovery sources and loads. For a heat recovery system to be economically feasible, the sources and loads should match as closely as possible. In order to determine the feasibility of a heat recovery system, it is necessary to perform a heat and mass balance for the facility and proposed heat recovery system.

It is important that the heating load that will absorb the heat is available at the same time as the waste heat. If there is a difference in the time that waste heat is available and when it is needed, if neither can be rescheduled, some form of thermal storage will be required. Thermal storage systems are discussed in Chapter 15. If the source and load cannot be rescheduled and thermal storage is not provided, the waste heat will be useless, regardless of its quantity and quality.

A suitable use for waste heat is critical to the success of any waste heat recovery project. No matter how much heat is recovered, it can only reduce energy consumption and costs if the energy is actually used at the facility. Typical uses of waste heat energy include:

- boiler feedwater heating
- make-up water heating
- combustion air heating
- heat for drying processes
- process heating
- domestic water heating
- building heating
- generating electrical power

In order to determine whether there is a suitable match of waste heat and heat loads, it is necessary to conduct a waste heat survey of the facility. Waste heat surveys usually would proceed in a manner similar to that illustrated in Figure 14-23.

Waste heat streams leaving a facility that are not part of a commercial product or service could include hot condensate, steam, flue gases, hot water and heated air. Waste heat streams may also include streams containing combustible fuels, such as paper by-products, sawdust, organic compounds, vapors and fumes.

After the identification of existing waste heat streams, the next step is to perform a heat balance on the system. Such a balance would include the determination of the mass flow rate, the temperature, pressure, and specific heat of each entering and exiting stream. From this balance, the quantity and quality of waste heat for each process may be determined. Similar heat balances should be performed for loads that could potentially use waste heat. The results of the heat balances on the waste heat sources and potential heat loads should be tabulated and sources matched with loads as closely as possible.

An accurate determination of the characteristics of waste heat streams and potential heat loads is important since capital costs for heat recovery systems tend to be high. In many cases, data for various processes is unattainable. In these cases, reasonably accurate estimates must be made.

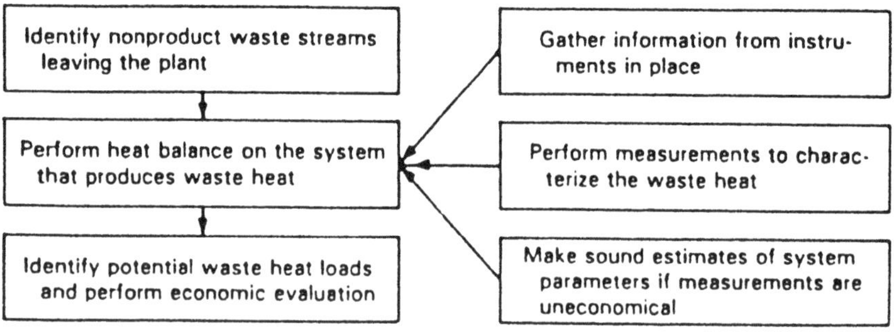

Figure 14-23. Typical sequence of events for a waste heat survey.

Chapter 15

Thermal Energy Storage Systems

Thermal energy storage systems store heat or cooling energy for use at a later time. These systems remove heat from or add heat to a storage medium for use at another time. Stored heat is typically energy removed from a heat recovery system for which the heating load is not coincidental with the time at which the heat is generated. Cooling thermal energy is typically cooling energy that has generated during off-peak energy periods for use during peak energy periods. Cool storage systems remove heat energy from a thermal storage medium during periods of low cooling demand. The stored cooling is later used to meet air conditioning or process cooling loads during periods of high cooling demand and/or periods of higher energy costs. Cool storage systems de-couple the refrigeration system from instantaneous cooling loads.

Often the first cost of central plant equipment, such as boilers, chillers, cooling towers and air handling units, can be reduced due to reductions in required capacities as the result of thermal energy storage.

15.1 BASIC PRINCIPLES OF THERMAL ENERGY STORAGE

Thermal energy storage systems use conventional heating and cooling equipment and a storage system to store thermal energy for later use. They can also store thermal energy that has been recovered from a process or from a cogeneration system. Thermal storage is frequently used to handle short-term peak thermal loads. Systems that include thermal storage can cost less to install and operate than non-storage systems, maintain better environments, offer greater flexibility in use, reduce the size of central plant equipment and save energy by transferring loads to more efficient equipment. Thermal storage materials typically include water, ice and eutectic salt phase change materials.

15.1.1 Advantages of Thermal Energy Storage

Although most thermal storage systems are cool storage systems, energy can be saved by storing heat. Heat may be reclaimed and stored for later use, which reduces the energy consumed to produce heat by boilers and other heating equipment. Energy may also be saved by recovering and storing waste heat generated by cogeneration systems in the production of mechanical or electrical power. Cogeneration systems are discussed in Chapter 13, and heat recovery systems are discussed in Chapter 14.

Frequently, larger facilities purchase electricity from utilities under some form of time-of-use rate schedules. Time-of-use schedules divide a 24-hour work-day into peak time and off-peak time. Utilities usually charge higher rates for electricity consumed during on-peak periods than for electricity consumed during off-peak periods. Utilities also often charge for the peak rate at which electricity is consumed during a billing period, usually a month. Such a charge is referred to as a demand charge. The peak demand during a billing period can affect the monthly demand charges for several subsequent billing periods, up to a year. High demand charges are enforced throughout the demand billing period even if there was only one brief occurrence of the highest demand.

Cool storage systems employ "off the shelf" equipment to extract heat from a storage medium during those daily periods when it is most cost-effective to do so (typically during nighttime hours). The cool storage medium is then used to absorb heat during peak demand periods using only pump and fan energy in the process. As a result, the peak demand is less than it otherwise would be. Using thermal storage to shift loads from peak demand periods to off-peak periods is often referred to as demand balancing or demand shedding.

Thermal energy storage systems may also reduce the size of central plant heating and cooling equipment, such as boilers and chillers. By generating heating and cooling during non-peak load hours, the thermal energy storage system can supplement the capacity of central plant equipment. Frequently, peak heating or cooling loads can be met by supplementing the capacity of the central plant equipment, thus allowing smaller central plant equipment. Often, the reduction in the size and cost of central plant equipment allows the installation of thermal energy systems at no more cost or less cost than non-storage systems.

Cooling thermal energy storage systems can also reduce the size of air and chilled water distribution systems, since some cool thermal energy storage systems produce chilled water at lower than standard chilled water temperatures. Low temperature air and chilled water systems with greater temperature rises can result in initial cost savings due to reduced air and water flow rates. Low temperature air systems can also produce higher comfort levels in HVAC systems since more humidity is removed from the air at lower temperatures.

Cool thermal storage can also be beneficial when existing cooling systems need to be expanded. The addition of thermal energy storage to an existing system can allow cooling capacity that is currently used to supplement refrigeration equipment during peak load periods, thus reducing the need for adding cooling capacity to the plant.

15.1.2 Applications of Thermal Energy Storage

Probably the largest application for thermal storage is off-peak cooling. In most cooling applications, both comfort and process cooling, the peak cooling load occurs during the summer and during the daytime. This time period is also the time during which most utilities charge the highest time-of-day energy charge and the highest demand charge. Energy cost savings can usually be achieved if the actual cooling is generated during off-peak daily time periods. Cool thermal storage systems may be used to shift the cooling from peak periods (usually during the day) to off-peak periods (usually at night). The refrigeration system is operated at night to generate cooling (usually chilled water or ice), which is stored. The stored cooling is then used during the daytime peak energy cost period. The stored cooling medium is then used during the peak energy cost period to supply part, or all, of the cooling necessary to meet the on-peak cooling load. This type of operation is often called load shifting. Figure 15-1 shows a typical load shifting cooling profile.

Cool thermal storage systems may also be used for load leveling. In this application, the cool thermal storage is used to provide a level or constant cooling load on the refrigeration system. During light cooling load periods (usually at night), the refrigeration system is operated at the same capacity, which is generally less than that required by the peak cooling load, and supplemental cooling is supplied by the thermal storage system. Figure 15-2 shows a typical cooling profile for a load leveling thermal storage application. When the load leveling cool storage mode is used, capacities of the storage equipment are selected so the peak cooling load can be met by continuous operation of the refrigeration equipment. This application minimizes the refrigeration compressor capacity requirements.

Cooling thermal storage systems may also be used to limit peak demand. This thermal energy storage strategy is similar to off-peak cooling. Demand limiting systems employ sophisticated techniques to control the refrigeration equipment, preventing its operation during peak demand periods. Figure 15-3 shows a cooling profile for a demand limited operating mode. System operation requires monitoring of the non-cooling loads and forecasting the day's non-cooling peak demand. Compared to the load

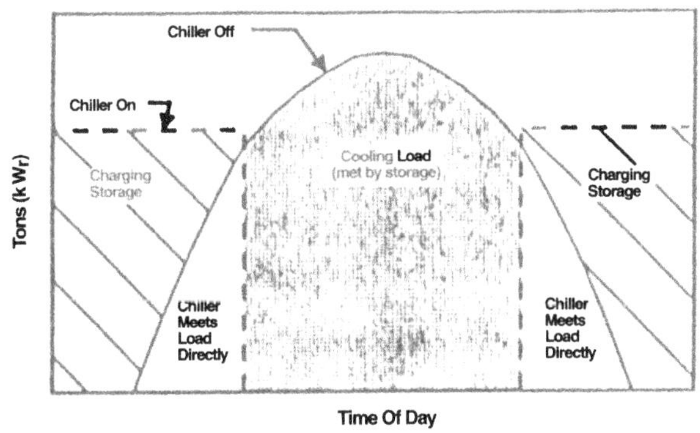

Figure 15-1. Load shifting cooling profile.

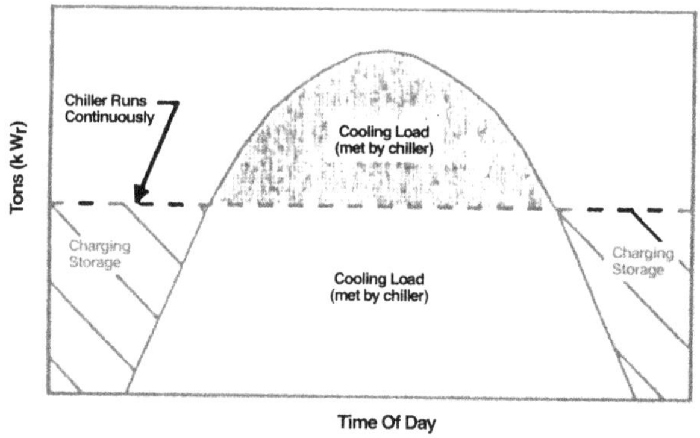

Figure 15-2. Cooling profile for load leveling mode.

leveling mode, a demand limited system does not result in as much refrigeration system compressor capacity reduction and requires larger capacity thermal storage systems.

Cool thermal storage systems are also frequently used in industrial/ process cooling applications. Refrigeration and process cooling applications typically require lower temperatures than comfort cooling applications. Low temperatures for freezing food, storing frozen food, etc., can be obtained by using low temperature thermal storage. Cool thermal storage systems can provide steady supply temperatures regardless of large, rapid changes in the cooling load or the return temperature. Cooling can be extracted very quickly from cool thermal storage systems to satisfy a very large cooling load and then be recharged slowly using a small refrigeration system.

Thermal storage systems may also be used for off-peak heating and to store solar heat. Some utilities provide incentives for off-peak water heating. This application is similar to off-peak cooling in which

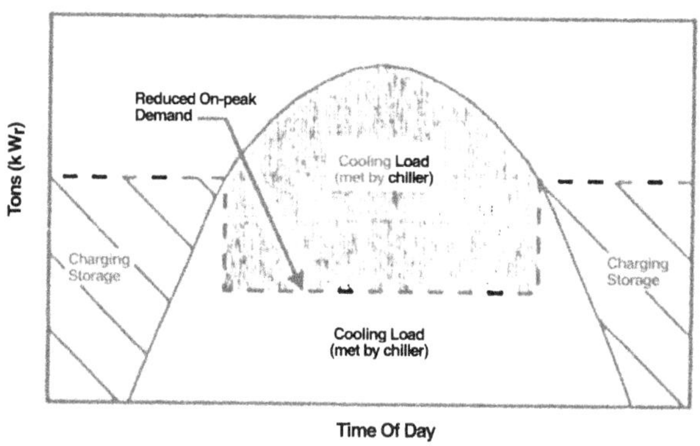

Figure 15-3. Demand limited cooling profile.

thermal energy is stored during off peak periods for use during on-peak periods. Applications of thermal heat storage also include the storage of heat collected by a solar energy system and heat recovered form heat generating processes, such as boilers and cogeneration systems.

15.2 SOURCES OF HEAT FOR THERMAL ENERGY STORAGE

There are a large number of potential sources of heat for thermal storage. The most common sources include waste heat recovery, cogeneration, chiller heat recovery and solar heating systems.

Waste heat recovery involves recovering energy for reuse, which would otherwise be discharged to the environment. Waste heat is rejected from a building or process because its temperature is too low for economical recovery or direct use. Heat may be recovered from heat exchangers and from heat pump systems. Heat exchangers are passive or active devices, which directly or indirectly transfer heat from a hotter exhaust stream to a cooler fluid stream. Heat pumps are heat exchange systems that transfer heat from a low temperature waste heat source to a higher temperature fluid stream. Typically, waste heat suitable for thermal storage must be in the form of a liquid, usually water. Additional information about heat recovery may be found in Chapter 14.

In the generation of power, either mechanical or electrical, considerable heat is rejected as the power is generated. Conventional power generation systems discharge the heat resulting from power generation to the atmosphere as waste heat. Cogeneration systems recover the heat for other uses. By capturing and applying the heat, or storing the heat for later use, cogeneration systems operate at efficiencies greater than those achieved when heat and power are produced in separate or distinct processes. Unless there is a coincidental use for the waste heat, it must be stored for later use. Cogeneration systems can generate large amounts of heat that may be stored for later use. Cogeneration systems are discussed in Chapter 13.

Chillers reject heat to the atmosphere with one of three rejection components: air cooled condensers, evaporative fluid coolers and cooling towers. Lager chillers typically use evaporative fluid coolers or open cooling towers. Heat rejected from vapor compression chillers is typically in the 85° to 100°F range. Absorption chillers use steam or hot water as a heating medium for the absorption process. Absorption chillers may also be direct-fired with fuels such as natural gas or fuel oil. Heat rejected from absorption chillers is typically available in thc 190° to 250°F range. Additional information on chillers may be found in Chapter 7.

In some areas, it is feasible to gather and use heat from solar collectors. Since sunlight is available only during the day, some form of thermal storage is required for nighttime heating and heating on days when there is no sunshine. Active solar systems use either a liquid or air as the collection fluid. Active systems absorb solar radiation with collectors and convey it to storage by circulating the collection fluid.

15.3 REFRIGERATION FOR THERMAL ENERGY STORAGE SYSTEMS

Refrigeration for thermal storage may be done directly or indirectly, with water or brine chillers. Direct refrigeration systems would be vapor compression types, in which the storage medium is cooled directly by the refrigerant as it is circulated through cooling coils or pipes. Indirect refrigeration systems would include vapor compression chillers and absorption chillers.

15.3.1 Direct Refrigeration Systems

Vapor compression refrigeration systems typically consist of a compressor, a condenser, an evaporator and some type of refrigerant flow control. A compressor is a mass mover that transfers refrigerant vapor from a low pressure area at its evaporating conditions to a high pressure area at condensing conditions. Refrigerant compressors used in thermal storage refrigeration systems are usually open reciprocating compressors and rotary screw compressors.

Open reciprocating compressors are available in sizes ranging from 20 to 300 HP. Capacity control is usually provided by loading or unloading cylinders. As a result, reciprocating compressors have a stepped response to varying refrigeration loads.

Rotary screw compressors have two rotors separated by an oil film to compress refrigerant gas. Capacity control is provide by a slide valve, which allows efficient, surge-free, infinitely variable capacity control from 100% down to about 10% of full capacity. Rotary screw compressors are available in sizes ranging from 100 to 1,500 HP.

Condensers are used in refrigeration systems to remove heat from the refrigerant after it has been compressed. There are three basic types of condensers: air-cooled, water-cooled and evaporative.

Air cooled condensers remove heat from the refrigerant by circulating air over an extended surface coil through which the refrigerant is circulated. The latent heat of the refrigerant is removed as it condenses to a saturated liquid by sensible heating of the air. The capacity of an air cooled condenser is a function of the ambient air temperature. Typically, they require fans with large horsepower motors to circulate large volumes of air.

Water cooled condensers are typically shell-and-tube heat exchangers in which heat is transferred from the refrigerant to water circulated through the condenser. Heat is then rejected from the cooling water to the atmosphere with a fluid cooler or cooling tower. Fluid coolers and cooling towers utilize evaporative cooling to improve heat transfer and efficiency.

The total energy required (pump horsepower plus cooling tower fan horsepower) to reject heat is less than for air cooled condensers at the same condensing temperature.

Refrigeration systems are discussed in more detail in Chapter 12. Cooling towers and fluid coolers are discussed in Chapter 8.

15.3.2 Chillers

Chillers are frequently used in cool storage applications. Chiller types include reciprocating compressor, screw compressor, centrifugal compressor, scroll compressor and absorption. Chillers can be used to cool water or brine that is circulated through the cool storage medium. Leaving water temperatures range from about 45°F down to 15°F for brine systems. Ammonia absorption chillers are capable of producing leaving brine temperatures as low as -51°F.

Chillers can have air cooled, water cooled or evaporatively cooled condensers. The capacity of air cooled condensers is limited by the ambient air dry bulb temperature. The capacity of chillers with water cooled condensers is limited by the ambient air wet bulb temperature, therefore, lower leaving temperatures can be achieved.

Chillers are discussed in more detail in Chapter 7.

15.4 OPERATING STRATEGIES FOR COOL THERMAL STORAGE SYSTEMS

There are two basic operating strategies for cool thermal storage: the full-storage (load-shifting) strategy and the partial-storage strategy. With a full-storage strategy, the storage capacity must be sufficient to meet the entire energy requirement of the cooling load, and the chiller capacity must be sufficient to recharge the storage during the off-peak period, usually at night. The chiller runs only during the off-peak period. The storage is used to fully decouple the operation of the chiller plant from the peak cooling load. A load shifting cooling load profile is shown in Figure 15-1. The full-storage strategy requires relatively large refrigeration and storage capacities. This strategy is best suited for applications where the on-peak cooling cycle duration is relatively short compared to the time required to recharge the cool storage. Full-storage systems are likely to be economically feasible only under one, or more, of the following conditions:

- Spikes in the peak load curve are of short duration.
- Time-of-use energy rates are based on short-duration on-peak energy periods.
- There are short overlaps between peak loads and peak energy periods.

- Large cash incentives are offered for using thermal storage
- High peak demand charges apply.

In a partial storage strategy, the peak cooling load is met partially by the cool storage and partially by a chiller. In this scenario, the chiller and the cool storage are used together to meet the load, and the chiller recharges the thermal storage when there is little or no cooling load. Under partial storage, the chiller runs many more hours than with a full storage strategy and, as a result, less demand reduction is obtained. However, partial storage is initially less expensive than a full storage system since less storage capacity is required and because smaller refrigeration capacity is required. Partial storage systems may be operated under a number of strategies, including load leveling, demand limiting, baseloading of chillers, sequencing of chillers and demand shifting. The operation of partial storage systems may also be optimized according to a chiller priority, storage priority, charging rate control and load prediction.

15.4.1 Load Leveling Mode

A load leveling thermal storage system is designed such that the load on the system chiller is a constant. A load leveling system typically operates with the refrigeration system running at full capacity for 24 hours. A typical cooling profile for a load leveling mode is shown in Figure 15-2. During peak cooling load period (usually during the daytime), part of the required cooling capacity is provided by the chiller operating at full cooling capacity, and the balance is provided by the thermal energy storage system. Rather than completely shifting the peak electrical demand for cooling, this approach levels the electrical demand for cooling over the design day. During low load periods (usually at night), the chiller operates at full cooling capacity to recharge the thermal storage system. The required available storage capacity must be sized so that the storage and refrigeration equipment meet the peak design day cooling.

15.4.2 Demand Limiting Mode

Demand limiting thermal storage systems are in-between load shifting systems and load leveling systems in terms of operation. A typical cooling profile for a demand limiting mode is shown in Figure 15-3. Operation in this mode reduces and limits the facility's electrical demand to a predetermined level. This level is normally slightly higher than the level peak demand imposed by the non-cooling electrical loads. In the demand-limiting mode, the refrigeration equipment operates at a reduced capacity or demand level during the on-peak period. In some applications, the refrigeration equipment is controlled to keep the facility demand at the billing meter at a given level. The storage capacity for a demand limiting mode is sized so that the simultaneous operation of the storage and refrigeration equipment does not cause the facility's total electrical demand to exceed a predetermined maximum level. To effectively keep the total allowable maximum demand, demand-limiting partial storage requires the use of real-time controllers to monitor the facility's non-cooling loads and control the ratio of storage and chiller supplied cooling.

15.4.3 Baseloading of Chillers

When the refrigeration system has more than one chiller, baseloading of one, or more, chillers can often improve the economics of cool thermal storage. This mode is most suitable for applications with relatively flat load profiles and those using ice storage technology. In this mode, the most efficient chiller, or chillers, is used to meet the constant component of a facility's load. A smaller chiller is then used to level, or shift, the remaining load. A load profile for baseloading operation with partial storage is shown in Figure 15-4.

Frequently, a separate chiller is used to meet off-peak loads while a storage chiller is used to charge the cool storage. The use of multiple chillers generally increases the flexibility of a cool storage system in meeting various load conditions that may occur.

15.4.4 Chiller Priority Operation

Under a chiller priority operating strategy, the chiller is used to directly meet as much of the cooling load as possible. This approach is most applicable to situations where the cost of stored cooling energy is higher than the cost of direct cooling, and where large utility cost savings are achieved through reductions in on-peak demand.

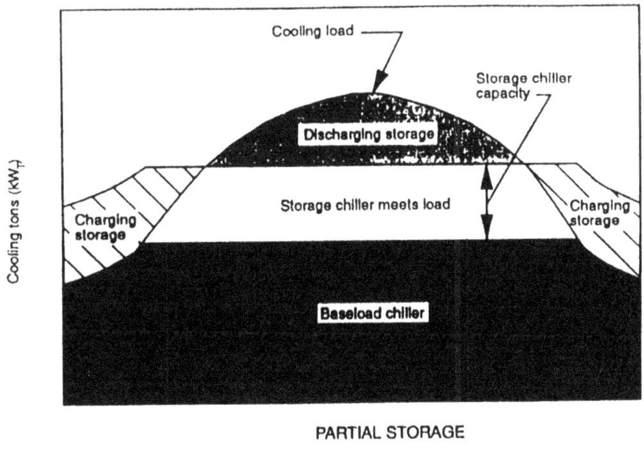

Figure 15-4. Load profile for baseloading operation with partial storage.

15.4.5 Storage Priority Operation

Cool thermal storage systems may also be operated under a storage priority strategy. The storage priority strategy uses stored cool thermal energy to meet as much of the cooling load as possible. Direct cooling from chillers is provided only when the cooling load exceeds the total stored cooling capacity. Under a storage priority, the chiller is sized such that it has just enough capacity for later use in the daily cycle. Operation of a thermal storage system under a storage priority strategy usually requires a more complex control sequence than operation under a chiller priority strategy. Storage priority operation is typically most feasible where the cost of stored cooling energy is lower than the cost of direct cooling.

15.5 THERMAL ENERGY STORAGE MEDIA

There are several types of thermal storage media. Most are commonly found materials such as water and ice. Other media include various types of phase change materials (PCMs). Phase change materials are those materials that exhibit properties, such as melting points, that lend themselves to thermal energy storage. Some materials are suitable only for cool thermal storage, while others are suitable for both warm and cool thermal storage. Desirable characteristics of thermal storage media include the following:

- commonly available
- low cost
- environmentally benign
- non-flammable
- non-explosive
- non-toxic
- compatible with commonly used materials
- non-corrosive
- inert
- well-documented physical properties
- high density
- high specific heat (for sensible heat storage)
- high heat of fusion (for latent heat storage)
- high heat transfer characteristics
- storage at ambient pressure
- characteristics unchanged over long use

The most commonly used methods for storing thermal energy include water storage, ice harvesting, external melt ice-on-coil, internal melt ice-on-coil, encapsulated ice and eutectic salt phase change materials.

15.5.1 Water Storage

Water is the most commonly used material for thermal storage. It can be used for both warm thermal storage and cool thermal storage. Water has the highest specific heat (1 Btu/lb-°F) of all common materials, which make it well suited for thermal storage. Typically, water storage systems are easily interfaced with other heating and cooling systems. In chilled water thermal storage systems, the chillers are usually not required to produce water any colder than standard chilled water systems. Chillers providing chilled water to existing chilled water systems may be used to supply chilled water for thermal storage.

In order for a chilled water storage system to operate at optimum efficiency, the storage system should maximize stored cooling capacity by maintaining a high chilled water temperature difference between the supply water and the return water. To ensure maximum temperature differentials, it is necessary to prevent the warm return water from mixing with the stored chilled water as much as possible. Although the concept of not allowing warm return water to mix with the stored chilled water seems practical, it becomes complicated in actual practice. There are a number of methods for reducing mixing, including diffusers, multiple tanks, diaphragms and labyrinth baffles.

The density of water is directly related to its temperature. As a result, cold water between 39.2°F and 42°F will collect and stabilize in the lowest areas of a storage tank. Warm water between 50°F and 65°F will collect near the top of a tank . A stable, sharply defined thermocline prevents cold water from mixing with the warm water above. A typical stratification temperature profile is shown in Figure 15-5.

Since it is important to prevent the warm water from mixing with the cold water, it is necessary that the returning warm water not mix with the

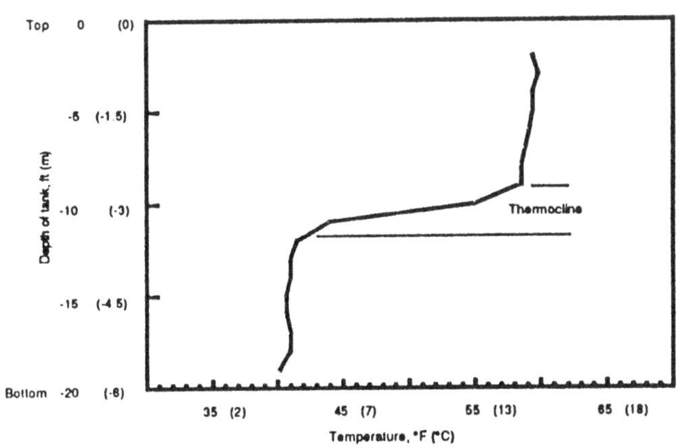

Figure 15-5. Typical stratification temperature profile.

cold as it enters the storage tank. Diffusers are frequently used to reduce the mixing of warm and cold water as it flows into and out of the tank. With a properly designed diffuser, water gently flows into and out of the tank. As a result, flow turbulence is minimized and the thermocline is undisturbed. A tank with an H-style diffuser pattern is shown in Figure 15-6.

A high temperature difference between the chilled water supply and return in a thermal storage system may also be accomplished with multiple storage tanks. Multiple tank systems use an extra tank, which is empty, to separate the warm return water from the cold leaving water. The warm return water is stored in one tank, or tanks, and the cold supply water is stored in a separate tank, or tanks. An extra tank, which is empty, is also provided for the warm water to flow into as it returns. The tank from which the cold water is flowing is allowed to empty while the extra return tank is filled. Multiple tank storage systems require relatively complex piping arrangements and flow controls to properly direct the flow to the appropriate tank.

A high temperature difference in thermal storage tanks may also be maintained with membranes or diaphragms within the storage tank. These storage tanks have a flexible membrane, or rigid, moveable diaphragm mounted inside the tank. The membrane or diaphragm is used to separate the warm return water from the stored cold water. The membrane or diaphragm eliminates the mixing of the warm and cold water without the need for diffusers. The first cost of tanks with membranes is relatively high and there is a risk of a tear in the membrane.

Mixing of warm and cold water may also be reduced by a labyrinth and baffle system. Labyrinth and baffle systems separate the storage volume into multiple compartments separated by partial walls or baffles. A labyrinth tank is shown in Figure 15-7. Due to flow turbulence, density currents,

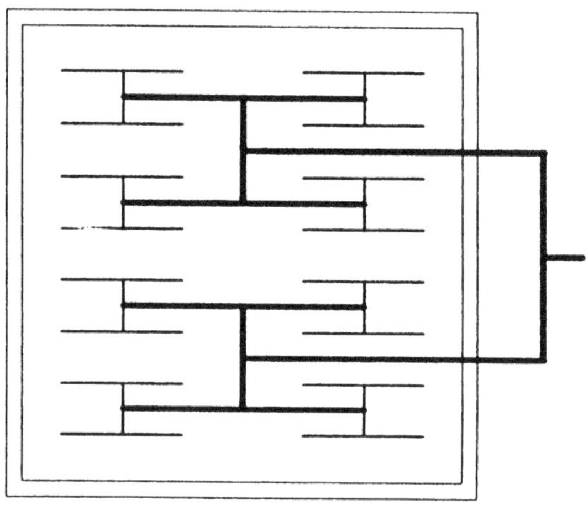

Figure 15-6. Tank with H-style diffuser.

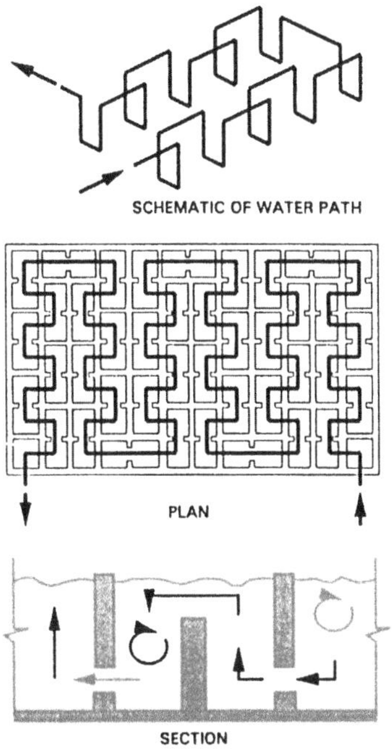

Figure 15-7. Labyrinth tank flow pattern.

velocity currents, thermal conduction and wall-water interaction, significant mixing of warm and cold water occurs in labyrinth and baffle tanks.

15.5.2 Ice Harvesting

Ice harvesting cool thermal storage systems are "dynamic" storage systems. They can simultaneously generate ice for storage and provide chilled water for cooling. These systems separate, or decouple, the formation of ice from the storage of ice as cool thermal energy.

Ice is typically formed on both sides of a hollow flat plate, or on the outside or inside (or both) of a cylindrical evaporator surface. Ice is generated on the evaporator surfaces by circulating water at 32°F over the surface of the evaporator. These systems make ice in plate, chunk, crushed or slush form. The ice is then stored in large tanks or bins. Typically, the evaporators are arranged in vertical banks above the storage tank or bin, as shown in Figure 15-8.

In the ice making mode, liquid refrigerant flows through the inside of the evaporator while water is circulated over the outside, as shown in Figure 15-9. Ice is formed to a thickness between 0.25 and 0.40 inches. The ice that has formed is then harvested, usually by introducing hot refrigerant gas into the evaporator, as shown in Figure 15-10. The refrigerant gas warms the evaporator, which breaks the bond between

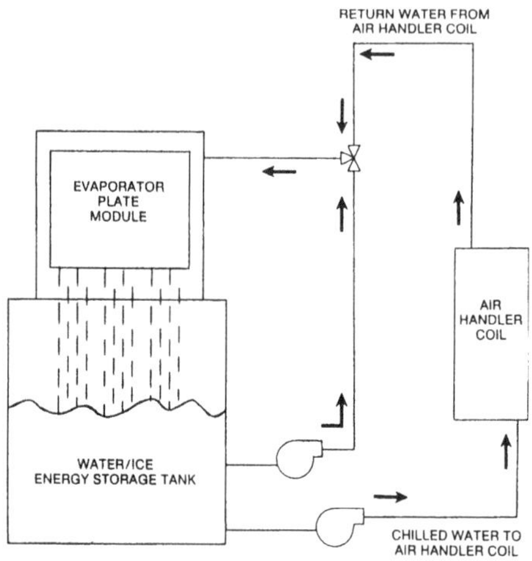

Figure 15-8. Typical plate type ice harvesting system.

the ice and the evaporator surface and allows the ice to drop into the storage tank below.

When the ice drops into the tank, it mixes with water in the tank to form a slurry of ice and water. As the ice melts in the water, the water is cooled to near 32°F, its freezing temperature. The water, which has been chilled by the ice, is then circulated to the cooling load where it is warmed by the load, and then returned to the storage tank. Ice harvesting systems can generate ice during off-peak periods for use during on-peak periods. They can also both generate ice and provide chilled water during either period. Ice harvesting systems can be simultaneously charged and discharged.

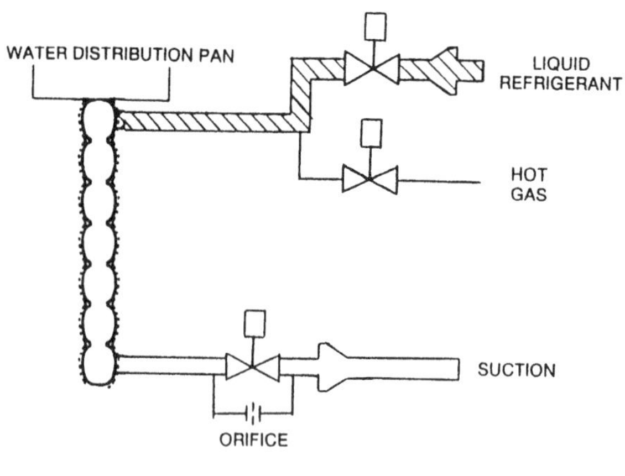

Figure 15-9. Ice making mode.

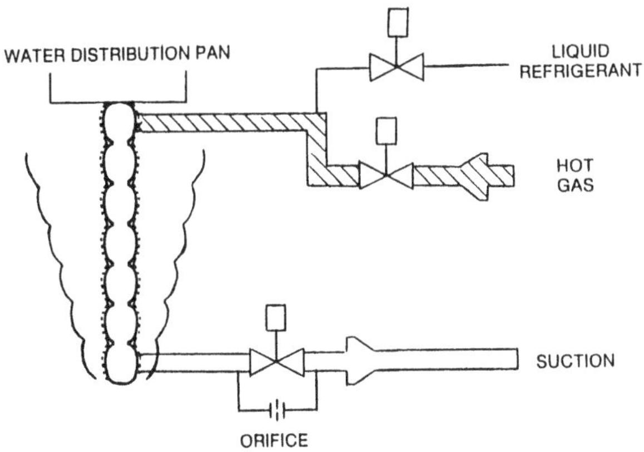

Figure 15-10. Ice harvesting mode.

Ice harvesting systems can be operated in any of the three basic operating strategies: full storage, partial storage load leveling or partial storage demand limiting. Base loading of non-storage chillers is also an effective strategy for ice harvesting thermal storage systems. A high efficiency chiller should be used to meet a constant load, while the stored cooling from the ice harvester should be used to reduce peak demand.

Unlike most other ice storage methods, the efficiency of an ice harvester refrigeration system is not affected by the amount of ice stored in the storage tank. Since ice is built on the evaporator surfaces, the refrigerant has a relatively high and nearly constant suction temperature. However, the losses associated with the defrost cycle tend to reduce and offset some of the efficiency gained from the constant and higher suction temperature.

15.5.3 External Melt Ice-on-Coil

External melt ice-on-coil storage systems are refrigerant-fed ice builders consisting of refrigerant-fed or brine-fed coils inside a storage tank filled with water. A schematic diagram of an external melt ice-on-coil system is shown in Figure 15-11. Ice is formed on the outer surface of submerged pipes or tubes through which a refrigerant or secondary coolant (brine) is circulated. External melt ice-on-coil storage systems are also referred to as ice builders.

Direct refrigerant external melt ice-on-coil systems are widely used in industrial applications, such as dairies and food processing plants. A direct refrigerant system is more efficient than a secondary coolant system, since there is one less step of heat transfer between the evaporating equipment and the ice building surface. One disadvantage of direct refrigerant systems, however, is that the system requires a large amount of refrigerant, which can be costly.

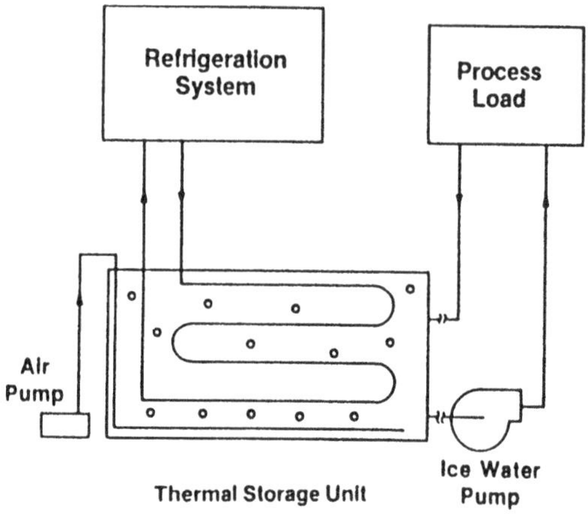

Figure 15-11. Schematic of external melt ice-on-coil system.

Secondary coolant-based systems use standard chillers to cool a brine (e.g., 25% ethylene glycol and 75% water) for ice making temperatures of 18 – 26°F. The secondary coolant has the advantage of significantly reducing the amount of refrigerant required by the system.

External melt systems are well suited for any of the three basic operating strategies: full storage, partial load leveling and partial storage demand limiting. Storage priority control is generally recommended for external melt systems. These strategies usually require some method of estimating or predicting the load for the upcoming on-peak period to provide the amount of cooling from the supplementary chiller.

To maximize the efficiency and minimize energy consumption, all ice should be melted each day so the charge cycle for the next day begins with bare pipes. When operating an external melt ice-on-coil system, it is important (1) to limit the ice thickness (and resulting excess compressor energy) during the ice building cycle; and (2) to minimize the bridging of ice between individual tubes in the ice bank. Bridging must be avoided because it restricts the free circulation of water during the discharge cycle.

15.5.4 Internal Melt Ice-on-Coil

Internal melt ice-on-coil thermal storage systems have coils or tubing located inside a water tank. The coils occupy about 10% of the tank volume; another 10% volume of the tank is left empty to allow for expansion of the water when it freezes (the rest of the tank contains water). An internal melt ice-on-coil thermal storage tank is shown in Figure 15-12. These systems use a secondary coolant as the charging and discharging heat transfer fluid. The secondary coolant is typically a brine solution, such as 25% ethylene glycol and 75% water.

The secondary coolant is cooled by a liquid chiller and circulated through the coils to freeze the water in the storage tank. To build ice, the chiller cools

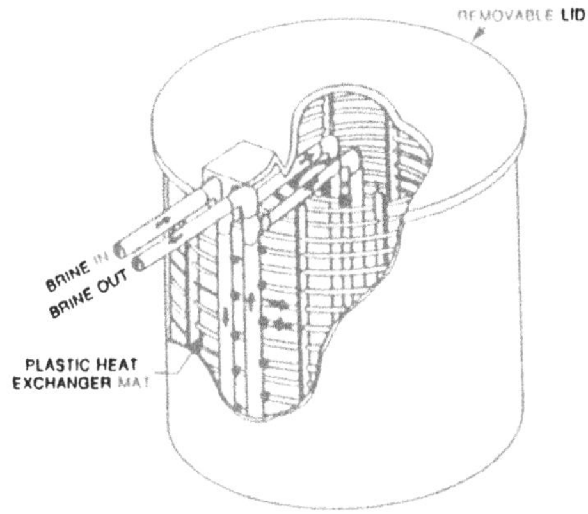

Figure 15-12. Internal melt ice-on-coil storage tank.

the secondary coolant to 22 – 26°F and ice forms on the outside of the tubes. A standard chiller can provide the refrigeration for these systems.

After the ice is built during the off-peak charging period, the system is discharged by circulating water through the tubes. Warm return water, or brine, from the chilled water distribution system flows through the tubes, melting the ice from the inside and reducing the temperature of the circulating fluid.

Internal melt ice-on-coil systems usually are configured with the ice storage tanks piped in series with the chiller. If the chiller is located upstream of the tank, as shown in Figure 15-13, and a partial storage operating strategy is employed, warm return fluid from the cooling load is pre-cooled by the chiller before entering the ice storage tank. This

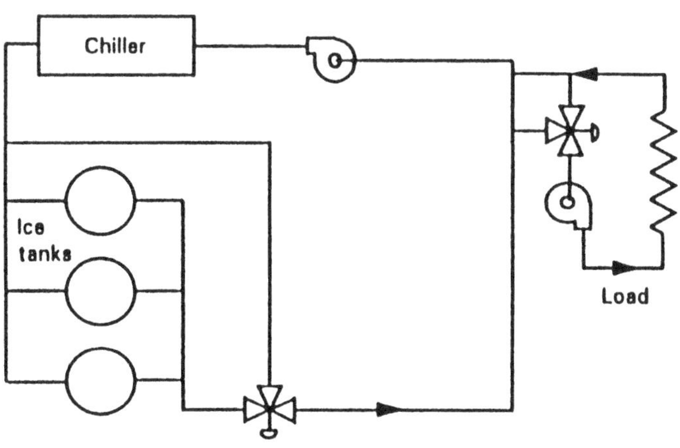

Figure 15-13. Chiller upstream configuration.

arrangement results in more efficient chiller operation due to the higher operating temperatures.

If the chiller is located downstream of the ice storage tank, as shown in Figure 15-14, warm return fluid is cooled by the ice in the storage tank before it enters the chiller where it is cooled to a lower temperature. This arrangement results in a higher useable storage capacity, in addition to a constant supply water temperature from the system. One disadvantage of this arrangement is that the chiller operates at a lower discharge temperature, which results in less efficient chiller operation.

Internal melt ice-on-coil storage systems are suitable for each of the three basic operating strategies: full storage, partial storage load leveling or partial storage demand limiting. Internal melt systems with the chiller upstream configuration are well suited to a chiller priority strategy. A storage priority is generally more complicated and difficult to implement.

It should be noted that the thermal output of internal melt ice-on-coil systems varies with time. If the discharge flow rate is held constant, the discharge temperature will steadily increase. Conversely, if the discharge temperature is to remain constant , the discharge flow rate must steadily be reduced. The tendency for the thermal output of internal melt systems to decrease as the tank is discharged results from the insulating effect of the water around the tube as the ice melts. The ability of the storage system to meet the load during the last several hours of the discharge cycle is generally most important in determining the required storage capacity of the system. The storage capacity of the system must be selected so that the required discharge temperature and discharge flow rate can be maintained throughout the discharge cycle.

15.5.5 Encapsulated Ice Systems

Encapsulated ice storage systems provide cool thermal storage by employing plastic containers filled with deionized water and an ice-

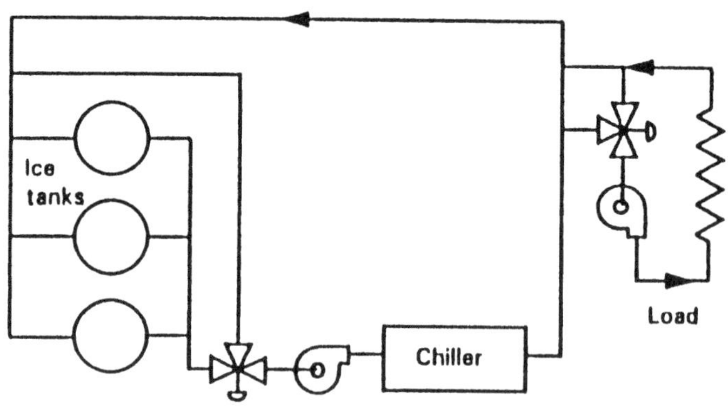

Figure 15-14. Chiller downstream arrangement.

nucleating agent. The containers with the water are submerged in a secondary coolant, such as an ethylene glycol brine solution, which is cooled to 24 – 26°F by a liquid chiller. The water filled containers and the secondary coolant are contained inside a storage tank. The secondary coolant is circulated through the tank and over the water filled containers to charge or discharge the storage system. The secondary coolant freezes the water in the containers during the charging cycle. The water inside the containers is thawed during the discharge cycle.

The plastic containers are typically available in two shapes: rectangular and dimpled spheres, as shown in Figure 15-15. Rectangular containers are available in 1.1-gallon and 4.5-gallon sizes. Spherical containers are available with a diameter of about 4 inches. Encapsulated ice containers are made of high density polyethylene plastic, which can allow for expansion and contraction as the water in the containers freezes and thaws.

The most common arrangement for chillers in an encapsulated ice storage system is with the chiller upstream of the storage tank. The chiller pre-cools the return water before it enters the thermal storage tank. Since the circulated water temperature is higher, the chiller operates more efficiently than if it were located downstream of the ice storage tanks.

When the chiller is located downstream of the ice storage tank, the return water is pre-cooled by the ice before entering the chiller. Although this arrangement results in a higher useable storage capacity and a constant discharge temperature, the chiller will operate at a lower efficiency due to the lower entering water temperature.

Encapsulated ice storage systems are applicable to any of the three operating strategies: full storage, partial storage load leveling or partial storage demand limiting. Chiller priority control is very compatible with encapsulated ice storage systems when the chiller is located upstream of the ice storage tank. When the cooling load is less than the chiller capacity, it is possible to bypass the storage tanks. Chiller priority may also be used when the chiller is located downstream of the ice storage tank.

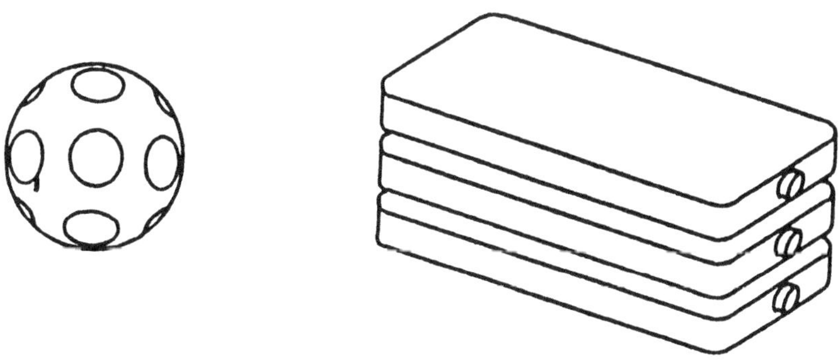

Figure 15-15. Encapsulated ice containers.

Storage priority control strategies are more difficult to apply. A load estimation or prediction algorithm is usually necessary to determine the extent of the required chiller cooling during the on-peak cycle. As a result, the chiller capacity is limited so that most of the cooling load is met by the thermal storage system.

For a constant water discharge temperature, encapsulated ice storage systems will have a decreasing discharge flow rate. If a constant discharge flow rate is maintained, the discharge water temperature will steadily increase. This characteristic results from the decreasing area of ice in contact with the walls of the container as the ice melts. The last several hours of the discharge cycle are the most critical in determining the storage capacity of an encapsulated ice storage system.

15.5.6 Eutectic Salt Phase-Change Materials

Eutectic salts are frequently used for both cool thermal storage and warm thermal storage. An eutectic salt is a material that changes phase from a liquid to a solid at a specific temperature, depending upon its chemical mixture.

Eutectic salts typically have a latent heat of fusion around 40 Btu/lb. The latent heat required for phase change of the salt reduces the storage volume by about 66% of that required for an equivalent capacity water storage system. The advantages of using eutectic salts as a thermal storage medium include:

- Eutectic salts can utilize existing chillers or heat generating equipment for generating thermal storage due to the similarity in temperatures.
- Eutectic salt storage requires less space than that required by water systems.
- Salts have higher storage efficiencies than water.

There are a number of chemical mixtures used for eutectic salts in thermal storage systems. Cool thermal storage systems frequently use a mixture of inorganic salts, water, nucleating agents and stabilizing agents. This mixture melts and freezes at 47° F. The salt mixture is contained in encapsulated rectangular plastic containers, as shown in Figure 15-16. The containers with the salt solution are stacked in a storage tank. Water is used as a heat transfer fluid and is circulated through the tank and over the containers to carry heat to or from the eutectic salt. Typical discharge temperatures are around 48 – 50°F, which is relatively high for most chilled water cooling systems.

There are two common salt hydrates used for heat storage: sodium thiosulfate pentahydrate and sodium sulfate decahydrate. Sodium thiosulfate forms salt hydrates, which have melting temperatures between 118 and 126°F. The density of the solid at 77°F is 108 lb/ft^3 and it has an average

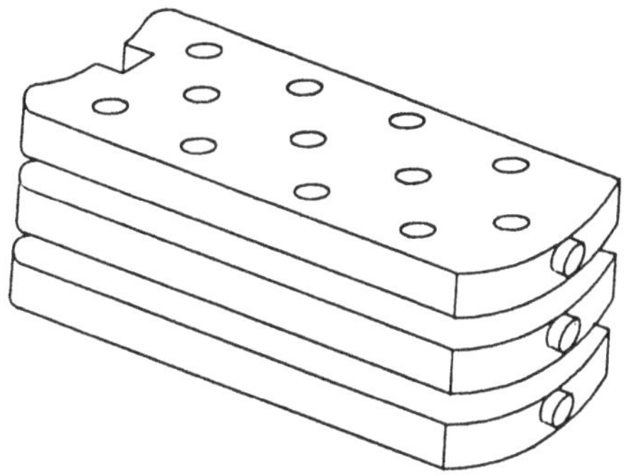

Figure 15-16. Stacked eutectic salt storage containers.

heat of fusion around 90 Btu/lb. Sodium sulfate is a relatively low cost chemical material suitable for heat storage. It has a melting point of 90°F and has a relatively high heat of fusion of 108 Btu/lb.

Cool thermal storage systems can have the chiller located upstream or downstream of the storage tank. The chiller upstream configuration can be used for full storage operating strategies and no additional cooling, other than from storage, is required. Since the water discharge temperature from a storage tank with eutectic salts is relatively high, the chiller is frequently located downstream of the storage. In this arrangement, the chiller is used to cool the water to the required supply temperature.

Storage systems with eutectic salts may be used for any of the three operating strategies: full storage, partial storage load leveling and partial storage demand limiting. For full storage applications, it must be noted that the water discharge temperature will be 48 – 50°F, which is relatively high for most chilled water systems. In many cases, however, non-peak loads can be satisfied by the higher water temperatures. In situations where lower temperatures are required, a chiller must be provided downstream of the thermal storage. Either chiller priority or storage priority strategies can be used to control storage system operation at less than peak loads.

15.6 QUICK SIZE ESTIMATION FOR THERMAL STORAGE SYSTEMS

This section presents quick sizing equations, which may be utilized for a preliminary feasibility analysis. The equations may be used to estimate the required size of a chiller and the required thermal storage capacity for a given application. Information for performing a more detailed analysis may be found in Dorgan and Elleson (1994).

In order to estimate the required capacity of a chiller for a storage system, both a design load profile for the cooling system and the utility rate schedule for the facility are required. A design load profile may require a detailed hourly analysis of the cooling load. Occasionally, the load profile can be developed based upon the projected peak load and a typical profile shape for the type of facility or load under consideration. Estimates of the chiller capacity and thermal storage capacity may be calculated based upon the requirement that the total integrated capacity of the refrigeration plant must be equal to the total integrated cooling load, plus any losses.

$$q_l = q_c \qquad \text{Eq. (15.1)}$$

where q_l = integrated cooling load, Btu/cycle
q_c = integrated chiller cooling capacity, Btu/cycle

The total integrated chiller capacity may be estimated by multiplying the chiller cooling capacity by the number of hours the chiller is available during a cycle. The total integrated chiller capacity is the summation of the chiller cooling capacity in each operating mode times the number of hours in that mode, and is expressed by Equation (15.2).

$$q_{ch} = q_{chrg}\, t_{chrg} + q_{D-on}\, t_{D-on} + q_{D-off}\, t_{D-off} \qquad \text{Eq. (15.2)}$$

where q_{ch} = chiller capacity, Btu/cycle
q_{chrg} = capacity when charging storage, Btu/hr
t_{chrg} = time charging storage, hours
$q_{D\text{-}on}$ = capacity when direct cooling during peak period, Btu/hr
$t_{D\text{-}on}$ = time for direct cooling during on-peak period, hours
$q_{D\text{-}off}$ = capacity when direct cooling during off-peak period, Btu/hr
$t_{D\text{-}off}$ = time for direct cooling during off-peak period, hours

If Equations (15.1) and (15.2) are combined, a general expression for the required cooling capacity is obtained. Combining and rearranging Equations (15.1) and (15.2) results in Equation (15.3). The chiller capacity for each mode is expressed as a percentage of its nominal capacity.

$$q_{ch} = \frac{q_L}{t_{chr}\, R_{ch} + t_{D=on}\, R_{D=on} + t_{D-off}\, R_{D-off}} \qquad \text{Eq. (15.3)}$$

where q_{ch} = chiller capacity, Btu/cycle
q_L = total cooling load, Btu/cycle
t_{chr} = time charging storage, hours
R_{ch} = capacity ratio when charging storage
$t_{D=on}$ = time for direct cooling during on-peak period, hours
$R_{D=on}$ = capacity ratio when cooling during on-peak period
$t_{D\text{-}off}$ = time for direct cooling during off-peak period, hours
$R_{D\text{-}off}$ = capacity ratio when cooling during off-peak period

If the resulting nominal chiller capacity is greater than the cooling load for any hour in the direct cooling mode, the chiller capacity must be recalculated. For a full storage system, the chiller cooling capacity is zero during the on-peak period since the chiller is off. For a load leveling system, the direct cooling capacity is the same during on- and off-peak periods. For ice harvesting systems, the ice making capacity is typically assumed to be the nominal capacity, and the direct cooling capacity ratio is approximately 1.2 : 1.3. For other ice storage systems, the chiller ice making capacity is usually about 60 – 70% of the nominal capacity.

The required storage capacity for a cool thermal storage system may be estimated by determining the cooling provided directly by the chiller, as well as by determining the load that is either met directly by the chiller or from storage while the storage is being charged. The required storage capacity may be calculated using Equation (15.4).

$$q_{stor} = q_L - (q_{D-on}\, t_{D-on} + q_{D-off}\, t_{D-off} + q_{D-chrg}\, t_{D-chrg}) \qquad \text{Eq. (15.4)}$$

where q_{stor} = required storage capacity, Btu/cycle
q_L = total cooling load, Btu/cycle
$q_{D\text{-}on}$ = capacity when direct cooling during peak period, Btu/hr
$t_{D\text{-}on}$ = time for direct cooling during on-peak period, hours
$q_{D\text{-}off}$ = capacity when direct cooling during off-peak period, Btu/hr
$t_{D\text{-}off}$ = time for direct cooling during off-peak period, hours
$q_{D\text{-}chrg}$ = direct cooling with simultaneous charging, Btu/hr
$t_{D\text{-}chrg}$ = time for direct cooling with simultaneous charging, hours

The total direct cooling capacity is typically equal to the time available for direct cooling in hours times the chiller cooling capacity for direct cooling. The storage capacity calculated by Equation (15.4) is the nominal storage required. The actual storage capacity must be determined by a detailed analysis of the charge and discharge characteristics of the selected storage type and their interaction with the load, as well as the performance of the chiller for each hour of the cycle.

Chapter 16

Energy Management and Control Systems

The purpose of a control system is to regulate energy consuming equipment and systems such that they operate at capacities required to handle loads at any given time without excessive overshooting or undershooting of the actual load. The control systems should automatically adjust the output of the energy consuming equipment or system to reasonably match the actual load. If the energy consuming equipment and systems are not properly controlled, the process being controlled will not operate optimally and energy will be wasted. Typically, control systems maintain temperatures, humidity levels, flow rates, power consumption, and lighting levels to meet loads and operate systems for optimum energy usage.

16.1 ENERGY MANAGEMENT AND CONTROL SYSTEM BASICS

Control systems may either be pneumatic, electric or electronic. Most energy management systems are computer-based or microprocessor-based, direct digital control (DDC) electronic systems. Computerized energy management and control systems (EMCS) provide an excellent means of controlling energy consuming systems and equipment, as well as an effective means of reducing energy costs. Energy management systems vary from simple time switches to computer- or microprocessor-based data and operations systems. The more sophisticated systems reduce operating costs, optimize performance, as well as provide programmed maintenance, security/fire detection and process control.

16.1.1 Definition and Terms Used in Energy Management and Control Systems

Prior to discussing energy management and control systems, it is best to have an understanding of the terminology used to describe these systems. The following are frequently used terms and definitions as they relate to energy management and control systems.

16.1.2 Types of Control Action

Generally speaking, there are two basic methods for providing automatic control: closed loop and open loop. A closed loop (feedback) control system senses changes in the controlled variable (such as air temperature) and causes the controlled device to change and take corrective action based on the measurement of the controlled variable. The controlling or modifying action continues to take place until the output (controlled) variable is brought to within an acceptable level. This method is error-sensitive and self-correcting. The acceptable level is probably not the exact desired level (or set point) of the variable, but is reasonably close to the desired level of the variable. The difference between the set point and the actual level of the variable is the offset. The closed loop control method is the method most commonly used to control energy-consuming systems. A closed loop is one in which all parts have an effect on the next step in the loop and are affected by the action of the previous step.

An open loop system is one in which one or more of the steps has no effect or action imposed on the following step, or is not affected by other steps in the loop. For open loop (feed forward) control systems, there is no feedback of information from the controlled variable to adjust the controller. Open loop control systems anticipate the effect that a change on the controller will have on the output variable. An example of an open loop control system would be an outdoor air thermostat, which resets the control point of a heating system controller based on the outdoor air temperature. The temperature inside the building is not sensed and the heat output is not adjusted accordingly. The only variable that affects the heat output is the outdoor air temperature.

Control systems can operate in several different modes. A mode of control is the manner in which the control system makes changes and responds to a change in the controlled variable. It relates the operation of the final control element to the measurement signal from the sensor. The type of control mode determines the type of response that the control system has as a result of a change in the controlled variable. The basic modes of control are two position, multi-position, floating control and modulating control.

The two position (on-off) control is the simplest mode. The controller output has only two positions: either on or off, as shown in Figure 16-1. When applied to a valve or damper, this means either fully open or fully

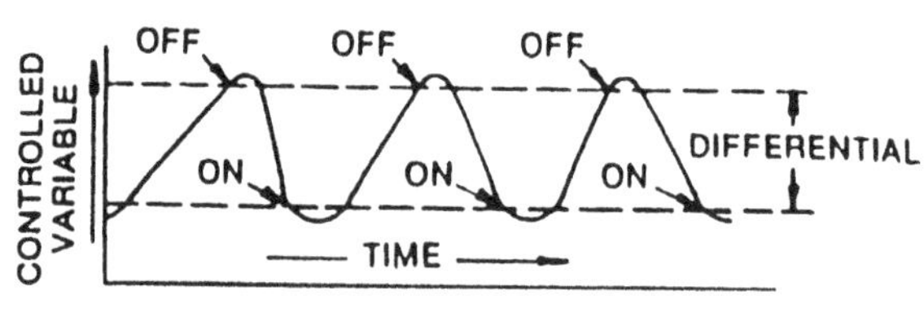

Figure 16-1. Two-position control.

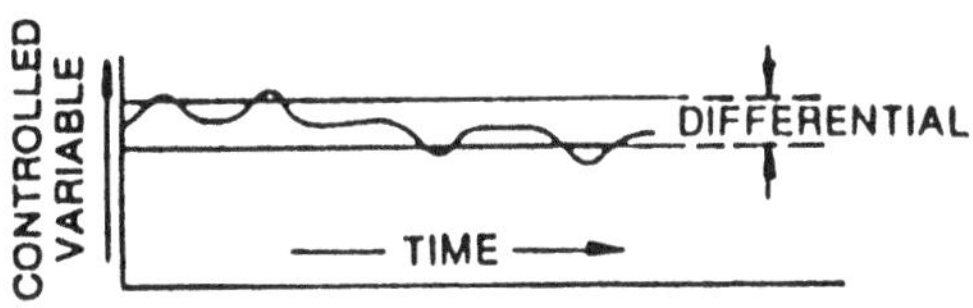

Figure 16-2. Floating control showing variations in controlled variable as load changes.

closed. There are no intermediate positions. When the controlled variable deviates a pre-set amount from the set point, the controller is actuated to either of its extreme positions—full on or full off. This mode allows the controlled variable to vary over a wide range of out put. The output is similar to a step function between two settings as opposed to a near steady condition.

The multi-position (multistage) mode is similar to the two position except there are more intermediate positions or stages. Multistage control is used when the range between full on and full off is too great. The multistage control divides the range into multiple stages or steps. Each stage is either on or off; however, the stage is much smaller than the range between full on and off. An example of multistage control is a heater that can come on in stages of 25%, 50%, 75% and 100% of full capacity. The greater the number of steps, the smoother the operation.

Floating control is different from two position or multi-position control in that the final control element may be moved to any position between the extremes (full on to full off). The controller has only two positions, with a neutral or dead band in between as shown in Figure 16-2. When the controlled variable changes enough, the controller moves to one or the other position depending on the direction of the change in the controlled variable. When the controlled variable returns to its neutral (dead band) range, the controller moves to the off position. This mode of control is called floating because the controller is off when the controlled variable is floating between its two positions.

Modulating (proportional) control is similar to floating control except that the controlled variable may assume any position between its extremes. The controller may also assume any position in response to changes in the controlled variable, as shown in Figure 16-3. There is no neutral zone or dead band. Any changes in the controlled variable will cause the controller

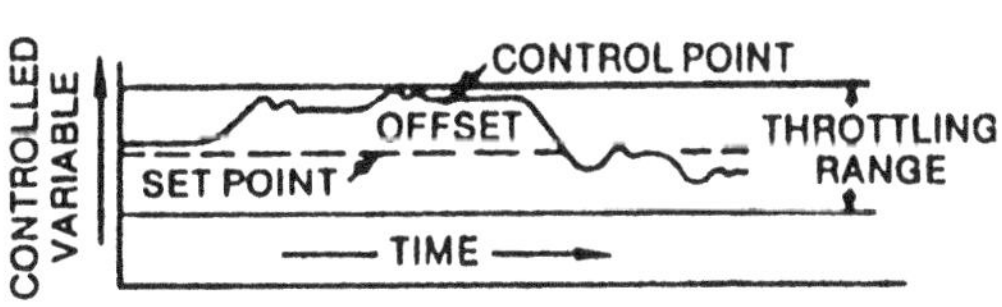

Figure 16-3. Proportional control showing variations in controlled variable as load changes.

to respond. For each movement in the controller, there is a proportional change in the controlled variable. These movements can occur as frequently as changes in the controlled variable occur. Modulating control produces a linear relationship between the change in the controlled variable and the output of the controller to change the position of the controlled device. A proportional controller can be described mathematically by Equation (16.1).

$$V_P = K_P e + V_O \qquad \text{Eq. (16.1)}$$

where V_P = output of proportional controller
K_P = proportional gain (proportional to throttling range)
e = offset or error signal
V_O = offset adjustment parameter

Incremental action control is a variation of floating control. Incremental action varies the pulse action to open or close an actuator depending on how close the controlled variable is to the set point. As the controlled variable approaches the set point, the pulses become shorter. As a result, incremental action provides closer control than floating control.

Proportional plus integral (PI) control adds another component to the control action that eliminates the offset and improves a simple proportional control. Proportional plus integral control is shown in Figure 16-4. A correcting element may be arranged to remain stationary when the control medium is at the desired value and move in a corrective manner at an increasing speed proportional to the deviation from the set point. This form of floating action with variable speed is known as integral action. The reset may be described by Equation (16.2).

$$V_P = K_P e + K_i \int e\,d\theta + V_O \qquad \text{Eq. (16.2)}$$

where V_P = output of controller
K_i = integral gain

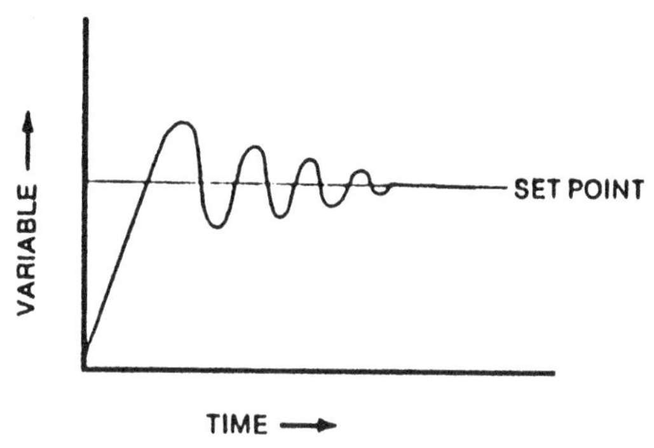

Figure 16-4. Proportional plus integral (PI) control.

θ = time
e = error
V_O = offset adjustment parameter

Proportional-integral-derivative (PID) control adds a derivative term that varies with the value of the derivative of the error. In derivative control, a correcting element is set such that its speed of operation is proportional to the rate of change of the controlled variable. This action is used to eliminate overshoot during a fast load change. Adding the derivative term gives some anticipatory action to the controller, which results in a faster response and greater stability. Equation (16.3) is the equation for proportional-integral-derivative control action.

$$V_P = K_P e + K_i \int e d\theta + K_a \frac{de}{d\theta} + V_a \qquad \text{Eq. (16.3)}$$

where K_a = the derivative gain of the controller
$de/d\theta$ = the time derivative of error

16.2 CONTROL SYSTEM COMPONENTS

The most basic elements or components that comprise an automatic control system are a sensor, a controller and a controlled device. These three components can make a complete control system, however, most control systems are comprised of many components and accessories.

16.2.1 Sensors

A sensor is a device in the control system that measures the value of the controlled variable (for example, air temperature). If the controlled variable changes, it causes a change in some physical or electrical property of the primary sensing element. The change in property of the sensing element is then translated or amplified by a mechanical or electrical signal. The signal is then transmitted through the controller by the control system. Typically, sensors are used to measure the temperature, pressure, humidity or flow of the controlled variable.

Temperature sensors are used extensively in control systems. Temperature sensors operate on one of three physical characteristics, which are:

- A change in physical dimension as the result of thermal expansion.
- A change in the state of a vapor or a liquid.
- A change in the electrical properties of a substance due to a change in temperature.

Typical temperature sensing elements include a bimetal element; a vapor filled bellows; a liquid filled, gas filled or refrigerant filled bulb and capillary; and metals that change in temperature. Two dissimilar metals may also be used to produce a voltage change due to a change in temperature.

Bimetal temperature sensors are the most common type used in control systems. The element is made up of two strips of different metals. Each metal has a different coefficient of thermal expansion. The two metals are bonded together to form a strip or a spiral element. As the surrounding temperature changes, the temperature of the element changes, causing the two metals to expand or contract differently. In the case of the bimetal strip, the strip bends. The amount of the bend and direction is a function of the direction and amount of temperature change. The spiral element (coil) tends to either coil tighter or straighten as the result of a temperature change. Bimetal temperature sensors can be used for two-position or modulating control.

A rod and tube sensor is another type of bimetal temperature sensor. It consists of a rod that is made of a material with a low coefficient of thermal expansion and a tube with a high coefficient. The rod is attached at one end of the tube. As the tube changes length due to temperature changes, the free end of the rod moves. The movement is a function of the temperature change. Rod and tube type temperature sensors are frequently used as insertion type sensors.

A sealed bellows type temperature sensor has a bellows type element. The bellows is filled with a vapor, gas or a liquid depending on the expected temperature range for the sensor. Temperature changes cause the fluid inside the bellows to expand or contract with the temperature change, which results in a force or movement of the bellows. Bellows type temperature sensors are frequently used in room thermostats.

A remote bulb and bellows or diaphragm type sensor is similar to the bellows type temperature sensor. This type of sensor has a sensor bulb and a capillary tube that are filled with fluid that expands or contracts in response to temperature changes. The bulb is typically inserted into the material where the temperature is to be sensed. A bellows or diaphragm is connected to the other end of the capillary tube. Temperature changes sensed by the bulb result in a pressure change in the fluid, which is transmitted through the capillary tube to the bellows or diaphragm where the fluid pressure is converted to a force or movement.

A thermistor is a temperature sensing device that changes electrical resistance with changes in temperature. The thermistor is made of a semiconductor material that is temperature sensitive. The material typically has a resistance of about 3,000 Ω at a temperature of 0°C. Small changes in temperature result in large changes in resistance, which makes thermistors quite temperature sensitive. Thermistors are also relatively low in cost and are readily used in DDC control systems.

Resistance temperature detectors (RTD) are made of fine wire wound into a tight coil. They are similar to thermistors in that they change resistance in response to a temperature change. The RTD sensing element is available in several different forms. They are suitable for sensing surface temperatures. For immersion sensing, the coil may be encased in a stainless steel bulb for protection. Normally, the RTD is more reliable and more accurate than a thermistor.

A thermocouple is a junction to two dissimilar metals that produce an electrical voltage change in response to changes in temperature. The voltage is a function of the two materials and the temperature of the junction. If the other ends of the conductors are connected, a circuit is formed, which causes a flow of current. The other end of the thermocouple must be kept at a constant temperature. The constant temperature is referred to as the cold junction. Thermocouples are frequently used to sense a flame in a pilot light.

Hygroscopic (moisture-absorbing) materials are used for humidity sensors. These materials absorb or reject moisture until a balance is reached with the surrounding air. There are two types of humidity sensors (hygrometers) used in control systems: mechanical and electronic.

Mechanical hygrometers use materials that expand or contract based on their moisture content. As the material absorbs moisture, it expands (stretches or swells) and as it looses moisture, it shrinks. The change in size is detected by a mechanical linkage, which converts the motion to a pneumatic or electrical signal.

Many hygrometers have humidity sensing elements that are made of organic materials, such as hair, wood fibers, paper and cotton. Non-organic materials include nylon and other synthetic fabrics.

Electronic hygrometers use materials that change resistance or capacitance in response to changes in moisture level. The resistance type uses a grid that is coated with a hygroscopic substance such as lithium chloride. As moisture is absorbed by the hygroscopic substance, a high resistance electric circuit is created. The resistance is a function of the level of moisture. The capacitance type is a membrane made of a non-conductive film stretched between two metal electrodes mounted in a perforated plastic capsule. As the membrane absorbs moisture, the capacitance between the electrodes increases or decreases.

Since pressure is a force exerted over an area, it is easily converted to motion or displacement. Pressure sensors convert a change in absolute, gage or differential pressure into mechanical motion with a bellows, diaphragm or Bourdon tube mechanism. Pressure sensors are available for measuring pressures from a few inches of water to several thousand psi.

Mechanical pressure sensors usually employ a flexible diaphragm, a bellows or a Bourdon tube. The flexible diaphragm type has a diaphragm that distorts with changes in pressure on one side. The flexible diaphragm is connected to a linkage, which converts the motion into a pneumatic or electronic signal. Bellows type pressure sensors have a bellows made of a corrugated cylinder that expands or contracts with changes in pressure differential between the inside and outside. The reference pressure is usually atmospheric. One end of the bellows is firmly anchored while the other end is allowed to move with changes in pressure. The moveable end is connected to a linkage, which converts the motion to a pneumatic or electronic signal.

Bourdon tube pressure sensors have a closed semi-circular tube. The end of the tube is connected to a linkage, which produces mechanical motion proportional to the pressure change. The pressure being measured fills the inside of the semi-circular tube. The pressure inside the tube tends to straighten the tube, which results in motion at the end. The greater the pressure inside the tube, the more the tube straightens. Bourdon tubes are frequently used in dial type pressure gages, but rarely in control systems.

Most pressure sensors are mechanical devices. Pressure is typically not measured directly with electronics. Typically, an electronic signal is produced from a transducer, which converts from a mechanical force or motion to an electrical signal.

Flow sensors are used frequently in control systems. It is often desirable to measure the flow rate of gases and liquids. Examples of applications for flow sensors in control systems include measuring the rate of airflow in ductwork, measuring water flow in heating and chilled water systems, and measuring the rate of steam flow in steam systems. Flow measuring devices in control systems include sail switches, paddle wheels, hot wire anemometers, pitot tubes and orifice plates. Other more sophisticated flow sensing devices include venturi meters, turbine meters, magnetic flow meters, vortex shedding meters and Doppler effect meters. The more commonly used flow measuring devices used in control systems are briefly discussed here.

Sail switches are typically used to verify that a fluid is flowing, but they cannot be used to measure the rate of fluid flow. A sail switch has a lightweight plate that is mounted in the fluid, perpendicular to the direction of flow. Since the sail switch offers a relatively large flat surface for the flowing fluid, the fluid moves the sail slightly in the direction of flow. At the other end of the sail, a switch is mounted, which is opened on closed depending upon whether flow is present or not.

A paddle wheel is similar to a sail switch except that a plate or paddle is used in lieu of a sail and the paddles are mounted on a wheel. As the fluid flows, the paddles cause the wheel to rotate. The greater the rate of fluid flow, the faster the paddle wheel rotates. The wheel has a magnetic measuring mechanism that measures the rotational speed of the wheel. These devices are used to measure both instantaneous flow rate and to calculate total flow. Total flow is the quantity of fluid that has flowed over a long period of time, like a month.

Hot wire anemometers are frequently used to measure airflow rates. It consists of a small electric resistance heater and a temperature sensor. As air flows past the heated anemometer, heat is lost to the air. The rate at which the heat is lost to the air is a function of the air velocity. The air velocity is determined by the amount of electric energy required to maintain a reference temperature. Hot wire anemometers require a reference to neutralize the effect of air temperature changes on the output signal.

Pitot tubes are also frequently used to measure air velocities in HVAC systems. A pitot tube is a double tube, which may be inserted into a duct to measure the velocity of the airflow. The pitot tube indirectly measures air velocity by measuring the total pressure and static pressure of the flowing air.

Orifice plates may be used to measure the flow rate of all types of fluids. The orifice plate consists of a plate with a precise hole or orifice in it through which the fluid flows. Based on the size of the orifice and the properties of the flowing fluid, the rate of the fluid flow may be determined by measuring the fluid pressure drop from one side of the orifice to the other. The accuracy of orifice plates drops off as the flow rate decreases below 50% of the design flow rate.

16.2.2 Controlled Devices

Earlier in this chapter a controlled device was defined as a device that reacts to signals received from a controller. A signal is measured by a sensor that sends a signal to a controller, which in turn signals the controlled device to change as necessary to maintain the desired set point. The controlled device adjusts the flow of the controlled agent in response to a signal from the sensor and controller.

Controlled devices that typically regulate or vary the flow of steam or water are called control valves. Controlled devices that control the flow of air are called dampers. The device that physically causes the control valve or damper to move is called an operator or actuator.

Automatic control valves are frequently used to control the temperature of air leaving a coil in an air handling unit or some other type of heat exchange component. The leaving air temperature is controlled by regulating the flow of steam, water, gas or other fluid to a heat exchange component.

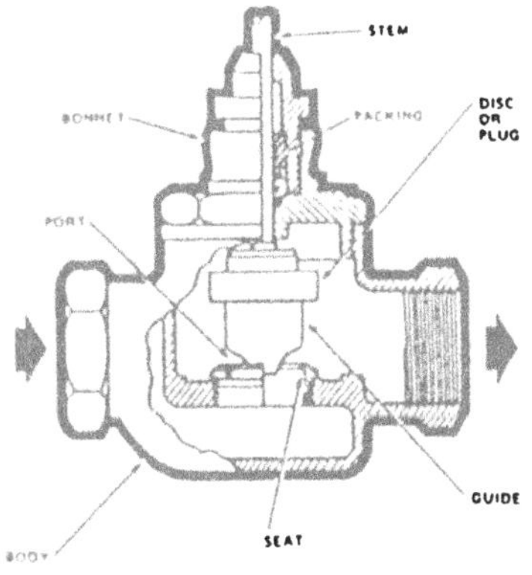

Figure 16-5. Components of a control valve.

A control valve acts as a variable orifice in a fluid flow line such as a pipe. A control valve itself typically consists of a body, a moveable disk or plug, a seat for the moveable disk, a stem to move the disk, and a bonnet with packing to prevent leakage from the valve. A typical control valve is shown in Figure 16-5.

In order for a control valve to move or be actuated, it is necessary that the valve have an actuator. Valve actuators may either be electric motors or pneumatic diaphragm type. A typical control valve with a pneumatic actuator is shown in Figure 16-6.

The control valves shown in Figures 16-5 and 16-6 are two way control valves. The fluid flows into one port and out another. A schematic for a two way control valve is shown in Figure16-7.

A two way control valve restricts the flow of fluid to the coil or heat exchange device by opening or closing in response to the load requirements. As the heating or cooling loads go down, the valve automatically closes, reducing the fluid flow to the heating or cooling coil. As the load increases, the valve automatically opens to increase the rate of flow. Two way valves are used to control the flow of steam to steam heating coils. Hydronic (heating water or chilled water) coils also frequently use two way control valves, depending upon the pumping/piping arrangement of the hydronic system. If two way valves are used in hydronic systems, it may be

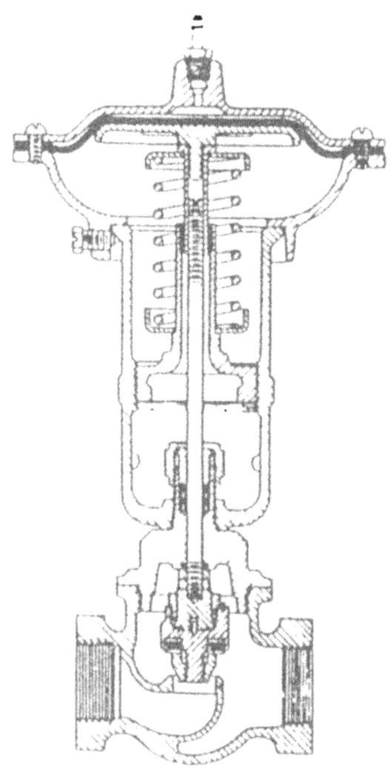

Figure 16-6. Control valve with pneumatic actuator.

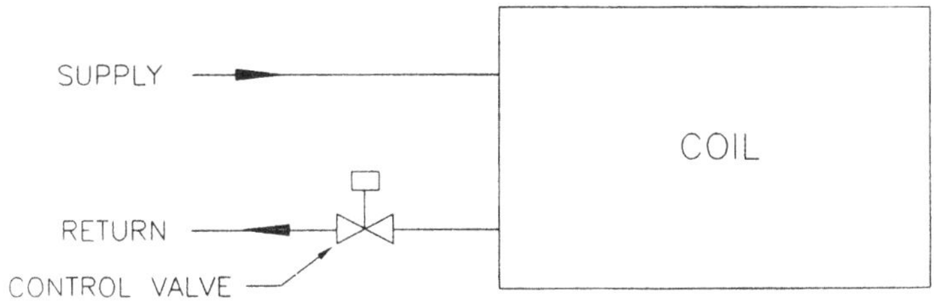

Figure 16-7. Schematic diagram for a two way control valve.

necessary to provide some means to relieve pressure in the system as the valves close and reduce total system flow.

Control valves may also be the three way type. Three way control valves have three ports. These valves are used on hydronic systems to relieve the pressure built up as flow through coils is reduced. They are not used for steam systems. The valve is piped to the hydronic coil such that water may either flow through the coil or be bypassed around the coil in response to the load requirements. Three way valves are available in two configurations, depending upon whether they are located in the coil inlet or outlet piping. A three way valve located in the coil inlet piping is referred to as a three way diverting valve. A three way diverting valve is shown in Figure16-8, and is shown schematically in Figure 16-9.

A three way diverting valve either allows the water to flow through the coil or diverts the water around the coil. The flow rate in the supply and return piping remains constant while the flow through the coil is varied.

Three way control valves may also be mixing type valves. Mixing valves are located on the discharge or outlet side of the coil piping and control the flow through the coil by mixing the flows. Since the flow rate in the piping to the coil remains constant, flow through the coil is varied by mixing water

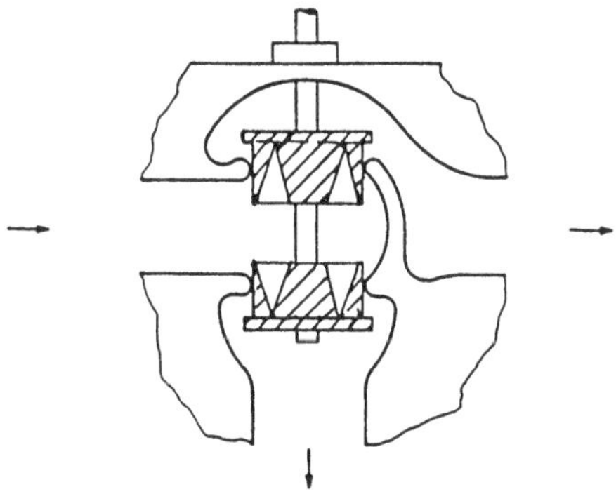

Figure 16-8. Three way diverting valve.

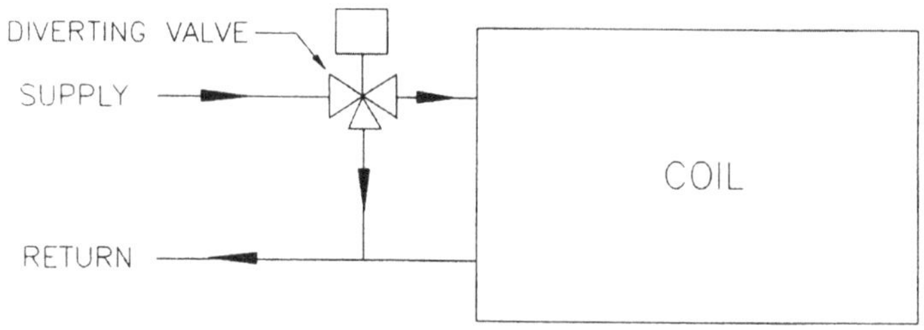

Figure 16-9. Schematic diagram for a three way diverting valve.

that has flowed through the coil with water bypassed around the coil in the proportions as required by the load. A typical three way mixing valve is shown in Figure 16-10, and the schematic for a three way mixing valve is shown in Figure 16-11.

The selection of control valves (two way or three way) should be based on the flow characteristics of the valve. The flow characteristics are the relationship between the flow rate and the valve stem travel (stroke) as the stem travel is varied between 0 and 100% of full travel and is based on a constant pressure drop through the valve. The three

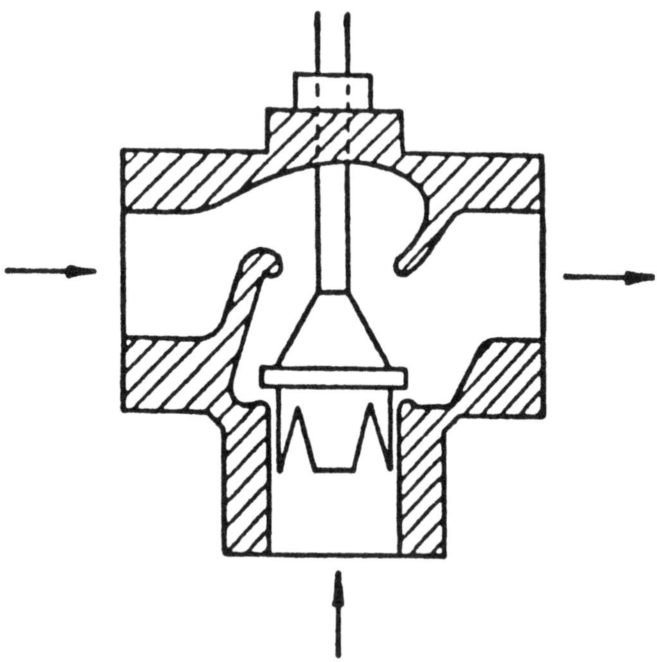

Figure 16-10. Three way mixing valve.

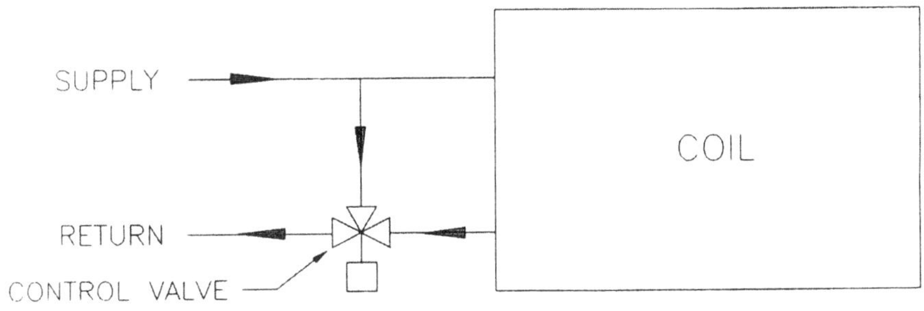

Figure 16-11. Schematic diagram for a three way mixing valve.

most common characteristics for control valves are quick opening, linear and equal percentage. These flow characteristics are shown graphically in Figure 16-12.

The quick opening valve approaches maximum flow quickly as the valve is opened. The slope of the percent flow to percent open curve is very steep initially and then levels off as the valve is opened farther. This type of valve is suitable for two position on-off applications.

Linear type control valves have a percent flow versus percent open curve that is a straight line as the name implies. The opening and flow are directly proportional if the pressure remains constant. Linear type control valves may be suitable for controlling the flow of steam, since the heat output of a steam coil or heat exchanger is directly proportional to the steam flow rate. This is due to the fact that the steam remains at the same temperature as it changes phase from a vapor to a liquid.

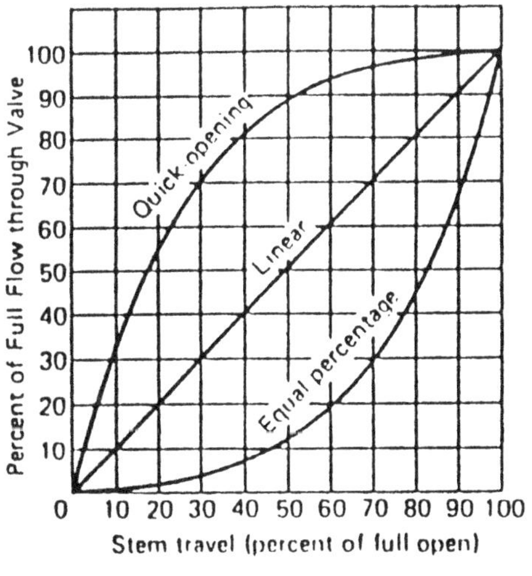

Figure 16-12. Control valve flow characteristics.

Equal percentage type control valves have a flow versus opening curve such that the flow increases exponentially as the valve opens. Each equal increment of opening increases the flow by an equal percentage over the previous value. This type of valve is most suitable for controlling the flow to heating water or chilled water coils since it has a lower ratio of flow increase in the region near full closure. The pressure characteristic of the coil provides a near linear temperature change to the valve control signal change.

Control valves in heating water and chilled water systems are also selected on their flow coefficient (CV). The flow coefficient is defined as "the flow rate of water at 60°F in gpm that will pass through a valve that is fully open at a pressure of 1 psig." The flow rate through a valve at pressure drops other than 1 psig may be calculated using Equation (16.4).

$$Q = C_V \sqrt{P} \qquad \text{Eq. (16.4)}$$

where Q = the flow rate, gpm
C_V = the valve flow coefficient
ΔP = the pressure drop through the valve, psig

Dampers are devices used to control the flow of air in an air distribution system. Dampers are made of one or more flat plates mounted in the airstream that rotate to control the airflow. Dampers accomplish airflow control by varying the resistance to the airflow. Automatic or motorized dampers may be used for modulating control or for two position control.

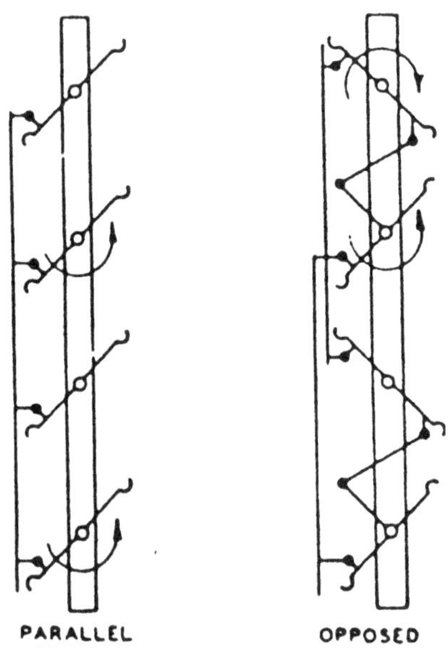

Figure 16-13. Typical multi-blade dampers.

Dampers may have a single large blade (butterfly damper) or may have multiple blades. Multi-blade dampers may be parallel or opposed blade type. Figure 16-13 shows typical multi-blade dampers.

Parallel blade dampers are best suited for two position control and for applications where two airstreams are mixed together. In mixing applications, parallel dampers tend to reduce stratification and improve mixing of two airsteams provided the dampers are properly placed. Better mixing occurs since the two airstreams can be directed toward each other and because parallel blade dampers tend to cause more turbulence in the airflow. Changes in blade rotation do not directly result in proportional changes in the rate of airflow. Their flow characteristics are non-linear.

Opposed blade dampers are constructed so that alternate blades rotate in opposite directions. They provide better control than parallel blade dampers and are used where tight shut-off is required. Opposed blade dampers have flow characteristics that are more linear than parallel blade dampers. The changes in damper rotation are more closely related to changes in airflow rate. This allows a closer control of airflow, especially at low rates. Opposed blade dampers are sometimes used as inlet vanes for fans used in VAV systems.

An important characteristic to consider when selecting dampers is the damper leakage. If dampers do not close tightly, air will leak past the dampers, which is frequently undesirable. Damper manufacturers will usually provide leakage information about their dampers. Damper leakage is expressed in one of two ways: (1) percentage of airflow through an open damper or (2) cfm of air per square foot of damper area at a given pressure differential. In both cases, the pressure differential across the damper must also be indicated.

Actuators are devices that are used to move a valve or damper in response to a signal from a controller. The energy used by an actuator to move a control valve or damper is usually dependent upon the type of energy used in the rest of the control system. Almost all actuators are electric or pneumatic. Electric actuators can operate on line voltage (120 V) or low voltage (24 V AC or DC). The most common types of actuators used in energy management and control systems are electric motors and actuators.

A solenoid is an electrical device with a magnetic coil and a moveable plunger. Solenoids are most commonly used for two position control, such as open or closed. Solenoids are typically used on small valves, usually less than two inches.

Electric motors may be used for modulating control as well as two position. The motor operates the valve or damper through a gear train and linkage. Motor operators are available in three different types:

- Unidirectional: Used for two position control.
- Spring return: Used for two position control in which the valve or damper is actuated to one position and a spring returns the valve or damper to the original position.

- Reversible: Used for floating and proportional control. The motor can operate in either direction and can stop in any position.

16.2.3 Controllers

Earlier a controller was defined as a device that sensed the value of a variable and compared it with a set point. A controller takes the input from a sensor, compares it with the desired set point, then regulates an output signal to cause a control action at the controlled device (control valve, automatic damper, etc). A controller and sensor can be combined in a single device such as a room thermostat, or it may be two separate devices. Controllers may be classified by the type of control action and type of energy used for the control signal. As described earlier, the control action may be two position or modulating. The term analog control suggests a continuous modulating signal. The term digital control suggests discrete on-off or two position control. The types of energy used by controllers are pneumatic and electric/electronic. When two separate pneumatic devices are used, the pneumatic controller is usually called a receiver-controller.

An electronic controller compares a signal from the sensor to a standard reference or set point to produce an electronic error signal. The error signal is then amplified to a high enough level to drive the controlled device. Electric/electronic controllers are available for all modes of control. For two position control, the controller output may be just an electrical contact that energizes or de-energizes the controlled device. For proportional control, a controller may have a switching circuit with a neutral zone where neither contact is made. Proportional control may also be such that a continuously or incrementally changing output signal is used to position an electrical actuator or controlled device.

A basic electronic controller is comprised of a null detector and amplifier with input adjustments and an output switching element to drive the controlled device. The null detector compares the difference between the sensor signal and the set point. When the two values are equal, there is a balance condition that results in zero (null) output. The amplifier multiplies the output from the null detector and produces an amplified error signal. The amount of amplification is called the gain of the amplifier. Some amplifiers can also differentiate or discriminate between signals of different polarity or phase. The discriminator determines whether the signal to the amplifier was the result of an increase or decrease in the controlled variable. The signal, therefore, can drive the controlled device in the proper direction to maintain the set point.

Direct digital controllers (DDC) are similar to electronic controllers except that a computer or microprocessor is used to provide control algorithms for the control system. DDC are different from electronic or pneumatic controllers in that the control algorithm is stored in the computer memory as a set of program instructions. The controller calculates the required output signal.

The controlled devices are then actuated to the controlled position through the hardware interface, which converts the digital signal from the computer to an analog signal that positions the actuator or energizes a relay. Based on the input data, the control algorithms stored in the computer memory make the required control decisions.

Pneumatic controllers produce a force or position output in order to produce a variable air pressure output signal. Pneumatic receiver-controllers are usually combined with pneumatic sensors. These controllers are usually classified as non-relay, or relay, and direct or reverse acting.

A pneumatic non-relay controller has a restrictor located in the air supply. It also has an air bleed nozzle. The sensor positions an air exhaust flapper, which controls the nozzle opening resulting in a variable air pressure output signal to the controlled device, such as a pneumatic actuator.

A relay type pneumatic controller actuates a relay device that amplifies the air volume available for control. This is accomplished, either directly or indirectly, with a nozzle and a flapper.

16.2.4 Control System Accessories

In addition to sensing devices, controlled devices and controllers, there are a number of accessories that are frequently required in order to have a complete control system. Some of the more common accessories found in control systems include transducers, switches, relays, transformers, step controllers, clocks and timers, air compressors, refrigerated air dryers, pneumatic tubing and control wiring.

A transducer is used to transform one form of energy to another. In control systems, transducers are used to convert electric input signals to pneumatic output or visa versa. Transducers may convert a proportional input signal to either a proportional output signal or a two position output signal. Pneumatic/electric (P/E) switches convert a pneumatic signal to an electric signal. Electro/pneumatic (E/P) switches convert electric signals to a pneumatic signal. P/E and E/P switches are examples of transducers.

Switches are devices typically used to turn another device on or off (two position), although they may be multi-position. They may be single pole to switch one device or they may have multiple poles to switch multiple devices. Pneumatic switches are typically manually operated devices, which are used to divert air from one circuit to another or for opening and closing air circuits. Electrical switches may be used to turn electrical devices on or off and to switch between electrical circuits.

A relay is a device used for passing a signal from one device to another or from one circuit to another. Relays may be pneumatic, electric or electronic. Pneumatic relays are actuated by the air pressure from a controller to perform several functions. Pneumatic relays are two position or proportional. Switching relays are pneumatically operated air valves that divert air from one circuit to another or opening/closing pneumatic circuits. Positive positioning relays are devices that ensure the accurate positioning of a controlled device, such as a valve or damper, in response to changes in pressure from a controller. Relays are also electric or electronic, which may be used to control electric devices.

Transformers are used in electrical systems to step the voltage in electrical systems up or down.

Amplifiers in control systems use solid state or vacuum tube components to amplify very small signals to a level where they are usable by the control system. Other types of transformers are used to step a signal down. For example, most electrical control systems require a transformer to step a 120 V AC power supply down to 24 V AC or DC for the control system.

Step controllers are used to operate a number of switches in a certain order or sequence by means of a proportional input signal. Step controllers can be either electric or pneumatic. They are frequently used to bring devices on in steps or stages in a predetermined order or sequence.

Timers and time clocks are used to turn components on and off or switch between circuits at predetermined times. An example of such a timer is a time clock on a setback thermostat that switches between day or night operation depending on the time of day. Timers can also be used for other time sequence functions. A time clock is a clockwork mechanism with a rotating dial on which trip tabs can be mounted. A program timer is an electronic timer with several switches that are opened or closed in a pre-set sequence.

16.3 DDC AND COMPUTER SYSTEMS

Computer-based energy management and control systems (EMCS) are frequently used to control energy-consuming systems and equipment. They provide an effective means of reducing energy consumption and costs while still maintaining the required comfort or process conditions. These computer-based systems can continuously scan the connected input points, output points and report unusual occurrences, while carrying out control functions automatically.

Transmission of information from data collecting points to a central monitoring panel is via a single or two wire systems using pulse coded messages. In some systems, the scan initiation is entirely from a central processor, which addresses each individual point in turn and processes the received data. Other systems have intelligent outstations (also known as data gathering panels or stand alone panels), which provide local control and data gathering. Systems with outstations address the control processor only if an alarm occurs or to transfer data. Software for central processing systems and systems with outstations is referred to as direct digital control (DDC). The software converts the digital output of the computer to an analog signal compatible with the control system hardware. A block diagram of a typical EMCS is shown in Figure 16-14.

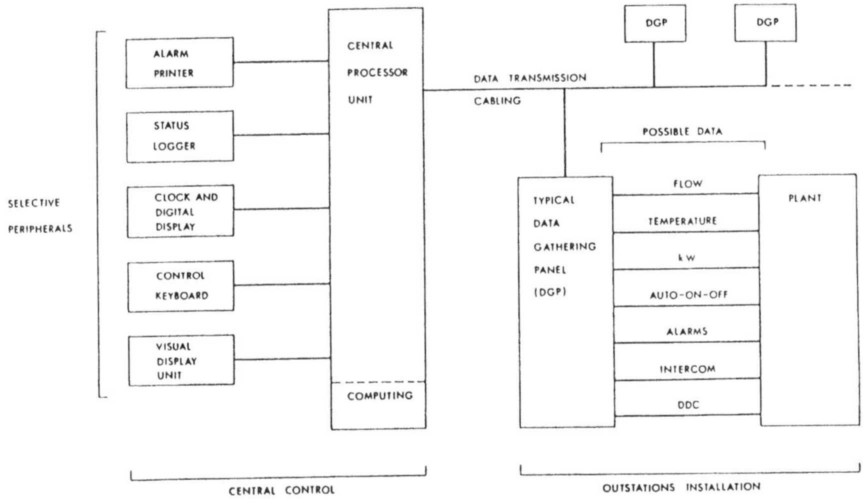

Figure 16-14. EMCS block diagram.

There are three basic types of components in an EMCS: inputs, outputs and intelligence. Input devices sense or collect information about the system or equipment and convey it to the intelligence in the system. At the other end, controlled output devices act to modify the operation of systems and equipment.

16.3.1 Data Input Devices

There are two basic types of input devices: analog inputs and digital inputs.

An analog input is any point that is able to read a continuously varying signal. Analog inputs are used for measuring such values as temperature, pressure, etc. The signals may be variations in voltage or current flow. The signal may also be in the form of a pulsed signal.

A digital input is any point that will accept only two states of information from the system, such as on-off or open-closed. Digital inputs are two position switching inputs and are usually used to indicate the status of a variable or indicate an alarm condition. Examples of digital input devices include dry contacts to sense relay status, fluid level sensors and pressure switches.

16.3.2 Output Devices

Output devices may also be divided into two basic types: analog and digital. Analog output devices and digital output devices have the same characteristics as the respective input device. Analog output devices are frequently used to drive valve or damper motors. Digital outputs almost always switch some type of relay.

16.3.3 Controlled Devices

Controlled devices were discussed earlier in this chapter. These devices include valves, dampers and relays that the output signal acts upon to affect the energy-consuming equipment and systems to be controlled.

16.3.4 EMCS Intelligence

EMCS intelligence includes all the system computer hardware, software and system communication. It consists of printed circuit boards, microprocessor, analog-to-ditital converters, digital-to-analog converters, power supplies, panels and communication system, which ties it all together as a system. See Figure 16-15.

The central microprocessor contains the memory and the software for automation and alarm functions and also provides an interface with the peripheral equipment. Included with the central processing unit (CPU) is the system software. The software is the opeating systems, databases, software controllers, control language and operator interfaces. The basis for the system intelligence is the microprocessor with its built-in hard-wired instruction set. The next step up is some form of higher language (such as C++) that the system manufacturer uses to build the operating system features. The operating system must be real-time and multi-tasking. It is stored in erasable program read-only memory (EPROM).

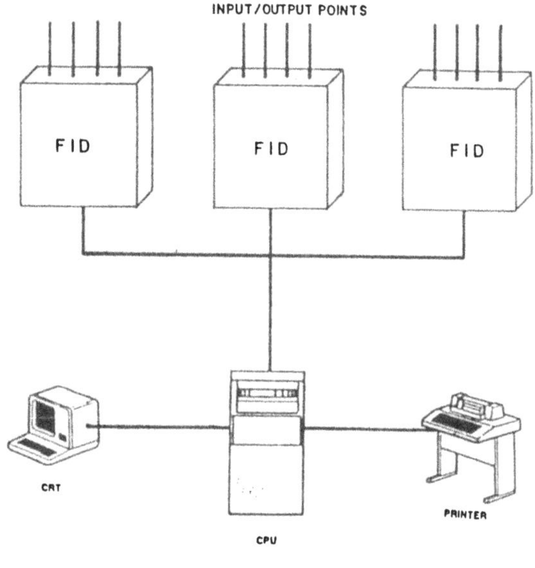

Figure 16-15. EMCS with Field Interface Devices (FID).

Operator interface includes such equipment as a personal computer (PC), a keyboard, a mouse, a video display and a printer. The keyboard and mouse provide a means for the operator to enter data, request data and to modify programs. Printers are used to provide a hard copy of data and historic information regarding the performance of the energy-consuming equipment or system.

Important functions of the system intelligence include software control, alarms, trends and trend graphing, totalization and scheduling. Software control provides the control of an output based upon a set point and an input. Almost all forms of control are some variation of three-term control: proportional, integral and derivative. All DDC control systems provide a method of tracking and reporting alarm conditions, which are conditions that are out of normal. Another important function of DDC systems is the ability to keep track continuously of what has happened with equipment and conditions where no one is there to observe the occurrence. This is referred to as trending or trend logging. Trends are useful in troubleshooting energy-consuming equipment and systems. Totalization is used for equipment that cycles during normal operation, such as a compressor or boiler. It is often useful to know the total accumulated time that it has operated for maintenance and troubleshooting. Finally, a function performed by DDC systems is scheduling. Time clocks are frequently used to schedule starting and stopping of equipment automatically. This is still a prime function of DDC systems. DDC systems provide a scheduling feature that allows for the creation of start/stop schedules and their use in programs.

Finally, DDC systems frequently have what is referred to as distributed intelligence. Systems with distributed intelligence have stand-alone panels (intelligent outstations or field interface devices) that have the full range of programming and communications required for the functioning of the system but can operate independently if communication to the rest of the system is lost. Stand-alone panels usually communicate on a peer-to-peer network, which means panels have equal priority, with some means of providing for regulating equitable access to the communication resources.

16.4 ENERGY MANAGEMENT AND CONTROL SYSTEM STRATEGIES

Computerized energy management and control systems can provide an effective means of controlling energy consumption and energy costs. In order for an EMCS to effectively control energy consumption, a control strategy that optimizes the energy performance of systems and equipment must be employed. This section reviews opportunities and control strategies associated with DDC systems for energy-consuming systems and equipment.

16.4.1 Basic Control Strategies

Almost all EMCS provide basic control strategies for energy-consuming equipment and systems. The basic control strategies include scheduling, duty cycling, demand limiting, optimal start/stop and temperature reset.

Time clocks are frequently used to turn equipment and systems on and off according to a set pattern. Although time clocks are simple and easy to use, there are several problems associated with time clocks, including:

- The on/off lugs sometimes loosen or fall off.
- Holidays, and other periods when systems can be set back, cannot easily be taken into account.
- Power failures require that time clocks be reset.
- There are a limited number of on/off cycles possible each day.
- It is time-consuming to change schedules on multiple time clocks.

Scheduling with an EMCS is very much the same as it is with time clocks; however, multiple start/stop settings can be accomplished very easily and accurately with an EMCS. Holidays, and other setback periods, can be entered into or calculated by an EMCS.

Most energy-consuming systems are designed for peak load conditions and frequently operate at excess capacity. As a result, these systems can sometimes be shut down for short periods each hour, typically, 15 minutes, without affecting process or comfort conditions. Shutting systems and equipment off for brief, predetermined periods of time is referred to as duty cycling. Duty cycling is easily accomplished with an EMCS.

Electrical utilities bill commercial customers based not only on the amount of energy consumed (KWh), but also on the peak demand (KW) for each month. In order to determine peak demand during a billing period, a utility establishes short periods of time called the demand interval (typically 15, 30 or 60 minutes). The billing demand is defined as the highest average demand recorded during any one demand period. EMCS with demand limiting capabilities utilize either pulses from a utility meter or current transformers to predict the facility demand during any demand interval. If the facility demand is predicted to exceed a user-entered set point, equipment is shed (shut-off) to reduce and control the demand.

Optimal start/stop strategies are frequently used for HVAC systems in buildings. Buildings with time clocks not controlled by an EMCS usually start HVAC equipment operation at a time early enough in the morning (such as three hours before occupancy time) to bring the temperature of the building up or down to the occupied set point temperature. EMCS with optimal start capabilities utilize indoor and outdoor temperature information, along with learned building characteristics, to vary the start time of HVAC equipment so that building temperatures reach the desired values just before occupancy occurs. Most EMCS have a "self-timing" capability, which allows the system to learn the characteristics of a building.

Temperature reset control is a function often provided by an EMCS. Air supply temperatures, chilled water temperatures or heating water temperatures may be reset easily based upon the load. The EMCS measures the load and automatically adjusts the supply temperature. The supply temperature is often reset according to the outside air temperature, especially for heating water systems.

16.4.2 Air Distribution Systems

Air distribution systems are also discussed in Chapter 9. In addition to the general control strategies listed above, there are additional control strategies, which are applicable to fans and air distribution systems, that can result in additional energy savings. An air distribution system can include terminal units (VAV terminals, etc.), air handling units, ductwork and controls. An air handling unit usually includes a cooling coil, dampers, fans, controls and some method of heating the air. Air distribution systems may be classified as one of two general types: constant volume and variable air volume (VAV).

Constant volume systems are characterized by the fact that the fan supplies air at a constant rate regardless of the heating or cooling load. Variations in the heating and cooling load are handled by varying the temperature of the air supplied to the load. Constant volume systems may be classified as either single zone or multi-zone.

A single zone air distribution system is shown in Figure 16-16 and is characterized by the fact that it supplies conditioned air to a single zone and is controlled in response to the temperature requirements of that zone as sensed by a temperature sensor in that zone. If air from the system is supplied to other zones with different heating or cooling loads, the system cannot adjust to meet the loads in the zone where the temperature sensor is not located. It can only adjust to the loads in the zone where the

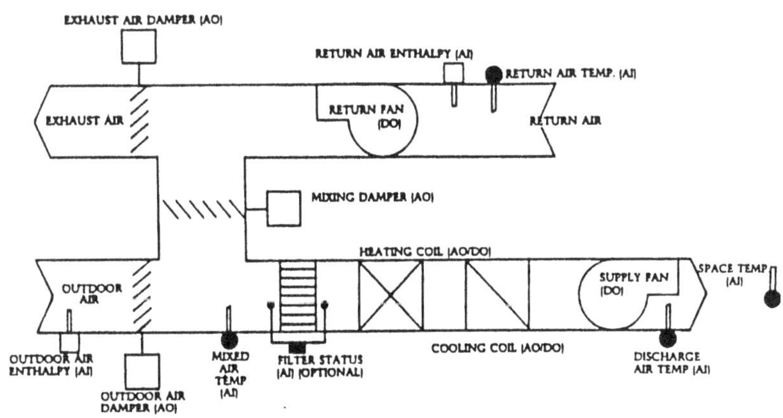

Figure 16-16. Single zone air handling unit control schematic.

temperature sensor is located. Some flexibility can be provided to a single zone constant volume system by providing terminal units with reheat for each individual zone. These systems are typically not very energy-efficient since a constant volume of air must first be cooled and then reheated at the zone to meet the zone heating or cooling load.

Multi-zone systems can supply a constant volume of air to zones with different heating or cooling loads. These systems usually supply a constant volume of air to a zone; however, the temperature of the supply air is varied by mixing cooled air with heated air to meet the load in the zone. Mixing can be accomplished inside the air handling unit, or a system with dual ducts (one with heated air and one with cooled air) can be used. The dual duct system mixes the hot and cold air at each zone to supply air at a temperature required to meet the zone heating or cooling load. Multi-zone air distribution systems are not very energy-efficient since a constant volume of air is either heated or cooled and then mixed to achieve the required supply air temperature.

The energy efficiency of constant volume systems can be improved with the addition of several control features and strategies. These control features and strategies include economizer operation, night setback/setup, pre-cooling, night purge and supply air temperature reset.

Air systems with economizer control can fully modulate return air, outside air and relief/exhaust air. The strategy employs the opening of the outside air damper beyond its minimum position if the heat content of the outside air is less than the heat content of the return air. This strategy can be based upon the total heat content of the air (enthalpy) or the dry bulb temperature of the air (sensible).

A night setback/setup strategy involves resetting the space temperature set points during unoccupied periods to appropriate levels, which will reduce the use of mechanical heating or cooling. Typical setback temperatures are 55°F for heating and 85°F for cooling.

A pre-cooling strategy starts the cooling process prior to occupancy to reduce the temperature in the building or process to at or below the normally required set point temperature. Once this level is reached, the cooling is shut off and the temperature is allowed to gradually rise or "coast" before the cooling is turned back on. Benefits of this strategy include a reduction in demand and the use of lower cost off-peak electrical power to cool the load.

The night purge strategy causes the air handling system to operate in a maximum outside air mode during unoccupied time periods in the cooling season whenever the outside air dry bulb temperature or enthalpy is lower than that of the space or load. The benefit of this strategy is the ability to cool the internal mass of a building or load with outside air, instead of mechanical cooling in anticipation of the cooling load that will occur later.

The supply air temperature reset strategy is based upon resetting the temperature according to the cooling load. The benefits of resetting the supply air temperature for constant volume systems can be quite significant. Increasing the supply air temperature set point for cooling

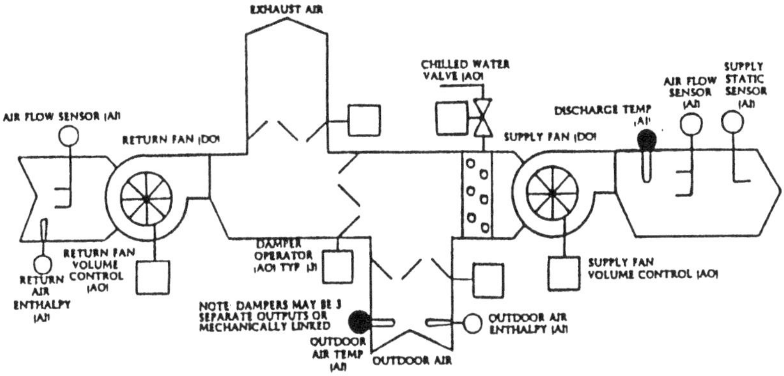

Figure 16-17. VAV air handling unit control schematic.

reduces both the cooling load and the amount of energy required for reheat; however, it does not change the amount of energy required to drive the fan. In general, the set point temperature for a constant air volume system could be set at the highest value that will satisfy the cooling load. Resetting the supply air temperature upward, however, may result in higher humidity levels in conditioned spaces.

Variable air volume (VAV) systems vary the rate of airflow based upon the cooling load. The air is supplied at a constant temperature. The airflow rate increases as the cooling load increases and vica versa. Airflow through the air handling unit fan may be modulated by using dampers on the outlet side of the fan, inlet vanes on the fan, vane-axial fans with controllable pitch fan blades, or variable speed control of the fan motor. Typically, the inputs to any of these controlled devices are modulated in order to maintain a static pressure set point in the duct. A VAV air handling unit control schematic is shown in Figure 16-17. The pressure sensor should be located at a point in the supply air ductwork such that the established set point will ensure proper operation of the zone variable air volume terminals under varying load (supply airflow) conditions. Significant fan energy savings are possible if the static pressure set point is reset so that at least one of the VAV terminals remains open. With this strategy, the airflow resistance remains relatively constant.

16.4.3 Chilled Water Systems

Chillers and chilled water systems were discussed in Chapter 7. Chilled water systems consist of chillers, pumps, piping, valves and controls. Chilled water piping loops may be either primary or primary/secondary. Pumping and piping systems were discussed in Chapter 6.

A chiller plant consists of one, or more, chillers pumps, piping and accessories. Most chillers come with their own packaged control system. Multiple chillers are typically arranged in a parallel piping arrangement with dedicated pumps. Individual feedback controllers start the dedicated chilled water pump and adjust the cooling capacity of each chiller in order to maintain a specified chilled water temperature. Control strategies for chillers and chilled water plants include optimum start/stop, soft start, multiple chiller lead/lag control, chilled water reset and condenser water reset.

The optimum start/stop strategy for chillers is most important when cooling loads vary and when operation is far from design conditions for a significant period of time. This operating strategy reduces energy consumption and costs by minimizing equipment on time while still satisfying the cooling load during on-peak periods. The optimum start/stop function monitors outside air temperature, relative humidity, cooling load temperatures, time of day, etc., and makes decisions as to the latest possible start time for a chiller. Conversely, the optimum stop determines the earliest time that equipment can be shut off while still satisfying the cooling load. Optimum start and stop calculations are updated on each scan of the microprocessor, typically a maximum of every 10 seconds. Therefore, any sudden changes in environmental conditions or in load conditions would be reflected in the calculated start/stop times.

As a chiller is started, conventional controls will permit it to run at full load until the chilled water temperature is brought down to the set point. This results in a brief high electrical current situation due to motor in-rush current, plus up to 20 minutes of full load operation while the water in the distribution system is brought down to the set point temperature. This type of operation may set a demand peak for the billing period and high electrical demand charges. These high demand charges can be avoided by soft starting of the chiller or "ramping it up" as it is started. The energy management system software limits the maximum allowable electrical demand during pull-down. The strategy typically involves starting a chiller at 50% capacity, and incrementally increasing the capacity over a user-definable ramp-up time such that the capacity increases from 50% to the maximum required capacity over a period of time.

Multiple chillers are typically configured in a parallel manner and usually controlled to give identical water supply temperatures. In general, controlling for identical set point temperatures is the best and simplest strategy. With this operating strategy, the relative loading on operating chillers is controlled by the relative chilled water flow rates.

Chiller sequencing defines the order that, and conditions where, chillers are brought on-line. In general, chillers should be brought on-line at conditions where the total power consumption (including pumps, tower fans or condenser fans) of operating with the additional chiller would be less than without. The effect of chiller and pump sequencing is shown graphically in Figure 16-18. EMCS software selects the combination of chillers and pumps that would optimize chilled water systems performance and minimize energy consumption. The following logic may be used for sequencing multiple chillers.

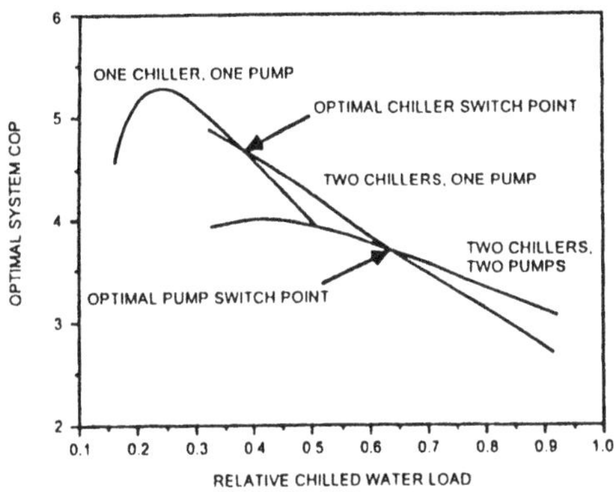

Figure 16-18. Effect of chiller and pump sequencing on optimal performance.

1. Evaluate the time-averaged values of the chilled water supply temperature and overall cooling load over a fixed time interval (e.g., five minutes).
2. If the chilled water supply temperature is significantly greater than the set point (e.g., 1°F), then bring the next chiller on-line.
3. Determine the time-averaged position of the cooling coil water valves and corresponding discharge air temperature for "representative" air handlers or cooling loads.
4. If the chilled water supply set point is at its lower limit and more than one valve is saturated at 100% open and their corresponding discharge air temperatures are significantly greater than set point (e.g., 1°F), then bring another chiller/pump combination on-line. Otherwise, go to step 5.
5. If the cooling load is significantly less (e.g., 10%) than the value associated with the first time interval after the last chiller was brought on-line, then take that chiller off-line.

Most chillers are operated such that they supply chilled water at a constant discharge temperature regardless of the cooling load. Significant chiller energy can be saved by resetting the chilled water supply temperature upward as the cooling load decreases. The EMCS software can reset the chilled water temperature to a maximum value while still satisfying the cooling load and maintaining comfort levels. There are three methods commonly used to reset chilled water supply temperatures: average system load, outside air temperature and the greatest zone demand.

Resetting the chilled water supply temperature according to the average system load involves sensing the temperature difference between the return water and the supply water. Inputs required for this control strategy are the chilled water supply temperature, chilled water return temperature and chilled water flow rate. This strategy employs a user selectable reset ratio, for example, for each one degree decrease in the temperature difference, the supply water temperature is reset upward one degree.

An outside air temperature strategy involves resetting the chilled water supply temperature as a function of the outside air enthalpy. This strategy assumes that the cooling load is directly proportional to the outside environmental conditions. As the outdoor air enthalpy increases, the chilled water temperature is decreased.

The most accurate method for resetting the chilled water supply temperature is by the greatest zone or load demand. This strategy resets the supply water temperature as a function of the highest zone temperature or actual cooling load valve position. The control valve calling for the most cooling controls the temperature of the chilled water. The chilled water temperature is reset upward until the valve with the greatest cooling load is fully open. The representative control valves should be chosen to include load diversity at all times and ensure reliable data.

The benefits on chiller energy consumption, which would result from resetting the condenser water temperature, was discussed in Chapter 7. With this strategy, the EMCS software optimizes the evaporation effect on the cooling tower to deliver condenser water to the chiller at an appropriately low temperature. Cooling towers controls are typically set up to provide a constant water temperature, typically 85°F. In relation to the energy consumed by the chiller, a cooling tower consumes comparatively little energy. Under this strategy, the EMCS software analyzes the outside air wet bulb temperature and controls the tower fan(s) and/or bypass control valve and/or tower dampers.

16.4.4 Cooling Towers

Cooling towers and fluid coolers were discussed in Chapter 8. Both cooling towers and fluid coolers reject heat to the atmosphere by circulating warm water from the condenser of a chiller. Basically, a cooling tower is an open device that rejects heat by bringing the condenser water into direct contact with ambient air. Fluid coolers are closed circuit devices with coils or heat exchangers, which transfer heat from the condenser water to the ambient air. The condenser does not come into direct contact with the ambient air.

Cooling tower water leaves the tower and returns to the chiller condenser at a set temperature, typically about 85°F. In order to supply water at a constant temperature, some form of cooling tower capacity control must be provided to adjust for varying ambient conditions and varying water temperatures from the chiller. Basically, there are two methods of controlling the capacity of a cooling tower: varying the flow of water through the tower and varying the flow of ambient air through the tower. Capacity control may also be a combination of both water side and air side control.

Water side control is achieved by either circulating the water through the tower or bypassing some of the water around it while the tower fans

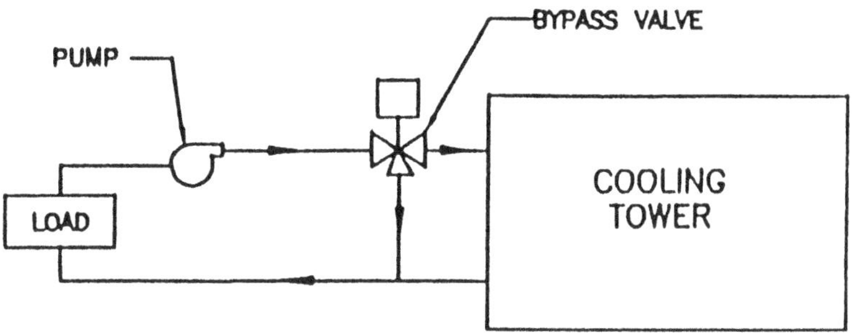

Figure 16-19. Cooling tower bypass control.

operate at a constant air volume. A schematic of a cooling tower bypass control is shown in Figure 16-19. The bypass flow control valve may be either a diverting valve, as shown in Figure 16-19, or a mixing valve. Variable flow control may also be provided by a two way valve; however, some form of capacity control, such as a variable frequency drive, is usually required for the pump to prevent overpressure. Water side capacity control tends to not be as energy-efficient as other means of control. First, no energy savings results since both the fan motors and pump motors are operating at full load. Second, the spray efficiency of the cooling tower spray nozzles is reduced at lower flow rates. This results in less water to air heat transfer in addition to reduced water flow rates.

There are three methods of providing air side capacity control for cooling towers: sequenced switching of fans, damper control and fan speed control. For cooling towers large enough to have multiple fans, sequenced switching of fans provides an effective method of capacity control. Fan motors are merely switched on or off in response to a tower supply water temperature sensor in the discharge piping of the cooling tower. Damper control is suitable for smaller cooling towers. A damper on the discharge side of the fan opens and closes to control the airflow rate and the tower leaving water temperature. Energy savings with fan dampers may be insignificant since the fan and motor are constantly operating at full speed. Fan speed control probably provides the most efficient method of tower capacity control. Cooling tower fan speed may be controlled by multi-speed gear transmissions or by variable frequency drives. The optimum number of cells operating and individual fan speeds depend upon the system characteristics and ambient conditions. The rules for bringing cell fans on-line are as follows:

- All variable speed fans: Operate all cells with fans at equal speeds.
- Multi-speed fans: Increment lowest speed fans first when adding tower capacity. Reverse for removing capacity.
- Variable speed/multi-speed fans: Operate all cells with variable speed fans at equal speeds. Increment lowest speed fans first when adding

tower capacity with multi-speed fans. Add multi-speed fan capacity when variable speed fan speeds match the fan speed associated with the next multi-speed fan increment to be added.

Similarly, for removing tower capacity, the highest speeds are the first to be reduced and the sequences defined above are reversed.

16.4.5 Pumping Systems

Hydronic and pumping systems were discussed in Chapter 6. The energy savings associated with two way control valves was discussed. Variable speed, or sequencing multiple pumps with different capacities, can provide an effective method for saving pumping energy over constant volume pumping systems. However, the problem of sequencing multiple pumps involves determining both the order and point at which pumps should be brought on-line and off-line.

Pumps should be brought on-line in an order that allows a continuous variation in flow rate and maximized operating efficiency of the pumps at each point for the specific pressure loss characteristics. For a combination of fixed speed and variable speed pumps, at least one variable speed pump should be brought on-line prior to any fixed speed pumps.

An additional pump should be brought on-line whenever the current set of pumps are operating at full capacity and can no longer satisfy the differential pressure set point. This situation can be detected by monitoring the differential pressure or controller output signal. Insufficient pump capacity leads to extended periods with differential pressures that are less than the set point and controller output that is saturated at 100%. A pump may be taken off-line whenever the remaining pumps have sufficient capacity to maintain the differential set point. This condition can be determined by comparing the current (time-averaged) controller output with the controller output (time-averaged) at the point just after the last pump was brought on-line. The pump can be taken off-line when the current is less than the switch point value by a specified deadband (e.g., 5%).

16.4.6 Boiler and Boiler Systems

Most package boiler systems are supplied with factory mounted control packages. Boilers were discussed in Chapter 4. These controls typically use a mechanical linkage to control combustion. The normal method of control uses a simple jackshaft, which is a steel shaft mounted between bearings across the front of the boiler. Arms connected to the jackshaft move as the shaft rotates and positions an air damper and fuel valve(s). Frequently, cams are inserted in the linkage to adjust the relationship between the air and fuel during installation.

Reductions in excess oxygen reduce stack losses directly. In addition, it reduces the flue gas volume and consequently the velocity through the boiler. This reduction in velocity provides more time for the heat to be removed from the flue gases, thereby reducing stack temperatures and reducing the temperature between the entering combustion air and exiting flue gases. This reduces stack losses and increases combustion efficiency. EMCS software can increase boiler efficiency by more accurately sensing stack temperatures and more precisely controlling combustion air volume.

Automatic oxygen trim control strategy modulates the inlet damper to maintain the proper excess air level for the various firing rates of the boiler. The oxygen level and firing rate are monitored. Depending upon the firing rate, an oxygen set point is calculated. The inlet air damper is then modulated to maintain this oxygen set point. In addition, the rate of change in the firing rate or the boiler load can be determined to enable the system to quickly anticipate and adjust the trim damper position. This feed-forward action allows the oxygen set point to be readjusted to match the firing rate without resultant oxygen set point droop that would be present through strict feedback control.

EMCS software can also improve boiler efficiency with several control strategies, including optimum start/stop, soft start and heating water reset. The optimum start/stop strategy reduces energy consumption by minimizing boiler operating, while supplying the necessary heat to satisfy the heating load. The optimum start function monitors the outside air temperature, temperature of the heating load, time of day, occupancy and operational schedules, to make decisions as to the latest possible start time for the boiler system, which will provide sufficient heat to meet the heating load.

The soft start strategy prevents thermal shock of the tube bundle in fire tube boilers. It also limits electrical demand in electric boilers. For fire tube boilers, thermal shock can be avoided in two different ways. One strategy calls for the heating water loop to be maintained at a certain minimum temperature, regardless of the load. The other strategy is to have the boiler go through a number of "soak" cycles upon start-up. With this strategy, the boiler is started and allowed to run for a predetermined period of time or until the heating water temperature rises above a certain predetermined level. At this point, the boiler is shut down for a period of time to let the heat "soak" into the tube bundle for a safe expansion of the various components. After the "soak" period, the boiler is refired for another period and then again shut off for a "soak" period. This cycle is continued until the heating water temperature reaches a level at which thermal shock is no longer a possibility.

The heating water reset strategy varies the heating water temperature to the heating load. The heating water reset function monitors one, or more, indicators of the heating load, which may include outside air temperature, load mass temperature, temperature difference between the heating water supply and return, and the space temperature. The appropriate indicator is chosen with respect to the type of heating load. The EMCS reset program calculates the heating load and resets the heating water supply set point.

16.4.7 Lighting System Controls

Lighting systems were discussed in Chapter 10. There are two general categories of lighting controls: centralized controls and local controls. Centralized controls are used in facilities where it is desirable to control large areas on the same schedule. With centralized controls, it is not possible to have individual differences within large work areas of the facility. Local controls are designed to affect only specific areas of a facility.

These two general categories of controls can be combined with an EMCS to create a hybrid control system to meet the needs of a facility, while controlling energy consumption. A timer in the EMCS software can activate the centralized control of large areas on a time-of-day basis. Personalized control, such as local switches or occupancy sensors, can be used to control lights in specific local areas of a facility. EMCS software can also sense daylighting and reduce electrical lighting in local areas when not required.

16.4.8 Thermal Storage System Control

Thermal storage systems were discussed in Chapter 15. Thermal storage systems can save significant amounts of energy by storing waste heat recovered from heat generating systems, such as cogeneration systems, and by providing cooling during peak electrical demand periods.

Energy management and control for heat storage systems is relatively straightforward. These controls usually consist of temperature sensors and flow control valves. Temperature sensors and flow control valves were discussed earlier in this chapter.

For cool thermal storage systems, the choice of control strategy results from a trade-off between system performance (i.e., operating cost) and implementation cost (i.e., initial cost). Chiller priority control has the lowest implementation cost, but generally results in higher operating cost. Storage priority strategies result in better performance, but require the use of a forecaster and measurement of state of change for storage.

For ice thermal storage, ice making should be initiated when the cooling load is low and off-peak electrical rates are in effect. During the ice making period, the chiller should operate at full capacity. Cooling plants for ice storage generally operate most efficiently at full load due to auxiliaries and characteristics of ice making chillers. With feedback control of the chilled water/glycol supply temperature, full capacity control is accomplished by establishing a low enough set point to ensure this condition (i.e., 20°F).

For internal melt ice storage tanks, the chiller should operate until the tank reaches its maximum state of charge or the charging period (e.g., off-peak period) ends. This strategy ensures that sufficient ice will be available for the next day without the need for a forecaster. Typically, only a small heat transfer penalty is associated with restoring a partially discharged, internal melt storage tank to full charge. For this type of storage device, the charging cycle always starts with high transfer effectiveness because water surrounds the tubes regardless of the amount of ice melted. The heat transfer effectiveness drops gradually until the new ice formations intersect with old formations, at which point the tank is fully recharged.

External ice melt storage tanks have a more significant heat transfer penalty associated with recharging after a partial discharge because ice forms on the outside of existing formations during charging. In this case, it is desirable from an energy efficiency standpoint to fully discharge the tank each day and only recharge as necessary to meet the next day's cooling requirements. To ensure that adequate ice is available, the maximum possible storage capacity needed for the next day depends on the discharge strategy used and the cooling load.

There are three strategies for use with utility structures having on-peak and off-peak energy demand charges: (1) chiller priority control, (2) storage priority, load limiting control and (3) a rule-based strategy that uses both chiller priority and load limiting strategies.

For chiller priority discharge strategy, the chiller operates at full cooling capacity (or less if sufficient to meet the load) during the storage discharge mode and matches the difference between the cooling load requirement and the chiller capacity. This strategy is easy to implement and does not require a load forecast. It works well for design conditions, but can result in relatively high demand and energy costs at full capacity during the on-peak period.

For load limiting storage priority control, several storage priority control approaches ensure that storage is not depleted prematurely. A storage priority control strategy, termed load limiting control, tends to minimize the peak cooling plant power demand.

Rule-based control combines elements of chiller priority and storage priority strategies, along with a demand limiting algorithm to achieve near-optimal control. The demand limiting algorithm requires measurement of the total facility electrical use.

Appendix A

Climatic Conditions for the United States and Canada

APPENDIX A. Climatic Design Criteria (Copyright 1993 by the American Society of Heating, Refrigerating, and Air Conditioning Engineers. Inc. from the ASHRAE Handbook-Fundamentals. Used by permission)

				Winter,[b] °F		Summer,[c] °F							Prevailing Wind		Temp., °F	
Col. 1	Col. 2	Col. 3	Col. 4	Col. 5		Col. 6			Col.7	Col. 8			Col. 9		Col.10	
State and Station[a]	Lat.	Long.	Elev.	Design Dry-Bulb		Design Dry-Bulb and Mean Coincident Wet-Bulb			Mean Daily	Design Wet-bulb			Winter	Summer	Median of Annual Extr.	
	° 'N	° 'W	Feet	99%	97.5%	1%	2.5%	5%	Range	1%	2.5%	5%	Knots[4]		Max.	Min.
ALABAMA																
Alexander City	32 57	85 57	660	18	22	96/77	93/76	91/76	21	79	78	78				
Anniston AP	33 35	85 51	599	18	22	97/77	94/76	92/76	21	79	78	78	SW 5	SW	98.4	12.4
Auburn	32 36	85 30	652	18	22	96/77	93/76	91/76	21	79	78	78			99.8	14.6
Birmingham AP	33 34	86 45	620	17	21	96/74	94/75	92/74	21	78	77	76	NNW 8	WNW	98.5	12.9
Decatur	34 37	86 59	580	11	16	95/75	93/74	91/74	22	78	77	76				
Dothan AP	31 19	85 27	374	23	27	94/76	92/76	91/76	20	80	79	78				
Florence AP	34 48	87 40	581	17	21	97/74	94/74	92/74	22	78	77	76	NW 7	NW		
Gadsden	34 01	86 00	554	16	20	96/75	94/75	92/74	22	78	77	76	NNW 8	WNW		
Huntsville AP	34 42	86 35	606	11	16	95/75	93/74	91/74	23	78	77	76	N 9	SW		
Mobile AP	30 41	88 15	211	25	29	95/77	93/77	91/76	18	80	79	78	N 10	N		
Mobile Co	30 40	88 15	211	25	29	95/77	93/77	91/76	16	80	79	78			97.92	2.3
Montgomery AP	32 23	86 22	169	22	25	96/76	95/76	93/76	21	79	79	78	NW 7	W	98.9	18.2
Selma, Craig AFB	32 20	87 59	166	22	26	97/78	95/77	93/77	21	81	80	79	N 9	SW	100.1	17.6
Talladega	33 27	86 06	565	18	22	97/77	94/76	92/76	21	79	78	78			99.6	11.2
Tuscaloosa AP	33 13	87 37	169	20	23	96/75	96/76	94/76	22	79	78	77	N 5	WNW		
ALASKA																
Anchorage AP	61 10	150 01	114	-23	-18	71/59	68/58	66/56	15	60	59	57	SE 3	WNW		
Barrow	71 18	156 47	31	-45	-41	57/53	53/50	49/47	12	54	50	47	SW 8	SE		
Fairbanks AP	64 49	147 52	436	-51	-47	82/62	78/60	75/39	24	64	62	60	N 5	S		
Juneau AP	58 22	134 35	12	-4	1	74/60	70/58	67/37	15	61	59	58	N 7	W		
Kodiak	57 45	152 29	73	10	13	69/58	65/56	62/55	10	60	58	56	WNW 14	NW		

Nome AP	64 30	165 26	13	-31	-27	66/57	62/55	59/54	10	58	56	55	N 4	W		
ARIZONA																
Douglas AP	31 27	109 36	4098	27	31	98/63	95/63	93/63	31	70	69	68			104.4	14.0
Flagstaff AP	35 08	111 40	7006	-2	4	84/55	82/55	80/54	31	61	60	59	NE 5	SW	90.0	-11.6
Fort Huachuca AP	31 35	110 20	4664	24	28	95/62	92/62	90/62	27	69	68	67	SW 5	W		
Kingman AP	35 12	114 01	3539	18	25	103/65	100/64	97/64	30	70	69	69				
Nogales	31 21	110 55	3800	28	32	99/64	96/64	94/64	31	71	70	69	SW 5	W		
Phoenix AP	33 26	112 01	1112	31	34	109/71	107/71	105/71	27	76	75	75	E 4	W	112.8	26.7
Prescott AP	34 59	112 26	5010	4	9	96/61	94/60	92/60	30	66	65	64				
Tucson AP	32 07	110 56	2558	28	32	104/66	102/66	100/66	26	72	71	71	SE 6	WNW	108.9	0.3
Winslow AP	35 01	110 44	4895	5	10	97/61	95/60	93/60	32	66	65	64	SW 6	WSW	102.7	-0.4
Yuma AP	32 39	114 37	213	36	39	111/72	109/72	107/71	27	79	78	77	NNE 6	WSW	114.8	30.8
ARKANSAS																
Blytheville AFB	35 57	89 57	264	10	15	96/78	94/77	91/76	21	81	80	78	N 8	SSW		
Camden	33 36	92 49	116	18	23	98/76	96/76	94/76	21	80	79	78				
El Dorado AP	33 13	92 49	277	18	23	98/76	96/76	94/76	21	80	79	78	S 6	SE	101.0	13.9
Fayetteville AP	36 00	94 10	1251	7	12	97/72	94/73	92/73	23	77	76	75	NE 9	SSW	99.4	-0.4
Fort Smith AP	35 20	94 22	463	12	17	101/75	98/76	95/76	24	80	79	78	NW 8	SW	101.9	7.0
Hot Springs	34 29	93 06	535	17	23	101/77	97/77	94/77	22	80	79	78	N 8	SW	103.0	10.6
Jonesboro	35 50	90 42	345	10	15	96/78	94/77	91/76	21	81	80	78			101.7	7.3
Little Rock AP	34 44	92 14	257	15	20	99/76	96/77	94/77	22	80	79	78	N 9	SSW	99.0	11.2
Pine Bluff AP	34 18	92 05	241	16	22	100/78	97/77	95/78	22	81	80	80	N 7	SW	102.2	13.1
Texarkana AP	33 27	93 59	389	18	23	98/76	96/77	93/76	21	80	79	78	WNW 9	SSW	104.8	14.0
CALIFORNIA																
Bakersfield AP	35 25	119 03	475	30	32	104/70	101/69	98/68	32	73	71	70	ENE 5	WNW	109.8	25.3
Barstow AP	34 51	116 47	1927	26	29	106/68	104/68	102/67	37	73	71	70	WNW 7	W	110.4	17.4
Blythe AP	33 37	114 43	395	30	33	112/71	110/71	108/70	28	75	75	74			116.8	24.1
Burbank AP	34 12	118 21	775	37	39	95/68	91/68	88/67	25	71	70	69	NW 3	S		
Chico	39 48	121 51	238	28	30	103/69	101/68	98/67	36	71	70	68	NW 5	SSE	109.0	22.6
Concord	37 58	121 59	200	24	27	100/69	97/68	94/67	32	71	70	68	WNW 5	NW		

APPENDIX A. (Continued)

Col. 1	Col. 2	Col. 3	Col. 4	Col. 5 Winter,[b] °F		Col. 6 Summer,[c] °F			Col.7	Col. 8			Col. 9 Prevailing Wind		Col.10 Temp., °F	
State and Station[a]	Lat.	Long.	Elev.	Design Dry-Bulb		Design Dry-Bulb and Mean Coincident Wet-Bulb			Mean Daily	Design Wet-bulb			Winter	Summer	Median of Annual Extr.	
	° ′N	° ′W	Feet	99%	97.5%	1%	2.5%	5%	Range	1%	2.5%	5%	Knots[4]		Max.	Min.
Covina	34 05	117 52	575	32	35	98/69	95/68	92/67	31	73	71	70				
Crescent City AP	41 46	124 12	40	31	33	68/60	65/59	63/58	18	62	60	59				
Downey	33 56	118 08	116	37	40	93/70	89/70	86/69	22	72	71	70				
El Cajon	32 49	116 58	367	42	44	83/69	80/69	78/68	30	71	70	68				
El Centro AP	32 49	115 40	-43	35	38	112/74	110/74	108/74	34	81	80	78	W 6	SE		
Escondido	33 07	117 05	660	39	41	89/68	85/68	82/68	30	71	70	69				
Eureka, Arcata AP	40 59	124 06	218	31	33	68/60	65/59	63/58	11	62	60	59	E 5	NW	75.8	29.7
Fairfield, Travis AFB	38 16	121 56	62	29	32	99/68	95/67	91/66	34	70	68	67	N 5	WSW		
Fresno AP	36 46	119 43	328	28	30	102/70	100/69	97/68	34	72	71	70	E 4	WNW	108.7	25.8
Hamilton AFB	38 04	122 30	3	30	32	89/68	84/66	80/65	28	72	69	67	N 4	SE		
Laguna Beach	33 33	117 47	35	41	43	83/68	80/68	77/67	18	70	69	68				
Livermore	37 42	121 57	545	24	27	100/69	97/68	93/67	24	71	70	68	WNW 4	NW		
Lompoc, Vandenberg AFB	34 43	120 34	368	35	38	75/61	70/61	67/60	20	63	61	60	ESE 5	NW		
Long Beach AP	33 49	118 09	30	41	43	83/68	80/68	77/67	22	70	69	68	NW 4	WNW		
Long Beach AP	33 56	118 24	97	41	43	83/68	80/68	77/67	15	70	69	68	E 4	WSW		
Los Angeles Co	34 03	118 14	270	37	40	93/70	89/70	86/69	20	72	71	70	NW 4	NW	98.1	35.9
Merced, Castle AFB	37 23	120 34	188	29	31	102/70	99/69	96/68	36	72	71	70	ESE 4	NW		
Modesto	37 39	121 00	91	28	30	101/69	98/68	95/67	36	71	70	69			105.8	26.2
Monterey	36 36	121 54	39	35	38	75/63	71/61	68/61	20	64	62	61	SE 4	NW		
Napa	38 13	122 17	56	30	32	100/69	96/68	92/67	30	71	69	68			103.1	25.8
Needles AP	34 46	114 37	913	30	33	112/71	110/71	108/70	27	75	75	74			116.4	26.7

Oakland AP	37 49	122 19	5	34	36	85/64	80/63	75/62	19	66	64	63	E 5	WNW	93.0	31.8
Oceanside	33 14	117 25	26	41	43	83/68	80/68	77/67	13	70	69	68				
Ontario	34 03	117 36	952	31	33	102/70	99/69	96/67	36	74	72	71	E 4	WSW		
Oxnard	34 12	119 11	49	34	36	83/66	80/64	77/63	19	70	68	67				
Palmdale AP	34 38	118 06	2542	18	22	103/65	101/65	98/64	35	69	67	66	SW 5	WSW		
Palm Springs	33 49	116 32	411	33	35	112/71	110/70	108/70	35	76	74	73				
Pasadena	34 09	118 09	864	32	35	98/69	95/68	92/67	29	73	71	70			102.8	30.4
Petaluma	38 14	122 38	16	26	29	94/68	90/66	87/65	31	72	70	68			102.0	24.2
Pomona Co	34 03	117 45	934	28	30	102/70	99/69	95/68	36	74	72	71	E 4	W	105.7	26.2
Redding AP	40 31	122 18	495	29	31	105/68	102/67	100/66	32	71	69	68			109.2	26.0
Redlands	34 03	117 11	1318	31	33	102/70	99/69	96/68	33	74	72	71			106.7	27.1
Richmond	37 56	122 21	55	34	36	85/64	80/63	75/62	17	66	64	63				
Riverside, March AFB	33 54	117 15	1532	29	32	100/68	98/68	95/67	37	72	71	70	N 4	NW	107.6	26.6
Sacramento AP	38 31	121 30	17	30	32	101/70	98/70	94/69	36	72	71	70	NNW 6	SW	105.1	27.6
Salinas AP	36 40	121 36	75	30	32	74/61	70/60	67/59	24	62	61	59				
San Bernardino, Norton AFB	34 08	117 16	1125	31	33	102/70	99/69	96/68	38	74	72	71	E 3	W	109.3	25.3
San Diego AP	32 44	117 10	13	42	44	83/69	80/69	78/68	12	71	70	68	NE 3	WNW	91.2	37.4
San Fernando	34 17	118 28	965	37	39	95/68	91/68	88/67	38	71	70	69				
San Francisco AP	37 37	122 23	8	35	38	82/64	77/63	73/62	20	65	64	62	S 5	NW		
San Francisco Co	37 46	122 26	72	38	40	74/63	71/62	69/61	14	64	62	61	W 5	W	91.3	35.9
San Jose AP	37 22	121 56	56	34	36	85/66	81/65	77/64	26	68	67	65	SE 4	NNW	98.6	28.2
San Luis Obispo	35 20	120 43	250	33	35	92/69	88/70	84/69	26	73	71	70	E 4	W	99.8	29.3
Santa Ana AP	33 45	117 52	115	37	39	89/69	85/68	82/68	28	71	70	69	E 3	SW	101.0	29.9
Santa Barbara AP	34 26	119 50	10	34	36	81/67	77/66	75/65	24	68	67	66	NE 3	SW	97.1	31.7
Santa Cruz	36 59	122 01	125	35	38	75/63	71/61	68/61	28	64	62	61			97.5	26.8
Santa Maria AP	34 54	120 27	236	31	33	81/64	76/63	73/62	23	65	64	63	E 4	WNW		
Santa Monica Co	34 01	118 29	64	41	43	83/68	80/68	77/67	16	70	69	68				
Santa Paula	34 21	119 05	263	33	35	90/68	86/67	84/66	36	71	69	68				
Santa Rosa	38 31	122 49	125	27	29	99/68	95/67	91/66	34	70	68	67	N 5	SE	102.5	23.4
Stockton AP	37 54	121 15	22	28	30	100/69	97/68	94/67	37	71	70	68	WNW 4	NW	104.1	24.5
Ukiah	39 09	123 12	623	27	29	99/69	95/68	91/67	40	70	68	67			108.1	21.6
Visalia	36 20	119 18	325	28	30	102/70	100/69	97/68	38	72	71	70			108.4	25.1

APPENDIX A. (Continued)

				Winter,[b] °F		Summer,[c] °F							Prevailing Wind		Temp., °F	
Col. 1	Col. 2	Col. 3	Col. 4	Col. 5		Col. 6			Col.7	Col. 8			Col. 9		Col.10	
State and Station[a]	Lat.	Long.	Elev.	Design Dry-Bulb		Design Dry-Bulb and Mean Coincident Wet-Bulb			Mean Daily	Design Wet-bulb			Winter	Summer	Median of Annual Extr.	
	° ′N	° ′W	Feet	99%	97.5%	1%	2.5%	5%	Range	1%	2.5%	5%	Knots[4]		Max.	Min.
Yreka	41 43	122 38	2625	13	17	95/65	92/64	89/63	38	67	65	64			102.8	7.1
Yuba City	39 08	121 36	80	29	31	104/68	101/67	99/66	36	71	69	68				
COLORADO																
Alamosa AP	37 27	105 52	7537	-21	-16	84/57	82/57	80/57	35	62	61	60				
Boulder	40 00	105 16	5445	2	8	93/59	91/59	89/59	27	64	63	62			96.0	-8.4
Colorado Springs AP	38 49	104 43	6145	-3	2	91/58	88/57	86/57	30	63	62	61	N 9	S	92.3	-12.1
Denver AP	39 45	104 52	5283	-5	1	93/59	91/59	89/59	28	64	63	62	S 8	SE	96.8	-10.4
Durango	37 17	107 53	6550	-1	4	89/59	87/59	85/59	30	64	63	62			92.4	-11.2
Fort Collins	40 35	105 05	4999	-10	-4	93/59	91/59	89/59	28	64	63	62			95.2	-18.1
Grand Junction AP	39 07	108 32	4843	2	7	96/59	94/59	92/59	29	64	63	62	ESE 5	WNW	99.9	-3.4
Greeley	40 26	104 38	4648	-11	-5	96/60	94/60	92/60	29	65	64	63				
Lajunta AP	38 03	103 30	4160	-3	3	100/68	98/68	95/67	31	72	70	69	W 8	S		
Leadville	39 15	106 18	10155	-8	-4	84/52	81/51	78/50	30	56	55	54			79.7	-17.8
Pueblo AP	38 18	104 29	4641	-7	0	97/61	95/61	92/61	31	67	66	65	W 5	SE	100.5	-12.2
Sterling	40 37	103 12	3939	-7	-2	95/62	93/62	90/62	30	67	66	65			100.3	-15.4
Trinidad AP	37 15	104 20	5740	-2	3	93/61	91/61	89/61	32	66	65	64	W 7	WSW	96.8	-10.5
CONNECTICUT																
Bridgeport AP	41 11	73 11	25	6	9	86/73	84/71	81/70	18	75	74	73	NNW 13	WSW		
Hartford, Brainard Field	41 44	72 39	19	3	7	91/74	88/73	85/72	22	77	75	74	N 5	SSW	95.7	-4.4

New Haven AP	41 19	73 55	6	3	7	89/75	84/73	82/72	17	76	75	74	NNE 7	SW	93.0	-0.2
New London	41 21	72 06	59	5	9	88/73	85/72	83/71	16	76	75	74				
Norwalk	41 07	73 25	37	6	9	86/73	84/71	81/70	19	75	74	73				
Norwich	42 32	72 04	20	3	7	89/75	86/73	83/72	18	76	75	74				
Waterbury	41 35	73 04	843	-4	2	88/63	85/71	82/70	21	75	74	72	N 8	SW		
Windsor	41 56	72 41	169	0	4	91/74	88/72	85/71	22				N 8	SW		
Locks, Bradley Fld										76	75	73				
DELAWARE																
Dover AFB	39 08	75 28	28	11	15	92/75	90/75	87/74	18	79	77	76	W 9	SW	97.0	7.0
Wilmington AP	39 40	75 36	74	10	14	92/74	69/74	87/73	20	77	76	75	WNW 9	WSW	95.4	4.9
DISTRICT OF COLUMBIA																
Andrews AFB	38 5	76 5	279	10	14	92/74	90/74	87/73	18	78	76	75				
Washington. National AP	38 51	77 02	14	14	17	93/75	91/74	89/74	18	78	77	76	WNW 11	S	97.6	7.4
FLORIDA																
Brlle Glade	26 39	80 39	16	41	44	92/76	91/76	89/76	16	79	78	78			94.7	30.9
Cape Kennedy AP	28 29	80 34	16	35	38	90/78	88/78	87/78	15	80	79	79				
Daytona Beach AP	29 11	81 03	31	32	35	92/78	90/77	88/77	15	80	79	78	NW 8			
E Fort Lauderdale	26 04	80 09	10	42	46	92/78	91/78	90/78	15	80	79	79	NW 9	ESE		
Fort Myers AP	26 35	81 52	15	41	44	93/78	92/78	91/77	18	80	79	79	NNE 7	W	94.9	34.9
Fort Pierce	27 28	80 21	25	38	42	91/78	90/78	89/78	15	80	79	79			96.1	34.0
Gainesville AP	29 41	82 16	152	28	31	95/77	93/77	92/77	18	80	79	78	W 6	W	97.8	23.3
Jacksonville AP	30 30	81 42	26	29	32	96/77	94/77	92/76	19	79	79	78	NW 7	SW	97.5	25.4
Key West AP	24 33	81 45	4	55	57	90/78	90/78	89/78	09	80	79	79	NNE 12	SE	92.0	51.5
Lakeland Co	28 02	81 57	214	39	41	93/76	91/76	89/76	17	79	78	78	NNW 9	SSW		
Miami AP	25 48	80 16	7	44	47	91/77	90/77	89/77	15	79	79	78	NNW 8	SE	92.5	39.0

APPENDIX A. (Continued)

Col. 1	Col. 2	Col. 3	Col. 4	Col. 5 Winter,[b] °F		Col. 6 Summer,[c] °F			Col.7	Col. 8			Col. 9 Prevailing Wind		Col.10 Temp., °F	
State and Station[a]	Lat.	Long.	Elev.	Design Dry-Bulb		Design Dry-Bulb and Mean Coincident Wet-Bulb			Mean Daily	Design Wet-bulb			Winter	Summer	Median of Annual Extr.	
	° ′N	° ′W	Feet	99%	97.5%	1%	2.5%	5%	Range	1%	2.5%	5%	Knots[4]		Max.	Min.
Miami Beach Co	25 47	80 17	10	45	48	90/77	89/77	88/77	10	79	79	78				
Ocala	29 11	82 08	89	31	34	95/77	93/77	92/76	18	80	79	78			98.6	24.8
Orlando AP	28 33	81 23	100	35	38	94/76	93/76	91/76	17	79	78	78	NNW 9	SSW		
Panama City, Tyndall AFB	30 04	85 35	18	29	33	92/78	90/77	89/77	14	81	80	79	N 8	WSW		
Pensacola Co	30 25	87 13	56	25	29	94/77	93/77	91/77	14	80	79	79	NNE 7	SW	96.3	23.3
St. Augustine	29 58	81 20	10	31	35	92/78	89/78	87/78	16	80	79	79	NW 7	W	97.6	25.8
St. Petersburg	27 46	82 80	35	36	40	92/77	91/77	90/76	16	79	79	78	N 8	W	94.8	35.6
Sanford	28 46	81 17	89	35	38	94/76	93/76	91/76	17	79	78	78				
Sarasota	27 23	82 33	26	39	42	93/77	92/77	90/76	17	79	79	78				
Tallahassee AP	30 23	84 22	55	27	30	94/77	92/76	90/76	19	79	78	78	NW 6	NW	97.6	20.9
Tampa AP	27 58	82 32	19	36	40	92/77	91/77	90/76	17	79	79	78	N 8	W	95.0	31.5
West Palm Beach Ap	26 41	80 06	15	41	45	92/78	91/78	90/78	16	80	79	79	NW 9	ESE		
GEORGIA																
Albany, Turner AFB	31 36	84 05	223	25	29	97/27	95/76	93/76	20	80	79	78	N 7	W	100.6	19.9
Americus	32 03	84 14	456	21	25	97/77	94/76	92/73	20	79	78	77			100.4	16.5
Athens	33 57	83 19	802	18	22	94/74	92/74	90/74	21	78	77	76	NW 9	WNW	98.7	13.5
Atlanta AP	33 39	84 26	1010	17	22	94/74	92/74	90/73	19	77	76	75	NW 11	NW	95.7	11.9
Augusta AP	33 22	81 58	145	20	23	97/77	95/76	93/76	19	80	79	78	W 4	WSW	99.0	17.5
Brunswick	31 15	81 29	25	29	32	92/78	89/78	87/78	18	80	79	79			99.3	24.7

Columbus, Lawson AFB	32 31	84 56	242	21	24	95/76	93/76	91/75	21	79	78	77	NW 8	W		
Dalton	34 34	84 57	720	17	22	94/76	93/76	91/76	22	79	78	77				
Dublin	32 20	82 54	215	21	25	96/77	93/76	91/75	20	79	78	77			101.0	16.7
Gainesville	34 11	83 41	50	24	27	96/77	93/77	91/77	20	80	79	78	WNW 7	SW	98.7	21.9
Griffin	33 13	84 16	981	18	22	93/76	90/75	88/74	21	78	77	76				
LaGrange	33 01	85 04	709	19	23	94/76	91/75	89/74	21	78	77	76			97.7	11.9
Macon AP	32 42	83 39	354	21	25	96/77	93/76	91/75	22	79	78	77	NW 8	WNW	99.6	16.9
Marietta, Dobbins AFB	33 55	84 31	1068	17	21	94/74	92/74	90/74	21	78	77	76	NNW 12	NW		
Savannah	32 08	81 12	50	24	27	96/77	93/77	91/77	20	80	79	78	WNW 7	SW	98.7	21.9
Valdosta-Moody AFB	30 58	83 12	233	28	31	96/77	94/77	92/76	20	80	79	78	WNW 6	W		
Waycross	31 15	82 24	148	26	29	96/77	94/77	91/76	20	80	79	78			100.0	19.5
HAWAII																
Hilo AP	19 43	155 05	36	61	62	84/73	83/72	82/72	15	75	74	74	SW 6	NE		
Honolulu AP	21 20	157 55	13	62	63	87/73	86/73	85/72	12	76	75	74	ENE 12	ENE		
Kaneohe Bay MCAS	21 27	157 46	18	65	66	85/75	84/74	83/74	12	76	76	75	NNE 9	NE		
Wahiawa	21 03	158 02	900	58	59	86/73	83/72	84/72	14	75	74	73	WNW 5	E		
IDAHO																
Boise AP	43 34	116 13	2838	3	10	96/65	94/64	91/64	31	68	66	65	SE 6	NW	103.2	0.6
Burley	42 32	113 46	4156	-3	2	99/62	95/61	92/66	35	64	63	61			98.6	-8.3
Coeur D'Alene AP	47 46	116 49	2972	-8	-1	89/62	86/61	83/60	31	64	63	61			99.9	-4.5
Idaho Falls AP	43 31	112 04	4741	-11	-6	89/61	87/61	84/59	38	65	63	61	N 9	S	96.2	-16.0
Lewiston AP	46 23	117 01	1413	-1	6	96/65	93/64	90/63	32	67	66	64	W 3	WNW	105.9	2.7
Moscow	46 44	116 58	2660	-7	0	90/63	87/62	84/61	32	65	64	62			98.0	-5.9
Mountain Home AFB	43 02	115 54	2996	6	12	99/64	97/63	94/62	36	66	65	63	ESE 7	NW	103.2	-6.5
Pocatello AP	42 55	112 36	4454	-8	-1	94/61	91/60	89/59	35	64	63	61	NE 5	W	97.9	-11.4
Twin Falls AP	42 29	114 29	4150	-3	2	99/62	95/61	92/60	34	64	63	61	SE 6	NW	100.9	-5.1

APPENDIX A. (Continued)

Col. 1	Col. 2	Col. 3	Col. 4	Col. 5 Winter,[b] °F		Col. 6 Summer,[c] °F			Col.7	Col. 8			Col. 9 Prevailing Wind		Col.10 Temp., °F	
State and Station[a]	Lat.	Long.	Elev.	Design Dry-Bulb		Design Dry-Bulb and Mean Coincident Wet-Bulb			Mean Daily	Design Wet-bulb			Winter	Summer	Median of Annual Extr.	
	° 'N	° 'W	Feet	99%	97.5%	1%	2.5%	5%	Range	1%	2.5%	5%	Knots[4]		Max.	Min.
ILLINOIS																
Aurora	41 45	88 20	744	-6	-1	93/76	91/76	88/75	20	79	78	76			96.7	-13.0
Belleville, Scott AFB	38 33	89 51	453	1	6	94/76	92/76	89/75	21	79	78	76	WNW 8	S		
Bloomington	40 29	88 57	876	-6	-2	92/75	90/74	88/73	21	78	76	75			98.4	-9.6
Carbondale	37 47	89 15	417	2	7	95/77	93/77	90/76	21	80	79	77			100.9	-0.8
Champaign/ Urbana	40 02	88 17	777	-3	2	95/75	92/74	90/73	21	78	77	75				
Chicago, Midway AP	41 47	87 45	607	-5	0	94/74	91/73	88/72	20	77	75	74	NW 11	SW		
Chicago. O'Hare AP	41 59	87 54	658	-8	-4	91/74	89/74	86/72	20	77	76	74	WNW 9	SW		
Chicago Co	41 53	87 38	590	-3	2	94/75	91/74	88/73	15	79	77	75			96.1	-8.3
Danville	40 12	87 36	695	-4	1	93/75	90/74	88/73	21	78	77	75	W 10	SSW	98.2	-8.4
Decatur	39 50	88 52	679	-3	2	94/75	91/74	88/73	21	78	77	75	NW 10	SW	99.0	-8.1
Dixon	41 50	89 29	696	-7	-2	93/75	90/74	88/73	23	78	77	75			97.5	-13.5
Elgin	42 02	88 16	758	-7	-2	91/75	88/74	86/73	21	78	77	75				
Freeport	42 18	89 37	780	-9	-4	91/74	89/73	87/72	24	77	76	74				
Galesburg	40 56	90 26	764	-7	-2	93/75	91/75	88/74	22	78	77	75	WNW 8	SW		
Joliet	41 31	88 10	582	-5	0	93/75	90/74	88/73	20	78	77	75	NW 11	SW		
Kankakee	41 05	87 55	625	-4	1	93/75	90/74	88/73	21	78	77	75				
La Salle/Peru	41 19	89 06	520	-7	-2	93/75	91/75	88/74	22	78	77	75				
Macomb	40 28	90 40	702	-5	0	95/76	92/76	89/75	22	79	78	76				
Moline AP	41 27	90 31	582	-9	-4	93/75	91/75	88/74	23	78	77	75	WNW 8	SW	96.8	-12.7

Mt Vernon	38 19	88 52	479	0	5	95/76	92/75	89/74	21	79	78	76			100.5	-2.9
Peoria AP	40 40	89 41	652	-8	-4	91/75	89/74	87/73	22	78	76	75	WNW 8	SW	98.0	-10.9
Quincy AP	39 57	91 12	769	-2	3	96/76	93/76	90/76	22	80	78	77	NW 11	SSW	101.1	-6.7
Rantoul, Chanute AFB	40 18	88 08	753	-4	1	94/75	91/74	89/73	21	78	77	75	W 10	SSW		
Rockford	42 21	89 03	741	-9	-4	91/74	89/73	87/72	24	77	76	74			97.4	-13.8
Springfield AP	39 50	89 40	588	-3	2	94/75	92/74	89/74	21	79	77	76	NW 10	SW	98.1	-7.2
Waukegan	42 21	87 53	700	-6	-3	92/76	89/74	87/73	21	78	76	75			96.5	-10.6
INDIANA																
Anderson	40 06	85 37	919	0	6	95/76	92/75	89/74	22	79	78	76	W 9	SW	95.1	-6.0
Bedford	38 51	86 30	670	0	5	95/76	92/75	89/74	22	79	78	76			97.5	-4.4
Bloomington	39 08	86 37	847	0	5	95/76	92/75	89/74	22	79	78	76	W 9	SW	97.8	-4.6
Columbus, Bakalar AFB	39 16	85 54	651	3	7	95/76	92/75	90/74	22	79	78	76	W 9	SW	98.3	-6.4
Crawfrodsville	40 03	86 54	679	-2	3	94/75	91/74	88/73	22	79	77	76			98.4	-7.6
Evansville AP	38 03	87 32	381	4	9	95/76	93/75	91/75	22	79	78	77	NW 9	SW	98.2	0.2
Fort Wayne AP	41 00	85 12	791	-4	1	92/73	89/72	87/72	24	77	75	74	WSW 10	SW		
Goshen AP	41 32	85 48	827	-3	1	91/73	89/73	86/72	23	77	75	74			96.8	-10.5
Hobart	41 32	87 15	600	-4	2	91/73	88/73	85/72	21	77	75	74			98.5	-8.5
Huntington	40 53	85 30	802	-4	1	92/73	89/72	87/72	23	77	75	74			96.9	-8.1
Indianapolis AP	39 44	86 17	792	-2	2	92/74	90/74	87/73	22	78	76	75	WNW 10	SW	96	-7
Jeffersonville	38 17	85 45	455	5	10	95/74	93/74	90/74	23	79	77	76			98	2
Kokomo	40 25	86 03	855	-4	0	91/74	90/73	88/73	22	77	75	74			98.2	-7.5
Lafayette	40 2	86 5	600	-3	3	94/74	91/73	88/73	22	78	76	75				
La Porte	41 36	86 43	810	-3	3	93/74	90/74	87/73	22	78	76	75			98.1	-10.5
Marion	40 29	85 41	859	-4	0	91/74	90/73	88/73	23	77	75	74			97.0	-8.6
Muncie	40 11	85 21	957	-3	2	92/74	90/73	87/73	22	76	76	75				
Peru, Grissom AFB	40 39	86 09	813	-6	-1	90/74	88/73	86/73	22	77	75	74	W 10	SW		
Richmond AP	39 46	84 50	1141	-2	2	92/74	90/74	87/73	22	78	76	75			94.8	-8.5
Shelbyville	39 31	85 47	750	-1	3	93/74	91/74	88/73	22	78	76	75			97.7	-6.0
South Bend AP	41 42	86 19	773	-3	1	91/73	89/73	86/72	22	77	75	74	SW 11	SSW	962	-9.2
Terre Haute AP	39 27	87 18	585	-2	4	95/75	92/74	89/73	22	79	77	76	NNW 7	SSW	98.3	-4.9

APPENDIX A. (Continued)

				Winter,[b] °F		Summer,[c] °F							Prevailing Wind		Temp., °F	
Col. 1	Col. 2	Col. 3	Col. 4	Col. 5		Col. 6			Col.7	Col. 8			Col. 9		Col.10	
State and Station[a]	Lat.	Long.	Elev.	Design Dry-Bulb		Design Dry-Bulb and Mean Coincident Wet-Bulb			Mean Daily	Design Wet-bulb			Winter	Summer	Median of Annual Extr.	
	° ′N	° ′W	Feet	99%	97.5%	1%	2.5%	5%	Range	1%	2.5%	5%	Knots[4]		Max.	Min.
Valparaiso	41 31	87 02	801	-3	3	93/74	90/74	87/73	22	78	76	75			95.5	-11.0
Vincennes	38 41	87 32	420	1	6	95/75	92/74	90/73	22	79	77	76			100.3	-2.8
IOWA																
Ames	42 02	93 48	1099	-11	-6	93/75	90/74	87/73	23	78	76	75			97.4	-17.8
Burlington AP	40 47	91 07	692	-7	-3	94/74	91/75	88/73	22	78	77	75	NW 9	SSW	98.6	-11.0
Cedar Rapids AP	41 53	91 42	863	-10	-5	91/76	88/75	86/74	23	78	77	75	NW 9	S	97.7	-15.6
Clinton	41 50	90 13	595	-8	-3	92/75	90/75	87/74	23	78	77	75			97.5	-13.8
Council Bluffs	41 20	95 49	1210	-8	-3	94/76	91/75	88/74	22	78	77	75				
Des Moines AP	41 32	93 39	938	-10	-5	94/75	91/74	88/73	23	78	77	75	NW 11	S	98.2	-4.2
Dubuque	42 24	90 42	1056	-12	-7	90/74	88/73	86/72	22	77	75	74	N 10	SSW	95.2	-15.0
FortDodge	42 33	94 11	1162	-12	-7	91/74	88/74	86/72	23	77	75	74	NW 11	S	98.5	-19.1
Iowa City	41 38	93 33	661	-11	-6	92/76	89/76	87/74	22	80	78	76	NW 9	SSW	97.4	-15.2
Keokuk	40 24	91 24	574	-5	0	95/75	92/75	89/74	22	79	77	76			98.4	-8.8
Marshalltown	42 04	92 56	898	-32	-7	92/76	90/75	88/74	23	78	77	75			98.5	-13.4
Mason City AP	43 09	93 20	1213	-15	-11	90/74	88/74	85/72	24	77	75	74	NW 11	S	96.5	-21.7
Newton	41 41	93 02	936	-10	-5	94/75	91/74	88/73	23	78	77	75			98.2	-14.7
Ottumwam AP	41 06	92 27	840	-8	-4	94/75	91/74	88/73	22	78	77	75			99.1	-12.0
Sioux City AP	42 24	96 23	1095	-11	-7	95/74	92/74	89/73	24	78	77	75	NNW 9	S	99.9	-17.7
Waterloo	42 33	92 24	868	-15	-10	91/76	89/75	86/74	23	78	77	75	NW 9	S	97.7	-19.8
KANSAS																
Atchison	39 34	95 07	945	-2	2	96/77	93/76	91/76	23	81	79	77			100.5	-8.8
Chanute AP	37 40	95 29	981	3	7	100/74	97/74	94/74	23	78	77	76	NNW 11	SSW	102.8	-2.8

Dodge City AP	37 46	99 58	2582	0	5	100/69	97/69	95/69	25	74	73	71	N 12	SSW	102.9	-7.0
El Dorado	37 49	96 50	1282	3	7	100/72	98/73	96/73	24	77	76	75			103.5	-5.0
Emporia	38 20	96 12	1230	1	5	100/74	97/74	94/73	25	78	77	76			102.4	-6.4
Garden City AP	37 56	100 44	2880	-1	4	99/69	96/69	94/69	28	74	73	71				
Goodland AP	39 22	101 42	3654	-5	0	99/66	96/65	93/66	33	71	70	68	WSW 10	S	103.2	-10.4
Great Bend	38 21	98 52	1889	0	4	101/73	98/73	95/73	28	78	76	75				
Hutchinson AP	38 04	97 52	1542	4	8	102/72	97/72	97/72	28	77	75	74	N 14	S	105.3	-6.1
Liberal	37 03	100 58	2870	2	7	99/68	96/68	94/68	28	73	72	71			105.8	-3.8
Manhattan, Ft Riley	39 03	96 46	1065	-1	3	99/75	95/75	92/74	24	78	77	76	NNE 8	S	104.5	-8.6
Parsons	37 20	95 31	899	5	9	100/74	97/74	94/74	23	79	77	76	NNW 11	SSW		
Russell AP	38 52	98 49	1866	0	4	101/73	98/73	95/73	29	78	76	75				
Salina	38 48	97 39	1272	0	5	103/74	100/74	97/73	26	78	77	75	N 8	SSW		
Topeka AP	39 04	95 38	877	0	4	99/75	96/75	93/74	24	79	78	76	NNW 10	S	101.8	-6.4
Wichita AP	37 39	97 25	1321	3	7	101/72	98/73	96/73	23	77	76	75	NNW 12	SSW	102.5	-2.8
KENTUCKY																
Ashland	38 33	82 44	546	5	10	94/76	91/74	89/73	22	78	77	75	W 6	SW	97.4	0.8
Bowling Green AP	35 58	86 28	535	4	10	94/77	92/75	89/74	21	79	77	76			99.9	1.2
Corbin AP	36 57	84 06	1174	4	9	94/73	92/73	89/72	23	77	76	75				
Covington AP	39 03	84 40	869	1	6	92/73	90/72	88/72	22	77	75	74	W 9	SW		
Hopkinsville, Ft Campbell	36 40	87 29	571	4	10	94/77	92/75	89/74	21	79	77	76	N 6	W	100.1	-0.4
Lexington AP	38 02	84 36	966	3	8	93/73	91/73	88/72	22	77	76	75	WNW 9	SW	95.3	-0.5
Louisville AP	38 11	85 44	477	5	10	95/74	93/74	90/74	23	79	77	76	NW 8	SW	97.4	1.2
Madisonville	37 19	87 29	439	5	10	96/76	93/75	90/75	22	79	78	77				
Owensboro	37 45	87 10	407	5	10	97/76	94/75	91/75	23	79	78	77	NW 9	SW	98.0	-0.2
Paducah AP	37 04	88 46	413	7	12	98/76	95/75	92/75	20	79	78	77				
LOUISIANA																
Alexandria AP	31 24	92 18	92	23	27	95/77	94/77	92/77	20	80	79	78	N 7	S	100.1	-5.7
Baton Rouge AP	30 32	91 09	64	25	29	95/77	93/77	92/77	19	80	80	79	ENE 8	W	98.0	21.4
Bogalusa	30 47	89 52	103	24	28	95/77	93/77	92/77	19	80	80	79			99.3	20.2

APPENDIX A. (Continued)

Col. 1	Col. 2	Col. 3	Col. 4	Col. 5 Winter,[b] °F		Col. 6 Summer,[c] °F			Col.7	Col. 8			Col. 9 Prevailing Wind		Col.10 Temp., °F	
State and Station[a]	Lat.	Long.	Elev.	Design Dry-Bulb		Design Dry-Bulb and Mean Coincident Wet-Bulb			Mean Daily	Design Wet-bulb			Winter	Summer	Median of Annual Extr.	
	° 'N	° 'W	Feet	99%	97.5%	1%	2.5%	5%	Range	1%	2.5%	5%	Knots[4]		Max.	Min.
Houma	29 31	90 40	13	31	35	95/78	93/78	92/77	15	81	80	79			97.2	22.5
Lafayette AP	30 12	92 00	42	26	30	95/78	94/78	92/78	18	81	80	79	N 8	SW	98.2	22.6
Lake Charles AP	30 07	93 13	9	27	31	95/77	93/77	92/77	17	80	79	79	N 9	SSW	99.2	20.5
Minden	32 36	93 18	250	20	25	99/77	96/76	94/76	20	79	79	78			101.7	-4.9
Monroe AP	32 31	92 02	79	20	25	99/77	96/76	94/76	20	79	79	78	N 9	S	101.1	-5.9
Natchitoches	31 46	93 05	130	22	26	97/77	95/77	93/77	20	80	79	78				
New Orleans AP	29 59	90 15	4	29	33	93/78	92/78	90/77	16	81	80	79	NNE 9	SSW	96.3	27.7
Shreveport AP	32 28	93 49	254	20	25	99/77	96/76	94/76	20	79	79	78	N 9	S		
MAINE																
Augusta AP	44 19	69 48	353	-7	-3	88/13	85/70	82/68	22	74	72	70	NNE 10	WNW		
Bangot, Dow AFB	44 48	68 50	192	-11	-6	86/70	83/68	80/67	22	73	71	69	WNW 7	S		
Caribou AP	46 52	68 01	624	-18	-13	84/69	81/67	78/66	21	71	69	67	WSW 10	SW		
Lewiston	44 02	70 15	200	-7	-2	88/73	85/70	82/68	22	74	72	70			94.0	13.7
Millinocket AP	45 39	68 42	413	-13	-9	87/69	83/68	80/66	22	72	70	68	WNW 11	WNW	92.4	-23.0
Portland	43 39	70 19	43	-6	-1	87/72	84/71	81/69	22	74	72	70	W 7	S	93.5	-9.9
Waterville	44 32	69 40	302	-8	-4	87/72	84/69	81/68	22	74	72	70				
MARYLAND																
Baltimore AP	39 11	76 40	148	10	13	94/75	91/75	89/74	21	78	77	76	W 9	WSW		
Baltimore Co	39 20	76 25	20	14	17	92/77	89/76	87/75	17	80	78	76	WNW 9	S	97.9	7.2
Cumberland	39 37	78 46	790	6	10	92/75	89/74	87/74	22	77	76	75	WNW 10	W		
Frederick AP	39 27	77 25	313	8	12	94/76	91/75	88/74	22	78	77	76	N 9	WNW		
Hagerstown	39 42	77 44	704	8	12	94/75	91/74	89/74	22	77	76	75	WNW 10	W		

Salisbury	38 20	75 30	59	12	16	93/75	91/75	88/74	18	79	77	76			96.8	7.4
MASSACHUSETTS																
Boston AP	42 22	71 02	15	6	9	91/73	88/71	85/70	16	75	74	72	WNW 16	SW	95.7	-1.2
Clinton	42	42 24	7141	398	-2	2	90/72	87/71	84/69	17	75	73	72			91.7
Fall River	41 43	71 08	190	5	9	87/72	84/71	81/69	18	74	73	72	NW 10	SW	92.1	-1.10
Framingham	42 17	71 25	170	3	6	89/72	86/71	83/69	17	74	73	71			96.0	-7.7
Gloucester	42 35	70 41	10	2	5	89/73	86/71	83/70	15	75	74	72				
Greenfield	42 3	72 4	205	-7	-2	88/72	85/71	82/69	23	74	73	71				
Lawrence	42 42	71 10	57	-6	0	90/73	87/72	84/70	22	76	74	73	NW 8	WSW	95.2	-9.0
Lowell	42 39	71 19	90	-4	1	91/73	88/72	85/70	21	76	74	73			95.1	-8.5
New Bedford	41 41	70 58	79	5	9	85/72	82/71	80/69	19	74	73	72	NW 10	SW	91.4	2.2
Pittsfield AP	42 26	73 18	1194	-8	-3	87/71	84/70	81/68	23	73	72	70	NW 12	SW		
Springfield,	72 32	245	-5	0	90/72	87/71	84/69	19	75				SSW	95.7	-4.7	
Westover AFB										73	72	N 8				
42 12																
Taunton	41 54	71 04	20	5	9	89/73	86/72	83/70	18	75	74	73			92.9	-9.8
Worcester AP	42 16	71 52	986	0	4	87/71	84/70	81/68	18	73	72	70	W 14	W		
MICHIGAN																
Adrian	41 55	84 01	754	-1	3	91/73	88/72	85/71	23	76	75	73			97.2	-7.0
Alpena AP	45 04	83 26	610	-11	-6	89/70	85/70	83/69	27	73	72	70	W 5	SW	93.9	-14.8
Battle Creek AP	42 19	85 15	941	1	5	92/74	88/72	85/70	23	76	74	73	SW 8	SW		
Benton Harbor AP	42 08	86 26	643	1	5	91/72	88/72	85/70	20	75	74	72	SSW 8	WSW		
Detroit	42 25	83 01	619	3	6	91/73	88/72	86/71	20	76	74	73	W 11	SW	95.1	-2.6
Escanaba	45 44	87 05	607	-11	-7	87/70	83/69	80/68	17	73	71	69			88.8	-16.1
Flint AP	42 58	83 44	771	-4	1	90/73	87/71	85/71	25	76	74	72	SW 8	SW	95.3	-9.9
Grand Rapids AP	42 53	85 31	784	1	5	91/72	88/72	85/70	24	75	74	72	WNW 8	WSW	95.4	-5.6
Holland	42 42	86 06	678	2		88/72	86/71	83/70	22	75	73	72			94.1	-6.8
Jackson AP	42 16	84 28	1020	1	5	92/74	88/72	85/70	23	76	74	73			96.5	-7.8
Kalamazoo	42 17	85 36	955	1	5	92/74	88/72	85/70	23	76	74	73			95.9	-6.7
Lansing AP	42 47	84 36	873	-3	1	90/73	87/72	84/70	24	75	74	72	SW 12	W	94.6	-11.0
Marquette Co	46 34	87 24	735	-12	-8	84/70	81/69	77/66	18	72	70	68			94.5	-11.8

APPENDIX A. (Continued)

Col. 1	Col. 2	Col. 3	Col. 4	Col. 5 Winter,[b] °F		Col. 6 Summer,[c] °F			Col.7	Col. 8			Col. 9 Prevailing Wind		Col.10 Temp., °F	
State and Station[a]	Lat.	Long.	Elev.	Design Dry-Bulb		Design Dry-Bulb and Mean Coincident Wet-Bulb			Mean Daily	Design Wet-bulb			Winter	Summer	Median of Annual Extr.	
	°'N	°'W	Feet	99%	97.5%	1%	2.5%	5%	Range	1%	2.5%	5%	Knots[4]		Max.	Min.
Mt Pleasant	43 35	84 46	796	0	4	91/73	87/72	84/71	24	76	74	72			95.4	-11.1
Muskegon AP	43 10	86 14	625	2	6	86/72	84/70	82/71	21	75	73	72	E 8	SW		
Pontiac	42 40	83 25	981	0	4	90/73	87/72	85/71	21	76	74	73			95.0	-6.8
Port Huron	42 59	82 25	586	0	4	90/73	87/72	83/71	21	76	74	73	W 8	S		
Saginaw AP	43 32	84 05	667	0	4	91/73	87/72	84/71	23	76	74	72	WSW 7	SW	96.1	-7.6
Sault Ste. Marie AP	46 28	84 22	721	-12	-8	84/70	81/69	77/66	23	72	70	68	E 7	SW	89.8	-21.0
Traverse City AP	44 45	85 35	624	-3	1	89/72	86/71	83/69	22	75	73	71	SSW 9	SW	95.4	-10.7
Ypsilanti	42 14	83 32	716	1	5	92/72	89/71	86/70	22	75	74	72	SW 10	SW		
MINNESOTA																
Albert Lea	43 39	93 22	1220	-17	-12	90/74	87/72	84/71	24	77	75	73				
Alexandria AP	45 52	95 23	1430	-22	-16	91/72	88/72	85/70	24	76	74	72			95.1	-28.0
Bemidji AP	47 31	94 56	1389	-31	-26	88/69	85/69	81/67	24	73	71	69	N 8	S	94.5	-36.9
Brainerd	46 24	94 08	1227	-20	-26	90/73	87/71	84/69	24	75	73	71				
Duluth AP	46 50	92 11	1428	-21	-16	85/70	82/68	79/66	22	72	70	68	WNW 12	WSW	90.9	-27.4
Fairbault	44 18	93 16	940	17	-12	91/74	88/72	85/71	24	77	75	73			95.8	-24.3
Fergus Falls	46 16	96 04	1210	-21	-17	91/72	88/72	85/70	24	76	74	72			96.9	-27.8
International Falls AP	48 34	93 23	1179	-29	-25	85/68	83/68	80/66	26	71	70	68	N 9	S	93.4	-36.5
Mankato	44 09	93 59	1004	-17	-12	91/72	88/72	85/70	24	77	75	73				
Minneapolis/ St. Paul AP	44 53	93 13	834	-16	-12	92/75	89/73	86/71	22	77	75	73	NW 8	S	96.5	-22.0
Rochester AP	43 55	92 30	1297	-17	-12	90/74	87/72	84/71	24	77	75	73	NW 9	SSW		

St. Cloud AP	45 35	94 11	1043	-15	-11	91/74	88/72	85/70	24	76	74	72				
Virginia	47 30	92 33	1435	-25	-21	85/69	83/68	80/66	23	71	70	68			92.6	-33.0
Willmar	45 07	95 05	1128	-15	-11	91/74	88/72	85/71	24	76	74	72			96.8	-24.3
Winona	44 03	91 38	652	-14	-10	91/75	88/73	85/72	24	77	75	74				
MISSISSIPPI																
Biloxi, Keesler AFB	30 25	88 55	26	28	31	94/79	92/79	90/78	16	82	81	80	N 8	S	98.0	23.0
Clarksdale	34 12	90 34	178	14	19	96/77	94/77	92/76	21	80	79	78			100.9	13.2
Columbus AFB	33 39	88 27	219	15	20	95/77	93/77	91/76	22	80	79	78	N 7	W	101.6	12.7
Greenville AFB	33 29	90 59	138	15	20	95/77	93/77	91/76	21	80	79	78			99.5	14.9
Greenwood	33 30	90 05	148	15	20	95/77	93/77	91/76	21	80	79	70			100.6	15.3
Hattiesburg	31 16	89 15	148	24	27	96/78	94/77	92/77	21	81	80	79			99.9	18.2
Jackson AP	32 19	90 05	310	21	25	97/76	95/76	93/76	21	79	78	78	NNW 6	NW	99.8	16.0
Laurel	31 40	89 10	236	24	27	96/78	94/77	92/77	21	81	80	79			99.7	17.8
McComb AP	31 15	90 28	469	21	26	96/77	94/76	92/76	18	80	79	78				
Meridian AP	32 20	88 45	290	19	23	97/77	95/76	93/76	22	80	79	78	N 6	WSW	98.3	15.7
Natchez	31 33	91 23	195	23	27	96/78	94/78	92/77	21	81	80	79			98.4	18.4
Tupelo	34 16	88 46	361	14	19	96/77	94/77	92/76	22	80	79	78			100.7	11.8
Vicksburg Co	32 24	90 47	262	22	26	97/78	95/78	93/77	21	81	80	79			96.9	18.0
MISSOURI																
Cape Girardeau	37 14	89 35	351	8	13	98/76	95/75	92/75	21	79	78	77				
Columbia AP	38 58	92 22	778	-1	4	97/74	94/74	91/73	22	78	77	76	WNW 9	WSW	99.5	-6.2
Farmington AP	37 46	90 24	928	3	8	96/76	93/75	90/74	22	78	77	75			99.9	-2.1
Hannibal	39 42	91 21	489	-2	3	96/76	93/76	90/76	22	80	78	77	NNW 11	SSW	98.4	-7.6
Jefferson City	38 34	92 11	640	2	7	98/75	95/74	92/74	23	78	77	76			101.2	-6.1
Joplin AP	37 09	94 30	980	6	10	100/73	97/73	94/73	24	78	77	76	NNW 12	SSW		
Kansas City AP	39 07	94 35	791	2	6	99/75	96/74	93/74	20	78	77	76	NW 9	S	100.2	-4.3
Kirksville AP	40 06	92 33	964	-5	0	96/74	93/74	90/73	24	78	77	76			98.3	-10.8
Mexico	39 11	91 54	775	-1	4	97/74	94/74	91/73	22	78	77	76			101.2	-8.0
Moberly	39 24	92 26	850	-2	3	97/74	94/74	91/73	23	78	77	76				
Poplar BlufF	36 46	90 25	380	11	16	98/78	95/76	92/76	22	81	79	78				
Rolla	37 59	91 43	1204	3	9	94/77	91/75	89/74	22	78	77	76			99.4	-3.1

APPENDIX A. (Continued)

Col. 1	Col. 2	Col. 3	Col. 4	Col. 5 Winter,[b] °F		Col. 6 Summer,[c] °F			Col. 7	Col. 8			Col. 9 Prevailing Wind		Col. 10 Temp., °F	
State and Station[a]	Lat.	Long.	Elev.	Design Dry-Bulb		Design Dry-Bulb and Mean Coincident Wet-Bulb			Mean Daily	Design Wet-bulb			Winter	Summer	Median of Annual Extr.	
	° ′N	° ′W	Feet	99%	97.5%	1%	2.5%	5%	Range	1%	2.5%	5%	Knots[4]		Max.	Min.
St. Joseph AP	39 46	94 55	825	-3	2	96/77	93/76	91/76	23	81	79	77	NNW 9	S	100.6	-8.0
St. Louis AP	38 45	90 23	535	2	6	97/75	94/75	91/74	21	78	77	76	NW 9	WSW		
St. Louis Co	38 39	90 38	462	3	8	98/75	94/75	91/74	18	78	77	76	NW 6	S	99.1	-2.7
Sikeston	36 53	89 36	325	9	15	98/77	95/76	92/75	21	80	78	77				
Sedalia, Whiteman AFB	38 43	93 33	869	-1	4	95/76	92/76	90/75	22	79	78	76	NNW 7	SSW	100.0	-5.1
Sikeston	36 53	89 36	325	9	15	98/77	95/76	92/75	21	80	78	77				
Springfield AP	37 14	93 23	1268	3	9	96/73	93/74	91/74	23	78	77	75	NNW 10	S	97.2	-2.4
MONTANA																
Billings AP	45 48	108 32	3567	-15	-10	94/64	91/64	88/63	31	67	66	64	NE 9	SW	100.5	-19.1
Bozeman	45 47	111 09	4448	-20	-14	90/61	87/60	84/59	32	63	62	60			92.2	-23.2
Butte AP	45 57	112 30	5553	-24	-17	86/58	83/56	80/56	35	60	58	37	S 5	NW	91.8	-26.3
Cut Bank AP	48 37	112 22	3838	-25	-20	88/61	85/61	82/60	35	64	62	61			94.7	-30.9
Glasgow AP	48 25	106 32	2760	-22	-18	92/64	89/63	85/62	29	68	66	64	E 8	S		
Glendive	47 08	104 48	3476	-18	-13	95/66	92/64	89/62	29	69	67	65			103.3	-29.8
Great Falls AP	47 29	111 22	3662	-21	-15	91/60	88/60	85/59	28	64	62	60	SW 7	WSW	98.0	-25.1
Havre	48 34	109 40	2492	-18	-11	94/65	90/64	87/63	33	68	66	65			99.7	-31.3
Helena AP	46 36	112 00	3828	-21	-16	91/60	88/60	85/59	32	64	62	61	N 12	WNW	95.6	-23.7
Kalispell AP	48 18	114 16	2974	-14	-7	91/62	87/61	84/60	34	65	63	62			94.4	-16.8
Lewistown AP	47 04	109 27	4122	-22	-16	90/62	87/61	83/60	30	65	63	62	NW 9	NW	96.2	-27.7
Livingstown AP	45 42	110 26	4618	-20	-14	90/61	87/60	84/59	32	63	62	60			97.2	-21.2
Miles City AP	46 26	105 52	2634	-20	-15	98/66	95/66	92/65	30	70	68	67	NW 7	SE	103.6	-27.7
Missoula AP	46 55	114 05	3190	-13	-6	92/62	88/61	85/60	36	65	63	62	ESE 7	NW	98.6	-13.9

NEBRASKA																
Beatrice	40 16	96 45	1235	-5	-2	99/75	95/74	92/74	24	78	77	76			103.1	-11.3
Chadron AP	42 50	103 05	3313	-8	-3	97/66	94/65	91/65	30	71	69	68				
Columbus	41 28	97 20	1450	-6	-2	98/74	95/73	92/73	25	77	76	75				
Fremont	41 26	96 29	1200	-6	-2	98/75	95/74	92/74	22	78	77	76				
Grand Island AP	40 59	98 19	1860	-8	-3	97/72	94/71	91/71	28	75	74	73	NNW 10	S	103.3	-14.2
Hastings	40 36	98 26	1954	-7	-3	97/72	94/71	91/71	27	75	74	73	NNW 10	S	103.5	-10.7
Kearney	40 44	99 01	2132	-9	-4	96/71	93/70	90/70	28	74	73	72			102.9	-13.7
Lincoln Co	40 51	96 45	1180	-5	-2	99/75	95/74	92/74	24	78	77	76	N 8	S	102.0	-12.4
McCook	40 12	100 38	2768	-6	-2	98/69	95/69	91/69	28	74	72	71				
Norfolk	41 59	97 26	1551	-8	-4	97/74	93/74	90/73	30	78	77	75			102.0	-20.0
Norfolk Platte AP	41 08	100 41	2779	-8	-4	97/69	94/69	90/69	28	74	72	71	NW 9	SSE	100.8	-15.8
Omaha AP	41 18	95 54	977	-8	-3	94/76	91/75	88/74	22	78	77	75	NW 8	S	100.2	-13.2
Scottsbluff AP	41 52	103 36	3958	-8	-3	95/65	92/65	90/64	31	70	68	67	NW 9	SE	101.6	-18.9
Sidney AP	41 13	103 06	4399	-8	-3	95/65	92/65	90/64	31	70	68	67				
NEVADA																
Carson City	39 10	119 46	4675	4	9	94/60	91/59	89/58	42	63	61	60	SSW 3	WNW	99.2	-5.0
Elko AP	40 50	115 47	5050	-8	-2	94/59	92/59	90/58	42	63	62	60	E 4	SW		
Ely AP	39 17	114 51	6253	-10	-4	89/17	87/56	85/55	39	60	59	38	S 9	SSW		
Las Vega AP	36 05	115 10	2178	25	28	108/66	106/65	104/65	30	71	70	69	ENE 7	SW		
Lovelock AP	40 04	118 33	3903	8	12	98/63	96/63	93/62	42	66	65	64			103.0	-1.0
Reno AP	39 30	119 47	4404	5	10	95/61	92/60	90/59	45	64	62	61	SSW 3	WNW		
Reno Co	39 30	119 47	4408	6	11	96/61	93/60	91/59	45	64	62	61			98.9	0.2
Tonopah AP	38 04	117 05	5426	5	10	94/60	92/59	90/58	40	64	62	61	N 8	S		
Winnemucca AP	40 54	117 48	4301	-1	3	96/60	94/60	92/60	42	64	62	61	SE 10	W	100.1	-8.1
NEW HAMPSHIRE																
Berlin	44 03	71 01	1110	-14	-9	87/71	84/69	81/68	22	73	71	70			93.2	-24.7
Claremont	43 02	72 02	420	-9	-4	89/72	86/70	83/69	24	74	73	71				
Concord AP	43 12	71 30	342	-8	-3	90/72	87/70	84/69	26	74	73	71	NW 7	SW	94.8	-16.0
Keene	42 55	72 17	490	-12	-7	90/12	87/70	83/69	24	74	73	71			94.6	-18.9
Laconia	43 03	71 03	505	-10	-5	89/72	86/70	83/69	25	74	73	71				

APPENDIX A. (Continued)

Col. 1	Col. 2	Col. 3	Col. 4	Col. 5 Winter,[b] °F		Col. 6 Summer,[c] °F			Col.7	Col. 8			Col. 9 Prevailing Wind		Col.10 Temp., °F	
State and Station[a]	Lat.	Long.	Elev.	Design Dry-Bulb		Design Dry-Bulb and Mean Coincident Wet-Bulb			Mean Daily	Design Wet-bulb			Winter	Summer	Median of Annual Extr.	
	° ′N	° ′W	Feet	99%	97.5%	1%	2.5%	5%	Range	1%	2.5%	5%	Knots[4]		Max.	Min.
Manchester, Grenier AFB	42 56	71 26	233	-8	-3	91/72	88/71	85/70	24	75	74	72	N 11	SW	93.7	-12.6
Portsmouth, Pease AFB	43 04	70 49	101	-2	2	89/73	85/71	83/70	22	75	74	72	W 8	W		
NEW JERSEY																
Atlantic City Co	39 23	74 26	11	10	13	92/74	89/74	86/72	18	78	77	75	NW 11	WSW		
Long Branch	40 19	74 01	15	10	13	93/74	90/73	87/72	18	78	77	75			95.9	4.3
Newark AP	40 42	74 10	7	10	14	94/74	91/73	88/72	20	77	76	75	WNW 11	WSW		
New Brunswick	40 29	74 26	125	6	10	92/74	89/73	86/72	19	77	76	75				
Paterson	40 54	74 09	100	6	10	94/74	91/73	88/72	21	77	76	75				
Phillipsburg	40 41	75 11	180	1	6	92/73	89/72	86/71	21	76	75	74			97.4	-0.7
Trenton Co	40 13	74 46	56	11	14	91/75	88/74	85/73	19	78	76	75	W 9	SW	96.2	4.2
Vineland	39	29	75	00	112	8	11	91/75	89/74	86/73	19	78	76	75	r	
NEW MEXICO																
Alamagordo, Holloman AFB	32 51	106 06	4093	14	19	98/64	96/64	94/64	30	69	68	67				
Albuquerque AP	35 03	106 37	5311	12	16	96/61	94/61	92/61	27	66	65	64	N 7	W	98.1	5.1
Artesia	32 46	104 23	3320	13	19	103/67	100/67	97/67	30	72	71	70			105.5	3.7
Carlsbad AP	32 20	104 16	3293	13	19	103/67	100/67	97/67	28	72	71	70	N 6	SSE		
Clovis AP	34 23	103 19	4294	8	13	95/65	93/65	91/65	28	69	68	67			102.0	2.5
Farmington AP	36 44	108 14	5503	1	6	95/63	93/62	91/61	30	67	65	64	ENE 5	SW		
Gallup	35 31	108 47	6465	0	5	90/59	89/58	86/58	32	64	62	61				

Grants	35 10	107 54	6524	-1	4	89/59	88/58	85/57	32	64	62	61				
Hobbs AP	32 45	103 13	3690	13	18	101/66	99/66	97/66	29	71	70	69				
Las Cruces	32 18	106 55	4544	15	20	99/64	96/64	94/64	30	69	68	67	SE 5	SE		
Los Alamos	35 52	106 19	7410	5	9	89/60	87/60	85/60	32	62	61	60			89.8	- 2.3
Raton AP	36 45	104 30	6373	-4	1	91/60	89/60	87/60	34	65	64	63				
Roswell, Walker AFB	33 18	104 32	3676	13	18	100/66	98/66	96/66	33	71	70	69	N 6	SSE	103.0	2.7
Santa Fe Co	35 37	106 05	6307	6	10	90/61	88/61	86/61	28	63	62	61			90.1	- 1.2
Silver City AP	32 38	108 10	5442	5	10	95/61	94/60	91/60	30	66	64	63				
Socorro AP	34 03	106 53	4624	13	17	97/62	95/62	93/62	30	67	66	65				
Tucumcari AP	35 11	103 36	4039	8	13	99/66	97/66	95/65	28	70	69	68	NE 8	SW	102.7	1.1
NEW YORK																
Albany AP	42 45	73 48	275	-6	-1	91/73	88/72	85/70	23	75	74	72	WNW 8	S		
Albany Co	42 39	73 45	19	-4	1	91/73	88/72	85/70	20	75	74	72			95.2	-11.4
Auburn	42 54	76 32	715	-3	2	90/73	87/71	84/70	22	75	73	72			92.4	-9.5
Batavia	43 00	78 11	922	1	5	90/72	87/71	84/70	22	75	73	72			92.2	-7.5
Binghamton AP	42 13	75 59	1590	-2	1	86/71	83/69	81/68	20	73	72	70	WSW 10	WSW	92.9	-9.3
Buffalo AP	42 56	78 44	705	2	6	88/71	85/70	83/69	21	74	73	72	W 10	SW	90.0	-3.2
Cortland	42 36	76 11	1129	-5	0	88/71	85/71	82/70	23	74	73	71			93.8	-11.2
Dunkirk	42 29	79 16	692	4	9	88/73	85/72	83/71	18	75	74	72	SSW 10	WSW		
Elmira AP	42 10	76 54	955	-4	1	89/71	86/71	83/70	24	74	73	71			96.2	-6.7
Geneva	42 45	76 54	613	-3	2	90/73	87/71	84/70	22	75	73	72			96.1	-6.5
Glens Falls	43 20	73 37	328	-11	-5	88/72	85/71	82/69	23	74	73	71	NNW 6	S		
Gloversville	43 02	74 21	760	-8	-2	89/72	86/71	83/69	23	75	74	72			93.2	-14.6
Hornell	42 21	77 42	1325	-4	0	88/71	85/70	82/69	24	74	73	72				
Ithaca	42 27	76 29	928	-5	0	88/71	85/71	82/70	24	74	73	71	W 6	SW		
Jamestown	42 07	79 14	1390	-1	3	88/70	86/70	83/69	20	74	72	71	WSW 9	WSW		
Kingston	41 56	74 00	279	-3	2	91/73	88/72	85/70	22	76	74	73				
Lockport	43 09	79 15	638	4	7	89/74	86/72	84/71	21	76	74	73	N 9	SW	92.2	-4.8
Massena AP	44 56	74 51	207	-13	-8	86/70	83/69	80/68	20	73	72	70				
Newburgh, Stewart AFB	41 30	74 06	471	-1	4	90/73	88/72	85/70	21	76	74	73	W 10	W		

APPENDIX A. (Continued)

				Winter,[b] °F		Summer,[c] °F							Prevailing Wind		Temp., °F	
Col. 1	Col. 2	Col. 3	Col. 4	Col. 5		Col. 6			Col.7	Col. 8			Col. 9		Col.10	
State and Station[a]	Lat.	Long.	Elev.	Design Dry-Bulb		Design Dry-Bulb and Mean Coincident Wet-Bulb			Mean Daily	Design Wet-bulb			Winter	Summer	Median of Annual Extr.	
	° ′N	° ′W	Feet	99%	97.5%	1%	2.5%	5%	Range	1%	2.5%	5%	Knots[4]		Max.	Min.
NYC-Central Park	40 47	73 58	157	11	15	92/74	89/73	87/72	17	76	75	74			94.9	3.8
NYC-Kennedy AP	40 39	3 47	13	12	15	90/73	87/72	84/71	16	76	75	74	WNW 4	SSW		
NYC-La Guardia AP	40 46	73 54	11	11	15	92/74	89/73	87/72	16	76	75	74	WNW 15	SW		
Niagara Falls AP	43 06	79 57	590	4	7	89/74	86/72	84/71	20	76	74	73	W 9	SW		
Olean	42 14	78 22	2119	- 2	2	87/71	84/71	81/70	23	74	73	71				
Oneonta	42 31	75 04	1775	- 7	- 4	86/71	83/69	80/68	24	73	72	70				
Oswego Co	43 28	76 33	300	1	7	86/73	83/71	80/70	20	75	73	72	E 7	WSW	91.3	-7.4
Plattsburg AFB	44 39	73 28	235	-13	- 8	86/70	83/69	80/68	22	73	72	70	NW 6	SE		
Poughkeepsie	41 38	73 55	165	0	6	92/74	89/74	86/72	21	77	75	74	NNE 6	SSW	98.1	-5.6
Rochester AP	43 07	77 40	547	1	5	91/73	88/71	85/70	22	75	73	72	WSW 11	WSW		
Rome, Griffiss AFB	43 14	75 25	514	- 11	- 5	88/71	85/70	83/69	22	75	73	71	NW 5	W		
Schenectady	42 51	73 57	377	- 4	1	90/73	87/72	84/70	22	75	74	72	WNW 8	S		
Suffolk County AFB	40 51	72 38	67	7	10	86/72	83/71	80/70	16	76	74	73	NW 9	SW		
Syracuse AP	43 07	76 07	410	- 3	2	90/73	87/71	84/70	20	75	73	72	N 7	WNW	93.	-10.0
Utica	43 09	75 23	714	-12	- 6	88/73	85/71	82/70	22	75	73	71	NW 12	W		
Watertown	43 59	76 01	325	-11	- 6	86/73	83/71	81/70	20	75	73	72	E 7	WSW	91.7	-19.6
NORTH CAROLINA																
Asheville AP	35 26	82 32	2140	10	14	89/73	87/72	85/71	21	75	74	72	NNW 12	NNW	91.9	5.8
Charlotte AP	35 13	80 56	736	18	22	95/74	93/74	91/74	20	77	76	76	NNW 6	SW	97.8	12.6
Durham	35 52	78 47	434	16	20	94/75	92/75	90/75	20	78	77	76			98.9	9.6

Elizabeth City AP	36 16	76 11	12	12	19	93/78	91/77	89/76	18	80	78	78	NW 8	SW		
Fayetteville, Pope AFB	35 10	79 01	218	17	20	95/76	92/76	90/75	20	79	78	77	N 6	SSW	99.1	13.1
Goldsboro	35 20	77 58	109	18	21	94/77	91/76	89/75	18	79	78	77	N 8	SW	99.8	13.0
Greensboro AP	36 05	79 57	897	14	18	93/74	91/73	89/73	21	77	76	75	NE 7	SW	97.7	9.7
Greenville	35 37	77 25	75	18	21	93/77	91/76	89/75	19	79	78	77				
Henderson	36 22	78 25	480	12	15	95/77	92/76	90/76	20	79	78	77				
Hickory	35 45	81 23	1187	14	18	92/73	90/72	88/72	21	75	74	73			96.5	9.6
Jacksonville	34 50	77 37	95	20	24	92/78	90/78	88/77	18	80	79	78				
Lumberton	34 37	79 04	129	18	21	95/76	92/76	90/75	20	79	78	77				
New Bern AP	35 05	77 03	20	20	24	92/78	90/78	88/77	18	80	79	78			98.2	15.1
Raleigh/Durham AP	35 52	78 47	434	16	20	94/75	92/75	90/75	20	78	77	76	N 7	SW	97.7	12.2
Rocky Mount	35 58	77 48	121	18	21	94/77	91/76	89/75	19	79	78	77				
Wilmington AP	34 16	77 55	28	23	26	93/79	91/78	89/77	18	81	80	79	N 8	SW	96.9	18.2
Winston-Salem AP	36 08	80 13	969	16	20	94/74	91/73	89/73	20	76	75	74	NW 8	WSW		
NORTH DAKOTA																
Bismarck AP	46 46	100 45	1647	-23	-19	95/68	91/68	88/67	27	73	71	70	WNW 7	S	100.3	-31.5
Devils Lake	48 07	98 54	1450	-25	-21	91/69	88/68	85/66	25	73	71	69			97.5	-30.4
Dickinson AP	46 48	102 48	2585	-21	-17	94/68	90/66	87/65	25	71	69	68	WNW 12	SSE	101.0	-31.3
Fargo AP	46 54	96 48	896	-22	-18	92/73	89/71	85/69	25	76	74	72	SSE 11	S	97.3	-29.7
Grand Forks AP	47 57	97 24	911	-26	-22	91/70	87/70	84/68	25	74	72	70	N 8	S	97.6	-29.0
Jamestown AP	46 55	98 41	1492	-22	-18	94/70	90/69	87/68	26	74	74	71			101.3	-27.9
Minot AP	48 25	101 21	1668	-24	-20	92/68	89/67	86/65	25	72	70	68	WSW 10	S		
Williston	48 09	103 35	1876	-25	-21	91/68	88/67	85/65	25	72	70	68			99.7	-32.9
OHIO																
Akron, Canton AP	40 55	81 26	1208	1	6	89/72	86/71	84/70	21	75	73	72	SW 9	SW	94.4	-4.6
Ashtabula	41 51	80 48	690	4	9	88/73	85/72	83/71	18	75	74	72				
Athens	39 20	82 06	700	0	6	95/75	92/74	90/73	22	78	76	74				
Bowling Green	41 23	83 38	675	- 2	2	92/73	89/73	86/71	23	76	75	73			96.7	-7.3
Cambridge	40 04	81 35	807	1	7	93/75	90/74	87/73	23	78	76	75				

APPENDIX A. (Continued)

				Winter,[b] °F		Summer,[c] °F							Prevailing Wind		Temp., °F	
Col. 1	Col. 2	Col. 3	Col. 4	Col. 5		Col. 6			Col.7	Col. 8			Col. 9		Col.10	
State and Station[a]	Lat.	Long.	Elev.	Design Dry-Bulb		Design Dry-Bulb and Mean Coincident Wet-Bulb			Mean Daily	Design Wet-bulb			Winter	Summer	Median of Annual Extr.	
	° 'N	° 'W	Feet	99%	97.5%	1%	2.5%	5%	Range	1%	2.5%	5%	Knots[4]		Max.	Min.
Chillicothe	39 21	83 00	640	0	6	95/75	92/74	90/73	22	78	76	74	W 8	WSW	98.2	-2.1
Cincinnati Co	39 09	84 31	758	1	6	92/73	90/72	88/72	21	77	75	74	W 9	SW	97.2	-0.2
Cleveland AP	41 24	81 51	777	1	5	91/73	88/72	86/71	22	76	74	73	SW 12	N	94.7	-3.1
Columbus AP	40 00	82 53	812	0	5	92/73	90/73	87/72	24	77	75	74	W 8	SSW	96.0	-3.4
Dayon AP	39 54	84 13	1002	- 1	4	91/73	89/72	86/71	20	76	75	73	WNW 11	SW	96.6	-4.5
Defiance	41 17	84 23	700	- 1	4	94/74	91/73	88/72	24	77	76	74				
Findlay AP	41 01	83 40	804	2	3	92/74	90/73	87/72	24	77	76	74			97.4	-7.4
Fremont	41 20	83 07	600	- 3	1	90/73	88/73	85/71	24	76	75	73				
Hamilton	39 24	84 35	650	0	5	92/73	90/72	87/71	22	76	75	73			98.2	-2.8
Lancaster	39 44	82 38	860	0	5	93/74	91/73	88/72	23	77	75	74				
Lima	40 42	84 02	975	- 1	4	94/74	91/73	88/72	24	77	76	74	WNW 11	SW	96.0	-6.5
Mansfield AP	40 49	82 31	1295	0	5	90/73	87/72	85/72	22	76	74	73	W 8	SW	93.8	-10.7
Marion	40 36	83 10	920	0	5	93/74	91/73	88/72	23	77	76	74				
Middletown	39 31	84 25	635	0	5	92/73	90/72	87/71	22	76	75	73				
Newark	40 01	82 28	880	- 1	5	94/73	92/73	89/72	23	77	75	74	W 8	SSW	95.8	-6.8
Norwalk	41 16	82 37	670	- 3	1	90/73	88/73	85/71	22	76	75	73			97.3	-8.3
Portsmouth	38 45	82 55	540	5	10	95/76	92/74	89/73	22	78	77	75	W 8	SW	97.9	1.0
Sandusky Co	41 27	82 43	606	1	6	93/73	91/72	88/71	21	76	74	73			96.7	-1.9
Springfield	39 50	83 50	1052	- 1	3	91/74	89/73	87/72	21	77	76	74	W 7	W		
Steubenville	40 23	80 38	992	1	5	89/72	86/71	84/70	22	74	73	72				
Toledo AP	41 36	83 48	669	- 3	1	90/73	88/73	85/71	25	76	75	73	WSW 8	SW	95.4	-5.2
Warren	41 20	80 51	928	0	5	89/71	87/71	85/70	23	74	73	71				
Wooster	40 47	81 55	1020	1	6	89/72	86/71	84/70	22	75	73	72			94.0	-7.7
Youngstown AP	41 16	80 40	1178	- 1	4	88/71	86/71	84/70	23	74	73	71	SW 10	SW		

Zanesville AP	39 57	81 54	900	1	7	93/75	90/74	87/73	23	78	76	75	W 6	WSW		
OKLAHOMA																
Ada	34 47	96 41	1015	10	14	100/74	97/74	95/74	23	77	76	75				
Altus AFB	34 39	99 16	1378	11	16	102/73	100/73	98/73	25	77	76	75	N 10	S		
Ardmore	34 18	97 01	771	13	17	100/74	98/74	95/74	23	77	77	76				
Bartlesville	36 45	96 00	715	6	10	101/73	98/74	95/74	23	77	77	76				
Chickasha	35 03	97 55	1085	10	14	101/74	98/74	95/74	24	78	77	76				
Enid, Vance AFB	36 21	97 55	1307	9	13	103/74	100/74	97/74	24	79	77	76				
Lawton AP	34 34	98 25	1096	12	16	101/74	99/74	96/74	24	78	77	76				
McAlester	34 50	95 55	776	14	19	99/74	96/74	93/74	23	77	76	75	N 10	S		
Muskogee AP	35 40	95 22	610	10	15	101/74	98/75	95/75	23	79	78	77				
Norman	35 15	97 29	1181	9	13	99/74	96/74	94/74	24	77	76	75	N 10	S		
Oklahoma City AP	35 24	97 36	1285	9	13	100/74	97/74	95/73	23	78	77	76	N 14	SSW		
Ponca City	36 44	97 06	997	5	9	100/74	97/74	94/74	24	77	76	76				
Seminole	35 14	96 40	865	11	15	99/74	96/74	94/73	23	77	76	75				
Stillwater	36 10	97 05	984	8	13	100/74	96/74	93/74	24	77	76	75	N 12	SSW	103.7	1.6
Tulsa AP	36 12	95 54	650	8	13	101/74	98/75	95/75	22	79	78	77	N 11	SSW		
Woodward	36 36	99 31	2165	6	10	100/73	97/73	94/73	26	78	76	75			107.1	-1.3
OREGON																
Albany	44 38	123 07	230	18	22	92/67	89/66	86/65	31	69	67	66			97.5	16.6
Astoria AP	46 09	123 53	8	25	29	75/65	71/62	68/61	16	65	63	62	ESE 7	NNW		
Baker AP	44 50	117 49	3372	- 1	6	92/63	89/61	86/60	30	65	63	61			97.5	-6.8
Bend	44 04	121 19	3595	- 3	4	90/62	87/60	84/59	33	64	62	60			96.4	-5.8
Corvallis	44 30	123 17	246	18	22	92/67	89/66	86/65	31	69	67	66	N 6	N	98.5	17.1
Eugene AP	44 07	123 13	359	17	22	92/67	89/66	86/65	31	69	67	66	N 7	N		
Grants Pass	42 26	123 19	925	20	24	99/69	96/68	93/67	33	71	69	66	N 5	N	103.6	16.4
Klamath Falls AP	42 09	121 44	4092	4	9	90/61	87/60	84/59	36	63	61	60	N 4	W	96.3	0.9
Medford AP	42 22	122 52	1298	19	23	98/68	94/67	91/66	35	70	68	67	S 4	WMW	103.8	15.0
Pendleton AP	45 41	118 51	1482	- 2	5	97/65	93/64	90/62	29	66	65	63	NNW 6	WNW		
Portland AP	45 36	122 36	21	17	23	89/68	85/67	81/65	23	69	67	66	ESE 12	NW	96.6	18.3
Portland Co	45 32	122 40	75	18	24	90/68	86/67	82/65	21	69	67	66			97.6	20.5

APPENDIX A. (Continued)

Col. 1	Col. 2	Col. 3	Col. 4	Col. 5 Winter,[b] °F		Col. 6 Summer,[c] °F			Col.7	Col. 8			Col. 9 Prevailing Wind		Col.10 Temp., °F	
State and Station[a]	Lat.	Long.	Elev.	Design Dry-Bulb		Design Dry-Bulb and Mean Coincident Wet-Bulb			Mean Daily	Design Wet-bulb			Winter	Summer	Median of Annual Extr.	
	° ′N	° ′W	Feet	99%	97.5%	1%	2.5%	5%	Range	1%	2.5%	5%	Knots[4]		Max.	Min.
Roseburg AP	43 14	123 22	525	18	23	93/67	90/66	87/65	30	69	67	66			99.6	19.5
Salem AP	44 55	123 01	196	18	23	92/68	88/66	84/65	31	69	68	66	N 6	N	98.9	15.9
The Dalles	45 36	121 12	100	13	19	93/69	89/68	85/66	28	70	68	67			105.1	7.9
PENNSYLVANIA																
Allentown AP	40 39	75 26	387	4	9	92/73	88/72	86/72	22	76	75	73	W 11	SW		
Altoona Co	40 18	78 19	1504	0	5	90/72	87/71	84/70	23	74	73	72	WNW 11	WSW	93.7	-5.2
Butler	40 52	79 54	1100	1	6	90/73	87/72	85/71	22	75	74	73				
Chambersburg	39 56	77 38	640	4	8	93/75	90/74	87/73	23	77	76	75			97.4	-0.3
Erie AP	42 05	80 11	731	4	9	88/73	85/72	83/71	18	75	74	72	SSW 0	WSW	91.3	-2.2
Harrisburg AP	40 12	76 46	308	7	11	94/75	91/74	88/73	21	77	76	75	NW 11	WSW	96.5	3.7
Johnstown	40 19	78 50	2284	- 3	2	86/70	83/70	80/68	23	72	71	70	WNW 8	WSW	96.4	-1.8
Lancaster	40 07	76 18	403	4	8	93/75	90/74	87/73	22	77	76	75	NW 11	WSW		
Meadville	41 38	80 10	1065	0	4	88/71	85/70	83/69	21	73	72	71			93.2	-8.5
New Castle	41 01	80 22	825	2	7	91/73	88/72	86/71	23	75	74	73	WSW 10	WSW	94.7	-6.4
Philadelphia AP	39 53	75 15	5	10	14	93/75	90/74	87/72	21	77	76	75	WNW 10	WSW	96.4	5.9
Pittsburgh AP	40 30	80 13	1137	1	5	89/72	86/71	84/70	22	74	73	72	WSW 10	WSW		
Pittsburgh Co	40 27	80 00	1017	3	7	91/72	88/71	86/70	19	74	73	72			94.6	-1.1
Reading Co	40 20	75 38	266	9	23	92/73	89/72	86/72	19	76	75	73	W 11	SW	97.0	3.6
Scranton/ Wilkes-Barre	41 20	75 44	930	1	5	90/72	87/71	84/70	19	74	73	72	SW 8	WSW	94.8	-2.2
State College	40 48	77 52	1175	3	7	90/72	87/71	84/70	23	74	73	72	NNW 8	WSW	93.2	-3.6
Sunbury	40 53	76 46	446	2	7	92/73	89/72	86/70	22	75	74	73				
Uniontown	39 55	79 43	956	5	9	91/74	88/73	85/72	22	76	75	74			93.9	-2.5

Warren	42 51	79 08	2280	- 2	4	89/71	86/71	83/70	24	74	73	72			93.3	-10.7
West Chester	39 58	75 38	450	9	13	92/75	89/74	86/72	20	77	76	75				
Williamport AP	41 15	76 55	524	2	7	97/73	89/72	86/70	23	75	74	73	W 9	WSW	95.5	-4.6
York	39 55	76 45	390	8	12	94/75	91/74	88/73	22	77	76	73			97.0	-2.4
RHODE ISLAND																
Newport	42 30	71 20	10	5	9	88/73	85/72	82/70	16	76	75	73	WNW 10	SW		
Providence AP	41 44	71 26	51	5	9	89/73	86/72	83/70	19	75	74	73	WNW 11	SW	94.6	-0.5
SOUTH CAROLINA																
Anderson	34 30	82 43	774	19	23	94/74	92/74	90/74	21	77	76	75			99.5	13.3
Charleston AFB	32 54	80 02	45	24	27	93/78	91/78	89/77	18	81	80	79	NNE 9	SW		
Charleston Co	32 54	79 58	3	25	28	94/78	92/78	90/77	13	81	80	79			97.8	21.4
Columbia AP	33 57	81 07	213	20	24	97/76	95/75	93/75	22	79	78	77	W 6	SW	100.6	16.2
Florence AP	34 11	79 43	47	22	25	94/77	92/77	90/76	21	80	79	78	N 7	SW	99.5	16.5
Georgetown	33 23	79 17	14	23	26	92/79	90/78	88/77	18	81	80	79	N 7	SSW	98.2	19.1
Greenville AP	34 54	82 13	957	18	22	93/74	91/74	89/74	21	77	76	75	NW 8	SW	97.3	12.6
Greenwood	34 10	82 07	620	18	22	95/75	93/74	91/74	21	78	77	76			99.5	14.1
Orangeburg	33 30	80 52	260	20	24	97/76	95/75	93/75	20	79	78	77			101.2	18.0
Rock Hill	34 59	80 58	470	19	23	96/75	94/74	92/74	20	78	77	76				
Spartanburg AP	34 58	82 00	823	18	22	93/74	91/74	89/74	20	77	76	75			99.5	13.9
Sumter, Shaw AFB	33 54	80 22	169	22	25	95/77	92/76	90/75	21	79	78	77	NNE 6	W	100.0	15.4
SOUTH DAKOTA																
Aberdeen AP	45 27	98 26	1296	-19	-15	94/33	91/72	88/70	27	77	75	73	NNW 8	S	102.3	-28.1
Brookings	44 18	96 48	1637	-17	-13	95/73	92/72	89/71	25	77	75	73			97.8	-26.5
Huron AP	44 23	98 13	1281	-18	-14	96/73	93/72	90/71	28	77	75	73	NNW 8	S	101.5	-25.8
Mitchell	43 41	98 01	1346	-15	-10	96/72	93/71	90/70	28	76	75	73			103.0	-22.7
Pierre AP	44 23	100 17	1742	-15	-10	99/71	95/71	92/69	29	75	74	72	NW 11	SSE	105.7	-20.6
Rapid City AP	44 03	103 04	3162	-11	- 7	95/66	92/65	89/65	28	71	69	67	NNW 10	SSE	100.9	-19.0
Sioux Falls AP	43 34	96 44	1418	-15	-11	94/73	91/72	88/71	24	76	75	73	NW 8	S		
Watertown AP	44 55	97 09	1738	-19	-15	94/73	91/72	88/71	26	76	75	73			97.8	-26.5

APPENDIX A. (Continued)

Col. 1	Col. 2	Col. 3	Col. 4	Col. 5 Winter,[b] °F		Col. 6 Summer,[c] °F			Col.7	Col. 8			Col. 9 Prevailing Wind		Col.10 Temp., °F	
State and Station[a]	Lat.	Long.	Elev.	Design Dry-Bulb		Design Dry-Bulb and Mean Coincident Wet-Bulb			Mean Daily	Design Wet-bulb			Winter	Summer	Median of Annual Extr.	
	° 'N	° 'W	Feet	99%	97.5%	1%	2.5%	5%	Range	1%	2.5%	5%	Knots[4]		Max.	Min.
Yankton	42 55	97 23	1302	-13	- 7	94/73	91/72	88/71	25	77	76	74			100.8	-19.1
TENNESSEE																
Athens	35 26	84 35	940	13	18	95/74	92/73	90/73	22	77	76	75				
Bristol-Tri City AP	36 29	82 24	1507	9	14	91/72	89/72	87/71	22	75	75	73	WNW 6	SW		
Chattanooga AP	35 02	85 12	665	13	18	96/75	93/74	91/74	22	78	77	76	NNW 8	WSW	97.2	9.8
Clarksville	36 33	87 22	382	6	12	95/76	93/74	90/74	21	78	77	76			99.8	3.7
Columbia	35 38	87 02	690	10	15	97/75	94/74	91/74	21	78	77	76				
Dyersburg	36 01	89 24	344	10	15	96/78	94/77	91/76	21	81	80	78				
Greeneville	36 04	82 50	1319	11	16	92/73	90/72	88/72	22	76	75	74			95.6	0.8
Jackson AP	35 36	88 55	423	11	16	98/76	95/75	92/75	21	79	78	77			99.2	6.6
Knoxville AP	35 49	83 59	980	13	19	94/74	92/73	90/73	21	77	76	75	NE 8	W	96.0	7.0
Memphis AP	35 03	90 00	258	13	18	98/77	95/76	93/76	21	80	79	78	N 10	SW	97.9	10.4
Murfreesboro	34 55	86 28	600	9	14	97/75	94/74	91/74	22	78	77	76			97.7	4.5
Nashville AP	36 07	86 41	590	9	14	91/75	94/74	91/74	21	78	77	76	NW 8	WSW		
Tullahoma	35 23	86 05	1067	8	13	96/74	93/73	91/73	22	77	76	75	NW 8	WSW	96.7	3.7
TEXAS																
Abilene AP	32 25	99 41	1784	15	20	101/71	99/71	97/71	22	75	74	74	N 12	SSE	103.6	10.4
Alice AP	27 44	98 02	180	31	34	100/78	98/77	95/77	20	82	81	79			104.9	24.8
Amarillo AP	35 14	101 42	3604	6	11	98/67	95/67	93/67	26	71	70	70	N 11	S	100.8	0.9
Austin AP	30 18	97 42	597	24	28	100/74	98/74	97/74	22	78	77	77	N 11	S	101.6	19.7
Bay City	29 00	95 58	50	29	33	96/77	94/77	92/77	16	80	79	79				

Beaumont	29 57	94 01	16	27	31	95/79	93/78	91/78	19	81	80	80			99.7	23.5
Beeville	28 22	97 40	190	30	33	99/78	97/77	95/77	18	82	81	79	N 9	SSE	103.1	22.5
Big Spring AP	32 18	101 27	2598	16	20	100/69	97/69	95/69	26	74	73	72			105.3	10.7
Brownsville AP	25 54	97 26	19	35	39	94/77	93/77	92/77	18	80	79	79	NNW 13	SE	98.1	30.1
Brownwood	31 48	98 57	1386	18	22	101/73	99/73	96/73	22	77	76	75	N 9	S	105.3	13.0
Bryan AP	30 40	96 33	276	24	29	98/76	96/76	94/76	20	79	78	78				
Corpus Christi AP	27 46	97 30	41	31	35	95/78	94/78	92/78	19	80	80	79	N 12	SSE	97.0	27.2
Corsicana	32 05	96 28	425	20	25	100/75	98/75	96/75	21	79	78	77			104.2	15.2
Dallas AP	32 15	96 51	481	18	22	102/75	100/75	97/75	20	78	78	77	N 11	S		
Del Rio, Laughlin AFB	29 22	100 47	1081	26	31	100/73	98/73	97/73	24	79	77	76			103.8	23.0
Denton	33 12	97 06	630	17	22	101/74	99/74	97/74	22	78	77	76			104.5	
Eagle Pass	28 52	100 32	884	27	32	101/73	99/73	98/73	24	78	78	77	NNW 9	ESE	107.7	22.1
El Paso AP	31 48	106 24	3918	20	24	100/64	98/64	96/64	27	69	68	68	N 7	S	103.0	15.7
Fort Worth AP	32 50	97 03	537	17	22	101/74	99/74	97/74	22	78	77	76	NW 11	S	103.2	13.5
Galveston AP	29 18	94 48	7	31	36	90/79	89/79	88/78	10	81	80	80	N 15	S	93.9	27.5
Greenville	33 04	96 03	535	17	22	101/74	99/74	97/74	21	78	77	76				103.6
Harlingen	26 14	97 39	35	35	39	96/77	94/77	93/77	19	80	79	79	NNW 10	SSE	102.3	29.3
Houston AP	29 38	95 21	96	27	32	96/77	94/77	92/77	18	80	79	79	NNW 11	S	r	
Houston Co	29 59	95 22	108	28	33	97/77	95/77	93/77	18	80	79	79				99.0
Huntsville	30 43	95 33	494	22	27	100/75	98/75	96/75	20	78	78	77			100.8	18.7
Killeen, Robert Gray AAF	31 05	97 41	850	20	25	99/73	97/73	95/73	22	77	76	75	r			Min.
Lamesa	32 42	101 56	2965	13	17	99/69	96/69	94/69	26	73	72	71			105.5	
Laredo AFB	27 32	99 27	512	32	36	102/73	101/73	99/74	23	78	78	77	N 8	SE	r	
Longview	32 28	94 44	330	19	24	99/76	97/76	95/76	20	80	79	78	r			Min.
Lubbock AP	33 39	101 49	3254	10	15	98/69	96/69	94/69	26	73	72	71	NNE 10	SSE	r	
Lufkin AP	31 25	94 48	277	25	29	99/76	97/76	94/76	20	80	79	78	NNW 12	S		
McAllen	26 12	98 13	122	35	39	97/77	95/77	44/77	21	80	79	79				
Midland AP	31 57	102 11	2851	16	21	100/69	98/69	96/69	26	73	72	71	NE 9	SSE	103.6	10.8
Mineral Wells AP	32 47	98 04	930	17	22	101/74	99/74	97/74	22	78	77	76				
Palestine Co	31 47	95 38	600	23	27	100/76	98/76	96/76	20	79	79	78			101.2	16.3
Pampa	35 32	100 59	3250	7	12	99/67	96/67	94/67	26	71	70	70				

APPENDIX A. (Continued)

Col. 1	Col. 2	Col. 3	Col. 4	Col. 5 Winter,[b] °F		Col. 6 Summer,[c] °F			Col.7	Col. 8			Col. 9 Prevailing Wind		Col.10 Temp., °F	
State and Station[a]	Lat.	Long.	Elev.	Design Dry-Bulb		Design Dry-Bulb and Mean Coincident Wet-Bulb			Mean Daily	Design Wet-bulb			Winter	Summer	Median of Annual Extr.	
	° ′N	° ′W	Feet	99%	97.5%	1%	2.5%	5%	Range	1%	2.5%	5%	Knots[4]		Max.	Min.
Pecos	31 35	103 30	2610	16	21	100/69	98/69	96/69	27	73	72	71				
Plainview	34 11	101 42	3370	8	13	98/68	96/68	94/68	26	72	71	70			102.7	3.1
Port Arthur AP	29 57	94 01	16	27	31	95/79	93/78	91/78	19	81	80	80	N 9	S	97.7	24.0
San Angelo, Goodfellow AFB	31 26	100 24	1877	18	22	101/71	99/71	97/70	24	75	74	73	NNE 10	SSE		
San Antonio AP	29 32	98 28	788	25	30	99/72	97/73	96/73	19	77	76	76	N 8	SSE	101.3	21.1
Sherman, Perrin AFB	33 43	96 40	763	15	20	100/75	98/75	95/74	22	78	77	76	N 10	S	103.0	11.9
Snyder	32 43	100 55	2325	13	18	100/70	98/70	96/70	26	74	73	72				
Temple	31 06	97 21	700	22	27	100/74	99/74	97/74	22	78	77	77				
Tyler AP	32 21	95 16	530	19	24	99/76	97/76	95/76	21	80	79	78	NNE 23	S		
Vernon	34 10	99 18	1212	13	17	102/73	100/75	97/73	24	77	76	75				
Victoria AP	28 51	96 55	104	29	32	98/78	96/77	94/77	18	82	81	79			101.4	23.4
Waco AP	31 37	97 13	500	21	26	101/75	99/75	97/75	22	78	78	77				
Wichita Falls AP	33 58	98 29	994	14	18	103/73	101/73	98/73	24	77	76	75	NNW 12	S		
UTAH																
Cedar City AP	37 42	113 06	5617	- 2	5	93/60	91/60	89/59	32	65	63	62	SE 5	SW		
Logan	41 45	111 49	4785	- 3	2	93/62	91/61	88/60	33	65	64	63			95.5	-7.8
Moab	38 36	109 36	3965	6	11	100/60	98/60	96/60	30	65	64	63				
Ogden AP	41 12	112 01	4455	1	5	93/63	91/61	88/61	33	66	65	64	S 6	SW	99.5	-3.9
Price	39 37	110 50	5580	- 2	5	93/60	91/60	89/59	33	65	63	62				
Provo	40 13	111 43	4448	1	6	98/62	96/62	94/61	32	66	65	64	SE 5	SW		
Richfield	38 46	112 05	5270	- 2	5	93/60	91/60	89/59	34	65	63	62			98.1	-10.5

St George Co	37 02	113 31	2900	14	21	103/65	101/65	99/64	33	70	68	67			109.3	11.1
Salt Lake City AP	40 46	111 58	4220	3	8	97/62	95/62	92/61	32	66	65	64	SSE 6	N	99.4	-0.1
Vernal AP	40 27	109 31	5280	- 5	0	91/61	89/60	86/59	32	64	63	62				
VERMONT																
Barre	44 12	72 31	600	- 16	-11	84/71	81/69	78/68	23	73	71	70				
Burlington AP	44 28	73 09	332	- 12	- 7	88/72	85/70	82/69	23	74	72	71	E 7	SSW	92.4	-16.9
Rutland	43 36	72 58	620	-13	- 8	87/72	84/70	81/69	23	74	72	71			92.5	-17.5
VIRGINIA																
Charlottesville	38 02	78 31	870	14	18	94/74	91/74	88/73	23	77	76	75	NE 7	SW	97.4	8.0
Danville AP	36 34	79 20	590	14	16	94/74	92/73	90/73	21	77	76	75			100.1	9.2
Fredericksburg	38 18	77 28	100	10	14	96/76	93/75	90/74	21	78	77	76				
Harrisonburg	38 27	78 54	1370	12	16	93/72	91/72	88/71	23	75	74	73				
Lynchburg AP	37 20	79 12	916	12	16	93/74	90/74	88/73	21	77	76	75	NE 7	SW	97.2	7.6
Norfolk AP	36 54	76 12	22	20	22	93/77	91/76	89/76	18	79	78	77	NW 10	SW	97.2	15.3
Petersburg	37 11	77 31	194	14	17	95/76	92/76	90/75	20	79	78	77				
Richmond AP	37 30	77 20	164	14	17	95/76	92/76	90/75	21	79	78	77	N 6	SW	97.9	9.6
Roanoke AP	37 19	79 58	1193	12	16	93/72	91/72	88/71	23	75	74	73	NW 9	SW		
Staunton	38 16	78 54	1201	12	16	93/72	91/72	88/71	23	75	74	73	NW 9	SW	95.9	2.5
Winchester	39 12	78 10	760	6	10	93/75	90/74	88/74	21	77	76	75			97.3	3.7
WASHINGTON																
Aberdeen	46 59	12 49	12	25	28	80/65	77/62	73/61	16	65	63	62	ESE 6	NNW	91.9	19.3
Bellingham AP	48 48	12 32	158	10	15	81/67	77/65	74/63	19	68	65	63	NNE 15	WSW	87.4	10.3
Bremerton	47 34	12 40	162	21	25	82/65	78/64	75/62	20	66	64	63	E 8	N		
Ellensburg AP	47 02	12 31	1735	2	6	94/65	91/64	87/62	34	66	65	63				
Everett, Paine AFB	47 55	12 17	596	21	25	80/65	76/64	73/62	20	67	64	63	ESE 6	NNW	84.9	15.2
Kennewick	46 13	11 08	392	5	11	99/68	96/67	92/66	30	70	68	67			103.4	2.0
Longview	46 10	12 56	12	19	24	88/68	85/67	81/65	30	69	67	66	ESE 9	NW	96.0	14.8
Moses Lake, Larson AFB	47 12	11 19	1185	1	7	97/66	94/65	90/63	32	67	66	64	N 8	SSW		

APPENDIX A. (Continued)

Col. 1	Col. 2	Col. 3	Col. 4	Col. 5 Winter,[b] °F		Col. 6 Summer,[c] °F			Col.7	Col. 8			Col. 9 Prevailing Wind		Col.10 Temp., °F	
State and Station[a]	Lat.	Long.	Elev.	Design Dry-Bulb		Design Dry-Bulb and Mean Coincident Wet-Bulb			Mean Daily	Design Wet-bulb			Winter	Summer	Median of Annual Extr.	
	°′N	°′W	Feet	99%	97.5%	1%	2.5%	5%	Range	1%	2.5%	5%	Knots[4]		Max.	Min.
Olympia AP	46 58	12 54	215	16	22	87/66	83/65	79/64	32	67	66	64	NE 4	NE		
Port Angeles	48 07	12 26	99	24	27	72/62	69/61	67/60	18	64	62	61			83.5	19.4
Seattle-Boeing Field	47 32	12 18	23	21	26	84/68	81/66	77/65	24	69	67	65				
Seattle Co	47 39	12 18	20	22	27	85/68	82/66	78/65	19	69	67	65	N 7	N	90.2	22.0
Seattle-Tacoma AP	47 27	12 18	400	21	26	84/65	80/64	76/62	22	66	64	63	E 9	N	90.1	19.9
Spokane AP	47 38	11 31	2357	- 6	2	93/64	90/63	87/62	28	65	64	62	NE 6	SW	98.8	-4.9
Tacoma, McChord AFB	47 15	12 30	100	19	24	86/66	82/65	79/63	22	68	66	64	S 5	NNE	89.4	18.8
Walla Walla AP	46 06	11 17	1206	0	7	97/67	94/66	90/65	27	69	67	66	W 5	W	103.0	3.8
Wenatchee	47 25	12 19	632	7	11	99/67	96/66	92/64	32	68	67	65			101.1	1.0
Yakima AP	46 34	12 32	1052	- 2	5	96/65	93/65	89/63	36	68	66	65	W 5	NW		
WEST VIRGINIA																
Beckley	37 47	81 07	2504	- 2	4	83/71	81/69	79/69	22	73	71	70	WNW 9	WNW		
Bluefield AP	37 18	81 13	2867	- 2	4	83/71	81/69	79/69	22	73	71	70				
Charleston AP	38 22	81 36	939	7	11	92/74	90/73	87/72	20	76	75	74	SW 8	SW	97.2	2.9
Clarksburg	39 16	80 21	977	6	10	92/74	90/73	87/72	21	76	75	74				
Elkins AP	38 53	79 51	1948	1	6	86/72	84/70	82/70	22	74	72	71	WNW 9	WNW	90.6	-7.3
Huntington Co	38 25	82 30	565	5	10	94/76	91/74	89/73	22	78	77	75	W 6	SW	97.1	2.1
Martinsburg AP	39 24	77 59	556	6	10	93/75	90/74	88/74	21	77	76	75	WNW 10	W	99.0	1.1
Morgantown AP	39 39	79 55	1240	4	8	90/74	87/73	85/73	22	76	75	74				
Parkersburg Co	39 16	81 34	615	7	11	93/75	90/74	88/73	21	77	76	75	WSW 7	WSW	95.9	0.7

Wheeling	40 07	80 42	665	1	5	89/72	86/71	84/70	21	74	73	72	WSW 10	WSW	97.5	-0.6
WISCONSIN																
Appleton	44 15	88 23	730	-14	- 9	89/74	86/72	83/71	23	76	74	72			94.6	-16.2
Ashland	46 34	90 58	650	-21	-16	85/70	82/68	79/66	23	72	70	68			94.1	-26.8
Beloit	42 30	89 02	780	- 7	- 3	92/75	90/75	88/74	24	78	77	75				
Eau Claire AP	44 52	91 29	888	-15	-11	92/75	89/73	86/71	23	77	75	73				
Fond Du Lac	43 48	88 27	760	-12	- 8	89/74	86/72	84/71	23	76	74	72			96.0	-17.7
Green Bay AP	44 29	88 08	682	-13	- 9	88/74	85/72	83/71	23	76	74	72	W 8	SW	94.3	-17.9
La Crosse AP	43 52	91 15	651	-13	- 9	91/75	88/73	85/72	22	77	75	74	NW 10	S	95.7	-21.3
Madison AP	43 08	89 20	858	-11	- 7	91/74	88/73	85/71	22	77	75	73	NW 8	SW	93.6	-16.8
Manitowoc	44 06	87 41	660	-11	- 7	89/74	86/72	83/71	21	76	74	72			94.1	-13.7
Marinette	45 06	87 38	605	-15	-11	87/73	84/71	82/70	20	75	73	71			95.9	-15.8
Milwaukee AP	42 57	87 54	672	- 8	- 4	90/74	87/73	84/71	21	76	74	73	WNW 10	SSW		
Racine	42 43	87 51	730	- 6	- 2	91/75	88/73	85/72	21	77	75	74				
Sheboygan	43 45	87 43	648	-10	- 6	89/75	86/73	83/72	20	77	75	74			97.0	-12.4
Stevens Point	44 30	89 34	1079	-15	-11	92/75	89/73	86/71	23	77	75	73			95.3	-24.1
Waukesha	43 01	88 14	860	- 9	- 5	90/74	87/73	84/71	22	76	74	73			95.7	-14.3
Wausau AP	44 55	89 37	1196	-16	-12	91/74	88/72	85/70	23	76	74	72				
WYOMING																
Casper AP	42 55	10 28	5338	-11	- 5	92/58	90/57	87/57	31	63	61	60	NE 10	SW	97.3	-20.9
Cheyenne	41 09	10 49	6126	- 9	- 1	89/58	86/58	84/57	30	63	62	60	N 11	WNW	92.5	-15.9
Cody AP	44 33	10 04	4990	-19	-13	89/60	86/60	83/59	32	64	63	61			97.4	-21.9
Evanston	41 16	11 57	6780	- 9	- 3	86/55	84/55	82/54	32	59	58	57			89.2	-21.2
Lander AP	42 49	10 44	5563	-16	-11	91/61	88/61	85/60	32	64	63	61	E 5	NW	94.9	-22.6
Laramie AP	41 19	10 41	7266	-14	- 6	84/56	81/56	79/55	28	61	60	59				
Newcastle	43 51	10 13	4265	-17	-12	91/64	87/63	84/63	30	69	68	66			99.4	-19.0
Rawlins	41 48	10 12	6740	-12	- 4	86/57	83/57	81/56	40	62	61	60				
Rock Springs AP	41 36	10 04	6745	- 9	- 3	86/55	84/55	82/54	32	59	58	57	WSW 10	W		
Sheridan AP	44 46	10 58	3964	-14	- 8	94/62	91/62	88/61	32	66	65	63	NW 7	N	99.8	-23.6
Torrington	42 05	10 13	4098	-14	- 8	94/62	91/62	88/61	30	66	65	63			101.1	-20.7

APPENDIX A. (Continued)

				Winter,[b] °F		Summer,[c] °F							Prevailing Wind		Temp., °F	
Col. 1	Col. 2	Col. 3	Col. 4	Col. 5		Col. 6			Col.7	Col. 8			Col. 9		Col.10	
State and Station[a]	Lat.	Long.	Elev.	Design Dry-Bulb		Design Dry-Bulb and Mean Coincident Wet-Bulb			Mean Daily	Design Wet-bulb			Winter	Summer	Median of Annual Extr.	
	° ′N	° ′W	Feet	99%	97.5%	1%	2.5%	5%	Range	1%	2.5%	5%	Knots[4]		Max.	Min.
ALBERTA																
Calgary AP	51 06	114	01	3540	-27	-23	84/63	81/61	79/60		25		65 63	62	NNW	8SE
Edmonton AP	53 34	113	31	2219	-29	-25	85/66	82/65	79/63		23		68 66	65	E 9	SE
Grande Prairie AP	55 11	118	53	2190	-39	-33	83/64	80/63	78/61		23		66 64	62		
Jasper	52 53	118	04	3480	-31	-26	83/64	80/62	77/61		28		66 64	63		
Lethbridge AP	49 38	112	48	3018	-27	-22	90/65	87/64	84/63		28		68 66	65		
McMurray AP	56 39	111	13	1216	-41	-38	86/67	82/65	79/64		26		69 67	65		
Medicine Hat AP	50 01	110	43	2365	-29	-24	93/66	90/65	87/64		28		70 68	66		
Red Deer AP	52 11	113	54	2965	-31	-26	84/65	81/64	78/62		25		67 66	64		
BRITISH COLUMBIA																
Dawson Creek	55 44	120	11	2164	-37	-33	82/64	79/63	76/61		26		66 64	62		
Fort Nelson AP	58 50	122	35	1230	-43	-40	84/64	81/63	78/62		23		67 65	64		
Kamloops Co	50 43	120	25	1133	-21	-15	94/66	91/65	88/64		29		68 66	65		
Nanaimo	49 11	123	58	230	16	20	83/67	80/65	77/64		21		68 66	65		
New Westminister	49 13	122	54	50	14	18	84/68	81/67	78/66		19		69 68	66		
Penticton AP	49 28	119	36	1121	0	4	92/68	89/67	87/66		31		70 68	67		
Prince George AP	53 53	122	41	2218	-33	-28	84/64	80/62	77/61		26		66 64	62	N 11	N
Prince Rupert Co	54 17	130	23	170	- 2	2	64/59	63/57	61/56		12		60 58	57		
Trail	49 08	117	44	1400	- 5	0	92/66	89/65	86/64		33		68 67	65		
Vancouver AP	49 11	123	10	16	15	19	79/67	77/66	74/65		17		68 67	66	E 6	WNW
Victoria Co	48 25	123	19	228	20	23	77/64	73/62	70/60		16		64 62	60		

MANITOBA													
Brandon	49 52 99	59	1200	-30	-27	89/72	86/70	83/68	25	74	72	70	
Churchill AP	58 45 94	04	155	-41	-39	81/66	77/64	74/62	18	67	65	63	SE 11 S
Dauphin AP	51 06 100	03	999	-31	-28	87/71	84/70	81/68	23	74	72	70	
Flin Flon	54 46 101	51	1098	-41	-37	84/68	81/66	79/65	19	70	68	67	
Portage La Prairie AP	49 54 98	16	867	-28	-24	88/73	86/72	83/70	22	76	74	71	
The Pas AP	53 58 101	06	894	-37	-33	85/68	82/67	79/66	20	71	69	68	W 8 W
Winnipeg AP	49 54 97	14	786	-30	-27	89/73	86/71	84/70	22	75	73	71	W 8 S
NEW BRUNSWICK													
Campbellton Co	48 00 66	40	25	-8	-14	85/68	82/67	79/66	21	72	70	68	
Chatham AP	47 01 65	27	112	15	-10	89/69	85/68	82/67	22	72	71	69	
Edmundston Co	47 22 68	20	500	-21	-16	87/70	83/68	80/67	21	73	71	69	
Fredericton Ap	45 52 66	32	74	-16	-11	89/71	85/69	82/68	23	73	71	70	
Moncton AP	46 07 64	41	248	-12	- 8	85/70	82/69	79/67	23	72	71	69	
Saint John AP	45 19 65	53	352	-12	- 8	80/67	77/65	75/64	19	70	68	66	
NEWFOUNDLAND													
Corner Brook	48 58 57	57	15	- 5	0	76/64	73/63	71/62	17	67	66	65	
Gander AP	48 57 54	34	482	- 5	- 1	82/66	79/65	77/64	19	69	67	66	WNW 11SW
Goose Bay AP	53 19 60	25	144	-27	-24	85/66	81/64	77/63	19	68	66	64	N 9 SW
St John's AP	47 37 52	45	463	3	7	77/66	75/65	73/64	18	69	67	66	N 20 WSW
Stephenville AP	48 32 58	33	44	- 3	4	76/65	74/64	71/63	14	67	66	65	WNW 10S
NORTHWEST TERRITORIES													
Fort Smith AP	60 01 111	58	665	-49	-45	85/66	81/64	78/63	24	68	66	65	NW 4 S
Frobisher AP	63 45 68	33	68	-43	-41	66/53	63/51	59/50	14	54	52	51	NNW9NW
Inuvik	68 18 133	29	200	-56	-53	79/62	77/60	75/59	21	64	62	61	
Resolute AP	74 43 94	59	209	-50	-47	57/48	54/46	51/45	10	50	48	46	
Yellowknife AP	62 28 114	27	682	-49	-46	79/62	77/61	74/60	16	64	63	62	SSE 7 S
NOVA SCOTIA													
Amherst	45 49 64	13	65	-11	- 6	84/69	81/68	79/67	21	72 7068			

APPENDIX A. (Continued)

				Winter,[b] °F		Summer,[c] °F							Prevailing Wind		Temp., °F	
Col. 1	Col. 2	Col. 3	Col. 4	Col. 5		Col. 6			Col.7	Col. 8			Col. 9		Col.10	
State and Station[a]	Lat.	Long.	Elev.	Design Dry-Bulb		Design Dry-Bulb and Mean Coincident Wet-Bulb			Mean Daily	Design Wet-bulb			Winter	Summer	Median of Annual Extr.	
	° 'N	° 'W	Feet	99%	97.5%	1%	2.5%	5%	Range	1%	2.5%	5%	Knots[4]		Max.	Min.
Halifax AP	44 39	63	34	83	1	5	79/66	76/65	74/64		16	69	6766			
Kentville	45 03	64	36	40	- 3	1	85/69	83/68	80/67		22	72	7169			
New Glasgow	45 37	62	37	317	- 9	- 5	81/69	79/68	77/67		20	72	7069			
Sydney AP	46 10	60	03	197	- 1	3	82/69	80/68	77/66		19	71	7068			
Truro Co	45 22	63	16	131	- 8	- 5	82/70	80/69	78/68		22	73	7170			
Yarmouth AP	43 50	66	05	136	5	9	74/65	72/64	70/63		15	68	6665	NW 11	S	
ONTARIO																
Belleville	44 09	77	24	250	-11	- 7	86/73	84/72	82/71		20		75 74	73		
Chatham	42 24	82	12	600	0	3	89/74	87/73	85/72		19		76 75	74		
Cornwall	45 01	74	45	210	-13	- 9	89/73	87/72	84/71		21		75 74	72		
Hamilton	43 16	79	54	303	- 3	1	88/73	86/72	83/71		21		76 74	73		
Kapuskasing AP	49 25	82	28	752	-31	-28	86/70	83/69	80/67		23		72 70	69		
Kenora AP	49 48	94	22	1345	-32	-28	84/70	82/69	80/68		19		73 71	70		
Kingston	44 16	76	30	300	-11	- 7	87/73	84/72	82/71		20		75 74	73		
Kitchener	43 26	80	30	1125	- 6	- 2	88/73	85/72	83/71		23		75 74	72		
London AP	43 02	81	09	912	- 4	0	87/74	85/73	83/72		21		76 74	73		
North Bay AP	46 22	79	25	1210	-22	-18	84/68	81/67	79/66		20		71 70	68		
Oshawa	43 54	78	52	370	- 6	- 3	88/73	86/72	84/71		20		75 74	73		
Ottawa AP	45 19	75	40	413	-17	-13	90/72	87/71	84/70		21		75 73	72		
Owen Sound	44 34	80	55	597	- 6	- 2	84/71	82/70	80/69		21		73 72	70		
Peterborough	44 17	78	19	635	-13	- 9	87/72	85/71	83/70		21		75 73	72		
St Catharines	43 11	79	14	325	- 1	3	87/73	85/72	83/71		20		76 74	73		
Sarnia	42 58	82	22	625	0	3	88/73	86/72	84/71		19		76 74	73		

Sault Ste Marie AP	46	32	84	30	675	- 7	-13	85/71	82/69	79/68	22	73	71	70		
Sudbury AP	46	37	80	48	1121	-22	-19	86/69	83/67	81/66	22	72	70	68		
Thunder Bay AP	48	22	89	19	644	-27	-24	85/70	83/68	80/67	24	72	70	68	W 8	W
Timmins AP	48	34	81	22	965	-33	-29	87/69	84/68	81/66	25	72	70	68		
Toronto AP	43	41	79	38	578	- 5	- 1	90/73	87/72	85/71	20	75	74	73	N 10	SW
Windsor AP	42	16	82	58	637	0	4	90/74	88/73	86/72	20	77	75	74		
PRINCE EDWARD ISLAND																
Charlottetown AP	46	17	63	08	186	- 7	- 4	80/69	78/68	76/67	16	71	70	68		
Summerside AP	46	26	63	50	78	- 8	- 4	81/69	79/68	77/67	16	72	70	68		
QUEBEC																
Bagotville AP	48	20	71	00	536	-28	-23	87/70	83/68	80/67	21	72	70	68		
Chicoutimi	48	25	71	05	150	-26	-22	86/70	83/68	80/67	20	72	70	68		
Drummondville	45	53	72	29	270	-18	-14	88/72	85/71	82/69	21	75	73	71		
Granby	45	23	72	42	550	-19	- 14	88/72	85/71	53/70	21	75	73	72		
Hull	45	26	75	44	200	-18	-14	90/72	87/71	84/70	21	75	73	72		
Megantic AP	45	35	70	52	1362	-20	-16	86/71	83/70	81/69	20	74	72	71		
Montreal AP	45	28	73	45	98	-16	-10	88/73	85/72	83/71	17	75	74	72		
Quebec AP	46	48	71	23	245	-19	-14	87/72	84/70	81/68	20	74	72	70		
Rimouski	48	27	68	32	117	-16	-12	83/68	79/66	76/65	18	71	69	67		
St Jean	45	18	73	16	129	-15	-11	88/73	86/72	84/71	20	75	74	72		
St Jerome	45	48	74	01	556	-17	-13	88/72	86/71	83/70	23	75	73	72		
Sept.Iles AP	50	13	66	16	190	-26	-21	76/63	73/61	70/60	17	67	65	63		
Shawinigan	46	34	72	43	306	-18	-14	86/72	84/70	82/69	21	74	72	71		
Sherbrooke Co	45	24	71	54	595	-25	-21	86/72	84/71	81/69	20	74	73	71		
ThetFord Mines	46	04	71	19	1020	-19	-14	87/71	84/70	81/69	21	74	72	71		
Trois Rivieres	46	21	72	35	50	-17	-13	88/72	85/70	82/69	23	74	72	71		
Val D'or AP	48	03	77	47	1108	-32	-27	85/70	83/68	80/67	22	72	70	68		
Valleyfield	45	16	74	06	150	-14	-10	89/73	86/72	84/71	20	75	74	72		
SASKATCHEWAN																
Estevan AP	49	04	103	00	1884	-30	-25	92/70	89/68	86/67	26	72	70	69		

APPENDIX A. (Continued)

Col. 1	Col. 2	Col. 3	Col. 4	Col. 5 Winter,[b] °F		Col. 6 Summer,[c] °F			Col.7	Col. 8			Col. 9 Prevailing Wind		Col.10 Temp., °F	
State and Station[a]	Lat.	Long.	Elev.	Design Dry-Bulb		Design Dry-Bulb and Mean Coincident Wet-Bulb			Mean Daily	Design Wet-bulb			Winter	Summer	Median of Annual Extr.	
	° 'N	° 'W	Feet	99%	97.5%	1%	2.5%	5%	Range	1%	2.5%	5%	Knots[d]		Max.	Min.
Moose Jaw AP	50 20	105	33	1857	-29	-25	93/69	89/67	86/66		27		71 69	68		
North Battleford AP	52 46	108	15	1796	-33	-30	88/67	85/66	82/65		23		69 68	66		
Prince Albert AP	53 13	105	41	1414	-42	-35	87/67	84/66	81/65		25		70 68	67		
Regina AP	50 26	104	40	1884	-33	-29	91/69	88/68	84/67		26		72 70	68		
Saskatoon AP	52 10	106	41	1645	-35	-31	89/68	86/66	83/65		26		70 68	67		
Swift Current AP	50 17	107	41	2677	-28	-25	93/68	90/66	87/65		25		70 69	67		
Yorkton AP	51 16	102	28	1653	-35	-30	87/69	84/68	80/66		23		72 70	68		
YUKON TERRITORY																
Whitehorse AP	60 43	135	04	2289	-46	-43	80/59	77/58	74/56		22		61 59	58	NW 5 SE	

[a]AP, AFB. Following the station name designates airport or millitary airbase temperature observations. Co designates office locations within an urban area thar are affected by the surrounding area. Undersigned stations are semirural and may be compared to airport data.
[b]Winter design data are based on the 3-month period. December through February.
[c]Summer design data are based on the 4-month period, June through September.
[d]Mean wind speeds occurring coincidentally with the 99.59% dry-bulb winter design temperature.

Appendix B

STEAM TABLES

TABLE B.1 Properties of Saturated Steam Above Atmospheric Pressure

Gauge Pressure (psig)	Saturated Temperature (°F)	Specific Volume (ft^3/lb)	Heat Content (Btu/lb) Saturated Liquid	Heat Content (Btu/lb) Saturated Vapor	Latent Heat of Vaporization (Btu/lb)
0	212	26.8	180	1,150	970
1	215	24.3	183	1,151	967
2	218	23.0	186	1,153	965
3	222	21.8	190	1,154	963
4	224	20.7	193	1,155	961
5	227	19.8	195	1,156	959
6	230	18.9	198	1,157	958
7	232	18.1	200	1,158	956
8	235	17.4	203	1,158	955
9	237	16.7	205	1,159	953
10	239	16.1	208	1,160	952
12	244	15.0	212	1,161	949
15	250	13.6	218	1,164	945
18	255	12.5	224	1,165	941
20	259	11.1	227	1,166	939
25	267	10.4	236	1,169	933
30	274	9.4	243	1,171	926
40	287	7.7	256	1,175	919
50	298	6.6	267	1,178	911
60	307	5.8	277	1,181	903
75	320	4.9	290	1,184	893
100	338	3.9	308	1,190	882
125	353	3.2	324	1,193	867
150	366	2.8	337	1,196	858
175	378	2.4	350	1,198	848
200	388	2.1	361	1,199	838

Glossary

Absorption: The process in which refrigerant vapor is absorbed into a concentrated (strong) solution.

Acfm: The actual volume at specified conditions in ft^3/min (cfm).

Air Horsepower: The power developed by a fan in moving a given air rate against a given resistance.

Air Rate: Mass flow of dry air per square foot of cross-sectional area in the tower's heat transfer region per hour.

Air-to Fuel Ratio: The amount of air supplied in relationship to the amount of fuel supplied to a burner.

Alarm: An audible or visual signal from an energy management system that warns of an abnormal and critical operating condition.

Algorithm: A set of well-defined rules used in an energy management system to translate data into a desired system response.

Application Program: Custom designed software that allows an energy management system to perform the desired functions in a particular system.

Approach: Difference between the cold water temperature and either the ambient or entering wet bulb temperature.

ASCII: American Standard Code for information interchange. An 8-bit coded character set to be used for the general interchange of data among information processing systems, communications systems, process control systems and associated equipment.

Avoided Cost: Incremental cost for the electric utility to generate or purchase electricity that is avoided through the purchase of power.

Backup Power: Electric energy available from or to an electric utility during an unscheduled outage to replace energy ordinarily generated by the facility or the utility. Frequently referred to as standby power.

Ballast: A device used with fluorescent high intensity discharge lamps to provide the necessary circuit conditions for starting and operating.

Ballast Efficacy Factor: The ratio of the ballast factor expressed as a percentage of the power input, in watts, at specified test conditions.

Ballast Factor (BF): The ratio of a commercial ballast lamp lumen to a reference ballast lamp lumen, used to correct the lamp lumen output from rated to actual.

Baseload: Minimum electric or thermal load generated or supplied over one or more periods.

Baud: A unit of signal transmission speed representing the number of discrete signal events per second. For example, 300 baud = 30 characters per second.

Bit: An acronym for binary digit—the smallest unit of information that can be represented. A bit may be in one of two states, represented by the binary digits 0 and 1.

Bleed Off (Blowdown): The circulating water in the tower that is discharged to waste to help keep the dissolved solids concentration of the water below a maximum allowable limit.

Blowdown: Boiler water that is drained out of the boiler in order to maintain the desired dissolved solids in it.

Boiler Feedwater: Supplemental water fed into a boiler during operation to replenish the water that is converted to steam in it. Boiler feedwater can include condensate and treated make-up water.

Boiler Horsepower: The heat required to evaporate 34.5 pounds of water per hour at a temperature of 212°F into dry saturated steam at the same temperature; it is equal to 33,472 Btu/hr.

Boiler Rating: The heating capacity of a boiler expressed in boiler horsepower, Btu/hr or pounds of steam per hour.

Boiling Point: The temperature at which water begins to boil at a given pressure.

Bottoming Cycle: Cogeneration facility in which the energy input to the system is first applied to another thermal energy process; the rejected heat that emerges from the process is then used for power production.

Breakdown Torque: The maximum torque a motor will develop at rated voltage and frequency without an abrupt drop in speed.

Break-Even Temperature: The outdoor temperature at which the total heat losses from a conditioned space equal the internally generated heat gains.

Brine: A liquid used to transfer heat without changing phase, which is a mixture of water and other heat transfer liquids.

Burner: A device that mixes the fuel and air and provides a continuous source of ignition for the combustion process.

Bus: A circuit path (or parallel paths) over which data or instructions are transferred to all points in the computer system.

Byte: A group of eight bits.

Capacity: Load for which a generating unit, generation station or other electrical apparatus is rated.

Capacity of Compressor: The quantity of gas actually delivered when operating between specified inlet and discharge pressures. It is the volume of air measured at the conditions of pressure, temperature, gas composition and moisture content existing at the compressor inlet flange.

Cell: Smallest cooling tower subdivision, which can function as an independent unit with regard to air and water flow; it is bounded by exterior walls or partition walls.

Central Processing Unit (CPU): The communication, calculation and control center of an energy management system, it interprets data received from sensors, retrieves operation parameters from the systems memory, and transmits instructions to control points.

Charging: Storing cooling capacity by removing heat from a cool storage device, or storing heating capacity by adding heat to a heat storage device.
Charging Cycle: The period of time during which an energy storage system goes through a complete sequence of charging and discharging.
Chiller Priority: A control strategy for partial storage systems that uses the chiller to directly meet as much of the load as possible. Storage is used to supplement chiller operation only when the load exceeds the chiller capacity.
Coefficient of Performance (COP): An expression of the cycle efficiency of a refrigerating system; it is stated as the ratio of the heat absorbed in the refrigerated space to energy equivalent of the energy supplied to the compressor.
Cogeneration: Sequential production of electrical or mechanical energy and useful thermal energy from a single energy stream. To qualify as a cogeneration facility in the United States, a plant must meet certain efficiency standards.
Combined Cycle: A combination of a topping cycle and a bottoming cycle. Power is produced in a topping cycle, typically a gas turbine generator. Then, heat exhausted from the turbine is used to produce steam, which is subsequently expanded in a steam turbine to generate more electric or shaft power.
Combustion: The rapid oxidation of elements, which liberate heat during their combination with oxygen.
Combustion air: Air used for the combustion process.
Complete Combustion: A combustion process in which all hydrogen and carbon in the fuel are oxidized to H_2O and CO_2.
Compressibility: That property of a gas or a gas mixture that causes it to differ in volume from that of a perfect gas when each is under pressure and temperature conditions.
Compressibility Factor (Z): The ratio of the actual volume of the gas to the volume determined according to the perfect gas law.
Compression Efficiency: The ratio of the theoretical work requirement (using a stated process) to the actual work required to be done on the gas for compression and delivery.
Compression Ratio: The ratio of the absolute discharge pressure to the absolute intake pressure.
Compressor: A machine used to compress refrigerant vapor or air.
Computer: A programmable electronic device used for information storage, retrieval and processing; microprocessor, CPU fabricated onto a single integrated circuit (chip); microcomputer, a small microprocessor-based computer.
Concentrated (strong) solution: A solution with relatively little refrigerant and a high concentration of sorbent.
Condensate: Steam that has given up its latent heat and has condensed back to water.
Condensation: The liquefaction of a vapor caused when it gives up its latent heat.

Condenser: That part of a refrigerating system where the refrigerant is liquefied by the removal of heat.

Condenser Coil: A condenser constructed of pipe or tubing, not enclosed in a pressure vessel.

Condensing Unit: A combination of one or more power driven compressors, condensers, liquid receivers (when required) and regularly furnished accessories.

Continuous Combustion: Combustion process in which fuel is burned in a sustained manner as long as fuel and air are continuously fed to the combustion zone and the fuel-air mixture is within the flammability limits.

Control Agent (Medium): The medium that is adjusted by the controlled device

Control Point: The actual value of the controlled variable at which the instrument is controlling.

Control Sequence: Equipment operating procedure based upon loads or environmental conditions.

Control Strategy: A method for controlling the operation of energy consuming equipment in an energy-efficient manner.

Controlled Device (Actuator): A device that reacts to signals received from a controller.

Controlled Variable: The temperature, humidity, pressure or flow being controlled.

Controller: A device that compares the sensed values of a variable with a set point (the desired level of the controlled variable).

Cool Storage: The storage of thermal energy at temperatures below the nominal temperature required by the space or process.

Coproduction: Conversion of energy from a fuel (possibly including solid or other waste) into shaft horsepower (which may be used to generate electricity) and a second or additional useful form of energy. The process generally entails a series topping and bottoming arrangements of conversion of shaft power, as well as process or space heating. Cogeneration is a form of coproduction.

Critical Pressure, Critical Temperature, Critical Volume: A point on the saturation curve where a liquid and vapor have identical volume, density, and enthalpy, and there is no latent heat.

Crystallization: The freezing or conversion of the pumped liquid solution into a solid or slushy mass when it is excessively strong in sorbent at a given temperature.

Current: The flow of electrons in an electrical circuit, measured in amperes, as the result of an applied voltage.

Damper: A device used to vary the volume of air passing through it.

Deadband: In load control applications, a range in which loads are neither shed nor restored; or, in heating and cooling applications, a range in which neither heating nor cooling is applied. Also known as zero energy band.

Deareation: The process of removing air, dissolved oxygen and carbon dioxide for a boiler's supply water.

Demand: Rate at which electric energy is delivered at a given instant averaged over any designated time, generally over a period of less than one hour.
Demand Charge: See electrical power demand charge.
Demand, Electric: The rate at which electric energy is delivered to or by a system, part of a system or piece of equipment—expressed in kW, kilovoltamperes, or other suitable units—at a given instant, or averaged over any designated period.
Demand Limiting (electrical): A function that reduces demand by measuring incoming electrical power and turning off specified loads to keep consumption below a prescribed level during a demand interval. Also demand control, load shedding, peak demand limiting and peak demand control.
Demand Limiting (thermal storage): A partial storage operating strategy that limits the capacity of the refrigeration equipment during the on-peak period. The refrigeration equipment capacity may be limited based upon its cooling capacity, its electrical demand or the facility demand.
Density: The weight of a given volume of gas, usually expressed in lb/ft^3, at Standard Pressure and Temperature (STP) conditions.
Design Load Profile: The hour-by-hour sequence of heating or cooling loads over a complete charging cycle that must be met by a storage system.
Desorption (generation): The process in which heat is added to a dilute (weak) solution to drive refrigerant vapor from the solution.
Desuperheater: A device that injects liquid refrigerant into the hot gas bypass in order to reduce the superheat of the refrigerant.
Deviation: The difference between the set point value and the actual value of the controlled variable at any instant.
Dew Point: The temperature at which the vapor in a space (at a given pressure) will start to condense (form dew). Dew point of a gas mixture is the temperature at which the highest boiling point constituent will start to condense.
Digital Input: An "either/or" electrical impulse, which travels from a sensing device to a CPU; limited to reflecting a condition being on/off, open/closed or above/below a certain point.
Dilute (weak) solution: A solution with relatively large amounts of refrigerant and low concentration of sorbent.
Direct Acting: The output signal changes in the same direction as the controlled variable (increasing versus decreasing, etc.).
Direct Digital Control (DDC): A digital computer as the primary control loop.
Discharge Capacity: Total amount of usable cooling or heating discharge over a single cycle.
Discharging: Recovering useful energy by adding heat to a cool storage system or removing heat from a heat storage system.
Distributed Processing System: A system of multiple processors, each performing its own task, yet working together as a complete system to allow information sharing and coordinated control.

Discharge Pressure: The total gas pressure (static plus velocity) at the discharge flange of the compressor. Velocity pressure is usually considered only with dynamic compressors.

Discharge Temperature: The temperature at the discharge flange of the compressor.

Displacement: The net volume swept by the moving parts of a compressor in a unit of time, usually one minute. Displacement applies only to positive displacement compressors.

Dry Gas: Any gas or mixture of gases that contains no water vapor and/or in which all constituents are substantially above their respective saturation vapors at the existing temperature.

Drift: Circulating water lost from the tower as liquid droplets entrained in the exhaust air stream.

Drift Eliminators: An assembly of baffles or labyrinth passages through which the air passes prior to its exit from the tower, for the purpose of removing entrained water droplets from the exhaust air.

Dry Saturated Steam: Saturated steam containing no water in suspension.

Duty Cycling: A function that reduces consumption and demand by turning off electrical equipment for predetermined short periods of time during normal operating hours.

Economizer: A device that recovers waste heat by transferring heat from flue gases to warm incoming feedwater.

Economizer Control: The opening or closing of various HVAC dampers to cool a load with outside air while outside temperatures fall within a certain range, usually 60 – 75°F.

Efficacy: The ratio of light from a lamp (lumen) to the electrical power (watts) consumed, expressed as lumens per watt.

Encapsulated Ice: An ice storage technology that consists of plastic containers of water, which are alternately frozen and melted by the influence of glycol or other secondary coolants in which they are immersed.

Energy Charge: That portion of the billed charge for electric service based on the electric energy (kilowatt-hours) supplied, as contrasted with demand charge.

Energy Management System (EMS): A microprocessor, minicomputer or computer-based system whose primary function is to optimize energy use in a facility through the control of energy-using equipment.

Enthalpy: A thermal property of a substance, which consists of the internal energy of a substance plus energy in the form of work due to flow.

Enthalpy: A thermal property of a substance that consists of the internal energy of the substance, plus energy in the form of work due to flow.

Enthalpy: The total energy content of a fluid, including both heat and flow energy.

Enthalpy of evaporation h_{fg} (latent heat of evaporation): The amount of heat required to change a pound of boiling water at a given pressure to a pound of steam at the same pressure.

Enthalpy of the liquid h_f (sensible heat): The amount of heat in Btu required to raise the temperature of one pound of water from 32°F to the boiling point of steam at the same pressure.

Enthalpy of steam h_g (total heat): The combined enthalpy of liquid and vapor in the steam.

Entropy: A measure of the unavailability of energy in a substance.

Entropy: A measure of the unavailability of energy in a substance. All actual processes are irreversible and, therefore, occur with a decrease in the amount of energy available for doing work, i.e., with an increase in unavailable energy.

Evaporation: The vaporization of liquid refrigerant by heat supplied from a heat source, thus producing a cooling effect on the heat source.

Evaporator: That part of the refrigerating system designed to vaporize liquid refrigerant to produce refrigeration.

Evaporator Coil: An evaporator constructed of pipe or tubing, not enclosed in a pressure vessel.

Excess Air: The air supplied to the burner over and above that required for complete combustion.

External Melt Ice-on-Coil: Ice storage technology that forms ice by refrigerant or secondary coolant circulating within tubes or pipes, and melts by immersion in circulating water.

FERC Efficiency: The sum of the power output of the cogeneration system plus half of the useful thermal output divided by the energy input, with the energy input measured as the fuel's lower heating value.

Field Interface Device: In a distributed processing system, the unit that performs control actions in response to actions from the CPU, as well as senses environmental conditions.

Figure of Merit: The ratio of the integrated discharge capacity to the hypothetical maximum capacity for a single cycle or operation.

Fill: The portion of a cooling tower that constitutes its primary heat transfer surface.

Firetube: A boiler design type in which combustion gases flow inside the tubes and water flows outside the tubes.

Flame Temperature: The highest temperature produced in a combustion process.

Flash Point: The lowest temperature to which a fuel oil must be heated for its vapors to ignite in a flame.

Flash Steam: The rapid passing into steam of water at a high temperature to which the pressure it is under is reduced so that its temperature is above that of its boiling point for the reduced pressure.

Flume: A trough, which may be either totally enclosed or open at the top. Flumes are used in cooling towers to supply water to various sections of the distribution system.

Fluorescent Lamp: A low pressure mercury electric discharge lamp in which a (fluorescent) coating (phosphor) transforms ultraviolet energy into visible light.

Foot-candle (fc): A quantitative unit for measuring illuminance. The illumination on a surface 1 ft^2 on which there is an uniformly distributed flux of one lumen.

Fogging: A reference to the visibility and path of the effluent air stream after having exited the cooling tower.

Fouling: Any undesirable deposit on a heat transfer surface that increases the resistance to heat transmission.

Friction Head: The equivalent head needed to overcome the flow resistance of the pipe, valves and fittings in the pumping systems.

Full-load Torque: The torque necessary for a motor to produce the rated horsepower at full-load speed.

Full Storage (Load Shifting): A cool storage operation strategy that meets the entire on-peak load from storage.

Generating Efficiency: Electrical energy output from the engine or turbine generator divided by the energy input to the prime mover, in consistent units.

Grid: System of interconnected transmission lines, substations, and generating plants of one or more utilities.

Head: (The pressure exerted at the bottom of a pipe by a column of water or other liquid in a vertical pipe.) Head and pressure are interchangeable terms provided they are expressed in the same units.

Heat/Power Ratio: The heat to power ratio of the industrial plant (demand), or the rated heat to power ratio of the cogeneration cycle (capacity).

Heat Release Rate: The rate that describes the heat available per square foot of heat-absorbing surface in a furnace or per cubic foot of volume.

Heat Rate: Measure of generating station thermal efficiency, generally expressed in Btu per net kilowatt-hour or lb steam/kWh.

Heat Storage: The storage of thermal energy at temperatures above the nominal temperature required by the space or process.

Heating Value: Energy content in a fuel that is available as useful heat. The higher heating value (HHV) includes the energy transmitted to vapor formed during combustion, whereas the lower heating value deducts this energy because no work is performed.

High-intensity Discharge (HID): A group of lamps consisting of mercury, metal halide and high-pressure sodium.

High-pressure Sodium: A sodium vapor lamp in which light is produced by electricity passing through sodium and mercury vapors.

Horsepower, Brake: The total power input required, including all friction losses.

Horsepower, Indicated: The actual work required to compress and deliver a given gas quantity, including all thermodynamic, leakage and fluid friction losses. It does not include mechanical losses.

Horsepower, Theoretical: The work theoretically required to compress and deliver a given gas quantity in accordance with a specified process.

Humidity Ratio (absolute humidity): The weight of water vapor per unit of dry air.

Ice Harvester: A machine that manufactures ice on a cooling surface, then delivers it to storage.

Ice-on-Coil (Ice-on-Pipe): Ice storage technology that forms and stores ice on the outside of tubes or pipes.

Ideal Gas: A gas that follows the ideal gas law without deviation.

Ignition Temperature: The lowest temperature at which heat is generated by combustion faster than heat is lost to the surroundings and combustion becomes self-propagating.

Illuminance: The density of amount of light on a surface.

Incandescent Lamp: A lamp in which light is produced by heating a filament to incandescence.

Inlet Pressure: The total pressure (static plus velocity) at the inlet flange of the compressor. Velocity pressure is usually only a consideration for dynamic compressors.

Inlet Temperature: The temperature at the inlet flange of the compressor.

Instantaneous Storage Capacity: The rate at which usable cooling capacity can be discharged.

Interactive: Functions performed by a process where the machine prompts or otherwise assists an operator to program the device while it continues to perform all other tasks as scheduled.

Intercooling: The cooling of the gas between stages of compression to reduce the temperature, volume to be compressed in succeeding stage, to liquefy condensable vapors and to save power.

Internal Melt Ice-on-Coil: Ice storage technology that forms and melts ice by secondary coolant circulating within tubes or pipes.

Internal Process Heat: Heat from industrial activities and sources such as wastewater, boiler flue gas, steam, flue gas, coolants, exhaust air and some waste materials. This heat is normally wasted unless equipment is included to extract it for further use.

Interruptible Power: Electric energy supplied by an electric utility subject to interruption by the electric utility under specified conditions.

Latent Energy Storage (Latent Heat Storage): Heat storage or cool storage systems in which the energy stored is predominantly in the form of latent heat (usually heat of fusion) associated with a phase change (usually between solid and liquid states) in the storage medium.

Latent Heat: Heat that is associated with a phase change of a substance.

Latent Heat of Evaporation: The heat (Btu/lb) necessary to change 1 pound of liquid into vapor without raising its temperature.

Liquid Receiver: A vessel that is permanently connected to a refrigerating system by inlet and outlet pipes for storage of liquid refrigerant.

Load Leveling: A partial storage operating strategy that starts with the refrigeration equipment operating at full capacity for 24 hours on the design day. When the load is less than the chiller output, the excess is stored. When the load exceeds the chiller capacity, the additional requirement is discharged from storage.

Load Shedding: A function that turns off electrical loads according to a pre-planned priority schedule, primarily to reduce demand.

Local Control Loop: The controls for any system or sub-system, which will continue to function when the energy management and control microprocessor controller is non-operative.

Locked Rotor Torque: The minimum torque a motor will develop at rest when the rated voltage and frequency are applied.

Lumen: A measure of light output from a lamp.

Luminaire: A complete lighting unit (fixture) consisting of lamps and components to distribute the light (reflector, lens, louver) and to connect the lamps to the power supply.

Lux: A quantitative unit for measuring the illuminance. The illumination on a surface 1 meter square on which there is a uniformly distributed flux of one lumen.

Make-up: Water added to the circulated water system to replace water lost by evaporation, drift, windage, blowdown and leakage.

Mean Radiant Temperature: The uniform temperature of an imaginary enclosure, in which radiant heat transfer from a human body equals the radiant heat transfer in the actual non-unifom enclosure.

Mechanical Efficiency: The ratio, expressed in percent, of the indicated horsepower to actual shaft horsepower.

Memory: Any device that can store logic 1 and 0 bits in such a manner that a single bit or group of bits can be accessed and retrieved.

Mercury Lamp: A gaseous discharge lamp that is produced by the passage of electricity through a mercury vapor.

Metal Halide Lamp: A gaseous discharge lamp in which light is produced by the passage of electricity through a mixture of metallic vapor. These lamps contain mercury and argon, plus sodium and iodine.

Microprocessor: A central processing unit fabricated as one integrated circuit.

Modem: An acronym for Modulator/Demodulator. A hardware device used for changing digital information to and from an analog form to allow transmission over voice grade telephone circuits.

Modulating: The controlled device adjusts by increments or decrements in response to the control signal.

Modulating Control: A mode of automatic control in which the action of the final control element is proportional to the deviation of the controlled medium.

Motor Efficiency, Minimum: The minimum efficiency occurring in a population of motors from the same manufacturer and rating.

Motor Efficiency, Nominal: The median efficiency occurring in a population of motors of the same manufacturer and rating.

Multi-tasking: The procedure allowing a computer to perform a number of programs simultaneously under the management of the operating system.

Multiplexer: An energy management component that can reduce wiring costs in applications with a large number of loads; it transmits an encoded signal over one pair of twisted wires to remote stations, where the signal is decoded and acted upon.

Net Heat to Process: The difference between the thermal energy supplied to the industrial plant and the energy returned to the cogeneration system.

Net Positive Suction Head (NPSH): The absolute pressure of the pumped liquid at the pump suction nozzle.

Night Setback: A function that reduces use of mechanical heating or cooling systems when a building is occupied. Typically, reduced heating set points and raised cooling set points.

Object Code: A term used to describe the machine language version of a program.

Off-Peak Air Conditioning: An air conditioning system that uses cool storage.

Off-Peak Heating: A heating system that uses heat storage.

Off-Peak Period: The time of day at which utility rates for electrical demand, energy, or both, are reduced.

Offset: The difference between the set point of the controller and the control point of the controlled variable. Also called error or deviation.

On-Peak Period: The time of day at which a cool storage system must reduce electric demand. This period may or may not be coincident with the utility on-peak period, which is the time of day during which billing rates for electric demand, energy, or both, are increased.

Optimum Start/Stop: An energy management function that refines the start/stop program of energy-consuming equipment to get the latest possible start and the earliest possible stop of equipment.

Package Boiler: A boiler supplied with all of its components—burner, controls and auxiliary equipment—designed as a single package—and ready for on-site installation.

Partial Pressure: The absolute pressure exerted by that portion of a constituent gas in a mixture.

Partial Storage: A cool storage operating strategy that meets a portion of the on-peak cooling load from storage, with the remainder of the load met by operation of chilling equipment. Partial storage operating strategies can be further subdivided into load leveling and demand limiting operation.

Perfect Gas: See Ideal Gas.

Phase Change Material: A substance that undergoes a change of state while absorbing or rejecting thermal energy at a constant temperature.

Piston Displacement: The net volume of a reciprocating compressor displaced by the piston at rated machine speed, generally expressed in cfm. For single acting cylinders, it is the displacement of the compressing end only. For double acting cylinders, it is the total of both ends. For multi-stage compressors, it is the displacement of the first stage only.

Plume: The effluent mixture of heated air and water vapor (usually visible) discharged from a cooling tower.

Point: A single connected monitor control device (i.e., relay) or sensor (i.e., temperature, voltage transducer).

Power Factor: The ratio of total watts to the root-mean-square (RMS) volt amperes.

Primary Air: Air that is mixed with the fuel in a burner.

Primary Control Loop: An existing system, such as a thermostat or similar primary control circuit, which continues functioning during those periods when an energy management system does not have control of the circuit. Also local loop.

Prime Mover: A component of a cogeneration system that generates electric or shaft power. Typically, it is an engine or turbine.

Process Plant: The apparatus that is being controlled.

Products: The chemical constituents after a combustion process.

Program: A sequence of instructions causing the computer to perform a specified function.

Proportional Integral Derivative (PID) Control: A control technique that increases controlled variable control accuracy by measuring and reacting to the difference between the set point and actual condition, the length of time the difference has existed, and how quickly the actual controlled variable is changing.

Psychrometrics: The science involving the thermodynamic properties of moist air and the effect of atmospheric moisture on materials and human comfort.

Pull-up Torque: The minimum torque developed by a motor during the period of acceleration from rest until breakdown torque occurs.

Pulse Combustion: An acoustically resonant process that burns fuel in small, discrete fuel-air volumes in a very rapid series of combustions.

Quality: The mass fraction of vapor present in a mixture of vapor and liquid.

Quality: When steam exists as part liquid and part vapor at the saturation temperature, its quality is defined as the ratio of the mass of the vapor to the total mass.

Random Access Memory (RAM): Volatile semiconductor data storage device in which data may be stored or retrieved. Access time is effectively independent of data location.

Range: Difference between the hot water temperature and cold water temperature.

Ratchet: A utility billing mechanism by which the demand charge for the current month is based on the higher of two possibilities: either the peak demand for the present month or a percentage of the highest demand incurred during any of the previous 12 months.

Ratio of Specific Heats: The ratio of the specific heat at constant pressure to the specific heat at constant volume.

Reactants: The chemical constituents entering a combustion process.

Real Time: A situation in which a computer monitors, evaluates, reaches decisions and affects controls within the response time of the fastest phenomenon.

Receiver Controller: A device that accepts a signal produced by a transmitter and converts it for suitable operation of the controller device.

Refrigerant (absorption refrigeration): A volatile substance that (1) leaves the solution at the desorber or generator, (2) performs the refrigeration process at an evaporator, and (3) is reabsorbed at the absorber to complete the thermodynamic cycle.
Refrigerant (vapor compression refrigeration): The fluid used for heat transfer in a refrigerating system; the refrigerant absorbs heat and transfers it at a higher temperature and a higher pressure, usually with a change of state.
Refrigerating System: A combination of interconnected parts forming a closed circuit in which a refrigerant is circulated for the purpose of extracting, then rejecting, heat.
Reset: The process of automatically adjusting the control point of a controlled variable.
Resistance: The resistance to the flow of electrons through an electrical device, measured in ohms.
Reverse Acting: The output signal changing in the direction opposite to the change in the controlled variable.
Saturated Liquid: A liquid that exists at the saturation temperature and pressure.
Saturated Vapor: A vapor that exists at the saturation temperature and pressure.
Saturation (Dew Point) Temperature: The temperature at which water vapor begins to condense out of air.
Saturation Pressure: The pressure at which vapor and liquid exist in equilibrium at a given temperature.
Saturation Temperature: The temperature at which vaporization takes place at a given pressure; this pressure is called the saturation pressure.
Scfm: Volume flow rate at 14.696 psia and 60°F in ft^3/min (cfm).
Secondary Air: Air that diffuses into the flame from the atmosphere.
Sensible Energy Storage (Sensible Heat Storage): Heat storage or cool storage systems in which the energy stored is in the form of sensible heat associated with a temperature change in the storage medium.
Sensible Heat: Heat that causes a change in the temperature of a substance.
Sensible Heat: Heat that increases the temperature of objects and liquids.
Sensor: A device that measures the controlled variable and conveys representative values to a controller.
Sequencer: A mechanical or electrical device that may be used to initiate a series of events in a particular order.
Service Factor: The maximum overload that can be applied to general purpose motors and certain definite purpose motors without exceeding the temperature limitation of the insulation system.
Set Point: The desired value of the controlled variable.
Shaft Efficiency: Prime mover's shaft energy output divided by its energy input in equivalent, consistent units. For a steam turbine, input can be the thermal value of the steam. For a fuel-fired prime mover, it is the fuel input.

Slip: The internal leakage within a rotary compressor. It represents gas at least partially compressed but not delivered.

Slip RPM: The speed required of a rotary compressor to maintain a given discharge pressure supplying leakage only (zero actual output).

Solution: A liquid of at least two components, one soluble in the other.

Sorbent: The portion of the solution that transports the absorbed refrigerant through the processes that make up thermochemical compression.

Specific Volume: The volume of unit mass—it indicates the volumetric space that 1 pound of steam or water occupies.

Stability: The ability of a system to maintain a controlled variable at or near the control point without requiring the controller to "hunt" (change continuously from one extreme to another).

Stand-alone System: Performs all energy management controls functions independently without direction from another computer.

Standard Pressure and Temperature (STP): 14.696 psia and 60°F.

Standby Power: Electric energy supplied by a utility to a cogenerator, or vice-versa, during either one's unscheduled outage.

Static Discharge Head: The vertical distance, in feet, from the pump centerline to the point of free delivery of the liquid.

Static Pressure: The pressure that exists independent of any motion or velocity.

Static Suction Lift: The vertical distance, in feet, from the liquid supply level to the pump centerline; the pump being above the supply level.

Steam: Water in the vapor phase.

Steam Tables: Tables that list the properties of steam and condensate at various pressures.

Steam Trap: A device for allowing the passage of condensate and air but preventing the passage of steam.

Stoichiometric Combustion: A combustion process in which the fuel is reacted with the exact amount of oxygen required to oxidize all carbon, hydrogen and sulfur in the fuel to CO_2, H_2O and SO_2.

Storage Medium: A substance in which heating or cooling capacity is stored.

Storage Priority: Control strategy for partial storage systems that uses stored cooling to meet as much of the load as possible. Direct cooling is supplied by chillers if the daily load exceeds the total stored cooling capacity.

Stored Heat: Heat from external or recovered heat that is held in reserve for later use.

Subcooled Liquid: A liquid that has been cooled below its saturation point at a given pressure.

Superheated Steam: Steam heated above the temperature corresponding to its pressure.

Superheated Vapor: The temperature of a vapor above its saturation point at a given pressure.

Supplemental Thermal: Heat required when recovered engine heat is insufficient to meet thermal demands.

System Friction Curve: A plot of the friction head loss versus flow rate for a pumping system.
System Head Curve: A curve obtained by combining the system static head with the system friction curve.
Task Lighting: Lighting that provides illumination for specific visual functions and is directed to a specific surface or area.
Thermal Capacity: Maximum amount of instantaneous heat that a system can produce.
Thermal Shock: A cycle of temperature swings that result in failure of metal due to expansion and contraction.
Thermal Storage Device: Container plus all the contents of the container used for storing thermal energy. The transfer fluid and accessories, such as heat exchangers, agitators, circulating pumps, flow switching devices, valves and baffles that are integral to the thermal storage container are considered part of the storage device.
Thermocline: A layer of fluid in which the temperature and density gradient is greater than, and which separates, the cooler fluid below it and the warmer fluid above it.
Throttling Range: The amount of change in the controlled variable required to move the actuator of a controlled device from one end to another.
Time-of-day Scheduling: The process of turning a facility's electrical loads on and off according to the time of day, day of week, the month and the facilities occupancy patterns.
Transducer: A device that converts an input signal from one form to another, i.e., electrical input to a pneumatic output.
Topping Cycle: Cogeneration facility in which the energy input to the facility is first used to produce useful power, and the rejected heat from production is used for other purposes.
Total Dynamic Head: The sum of the friction head and the static discharge head.
Total Energy System (TES): Form of cogeneration in which all electrical and thermal energy needs are met by on-site systems. A total energy system can be completely isolated or switched over to a normally disconnected electrical utility system for backup.
Turndown Ratio: The ratio of the maximum to the minimum fuel-air mixture input rates at which the burner will operate satisfactorily.
Useable Temperature: The temperature or range of temperatures at which heat energy can be absorbed, rejected or stored for use within the system.
Variable Speed Drive: An electronic device that matches motor speed to that required for the application.
Velocity Pressure: Pressure that exists as a result of the motion of a fluid.
Viscosity: That property of a fluid which causes resistance to relative motion within the fluid.
Voltage (volts): The electrical pressure in an electrical circuit that causes current to flow.

Volumetric Efficiency: The ratio, in percent of the actual delivered capacity (measured at the inlet temperature and pressure and gas composition) to the piston displacement.

Waste Heat: Heat rejected from the building (or process) because its temperature is too low for economical recovery or direct use.

Watertube: A boiler design type in which combustion gases flow outside the tubes and water flows inside the tubes.

Wet Saturated Steam: Saturated steam containing some water particles in suspension.

Wheeling: Use of the transmission facilities of one systems to transmit gas or power for another system.

Windage: Water lost from a tower because of the effects of wind.

Zone: An area composed of a building, a portion of a building, or a group of buildings affected by a single device or piece of equipment.

References

CHAPTER ONE:

1. American Society of Heating, Refrigerating and Air Conditioning Engineers (ASHRAE), 1985, *ASHRAE Standard 100.3-1985, Energy Conservation in Existing Buildings, Atlanta, GA.*
2. American Society of Heating, Refrigerating and Air Conditioning Engineers (ASHRAE), 1989, *ASHRAE Standard 90.1-1989, Energy Efficient Design of New Buildings, Except Low-Rise Residential Buildings,* Atlanta, GA.
3. American Society of Heating, Refrigerating and Air Conditioning Engineers (ASHRAE), 1999, *ASHRAE Applications Handbook,* Atlanta, GA.
4. Oak Ridge Associated Universities, 1980, *Industrial Energy Use Data Book,* Oak Ridge, TN.
5. U.S. Department of Energy, 1990, *Architects and Engineers Guide to Energy Conservation in Existing Buildings, Volumes 1 and 2, Washington, D.C.*
6. U.S. Office of Technology Assessment, U.S. Department of Energy, 1983a, *Enhancing* the Competitiveness, *Efficiency, and Environmental Quality of American Industry Through Technology Partnerships, Washington, D.C.*
7. U.S. Office of Technology Assessments, 1983b, *Industrial Energy Use,* Washington, D.C.
8. Witte, Larry C., et. al., 1988, *Industrial Energy Management and Utilization, Hemisphere Publishing, New York, NY.*

CHAPTER TWO:

1. Albertson, M. L., Barton, J. R., and Simmons, D. B., (1960), *Fluid Mechanics for Engineers*, Prentice-Hall, Inc., Englewood Cliffs, NJ.
2. American Society of Heating, Refrigerating and Air Conditioning Engineers, Inc., (ASHRAE) 1997, *Fundamentals Handbook*, Atlanta, GA.
3. Chapman, A. J., 1974, *Heat Transfer*, Macmillan Publishing Co., New York, NY.
4. Holman, J. P., 1972, *Heat Transfer*, New York, NY, 1972.
5. Sontag, R. E., and Van Wylen, G. J., 1971, *Introduction to Thermodynamics, Classical and Statistical*, John Wiley & Sons, New York, NY.

CHAPTER THREE:

1. American Society of Heating, Refrigerating and Air Conditioning Engineers (ASHRAE), 1999, *HVAC Applications Handbook,* Atlanta, GA.

2. Kreider, K. G. and Mc Niel, M. B., 1977, Waste *Heat Management Guidebook, NBS Handbook 121,* U.S. Department of Commerce.
3. Newman, D. G., 1977, *Engineering Economic Analysis,* Engineering Press, San Jose, CA.
4. Reiter, S., 1985, *The Financial Evaluation of Energy Costs and Projects,* Van Nostrand Reinhold Company, New York, NY.
5. Witte, L. C., et. al., 1988, *Industrial Energy Management and Utilization,* Hemisphere Publishing Corporation, New York, NY.

CHAPTER FOUR:

1. American Society of Heating, Refrigerating and Air-Conditioning Engineers (ASHRAE), 1996, *HVAC Systems and Equipment Handbook,* Atlanta, GA.
2. American Society of Heating, Refrigerating and Air-Conditioning Engineers (ASHRAE), 1997, *Fundamentals Handbook*, Atlanta, GA.
3. Cleaver Brooks, 1993, *The Boiler Book,* Milwaukee, WI.
4. Kirts, R. E., 1984, *User's Guide to Automatic Boiler Controls,* Naval Civil Engineering Laboratory, Port Huenme, CA.
5. Payne, W. F., and Thompson, R. E., 1996, *Efficient Boiler Operations Sourcebook,* Fourth Edition, The Fairmount Press, Inc., Lilburn, GA.
6. Sontag, R. E. and Wylen, G. J., 1971, *Introduction to Thermodynamics: Classical and Statistical,* John Wiley & Sons, Inc., New York, NY.
7. Taplin, H., 1991, *Boiler Plant and Distribution System Optimization Manual,* The Fairmount Press, Inc., Lilburn, GA.
8. The Babcox & Wilcox Company, 1972, *Steam, Its Generation and Use,* New York, NY.
9. Thuman, A., 1988, *Guide to Improving Efficiency of Combustion Systems,* The Fairmount Press, Inc., Lilburn, GA.
10. Turner, W. C., 1993, *Energy Management Handbook,* Second Edition The Fairmount Press, Inc., Lilburn, GA.
11. U.S. Department of Energy, 1998, *Non-Chemical Technologies for Scale and Hardness Control,* Federal Energy Management Program, Technology Alert, Washington, D.C.
12. Witte, L. C., et. al., 1988, *Industrial Energy Management and Utilization,* Hemisphere Publishing Corp., New York, NY.

CHAPTER FIVE:

1. American Society of Heating, Refrigerating and Air-Conditioning Engineers (1993), *ASHRAE Standard 90.1, Energy Efficient Design of New Buildings Except Low-Rise Residential Buildings*, Atlanta, GA.
2. American Society of Heating, Refrigerating and Air-Conditioning Engineers (ASHRAE), 1996, *HVAC Systems and Equipment Handbook*, Atlanta, GA.

3. American Society of Heating, Refrigerating and Air-Conditioning Engineers (ASHRAE), 1997, *Fundamentals Handbook*, Atlanta, GA.
4. ITT Hoffman, 1994, *Steam Traps Engineering Data Manual*, Chicago, IL.
5. Makansi, J., 1985, *Managing Steam: An Engineering Guide to Commercial, Industrial, and Utility Systems*, Hemisphere Publishing Corporation, New York, NY.
6. Sontag, R. E., and W. G., J., 1971, *Introduction to Thermodynamics: Classical and Statistical*, John Wiley & Sons, Inc., New York, NY.
7. Spirax Sarco, 1985, *Steam Utilization Course*, Allentown, PA.
8. Taplin, H., 1991, *Boiler Plant and Distribution Systems Optimization Manual*, The Fairmont Press, Lilburn GA.
9. Turner, W. C., 1993, *Energy Management Handbook*, Second Edition, The Fairmont Press, Inc., Lilburn GA.
10. Witte, L. C., et al., 1988, *Industrial Energy Management and Utilization*, Hemisphere Publishing Corporation, New York, NY.
11. Yarway Corporation, 1984, Industrial Steam Trapping Handbook, Second Edition, Blue Bell, PA.

CHAPTER SIX:

1. American Society of Heating, Refrigerating and Air Conditioning Engineers (ASHRAE), 1996, *Systems and Equipment Handbook*, Atlanta, GA.
2. American Society of Heating, Refrigerating and Air Conditioning Engineers (ASHRAE), 1997, *Fundamentals Handbook*, Atlanta, GA.
3. Carlson, G. F., 1980, *Variable Volume Pumping Considerations*, Heating/Piping/Air Conditioning, April.
4. Coad, W. J., 1985, *Hydronic Energy Economics*, ASHRAE Journal, June.
5. Crane Company, 1974, *Flow of Fluids Through Valves, Fittings and Pipe, Technical Paper No. 410,* Chicago IL.
6. Garay, P. N., 1996, *Pump Application Deskbook*, Third Edition, The Fairmont Press, Inc., Atlanta, GA.
7. Hansen, E. G., 1995, *Parallel Operation of Variable Speed Pumps in Chilled Water Systems*, ASHRAE Journal, October.
8. Hicks, T. G., 1957, *Pump Selection and Application*, McGraw-Hill Book Company, Inc., New York, NY.
9. ITT Fluid Handling Division, 1967, *Pump & Systems Curve Data for Centrifugal Pump Selection and Application,* Morton Grove, IL.
10. Karassik, I. J., and McGuire, T. J., 1998, *Centrifugal Pumps*, Second Edition, Chapman & Hall, New York, NY.

CHAPTER SEVEN:

1. American Society of Heating, Refrigerating and Air-Conditioning Engineers, 1993, *ASHRAE Standard 90.1, Energy Efficient Design of New Buildings Except Low-Rise Residential Buildings, Atlanta, CA.*

2. American Society of Heating, Refrigerating and Air-Conditioning Engineers (ASHRAE), 1994, *Refrigeration Handbook*, Atlanta, GA.
3. American Society of Heating, Refrigerating and Air-Conditioning Engineers (ASHRAE), 1996, *HVAC Systems and Equipment Handbook*, Atlanta, GA.
4. Barr, R. A., 1986, *18 Ways to Improve Chiller Efficiency, Engineered Systems, September/October.*
5. Carrier Air Conditioning Company, 1969, *Carrier System Design Manual, Part 7, Refrigeration Equipment,* Syracuse, NY.
6. Katzel, J., 1992, *The Rediscovery of Absorption Chillers*, Plant Engineering, April 23.
7. Kozak, R. A., 1989, *Maintaining Centrifugal Chillers, Plant Engineering, December 21.*
8. Mull, T. E., 1997, *HVAC Principles and Applications Manual*, McGraw-Hill, Inc. New York, NY.
9. The Trane Company, 1974, *Trane Air Conditioning Manual*, La Crosse, WI.

CHAPTER EIGHT:

1. American Society of Heating, Refrigerating and Air-Conditioning Engineers (ASHRAE), 1992, *HVAC Systems and Equipment Handbook*, Atlanta, GA.
2. Becker, B. R. and Burdick, L. F., 1993, *Drift Eliminators and Cooling Tower Performance*, ASHRAE Journal, June.
3. Berger, R., 1987a, *Cooling Towers: The Energy Orphan*, Heating/Piping/Air Conditioning, January.
4. Berger, R., 1987b, *Cooling Tower Technology*, Heating/Piping/Air Conditioning, August 1987.
5. Goswami, D. and Reveitty, C. N., 1987, *Free Cooling by Cooling Tower Water*, ASHRAE Journal, January.
6. Hensley, J. C., 1981, *Cooling Tower Energy*, Heating/Piping/ Air Conditioning, October.
7. Hensley, J. C., 1985, *Cooling Tower Fundamentals*, The Marley Cooling Tower Company, Mission KS.
8. Katzel, J., 1989, *Fundamentals of Cooling Towers*, Plant Engineering, April.
9. Landhall, P., and Bugler, T., 1987, *Design for Efficient Cooling Tower Operation*, Counsulting-Specifying Engineer, August.
10. Leary, W. M., 1987, *Optimizing Cooling Tower Selections*, Heating/Piping/Air Conditioning, August.
11. Shelton, S. V., and Joyce, C. T., 1991, *Cooling Tower Optimization for Centrifugal Chillers*, ASHRAE Journal, June.
12. The Marley Cooling Tower Company, 1982, *The Application of Cooling Tower Water for Free Cooling*, Mission, KS.
13. The Marley Cooling Tower Company, 1983, *External Influences on Cooling Tower Performance*, Mission, KS.

14. The Marley Cooling Tower Company, 1983, *Automatic Variable Pitch Propeller Type Fans for Cooling Towers*, Mission KS.
15. U. S. Department of Energy, 1995, *Ozone Treatment for Cooling Towers*, Federal Energy Management Program Technology Alert, Washington, D. C.

CHAPTER NINE:

1. American Conference of Governmental Industrial Hygienists, Inc, (ACGIH), 1988, *Industrial Ventilation*, 20th Edition, Cincinnati, OH.
2. American Society of Heating, Refrigerating and Air Conditioning Engineers (ASHRAE), 1993, *ASHRAE Standard 90.1, Energy Efficient Design of New Buildings Except Low-Rise Residential Buildings*, Atlanta, GA.
3. American Society of Heating, Refrigerating and Air Conditioning Engineers (ASHRAE), 1995, *HVAC Applications Handbook*, Atlanta, GA.
4. American Society of Heating, Refrigerating and Air Conditioning Engineers (ASHRAE), 1996, *Systems and Equipment Handbook*, Atlanta, GA.
5. American Society of Heating, Refrigerating and Air Conditioning Engineers (ASHRAE), 1997, *Fundamentals Handbook*, Atlanta, GA.
6. Mull, T. E., 1997, *HVAC Principles and Applications Manual*, McGraw-Hill, Inc., New York, NY.
7. Watt, J. R. and Brown, W. K., 1997, *Evaporative Air Conditioning Handbook*, Third Edition, The Fairmont Press, Inc.

CHAPTER TEN:

1. American Society of Heating, Refrigerating and Air Conditioning Engineers (ASHRAE), 1993, *ASHRAE Standard 90.1, Energy Efficient Design of New Buildings Except Low-Rise Residential Buildings*, Atlanta, GA.
2. American Society of Heating, Refrigerating and Air Conditioning Engineers (ASHRAE), 1996, *Systems and Equipment Handbook*, Atlanta, GA.
3. Chen, K., 1994, *Energy Effective Industrial Illuminating Systems*, The Fairmount Press, Inc. Lilburn, GA.
4. Fitzgerald, A.E., et. al, 1967, *Basic Electrical Engineering*, McGraw-Hill Book Company, New York, NY.
5. Illuminating Engineering Society of North America (IES), 1981, *IES Lighting Handbook, Applications Volume*, New York, NY.
6. Illuminating Engineering Society of North America (IES), 1981, *IES Lighting Handbook, Reference Volume*, New York, NY.
7. The Institute of Electrical and Electronic Engineers (IEEE), 1984, *IEEE Recommended Practice of Energy Conservation and Cost Effective Planning in Industrial Facilities*, New York, NY.
8. Turner, W. C., 1993, *Energy Management Handbook*, Second Edition The Fairmount Press, Inc., Lilburn, GA.

9. Witte, L. C., et. al., 1988, *Industrial Energy Management and Utilization*, Hemisphere Publishing Corp., New York, NY.

CHAPTER ELEVEN:

1. American Society of Heating, Refrigerating and Air-Conditioning Engineers, *(ASHRAE), 1989, Standard 90.1 Energy Efficient Design of New Buildings Except Low-Rise Residential Buildings*, Atlanata, GA.
2. Carroll, Hatch & Associates, Inc., *Industrial Compressed Air System Energy Efficiency Guidebook*, Bonneville Power Administration, Portland OR.
3. Crane Company, 1974, *Flow of Fluids Through Valves, Fittings, and Pipe, Technical Paper 410*, New York, NY.
4. Loomis, A.W., 1980, *Compressed Air and Gas Data*, Ingersoll-Rand Company, Woodcliff Lake, NJ.
5. Mull, T. E., 1997, *HVAC Principles and Applications Manual*, McGraw-Hill, Inc. New York, NY.
6. Oviatt, M. D. and Miller, R.K., 1981, *Industrial Pneumatic Systems: Noise Control and Energy Conservation*, The Fairmont Press, Inc., Liliburn GA.
7. Talbott, E. M., 1992, *Compressed Air Systems; A Guidebook on Energy and Cost Savings*, The Fairmont Press, Inc., Lilburn GA.

CHAPTER TWELVE:

1. American Society of Heating, Refrigerating and Air Conditioning Engineers (ASHRAE), 1994, *Refrigeration Systems and Applications Handbook*, Atlanta, GA.
2. American Society of Heating, Refrigerating and Air Conditioning Engineers (ASHRAE), 1997, *Fundamentals Handbook,* Atlanta, GA.
3. Carrier Air Conditioning Company, 1969, *Carrier System Design Manual, Part 7 Refrigeration Equipment*, Syracuse, NY.
4. Cole, R. A., 1985, *Reduced Head Pressure Operation of Industrial Refrigeration Systems,* Heating/Piping/Air Conditioning, May.
5. Cole, R. A., 1988. *The Case for Refrigerant Liquid Subcooling,* Heating/Piping/Air Conditioning, December.
6. Dossat, R. J., 1997, *Principles of Refrigeration*, Prentice Hall, Inc., Upper Saddle River, NJ.
7. Electric Power Research Institute, (EPRI), 1987, *Commercial Cool Storage Design Guide*, EM-3981, Hemisphere Publishing Corporation, Washington, DC.
8. Tomczyk, J., 1997, *The Technology Keeps Improving, Engineered Systems,* May.
9. U.S. Department of Energy, 1985, *Refrigerant Subcooling, Technology for Improving Cooling System Performance*, Federal Energy Management Program Technology Alert, Washington, D.C.

10. U.S. Department of Energy, 1995, *Liquid Refrigerant Pumping, Technology for Improving Refrigeration Performance and Capacity*, Federal Energy Management Program Technology Alert, Washington, DC.

CHAPTER THIRTEEN:

1. American Society of Heating, Refrigerating and Air-Conditioning Engineers (ASHRAE), 1996, *Systems and Equipment Handbook,* Atlanta, GA.
2. Baumeister, T., 1967, *Marks Standard Handbook for Mechanical Engineers, 7th Edition, McGraw-Hill Book Company, New York, NY.*
3. Hu, D. S., 1985, *Cogeneration*, Reston Publishing Company, Inc., Reston, VA.
4. Institute of Electrical and Electronic Engineers (IEEE), 1984, *IEEE Recommended Practice for Energy Conservation and Cost-Effective Planning in Industrial Facilities,* New York, NY.
5. Orlando, J. A., 1996, *Cogeneration Design Guide,* American Society of Heating, Refrigerating and Air-Conditioning Engineers, Atlanta, GA.
6. Smith, J. R., 1996, *Capturing Energy Savings with Cogeneration*, Consulting-Specifying Engineer, July.
7. Sonntag, R. E. and Van Wylen, G. J., 1971, *Introduction to Thermodynamics: Classical and Statistical,* John Wiley & Sons, Inc., New York, NY.
8. Turner, W. C., 1993, *Energy Management Handbook, Second Edition,* The Fairmont Press, Inc., Lilburn, GA.

CHAPTER FOURTEEN:

1. American Society of Heating, Refrigerating and Air-Conditioning Engineers (ASHRAE), 1996, *Systems and Equipment Handbook,* Atlanta, GA.
2. Boyen, J. L., 1980, *Thermal Energy Recovery, Second Edition,* John Wiley & Sons, New York, NY.
3. Chapman, A. J., 1974, *Heat Transfer,* Macmillan Publishing Company, Inc., New York, NY.
4. Gilbert, J. S., 1987, *Heat Pump Strategies and Payoffs,* Consulting-Specifying Engineer, October.
5. Holman, J. P., 1972, *Heat Transfer, Third Edition,* McGraw-Hill Book Company, New York, NY.
6. Kreider, K. G. and McNeil, M. B., 1977, *Waste Heat Management Guidebook, NBS Handbook 121,* S Department of Commerce.
7. Taplin, H., 1991, *Boiler Plant and Distribution System Optimization Manual,* The Fairmount Press, Inc., Lilburn, GA.
8. Turner, W. C., 1993, *Energy Management Handbook,* Second Edition The Fairmount Press, Inc., Lilburn, GA.
9. Witte, L. C., et. al., 1988, *Industrial Energy Management and Utilization,* Hemisphere Publishing Corp., New York, NY.

CHAPTER FIFTEEN:

1. American Society of Heating, Refrigerating and Air-Conditioning Engineers (ASHRAE), 1977, *Applications of Solar Energy for Heating and Cooling of Buildings, ASHRAE GRP-170,* Atlanta, GA.
2. American Society of Heating, Refrigerating and Air-Conditioning Engineers (ASHRAE), 1995, *Applications Handbook,* Atlanta, GA.
3. American Society of Heating, Refrigerating and Air-Conditioning Engineers (ASHRAE), 1997, Fundamentals *Handbook,* Atlanta, GA.
4. Baltimore Aircoil Company, 1986, *A Guide to Ice Storage Systems, Design Bulletin E140/1-0A.,* Baltimore, MD.
5. Casey, C. W., et. al., 1995, *The Control of Ice Storage Systems*, ASHRAE Journal.
6. Dorgan, C. E. and Elleson, J. S., 1994, *Design Guide for Cool Thermal Storage,* American Society of Heating, Refrigerating and Air-Conditioning Engineers, Atlanta, GA.
7. Electric Power Research Institute, (EPRI), 1987, *Commercial Cool Storage Design Guide, EM-3981,* Hemisphere Publishing Corporation, Washington DC.
8. Knebel, D. E., 1995, *Current Trends in Thermal Storage*, Engineered Systems, January.
9. Paul Mueller Company, 1989, *Thermal Storage With Ice Harvesting,* Springfield, MO.
10. Turner, W. C., 1993, *Energy Management Handbook,* Second Edition The Fairmount Press, Inc., Lilburn, GA.

CHAPTER SIXTEEN:

1. American Society of Heating, Refrigerating and Air-Conditioning Engineers (ASHRAE), 1997, *Fundamentals Handbook,* Atlanta, GA.
2. American Society of Heating, Refrigerating and Air-Conditioning Engineers (ASHRAE), 1999, *Applications Handbook,* Atlanta, GA.
3. Andover Controls Corporation, 1990, *Specifying Engineers Guide,* Andover, MA.
4. Mull, T. E., 1997, *HVAC Principles and Applications Manual,* McGraw-Hill, Inc. New York, NY.
5. The Chartered Institution of Building Service Engineers, 1985, *Automatic Controls and Their Implications for System Design,* London, England.
6. Thuman, A., 1985, *Energy Management Systems Sourcebook,* The Fairmont Press, Inc., Atlanta, GA.
7. Zimmerman, Alex J., 1996, *Fundamentals of Direct Digital Control,* Heating/Piping/Air Conditioning, May.

Index